Name: _____

Safety Information Sheet

Locations of Safety Equipment

Local emergency telephone _____

1. Fire and Emergency Exits: _____

2. Eye wash fountain: _____

3. Fire extinguisher and type: _____

4. Fire blanket: _____

5. Safety shower: _____

Other Important Telephone Numbers

Yours _____   Instructor _____

Others _____

Microscale and Selected Macroscale Experiments for General and Advanced General Chemistry:

An Innovative Approach

Microscale and Selected Macroscale Experiments for General and Advanced General Chemistry:
An Innovative Approach
First Edition

Mono M. Singh
Merrimack College

Ronald M. Pike
Merrimack College

Zvi Szafran
Merrimack College

John Wiley & Sons, Inc.
New York • Chichester • Brisbane • Toronto • Singapore

CHEMISTRY EDITOR	Nedah Rose
ACQUISITIONS EDITOR	Joan Kalkut
MARKETING MANAGER	Catherine Faduska
PRODUCTION EDITOR	Publication Services and Erin Singletary
COVER DESIGNER	Carol C. Grobe
MANUFACTURING MANAGER	Susan Stetzer
ILLUSTRATION COORDINATOR	Jaime Perea

This book was set in 11/13 Times Roman by Publication Services and
printed and bound by Courier/Kendalville. The cover was printed by Phoenix Color Corp.

Library of Congress Cataloging in Publication Data:

CIP DATA
ISBN 0471-58596-3

Printed in the United States of America

10   9   8   7   6   5   4   3   2   1

Dedication

To Our Families

*To my brothers Amarjeet and Inderjeet and my sister Shabnam,
for their support while I was preparing for a career in chemistry,
and to my wife Shashi and children Yuvraj, Balraj, and Pamela.*

*To my wife Marilyn, children Dana and Jane, Gretchen and John and
grandchildren Benjamin, Erik, Daniel, Shane, Scott, Beth, and Allison.*

To the memory of my grandfather Bernard Dulzer, and to my wife Jill and son Mark.

Disclaimer

All experiments described in the text have been developed and tested by the authors. These experiments have been safely performed by students under the supervision of the authors. If they are performed in accordance with the methods developed in this text, the authors believe that the experiments are safe and that they provide a valuable learning experience. However, all duplication or performance of these experiments is conducted at one's own risk. The authors do not warrant or guarantee the safety of individuals performing these experiments. The authors hereby disclaim any liability for any loss or damage claimed to have resulted from or related in any way to the experiments, regardless of the form of action.

Preface

Why This Manual?

A large number of general chemistry laboratory manuals are available in the market. Then why are we writing this manual? There are several basic reasons:

1. Chemistry is an experimental science. Its home is in a laboratory and its major goal is to create good things (for example, medicines, home products, and food preservatives) for human life. The best way to learn chemistry is, obviously, to practice doing it in a laboratory. **This text is meant for those who want to learn chemistry by practicing it in the laboratory.**

2. The manual contains experiments that are meant for college-level chemistry courses and avoids, as far as practicable, repetitive material that is generally learned at the high-school level. However, reviewing some of the basic laboratory techniques is essential. **A good number of experiments are therefore provided in the text at a level that will be challenging to first-year college students and, at the same time, will serve as the link between what students have learned in high school and what is expected of them at the college level.**

3. This book is the outcome of a course that has been successfully offered at Merrimack College and elsewhere at the first-year chemistry level. The most difficult part of teaching this course has been the nonavailability of a suitable text for the laboratory. **To fill the void, we have developed this manual.**

4. There is a strong demand for educated chemists in the chemical manufacturing industry, in the pharmaceutical industry, in analytical and environmental laboratories, in other chemistry-related industries, and, finally, in the development of new, environmentally friendly processes. Good, solid training in laboratory practice—involving not only the minimum basic techniques supporting micro- and macroscale experimentation but also the use of modern instrumentation—is essential. **We believe that this laboratory manual provides such training.**

What Is the Scope of the Manual?

This manual contains experiments that can be adopted for a two-semester general chemistry course as well as for advanced general chemistry courses. Many experiments described in the text can also be used in inorganic, analytical, or quantitative analysis laboratories. **The choice and the level of experiments are such that they can accompany any general chemistry text.**

Why Microscale as Well as Macroscale (Normal Scale) Experiments?

Chemistry is a sword that cuts both ways. Nobody can deny that chemistry provides us with numerous products—such as medicines, plastics, pesticides, and fertilizers—that are desirable for a healthy and comfortable life. However, many such products are also responsible for pollution of the environment. Many of these problems are created by the abuse of the science of chemistry. The most pressing problem related to the practice of chemistry is the generation of chemical waste and its subsequent disposal.

One of the ways to handle the situation is to train future chemists so that, while they develop new, environmentally friendly products, they also become an integral part of the solution of the environmental problem. For example, using the newly coined concept of *maximum atom utilization,* the reactions are carried out in such a way that no residue is left as waste. A chemist should be able to develop a process where the chemical waste is totally eliminated.

Keeping these goals in view, we have adopted the microscale approach because we feel that it is an ethical way to practice chemistry. Merrimack College is one of the pioneering institutions in the nation to lead the microscale revolution. We perform all our laboratories under ***microscale conditions,*** with these advantages:

- Small amounts of chemicals and solvents are used, which **saves money**.
- This approach **saves laboratory time** by curtailing the time required for completing a reaction and subsequent manipulative steps. It also drastically reduces the instructor's time for setting up the laboratory.
- This approach prevents **fire hazards** and eliminates **accidents** related to chemical spills.
- Microscale work **eliminates chemical waste at the source**, serving the cause of pollution prevention (P2).
- The microscale approach keeps students and instructors healthy and effective by eliminating exposure to toxic chemicals and fumes.
- Experiments can be performed using chemicals that we otherwise would not have been able to use (see, for example, Experiment 6, where a student learns how to make a microscale mercury gas law tube).
- Most importantly, this training brings **a change in the way a chemist thinks and works**; it fosters the growth of a future chemist who is well prepared to keep the environment safe and healthy.

Though our emphasis is on the microscale approach, **we do believe that some macroscale or normal scale experiments must be performed to learn macro techniques**. Moreover, we do not perform microscale experiments just for the sake of doing them. Some experiments use chemicals that are not hazardous to the environment; in fact, they can be recycled and reused. For such experiments, alternative macroscale or normal scale procedures have been described (for example, see Experiments 10, 11, and 12). We recommend that students perform some of these normal scale experiments.

What Are the Distinctive Features of the Text?

1. There are several chapters that we consider to be attractive features of the book. For example, in **Chapter 3** we describe **procedures for students/instructors to construct a number of pieces of their own microscale equipment**.

Chapter 4 is written with the main purpose of encouraging students to keep notebooks and to write **laboratory reports**. The data sheets provided in the manual are attached to the report. At Merrimack College, our advanced students submit a written lab report that includes the prelaboratory report, problem sets, data sheets and graphs, spectra, and so on. They write independent reports (two pages) for each experiment.

Chapter 11 is dedicated to literature searching, undergraduate mini research projects, and the use of computers in the laboratory.

2. In using microscale techniques, **we did not sacrifice precision.** For example, in titrations, we recommend the use of a microburet instead of a Beral pipet (in the real world, a titration is not done by counting drops!).

3. The experiments are written in a less-structured fashion. The procedures are written in the style that students will find them in real life. As far as practicable, we have tried to avoid the "cookbook" style of writing. However, we feel that in the initial stages of training there is nothing wrong in following a cooking recipe. We believe that the beginning student **must** have some cookbook-style recipes for their work. **Without a good recipe we can have neither a good cook nor good food.**

4. A new approach has been introduced:

(*a*) A **prelaboratory report sheet** is given where students can develop a work plan. This encourages them to read the experiment before coming to the laboratory. The report is collected at the beginning of the lab. In addition, **prelab and postlab problem sheets** are also provided.

(*b*) To emphasize the importance of self-teaching and the independent thinking process, we have offered **additional independent projects** at the end of the majority of experiments. Some of these experiments are **open-ended and have been described very briefly**. After performing the regular laboratory, a student should be able to do these experiments with little or no help. In these projects, the student is required to **develop his or her own data and manipulation tables**. Students are asked to submit a written report at the end of the laboratory.

(*c*) **References to independent projects** have been provided. We expect that students (especially the advanced-level students) will search the literature before starting any independent project (see Chapter 11). We believe that this activity is a great self-teaching process.

(*d*) Many experiments in this manual are **interdependent**. For example, in the titration of vinegar (Experiment 10), we ask students to determine the density of vinegar using the micropycnometer method learned in Experiment 1. They also use the micropycnometers in Experiments 4, 10, and 34. Experiments 21, 24, and an independent project of Experiment 29 are also interconnected. Aspirin prepared in Experiment 32A is used in Experiment 19. Urea prepared in Experiment 32C is used in 32D. Such tools as TLC (Experiment 26) and the pH meter (Experiment 28) are used again in Experiments 33 and 35, respectively.

(*e*) Many experiments are **innovative**. Some examples follow:

Experiment 1: Construction and use of a micropycnometer for determining **accurately** the density of liquids and solutions is described. Further, determination of the density of a solid using **both volumetric and gravimetric methods**, followed by the determination of the thickness of a metal foil, has been included. Graphing and plotting of data are introduced in the very first experiment.

Experiment 9 and 32: We have developed a microscale technique for the **electrolytic preparation of iodoform and the determination of Faraday's constant (and Avogadro's number).**

Experiment 35: Reverse-phase TLC is used in determining the MW of polymers.

Experiments 17, 19 and 20: These are either sequential or interdependent experiments exploring new methods. For example, sunlight photochemistry is used in Experiment 20; isomerization of bis(glycinato)copper(II) compound is carried out in Experiment 19.

Experiment 36: This experiment explores the preparation and evaluation of a semiconductor material.

Experiments 27 and 37: These simple and easily performed experiments (NMR and IR) illustrate the versatile applications of instrumental methods of analysis at the undergraduate level.

5. In several experiments, we have developed the idea of the **Three R's: Recycle, Recover, and Reuse**. For example, in Experiment 18 students are asked to regenerate lead(II) carbonate for reuse.

6. The last chapter is devoted to **Undergraduate Mini Research** and how to implement it.

7. **Data collection tables** and **data manipulation tables** are included in the manual. The main purpose of providing these tables is to teach students how to **organize and manipulate data**.

8. **Hazard and caution instructions are highlighted inside boxes**.

9. In the beginning of the experimental procedure, **estimated time** to complete the experiment and the **main steps** involved in performing the assignment have been described. This will help each student to plan and to organize his or her laboratory work.

What Is the Experimental Format?

Each experiment is composed of several sections: objectives, prior reading, related experiments, introduction, general references, experimental section, independent projects, prelaboratory report sheet, prelaboratory problems, data sheets, and postlaboratory report.

An Instructor's Manual is also available. It describes the laboratory equipment and chemical compounds used and provides safety remarks and answers to problems.

These experiments have been tested by us and by our students. However, in any new endeavor, errors may creep in. If you find any corrections or suggestions for improving the text, we urge you to contact us or the publishers.

Acknowledgment

First, we wish to thank all the students who participated over the years in the laboratory during the development of the experiments described in this text. The following students deserve special recognition: Patricia Novelli, Nicole Casey, Stanley Mallory, and Stephanie Cook. We also gratefully acknowledge the help and the positive critiques from our colleagues at Merrimack College: Drs. J. David Davis, Stephen A. Leone, Jorge Ibanez, Kathleen C. Swallow, Diane Rigos, and Cynthia McGowan. The assistance of Mrs. Catherine Festa, Mrs. Susan Brien, and Mrs. Rita Fragala is also appreciated.

We gratefully acknowledge the financial assistance in the preparation of this manuscript from John Wiley & Sons, Inc. We also appreciate the help and the encouragement from our chemistry editor at Wiley, Mrs. Nedah Rose, during the development of this text. This text could not have materialized without her support and advice.

We also express our sincere thanks to the following reviewers, who provided many comments, thoughtful criticisms, and positive suggestions to improve the text.

Kenneth Brooks
New Mexico State U.

Warren Zemke
Wartburg College

Vahe Marganian
Bridgewater State College

Douglas Armstrong
Olivet Nazarene U.

Netkal Gowda
Western Illinois U.

Russell Baughman
Northeast Missouri State U.

Steven Wright
U. of Wisconsin
Stevens Point

Craig Jensen
U. of Hawaii

William Litchman
U. of New Mexico

Henry Hollinger
Rensselaer Polytechnic Inst.

Many of these experiments have been offered in numerous Microscale Chemistry Workshops over the past three years, both at the National Microscale Chemistry Center (NMC^2) located at Merrimack College, and at national and international meetings held in the United States and abroad. The participants in these workshops have assisted us in fine-tuning many of these experiments. In this regard, we wish to acknowledge support received from EPA, TURI, (MA), and NSF to our workshops at NMC^2.

Finally and most importantly, we thank our wives: Shashi, Marilyn, and Jill. Without their patience, support, and love, we could not have accomplished our goals.

North Andover, MA
June 14, 1994

Mono Mohan Singh
Ronald Marston Pike
Zvi Szafran

Table of Contents

Chapter 9 Organic Chemistry   605

CHAPTER **1** **Safety Precautions in the Laboratory**

"Safety First—Safety Last"

Objectives
- To be able to recognize hazards and risks in a laboratory and to know how to avoid them
- To learn and appreciate the importance of safety in a laboratory
- To know the locations of safety equipment and escape routes in case of emergencies

The first scheduled laboratory period in General Chemistry should be used for check-in and for a discussion of **laboratory safety rules** and the **proper disposal of chemical waste.** During this time, you will be shown the locations of common laboratory safety equipment, namely, the fire extinguisher, the fire blanket, fire exits, hoods, the eyewash facility, first aid box, nearest telephone, and the safety shower. Specific instructions and guidelines will be provided for the safe disposal of chemicals. For the purpose of laboratory safety, you will be asked to complete a safety form, printed on the inside page of the cover of this laboratory text. Keep this information handy for ready reference.

1.1 General Safety Rules[1–14]

Laboratory safety is of utmost concern. If not handled carefully, many chemicals and/or their combinations can be potentially dangerous. Similarly, many pieces of equipment used in a laboratory (such as electric wires, Bunsen burners, cables, batteries, high-voltage electric outlets, compressed gas cylinders) may also create hazardous situations.

One of the best ways to ensure safety in the laboratory is to minimize contact with all chemicals. The main way in which we promote safety in the laboratory is by using the microscale technique. Microscale chemistry reduces the amounts of chemicals used in the laboratory by a factor of 100 to 1000 from the traditional multigram scale. There are several advantages to doing this, many of which are safety related:

- The generation of chemical waste is drastically reduced. This saves money on disposal costs and prevents the environment from being polluted.
- Chances of chemical related fire or explosion are markedly reduced.
- Air quality in the laboratory is improved. Using smaller amounts of volatile compounds cuts down sharply on the amounts of chemicals present in the air, improving both the smell and the healthfulness of the laboratory.
- Exposure to chemicals is minimized.
- Because of the smaller amounts of chemicals used, the time required to manipulate and complete an experiment is considerably reduced, shortening the laboratory period. This also saves time and money. Furthermore, a larger variety of experiments, using a wider range of chemicals, may be performed.

Using microscale techniques certainly will minimize the risks. However, even though we use small amounts of material, some chemicals are still highly toxic; spills or splattering of a corrosive material still can occur; or a compound may still decompose to generate a noxious gas. Furthermore, in future work, a reaction may have to be scaled up to larger quantities. For these reasons, plus the fact that it just makes sense, each of us should be aware of several safety regulations concerning work in a chemical laboratory. As an individual, you have an obligation to protect yourself and your fellow workers.

There are four stages to maintaining a high standard of safety and a clean laboratory environment.

Stage 1: Before the Laboratory

Safety in the laboratory does not begin when one walks in the laboratory door. Three initial steps must be carried out before the experiment begins.

1. Read the directions of the experiment to be carried out carefully, *in advance* of the laboratory.
2. Think critically about what you are reading and write the experimental sequence as to the chemicals used and the arrangement of equipment. Especially note any safety warnings given in the procedure section. The safety warnings in this laboratory text are for the protection of you and your neighbors. Many people find it useful to prepare a flow chart for each experiment, listing each step of the laboratory procedure in sequence.
3. Check the toxicities of the chemicals involved. Safety data, such as found in Material Safety Data Sheets (MSDS) or the *Merck Index* (see Section 1.2), will be available in the laboratory or posted in the stock room area. Your instructor may also provide additional information on the chemicals to be used.

Stage 2: Dressing for the Laboratory

To enhance safety in the laboratory, you should consider several things with regard to what you wear.

1. **Safety goggles (Figure 1.1a) or other suitable protective glasses must be worn at all times** when experimentation is taking place *anywhere* in the laboratory. Any visitors to the laboratory must also have suitable eye protection. **Contact lenses should not be worn,** as corrosive fumes or chemicals may get underneath them and prevent effective irrigation, ventilation, and flushing of the eyes if an accident should occur.
2. Suitable clothing should be worn. Long hair should be tied back, and if ties or similar items of loosely-hanging clothing are worn, they should be tucked in in an appropriate manner. Clothing that offers protection against an accidental spill is most appropriate. **Laboratory aprons or coats are highly recommended** (Figure 1.1*b*). Similarly, clothing that leaves large areas of the body exposed (such as a cutoff T-shirt, short skirts, or shorts) should not be worn.
3. Never work barefoot in a laboratory. Since glass breakage is a common occurrence, your feet need proper protection. Obviously, open-toed shoes or sandals offer little protection to the feet from chemical spills. They should never be worn in the laboratory.

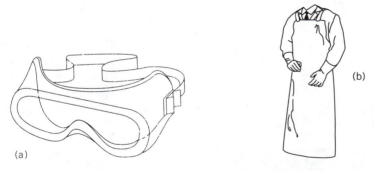

Figure 1.1 (a) Safety goggles (b) Laboratory apron

Stage 3: Inside the Laboratory—The Safety Rules

Specific rules and regulations have been developed based on the experience of those who have extensively studied the safety aspects of the laboratory. It is imperative that you learn these safety rules and follow them at all times. They will form a large part of your code of conduct in the laboratory.

1. Use common sense. Think before you act.
2. Don't rush, don't take short cuts. If you rush your work, at best you will get poor results. At worst, you will be a danger to yourself and to those around you.
3. Report any spill or accident *immediately* to your instructor.
4. Know the location and operation of emergency safety equipment in the laboratory from the first session, particularly the following:
   - Eyewash fountains (Figure 1.2)
   - Safety showers (Figure 1.3)
   - Fire extinguishers (Figure 1.4)
   - Nearest telephone
   - Fire blankets (Figure 1.5)
   - First aid kits
   - Fire exits
   - Spill clean-up kits
5. Smoking is absolutely forbidden in the laboratory. Volatile, flammable solvents can ignite easily and result in an explosion or fire. As far as practicable, avoid using an open flame. Severe burns can result from the use of an open flame in the vicinity of flammable solvents. The microscale technique markedly reduces the possibility of this aspect of potential injury, but we must always be on our guard, nevertheless.

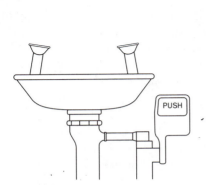

Figure 1.2 Eyewash fountain

Figure 1.3 Safety shower

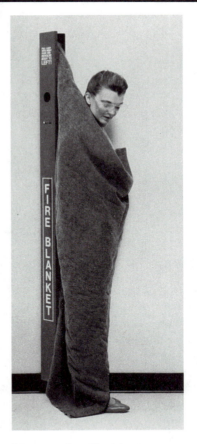

Figure 1.4 Fire extinguisher (courtesy of Fisher Scientific, 711 Forbes Ave., Pittsburgh, PA 15219)

Figure 1.5 Fire blanket (courtesy of Fisher Scientific, 711 Forbes Ave., Pittsburgh, PA 15219)

6. Working alone in a chemical laboratory is prohibited.
7. Minimize exposure to all chemicals. Handle all chemicals with respect. Follow the directions on the container, or those given to you by your instructor. Use a brush and a pan or a mop to clean up a spilled chemical. Use gloves before cleaning a spill. Never directly smell a chemical. Use your hand to waft a tiny amount of vapor toward your nose (Figure 1.6).
8. Dispose of chemicals properly, in the containers provided in the laboratory, according to the instructions given by the laboratory instructor. **Do not simply pour any chemicals in the sink.** Strict governmental regulations control the generation and the proper disposal of laboratory waste. Severe penalties are imposed on those who do not follow proper waste disposal procedures.
9. Never throw solids, loose papers, paper towels, or filter papers into the sink or drain. They can easily clog the drain, causing a flood.
10. Never *eat, chew, or drink* while in the laboratory. Do not put *pipets or your hands* in your mouth. Do not taste any chemicals. Some chemical compounds can be absorbed through the skin. In working with these chemicals, you must wear protective gloves and/or clothing. Your instructor and the experiment directions will provide information in such cases.

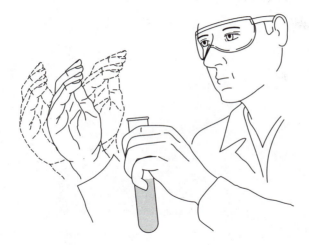

Figure 1.6 The proper way to "smell" chemical vapors

11. Much laboratory equipment consists of glassware. Follow proper procedures for handling this equipment. Ask your instructor for the correct technique. The common cause of laboratory cuts or lacerations is the careless handling of glass equipment. **Be particularly careful** when inserting thermometers, glass rods, or tubes into rubber stoppers. *Be sure the stopper is lubricated with water or glycerin* (see Chapter 3).

Report any incident of breakage or injury to your instructor immediately, no matter how insignificant it may seem at the time. Using a broom and a pan, clean up broken glass immediately. **Special containers are provided for the disposal of broken glass.**

12. Never carry out unauthorized experiments. At this stage of your chemical development, it is imperative for basic safety reasons that you follow the procedures given in the laboratory manual.

13. Keep your laboratory space clean. This rule also pertains to the balance area and where chemicals are dispensed. You or your fellow students could be burned or exposed to toxic chemicals if you do not clean up a spill.

14. Replace caps on containers immediately after use. An open container is an invitation for a spill. Furthermore, some reagents are very sensitive to moisture and may decompose if left open. **Never return an unused portion of a reagent to its original container.** Suitable disposal sites for chemical wastes will be provided in the laboratory.

15. Never heat a closed system. Always provide a vent to avoid an explosion. Provide a suitable trap for any toxic gases generated, such as sulfur dioxide, hydrogen chloride, chlorine, and the like. You will find directions in the experimental procedures.

16. Use a **HOOD** (Figure 1.7) when required. The experimental procedures in the text will indicate when a hood is to be used.

17. Do not engage in games or horseplay in the laboratory.

18. Never leave a burner flame unattended.

19. Always add acid to water (The letter *A* comes before *W*!). Always perform a dilution by adding the concentrated reagent to water. The solution may spatter back if this is done in the reverse order.

Figure 1.7 Chemical fume hood (courtesy of Fisher Scientific, 711 Forbes Ave., Pittsburgh, PA 15219)

20. Never remove any chemical substance or any equipment from the laboratory. Doing so is grounds for expulsion or other severe disciplinary action.
21. Never point a test tube that is being heated toward anyone.
22. When in doubt, do not hesitate to ask your instructor for help.

Stage 4: After the Laboratory is Over

Good housekeeping is an essential step toward safe and successful laboratory work. Before leaving the laboratory, you must do the following:

1. Always clean the area where you worked. Rinse all the glassware and clean any other equipment you used. Even a slight contamination may ruin a future experiment.
2. Return all chemicals and supplies to where they belong.
3. Clean the balance area and the area near other instruments that you have used.

1.2 Planning for Chemical Safety

There are many chemicals (some hazardous and some less so) that you may have to use in a laboratory. A chemical may enter your body through inhalation (breathing), eye contact, skin absorption or by ingestion. A chemical is considered hazardous if it meets one or more of the following criteria (see notes 1, 2):

- It is cancer causing, toxic, or a corrosive substance.
- It is an irritant, a strong sensitizer, an oxidizer, explosive, radioactive, inflammable or reactive, thereby posing a danger to health and the environment.
- It carries a low threshold limit value (TLV) assigned by the American Conference of Governmental Industrial Hygienists (ACGIH).
- It is listed under the Occupational Safety and Health Act.

Use of Material Safety Data Sheets

Many chemicals, as just pointed out, have dangers associated with their use. One way of combating these dangers is to use Material Safety Data Sheets (MSDS), which are provided by each manufacturer or vendor as *required by law* for the chemicals purchased and used in the laboratory. The information given relates to the risks involved when using a specific chemical. These sheets are available to you as a laboratory worker.

A typical MSDS is shown in Figure 1.8 for sodium chloride. The sheet is divided into several sections: Identification, Toxicity Hazards, Health Hazard Data, Physical Data, Fire and Explosion Data, Reactivity Data, and other data. The Identification section provides additional names by which the compound is known (salt, for example), the CAS (Chemical Abstract Service) number, and the Sigma or Aldrich catalog product number. The CAS number is especially useful, as one can access several databases using this number to obtain listings of papers and books in which this compound is discussed and used.

The Toxicity Hazards section contains results of studies detailing the toxicity of the compound in various animal and inhalation tests. Sodium chloride is a well-studied compound, so many such tests have been performed. Several common abbreviations are used:

HMN: Human
IVN: Intravenous (in the blood stream)
LC_{50}:   Lethal Concentration 50; concentration of a substance in air that
         can kill 50% of the test animals when administered
         as a single respiratory exposure
LD_{50}:   The lethal dose with which 50% of the test subjects will die
LD_{Lo}:   Lowest lethal dose
MUS: Mouse
ORL: Oral dose
SKN: Skin
TC_{Lo}:   Toxic concentration low
TD_{Lo}:   Toxic dose low

Thus, the listing ORL-RAT LD50: 3000 mg/kg indicates that when sodium chloride was given via oral dose to a test group of rats, the dose that would kill 50 percent of the rats was 3000 mg per kg of rat weight. If one could extrapolate directly from a rat to a human, an oral dose of 240 g of sodium chloride would kill half of a random group of 80 kg humans. Needless to say, this is well above the amount of sodium chloride one would expect to ingest inadvertently in a laboratory. We can conclude that sodium chloride is not very risky in this regard.

The Health Hazard Data section for sodium chloride indicates that inhalation, ingestion, or skin absorption may be harmful and that the material is irritating to mucous membranes and the upper respiratory tract. This may seem surprising for as "innocent" a material as salt, but it is certainly well enough known that salt will make the eyes sting and that prolonged exposure of the skin to salt water can be harmful. Although spilling a small amount of sodium chloride on the skin would not be harmful, this warning illustrates the general principle of trying to minimize contact with any chemical. The section also gives the treatment for having contact with salt in the eyes: flushing with water for at least 15 minutes.

The Fire and Explosion Hazard Data and Reactivity Data sections provide information about chemical incompatibilities and other chemical reaction dangers. We are told that sodium chloride does not combust and that it may react with strong oxidizing agents

MATERIAL SAFETY DATA SHEET

Sigma-Aldrich Corporation
1001 West Saint Paul Ave, Milwaukee, WI 53233 USA

July 1989 version

------------------ IDENTIFICATION ------------------
PRODUCT #: S9888   NAME: SODIUM CHLORIDE ACS REAGENT
CAS#: 7647-14-5
MF: CL1NA1

SYNONYMS
COMMON SALT * DENDRITIS * EXTRA FINE 200 SALT * EXTRA FINE 325 SALT* HALITE * H.G. BLENDING * NATRIUMCHLORID (GERMAN) * PUREX * ROCK SALT *
SALINE * SALT * SEA SALT * STERLING * TABLE SALT * TOP FLAKE * USP SODIUM CHLORIDE * WHITE CRYSTAL *

------------------ TOXICITY HAZARDS ------------------

RTECS NO: VZ4725000
SODIUM CHLORIDE

IRRITATION DATA
SKN-RBT 50 MG/24H MLD   BIOFX* 20-3/71
SKN-RBT 500 MG/24H MLD   28ZPAK -,7,72
EYE-RBT 100 MG MLD   BIOFX* 20-3/71
EYE-RBT 100 MG/24H MOD   28ZPAK -,7,72
EYE-RBT 10 MG MOD   TXAPA9 55,501,80

TOXICITY DATA
ORL-RAT LD50:3000 MG/KG   TXAPA9 20,57,71
ORL-MUS LD50:4000 MG/KG   FRPPA0 27,19,72
IPR-MUS LD50:6614 MG/KG   COREAF 256,1043,63
SCU-MUS LD50:3 GM/KG   ARZNAD 7, 445,57
IVN-MUS LD50:645 MG/KG   ARZNAD 7,445,57
ICV-MUS LD50:131 MG/KG   TYKNAQ 27,131,80

REVIEWS, STANDARDS, AND REGULATIONS
EPA GENETOX PROGRAM 1988, NEGATIVE: IN VITRO CYTOGENETICS-NONHUMAN; SPERM MORPHOLOGY-MOUSE
EPA GENETOX PROGRAM 1988, INCONCLUSIVE: MAMMALIAN MICRONUCLEUS
EPA TSCA CHEMICAL INVENTORY, 1986
EPA TSCA TEST SUBMISSION (TSCATS) DATA BASE, JANUARY 1989
MEETS CRITERIA FOR PROPOSED OSHA MEDICAL RECORDS RULE FEREAC 47,30420, 82

GET ORGAN DATA
MATERNAL EFFECTS (OVARIES, FALLOPIAN TUBES)
EFFECTS ON FERTILITY (PRE-IMPLANTATION MORTALITY)
EFFECTS ON FERTILITY (POST-IMPLANTATION MORTALITY)
EFFECTS ON FERTILITY (ABORTION)
EFFECTS ON EMBRYO OR FETUS (FETOTOXICITY)
EFFECTS ON EMBRYO OR FETUS (FETOTOXICITY)
SPECIFIC DEVELOPMENTAL ABNORMALITIES (MUSCULOSKELETAL SYSTEM)

------------------ HEALTH HAZARD DATA ------------------

ACUTE EFFECTS
MAY BE HARMFUL BY INHALATION, INGESTION, OR SKIN ABSORPTION.
CAUSES EYE IRRITATION.
CAUSES SKIN IRRITATION.
MATERIAL IS IRRITATING TO MUCOUS MEMBRANES AND UPPER RESPIRATORY TRACT.

FIRST AID
IN CASE OF CONTACT, IMMEDIATELY FLUSH EYES WITH COPIOUS AMOUNTS OF WATER FOR AT LEAST 15 MINUTES.
IN CASE OF CONTACT, IMMEDIATELY WASH SKIN WITH SOAP AND COPIOUS AMOUNTS OF WATER.
IF INHALED, REMOVE TO FRESH AIR. IF NOT BREATING GIVE ARTIFICIAL RESPIRATION. IF BREATING IS DIFFICULT,
GIVE OXYGEN. CALL A PHYSICIAN.

------------------ PHYSICAL DATA ------------------

MELTING PT: 801 C
SPECIFIC GRAVITY: 2.165

APPEARANCE AND ODOR
WHITE CRYSTALLINE POWDER

------------------ FIRE AND EXPLOSION HAZARD DATA ------------------

EXTINGUISHING MEDIA
NON-COMBUSTIBLE.
USE EXTINGUISHING MEDA APPROPRIATE TO SURROUNDING FIRE CONDITIONS.

SPECIAL FIREFIGHTING PROCEDURES
WEAR SELF-CONTAINED BREATHING APPARATUS AND PROTECTIVE CLOTHING TO PREVENT CONTACT WITH SKIN AND EYES.

------------------ REACTIVITY DATA ------------------

INCOMPATIBILITIES
STRONG OXIDIZING AGENTS
STRONG ACIDS

HAZARDOUS COMBUSTION OR DECOMPOSITION PRODUCTS
NATURE OF DECOMPOSITION PRODUCTS NOT KNOWN

------------------ SPILL OR LEAK PROCEDURES ------------------

STEPS TO BE TAKEN IF MATERIAL IS RELEASED OR SPILLED
WEAR RESPIRATOR, CHEMICAL SAFETY GOGGLES, RUBBER BOOTS AND HEAVY RUBBER GLOVES.
SWEEP UP, PLACE IN A BAG AND HOLD FOR WASTE DISPOSAL.
AVOID RAISING DUST.
VENTILATE AREA AND WASH SPILL SITE AFTER MATERIAL PICKUP IS COMPLETE.

WASTE DISPOSAL METHOD
FOR SMALL QUANTITIES: CAUTIOUSLY ADD TO A LARGE STIRRED EXCESS OF WATER. ADJUST THE PH TO NEUTRAL, SEPARATE ANY INSOLUBLE SOLIDS OR
LIQUIDS AND PACKAGE THEM FOR HAZARDOUS-WASTE DISPOSAL. FLUSH THE AQUEOUS SOLUTION DOWN THE DRAIN WITH PLENTY OF WATER. THE
HYDROLYSIS AND NEUTRALIZATION REACTIONS MAY GENERAL HEAT AND FUMES WHICH CAN BE CONTROLLED BY THE RATE OF ADDITION.
OBSERVCE ALL FEDERAL, STATE, AND LOCAL LAWS.

— PRECAUTIONS TO BE TAKEN IN HANDLING AND STORAGE —

CHEMICAL SAFETY GOGGLES.
USE PROTECTIVE CLOTHING, GLOVES AND MASK.
SAFETY SHOWER AND EYE BATH.
MECHANICAL EXHAUST REQUIRED.
DO NOT BREATHE DUST.
DO NOT GET IN EYES, ON SKIN, ON CLOTHING.
WASH THOROUGHLY AFTER HANDLING.
IRRITANT.
KEEP TIGHTLY CLOSED.
HYGROSCOPIC
STORE IN A COOL DRY PLACE.

------ ADDITIONAL PRECAUTIONS AND COMMENTS ------

SECTION 9 FOOTNOTES
REACTS VIOLENTLY WITH BROMINE TRIFLUORIDE AND LITHIUM.

THE ABOVE INFORMATION IS BELIEVED TO BE CORRECT BUT DOES NOT PURPORT TO BE ALL INCLUSIVE AND SHALL BE USED ONLY AS A GUIDE. SIGMA-ALDRICH
SHALL NOT BE HELD LIABLE FOR ANY DAMAGE RESULTING FROM HANDLING OR FROM CONTACT WITH THE ABOVE PRODUCT. SEE REVERSE SIDE OF INVOICE
OR PACKING SLIP FOR ADDITIONAL TERMS AND CONDITIONS OF SALE

Figure 1.8 Material safety data sheet: sodium chloride (reprinted with permission of Aldrich Chemical Co., Inc., Milwaukee, WI)

or strong acids. The steps to be taken if material is released or spilled generally refer to large industrial amounts. In this text specific information will be provided in the experimental procedures for materials with unusual handling characteristics.

Likewise, waste disposal methods generally refer to industrial quantities and are designed for materials that are less than pure. Pure sodium chloride added to water yields a neutral solution. This is not necessarily true of industrial grades of sodium chloride, so care should be indicated. The Handling and Storage section gives some advice on how to deal with the compound as well as recommendations about safety equipment that should be on hand (shower, eye bath).

Finally, the Additional Precautions and Comments section details specific dangers associated with this compound. Sodium chloride is known to react violently with lithium or bromine trifluoride under certain conditions. These materials must never be used in the same reaction step.

The MSDS may seem too detailed to you. This observation is certainly true in the microscale usage of sodium chloride, but keep in mind that these sheets are designed for many different kinds of use. A judicious reading of the sheets will provide the chemist with much useful information, and you will quickly learn what aspects of safety to focus in on. It is much better to have the detailed information and not to need it than to be in the opposite predicament.

The Fire Diamond

Labels on chemical containers exhibit a *fire diamond* that indicates the hazards of the chemical. The diamond is subdivided into four quadrants (Figure 1.9). Quadrant A (blue) refers to health hazards; quadrant B (red) deals with flammability; quadrant C (yellow) indicates reactivity or instability; and quadrant D (white) carries other specific hazards (such as radioactivity). This system uses a rating scale from 0 to 4, where 0 stands for the least danger and 4 indicates the most danger.

Position A: Health hazard (blue)

   0 = Ordinary combustible hazards in fire

   1 = Slightly hazardous

   2 = Hazardous

   3 = Extreme danger

   4 = Deadly

Position B: Flammability (red)

   0 = Will not burn

   1 = Will ignite if preheated

   2 = Will ignite if moderately heated

   3 = Will ignite at most ambient conditions

   4 = Burns readily at ambient conditions

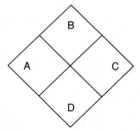

Figure 1.9 Fire diamond

Position C: Reactivity, instability (yellow)

0 = Stable and unreactive with water

1 = Unstable if heated

2 = Violent chemical change

3 = Shock and heat may detonate

4 = May detonate

Position D: Special hazard (white)

OX = Oxidizer, etc.

The Merck Index

Similar information to MSDS in a more compact form can be found in the *Merck Index* (Merck & Co., Rahway, NJ). This basic reference work gives the "bottom line" on the toxicity of chemicals, and their incompatibilities. In the case of sodium chloride (Figure 1.10), the *Index* lists under Human Toxicity: "Not generally considered poisonous. Accidental substitution of NaCl for lactose in baby formulas has caused fatal poisoning." Although the information in the *Merck Index* is not so complete as on the MSDS, it is generally sufficient for our purposes at the microscale laboratory level. The *Index* also supplies some interesting information about the common usage of the chemicals listed, with a special emphasis on medical usage. References to the chemical literature are also provided.

CRC Handbook of Chemistry and Physics

This reference work, which is updated each year, contains a wide range of data (in table form) in the areas of health, safety, and environmental protection. It also includes directions for the handling and disposal of laboratory chemicals. Most laboratories generally have copies of this work.

1.3 Fire Safety

Many solvents used in the general chemistry laboratory are nonflammable. However, solvents such as ethanol, methanol, diethyl ether, and acetone are flammable. (Appendix 7 gives safety data for common organic solvents, including fire safety data.) Chemical fires can also occur. There is also a small risk (due to the use of electronic equipment) of an electrical fire due to a short circuit, frayed electrical cord, or power surge.

If a fire should occur, the most important thing to remember is to keep calm. **The very first thing you do is notify your instructor immediately.** Several types of equipment for dealing with small fires are present in most laboratories. The most obvious is the fire extinguisher. A laboratory extinguisher (Figure 1.4) should weigh no more than 10 lb. It must be of convenient size to lift and employ rapidly. Ideally, there should be at least one fire extinguisher for every laboratory bench. Several types are available, the most common being of the dry chemical (bicarbonate powder under pressure) or compressed carbon dioxide type. Most small fire extinguishers are activated by pointing the nozzle toward the base of the fire and squeezing the handle. A jet of compressed powder or foam will then discharge

8430. Sodium Chloride, Salt; common salt. ClNa; mol wt 58.45. Cl 60.66%, Na 39.34%. NaCl. The article of commerce is also known as *table salt*, *rock salt* or *sea salt*. Occurs in nature as the mineral halite. Produced by mining (rock salt), by evaporation of brine from underground salt deposits and from sea water by solar evaporation: Faith, Keyes & Clark's *Industrial Chemicals*, F. A. Lowenheim, M. K. Moran, Eds. (Wiley-Interscience, New York, 4th ed., 1975) pp 733–730. Comprehensive monograph: D. W. Kaufmann, *Sodium Chloride*, ACS monograph Series no. 145 (Reinhold, New York, 1960) 743 pp.

Cubic, white crystals, granules, or powder; colorless and transparent or translucent when in large crystals. d 2.17. The salt of commerce usually contains some calcium and magnesium chlorides which absorb moisture and make it cake. mp 804° and begins to volatilize at a little above this temp. One gram dissolves in 2.8 ml water at 25°, in 2.6 ml boiling water, in 10 ml glycerol; very slightly sol in alcohol. Its soly in water is decreased by HCl and it is almost insol in concd HCl. Its aq soln is neutral. pH: 6.7–7.3 d of said aw soln at 25° is 1.202. A 23% aq soln of sodium chloride freezes at –20.5°C (5°F). LD_{50} orally in rates: 3.75 g/kg, Boyd, Shanas, *Arch. Int. Pharmacodyn. Ther.* **144**, 86 (1963).

Note: Blusalt, a brand of soldium chloride contg trace amounts of cobalt, iodine, iron, copper, manganese, zinc is used in farm animals.

Human Toxicity: Not generally considered poisonous. Accidental substitution of NaCl for lactose in baby formulas has caused fatal poisoning.

USE: Natural salt is the source of chlorine and of sodium as well as of all, or practically all, their compds, e.g., hydrochloric acid, chlorates, sodium carbonate, hydroxide, etc; for preserving foods, manuf soap, dyes—to salt them out; in freezing mixtures; for dyeing and pritning fabrics, glazing pottery, curing hides; metallurgy of tin and other metals.

THERAP CAT: Electrolyte replenisher, emetic; topical antiinflammatory.

THERAP CAT (VET): Essential nutrient factor. may be given orally as emetic, stomachic, laxative or to stimulate thirst (prevention of calculi). Intravenously as isotonic solution to raise blood volume, to combat dehydration. Locally as wound irrigant, rectal douche.

Figure 1.10 *Merck Index* listing: sodium chloride (reprinted by permission [10th ed., 1983] Merck & Co., Inc., Rahway, NJ)

from the nozzle, smothering the fire. In some cases, it may be necessary to pull a pin from the handle before it can be squeezed. Some fire extinguishers only operate when they are turned upside-down. It is imperative that each student be familiar with the proper use of the fire extinguishers located in the laboratory. In most cases of fire, you should leave the laboratory and allow the instructor to take the necessary steps.

Alternative ways exist for putting out small fires. Fires in small vessels can be extinguished by inverting a beaker or other such item over the burning vessel, thereby excluding oxygen. A second way of putting out such fires is by covering the vessel with soaking wet towels. Never use dry towels for this purpose. If a fire should occur, it is important to remove any flammable material from the vicinity immediately, especially containers of flammable solvents or gas cylinders. If fire comes in contact with these items, an explosion can occur.

Whenever a fire occurs, there is also an associated danger of inhalation of smoke or toxic fumes. This is potentially more dangerous than the fire itself. Thus, if the air is not fit to breathe, the fire should be abandoned, and the fire department called. Any persons overcome by fumes should be removed to a well ventilated area, and health professionals should be called immediately. If the fire is too large to contain, the area should be evacuated, and the fire department should be called. The person calling the fire department should tell the dispatcher the specific nature of the laboratory fire so that the firefighters can bring the proper equipment.

In the event that a person's clothing should catch on fire, all laboratories should be equipped with safety showers (Figure 1.3) and fire blankets (Figure 1.5). To activate a safety shower, merely stand beneath it and pull down on the lever or chain. Remain under the shower until you are thoroughly soaked. To use a fire blanket, grasp the rope or material at the end of the blanket and turn so that the blanket surrounds you tightly and smothers the fire.

Notes

1. Gorman, C., Ed. *Working Safely with Chemicals in the Laboratory;* Genium Publishing: Schenectady, NY, 1993.
2. Wulfman, D.S. *Materials Safety Manual for General Chemistry Lab Activities;* Genium Publishing: Schenectady, NY, 1992.

General References

1. Armour, M. *Hazardous Laboratory Chemicals Disposal Guide;* CRC Press: Boca Raton, FL, 1991.
2. Furr, A.K., Jr., Ed. *Handbook of Laboratory Safety,* 3rd ed.; CRC Press: Boca Raton, FL, 1990.
3. *Hazards in the Chemical Laboratory,* 4th ed.; D. Muir, Ed.; Royal Society of Chemistry: London, 1986.
4. *Informing Workers of Chemical Hazards: The OSHA Hazard Communication Standard;* American Chemical Society: Washington, DC, 1985.
5. Kaufman, J.A., Ed. *Waste Disposal in Academic Institutions;* Lewis Publishers: Chelsea, MI, 1990.
6. Kaufman, J.A. *J. Chem. Educ.* **1992,** *69,* 911 and *Laboratory Safety Guidelines: 40 Suggestions for Improving Laboratory Safety;* Curry College: Milton, MA.
7. Lenga, R.E., Ed. *The Sigma-Aldrich Library of Chemical Safety Data,* 2nd ed.; Sigma-Aldrich Corp.: Milwaukee, WI, 1987.
8. *Less Is Better (Laboratory Chemical Management for Waste Reduction),* 2nd ed.; American Chemical Society: Washington, DC, 1993.
9. Lide, D.R., Ed. *Handbook of Chemistry and Physics,* 71st ed.; CRC Press: Boca Raton, FL, 1990–1991.
10. *Merck Index of Chemicals and Drugs,* 11th ed.; Merck & Co.: Rahway, NJ, 1990.
11. *RCRA and Laboratories;* American Chemical Society: Washington, DC, 1986.
12. *Safety in Academic Chemistry Laboratories,* 5th ed.; American Chemical Society: Washington, DC, 1990.
13. Szafran, Z.; Singh, M.M.; Pike, R.M. *J. Chem. Educ.* **1989,** *66,* A263.
14. Szafran, Z.; Pike, R. M.; Foster, J.C. *Microscale General Chemistry Laboratory with Selected Macroscale Experiments;* Wiley: New York, 1993.

References 4, 8, 11, and 12 are available at minimal cost from the American Chemical Society. Reference 13 is available from the authors. Notes 1 and 2 and references 1, 5, and 6 are strongly recommended for the use of students (especially chemistry majors).

CHAPTER 2

Mathematical Methods and Manipulation of Data

Objectives

- To learn about the manipulation of numbers and how to treat experimental data

Chemistry is an experimental science. When we do experiments, we make measurements, which are subject to a degree of uncertainty or error. It is nearly impossible to measure a "true value" of a quantity. The best we can do is to use a reliable technique to measure a value. Our ability to draw conclusions from the experimental data depends on how accurately and precisely we make measurements (accuracy and precision are *not* the same thing), how we manipulate the numbers, and what the numbers mean. Some questions we might keep in mind include the following:

- What is the best way of expressing the data?
- How accurate and precise are our experimental measurements?
- How appropriate is the method we used? Would we get the same answer if we measured it another way?
- How many times must we repeat a measurement before we can be reasonably certain of the result?
- Once we are certain of the result, what is the best way to display the data so that we can see trends?

Each of these questions affects the quality of the results and the strength of interpretation of the results. We make efforts to evaluate our data (*i*) by comparing them with already known and established values (true values), (*ii*) by searching literature data collected by other scientists and comparing these with those we obtained, (*iii*) by expanding the capability of experimental techniques to avoid any source of uncertainty, and (*iv*) by applying statistical methods in manipulating the data we collected. However, the most important factor in doing an experiment is to collect the data as truthfully and reliably as possible and to report them as they are obtained without trying to "adapt the data" to the preconceived concept. In this chapter we deal with the relationship between the errors in experimental measurements and the degree of reliability of the results.

2.1 Accuracy, Precision, and Uncertainty[1–3]

Every measurement, presumably, has a "true value." At a given temperature, a block of wood has a definite length. A portion of a compound has a definite mass. The result we get when we measure the object, however, often depends on how we measure it. Suppose, for example, we have a certain quantity of a liquid. If we measure it with a 10 mL graduated cylinder (which is a rough way to measure things), we will get one answer. If we measure it with a 10 mL volumetric pipet (a more exact way to measure things), we will get a similar, but probably different result.

There is an uncertainty (error) associated with any measurement, due to the inherent limitations of the measuring device and the care with which it is used. Typically, the uncertainty is ±0.5 in the last calibrated digit.

Measuring Device	Uncertainty
Triple beam balance	±0.05 g
Analytical balance	±0.00005 g
Digital milligram balance	±0.001 g
100 mL graduated cylinder	±0.5 mL
10 mL graduated cylinder	±0.05 mL
25 mL buret	±0.02 mL
25 mL volumetric flask	±0.02 mL
1 mL graduated pipet (0.01 grad.)	±0.002 mL

Two broad categories of errors are associated with a measurement: **determinate** or **systematic error** and **indeterminate** or **random error.** Determinate errors have assignable causes. A determinate error is unidirectional with respect to the true value; it makes all of the data obtained by replicate measurements either high or low, but not both high and low. A determinate error can be caused by a personal error, by a faulty instrument or by a reagent error (impurity present in the reagent, wrong concentration, and so forth). Determinate errors can be detected and compensated for. On the other hand, an indeterminate error is caused by the limitations of physical measurement. A better experimental method may reduce the magnitude of the random error but cannot eliminate the error entirely. Most often statistical methods are used to reduce the magnitude of random errors.

Accuracy is defined as how close an experimentally determined value is to the "true value." Note that a "true value" has to be determined by a set of measurements and has been tested several times before being accepted as the "true value." In that sense, then, a "true value," which is supposed to be known with absolute correctness, does not exist. Suppose that the true value of the volume you were measuring was 1.00 mL, and you obtained values of 0.998, 0.999, 0.997, 1.001, and 1.002 mL using a graduated 2 mL volumetric pipet. Your values would be reasonably accurate, as they are all close to the true value of 1.00 mL. However, with a graduated cylinder, you might have obtained values of 0.95, 0.98, 1.03, 1.05, and 1.01 mL. These results would be less accurate, because the graduated cylinder is a less accurate way of measuring volumes than a volumetric pipet.

Precision refers to how reproducible your measurements are. It represents the agreement between two or more data being collected under identical conditions. Suppose that you had used a ruler to measure the length of a particular object 10 times, each time obtaining the result of 3.45 cm. The value 3.45 cm would be known with great precision. The result might not be at all accurate (if the ruler were warped or mismarked, for example), but it would be precise.

The precision of a group of data gives some indication of how many times an experiment must be repeated. Consider the following data, obtained by students A, B, and C, each of whom measured some phenomenon 12 times:

A	B	C
2.03	1.95	2.01
2.01	1.99	2.00
2.04	2.08	2.02
2.02	2.11	2.00
2.03	2.05	2.00
2.04	2.03	1.99
2.02	2.01	2.01
2.03	2.00	2.00
2.03	2.03	1.99
2.05	2.06	2.01
2.03	2.01	1.98
2.02	2.05	2.00

A simple way to visualize the precision of these measurements is to plot how many **times** (frequency) each value occurs (y axis) versus the value itself (x axis), and observe how tightly bunched (another way of saying precise) the readings are. The result is shown in Figure 2.1. We can see that a and c are more precise (more tightly bunched) than b. In cases a and c, a sufficient number of data points seem to have been gathered, as the data are fairly well centered about a mean value. No further measurements need to be made. In case b, more data are needed, as there are an insufficient number of data points to establish convergence.

In scientific measurements, we obviously strive for data that are both accurate and precise. How can we ensure that we obtain this sort of data? First, when we investigate a phenomenon, we will obtain several duplicate runs (helping our accuracy). Replicate measurements are made to ensure the reliability of the experimental data. Second, the phenomenon may be investigated in more than one way, in case one type of measurement does not give an accurate value. There have been many cases where results obtained in one manner have been invalidated owing to some external influence that the investigator did not take into consideration.

Before we can explore the mathematical methods of manipulating experimental data, we must first define a number of terms used in connection with the measurement and handling of data.

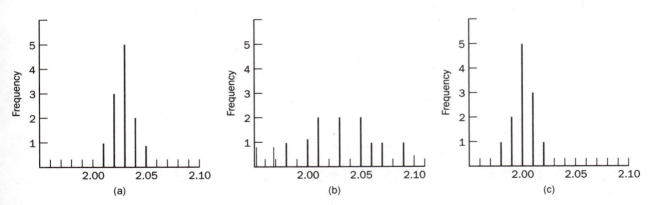

Figure 2.1 Frequency plot for tabulated data

The **mean** or the **average** (or **arithmetic mean**) of a set of data is defined as the sum of all the replicate measurements divided by the number of times the values have been measured. The mathematical expression of an average, x_{av} is

$$x_{av} = \frac{1}{N} \sum_{i=1}^{n} x_i$$

where Σ (pronounced *sigma*) is the symbol for the summation of N number of individual values, x_i.

The **median** m is that value of a set of data about which all other values are equally distributed. If we arrange a set of data from the highest to the lowest value, then the central value is the median value. In the given set of data, therefore, half of the values will be greater and half of the values will be smaller than the median value. The following example will illustrate these two definitions.

Example 2.1   The determination of the density of a substance is used to characterize the substance. A chemist has obtained the following values for the density of a liquid: 1.221, 1.432, 1.312, 1.223, and 1.181 g/mL. Calculate the mean (or the average) and the median of this set of data.

Answer

The average, x_{av}, is $[(1.221 + 1.312 + 1.223 + 1.432 + 1.181)\,g/mL]/5 = 1.2738\ g/mL = 1.274\ g/mL$ (rounding off). The median, m is chosen from 1.432 g/mL, 1.312 g/mL, 1.223 g/mL, 1.221 g/mL, and 1.181 g/mL. The median value is thus 1.223 g/mL. Note that two values are greater than the median and two values are smaller. In this example, we have used an odd number of readings (5), the median being the central value. In the case of an even number of values, select the two central values and average the two to find the median.

Precision can be expressed either by an **absolute method** or by a **relative method**. One of the absolute methods consists of expressing the deviation of individual values from the average without regard to sign. This is expressed as $|x_i - x_{av}|$. The higher the deviation, the lower the precision.

In example 2.1, the numerical differences between the mean (x_{av} = 1.274 g/mL) and the individual values (g/mL) are as follows:

x_i	$|x_i - x_{av}|$	Average deviation
1.221	0.053	$(0.053 + 0.038 + 0.051 + 0.158 + 0.093)/5$
1.312	0.038	$= 0.080$
1.223	0.051	
1.432	0.158	
1.181	0.093	

Precision can also be expressed as the **spread** or **range** (w) which is the difference between the highest and the lowest value of the set. In the given example the spread w is 0.251 g/mL.

One method of expressing the relative precision involves the calculation of relative deviation from the mean in percentage or in parts per thousand (ppt). For example, the following calculations are based upon the data in the table just shown. Relative deviation of the first result from the mean in percentage is calculated as follows:

$$\text{Percent relative deviation} = (|x_i - x_{av}|/x_{av}) \times 100$$
$$= 4.553\% = 4.6\%(\text{rounding off})$$

However, the most important method of expressing the precision is the **standard deviation** s and **the coefficient of variation**, v (see section 2.5). Accuracy is expressed as an **absolute error**. This is calculated as the difference between the experimental value x_i, and the "true value" x_t. This calculation requires that the true value of the quantity be known. The absolute error e of any determined value is given by the expression

$$e = x_i - x_t$$

Here, in contrast to precision, we will express the error with a positive or a negative sign to indicate the direction of the error. Suppose that the "true value" of the density of the liquid in example 2.1 is 1.220 g/mL. The accuracy of the first of the five values will be given by

$$e_1 = 1.221 - 1.220 = +0.001 \text{ g/mL}$$

Similarly, one can also calculate the absolute error of the mean as well as the relative absolute error in percentage or in parts per thousand. In every case, the result must be expressed with respect to its sign: positive (+) or negative (−).

In practice, absolute error expresses the error associated with a measured value. For example, if you deliver 4.50 mL of a solution from a graduated pipet that is known to have an error of + or −0.01 mL, then the reading must be expressed as 4.50 ± 0.01 mL. The relative uncertainty for a reading of 4.50 ± 0.01 mL is expressed as

$$\text{Relative error} = \text{absolute error}/\text{magnitude of measurement}$$
$$0.01 \text{ mL}/4.50 \text{ mL} = 0.002(\text{no units})$$
$$\text{Percent relative error} = 0.002 \times 100 = 0.2\%$$

2.2 Scientific Notation[3]

Numbers obtained in chemical measurements can span a very wide range of values. The mass of an oxygen molecule is 0.000000000000000000000053 g. The number of oxygen molecules in a commonly used quantity known as a mole is 602,300,000,000,000,000, 000,000. Neither of these numbers is particularly convenient to work with, as they contain too many zeros—it is too easy to write one zero too many or one too few. We can readily see that in chemistry it is not always convenient to express measurements using "plain" numbers.

Chemists use **scientific notation** to write very large or very small numbers. In scientific notation, numbers are written in the form $a.aaa \times 10^z$, as powers of 10. The number $a.aaa$ is called the **coefficient**, and z is called the **exponent**.

Consider the number 1000. This number is the product of $10 \times 10 \times 10$ and can therefore be written as 1×10^3. Note that one can do this step more easily (and avoid factoring the number) by moving the decimal point at the end of the number three places to the left until one obtains a number between 1 and 10. **The number of places to the left that the decimal point is moved is the value of the exponent**. Thus, we obtain the following representations in scientific notation for the numbers below:

$$145 = 1.45 \times 10^2 \qquad 95,134 = 9.5134 \times 10^4 \qquad 3,567,000 = 3.567 \times 10^6$$

If the number is less than one, the decimal point needs to be moved to the right until one obtains a number between 1 and 10. **The number of places to the right that the decimal point is moved is the negative of the value of the exponent**. Consider the number 0.0015. We need to move the decimal point three places to the right. Thus, in scientific notation, the number 0.0015 becomes 1.5×10^{-3}. The general rule is: **Whenever the decimal point is moved to the left, the exponent increases by one for each place moved. Whenever the decimal point is moved to the right, the exponent goes down by one for each place moved**.

When one is adding or subtracting numbers in scientific notation, all the numbers must have their exponents to the same power. Suppose that we wish to add the numbers 3.51×10^3 and 1.2×10^2. We must write both as having the same power, so we must either change the 3.51×10^3 to 35.1×10^2 by moving the decimal point one place to the right or, alternatively, write 1.2×10^2 as 0.12×10^3 by moving the decimal point one place to the left. Choosing the second option, we write

$$
\begin{array}{r}
3.51 \times 10^3 \\
+\,0.12 \times 10^3 \\
\hline
3.63 \times 10^3
\end{array}
$$

Note that *only the number* is added or subtracted—the exponent remains the same.

When one is multiplying numbers in scientific notation, the coefficients are multiplied and the exponents are added. Thus, when 2×10^5 is multiplied by 3×10^3, the result is 6×10^8. Similarly, when dividing numbers in scientific notation, the coefficients are divided and the exponents are subtracted. For example, when 9×10^6 is divided by 3×10^2, the result is 3×10^4.

2.3 Rounding Off

Suppose that the length of a block of wood is measured five times using a ruler calibrated to a tenth of a centimeter. The following results are obtained:

Reading #	Length, cm
1	120.54
2	120.58
3	120.55
4	120.53
5	120.54

$$\text{Average, } x_{\text{av}} = (1/5)(120.54 + 120.58 + 120.55 + 120.53 + 120.54) = 120.548 \,\text{cm}$$
$$= 120.55 \,\text{cm}$$

Note that the result has three decimal places, but none of the individual measurements had more than two. Further, the ruler was only calibrated to a tenth of a centimeter. How many decimal places should be displayed in the average? Since the readings were all given with two decimal places (tenths marked on the ruler and hundredths being estimated), the largest number of decimal places reported should be two.

We now need to round the number off to two decimal places. The rules for rounding off numbers are as follows:

1. Look at the digit to the right of the number of decimal places being retained. In this case, since we are retaining two decimal places, we look at the third digit to the right of the decimal point.
2. If the digit is less than five, drop it and all digits to its right.
3. If the digit is greater than five, drop it and all digits to its right, and increase the last retained digit by one.
4. If the last digit is a five:
    (a) If there are only zeros to its right, drop it and all digits to its right, and make the last retained digit even by increasing it or leaving it alone.
    (b) If there are nonzero numbers to its right, drop it and all digits to its right, and increase the last retained digit by one.

In the preceding example, we want to round the number 120.548 to two decimal places. We therefore look at the digit to the right of the last retained digit (last retained digit is a four, so we look at the eight. Since eight is greater than five (rule 3), we drop the eight, and increase the previous number by one. We therefore obtain the result 120.55 cm.

2.4 Significant Figures

We saw in the previous example that the formula for taking the average gave a result with three decimal places, but we only reported two of them (the third was an artifact of the mathematics—the division step—and did not correspond to the actual accuracy of the measurement). Properly, then, only two decimal places have any significance. The last digit was not significant and was dropped. The resulting average has five significant figures: three to the left of the decimal point, and two to the right. An interesting question that we might ask is, "Are all reported numbers significant?"

Many times, a chemist is asked to analyze other people's data. How do we know how many significant figures a number has in this case? For numbers *with decimal points*, the rule of thumb is that **all digits are significant except for zeros to the left of the leftmost nonzero digit** (i.e., 0.0015 has two significant figures, not 4 or 5).

Numbers *without decimal points* are a little trickier. If they end in zero(s) (e.g., 12,500), are the zeros significant? Normally, the assumption is that zeros of this type are not significant, unless some indication is made to the contrary. The most common ways to indicate a significant ending zero are to write a bar above the zero or to write the zero in boldface type (e.g., 12,5**0**0). An easier way to indicate significance is to use scientific notation. If the 10's place zero in 12,500 were significant, but not the 1's place zero, this number would be written as 1.250×10^4, making it clear that there are 4 significant figures. When one is reporting an average for several direct measurements, the number of significant figures in the average should not exceed the number of significant figures in the individual measurements.

In calculations, there are two major rules for significant figures, one for adding and subtracting, the other for multiplying and dividing. **When adding and subtracting numbers, retain all significant figures while calculating, then round off to the precision of the least precise value.**

Suppose we have recorded the relative amount of light absorbed by a series of solutions of known concentration. In performing calculations on these data, we must first add up the concentrations and absorbances (defined later in this chapter):

Concentration	1.24×10^{-4}	3.00×10^{-4}	6.00×10^{-4}	1.10×10^{-3}
Absorbances	0.010	0.026	0.53	1.1

Having converted the 1.10×10^{-3} to 11.0×10^{-4}, we add the concentrations, getting a total of 21.24×10^{-4}. The least precise of the values is 11.0×10^{-4} (one digit to the right of the decimal point), so the sum is rounded to match, resulting in 21.2×10^{-4} or 2.12×10^{-3}. Likewise, the sum of absorbances, properly rounded, is 1.7 (one digit to the right of the decimal point), not 1.666.

When one is subtracting, the answer may have fewer significant figures than the initial numbers. Say the mass of an empty filter crucible was 25.8367 g. It is used to collect a product, then dried and reweighed, and the final mass is 26.2391 g. These masses each have six significant figures, and each is precise to 0.0001. The difference is 0.4024 g, which has the same precision (four digits to the right of the decimal point).

> When one is multiplying and dividing, the number of significant figures in the answer is the same as the factor with the least number of significant figures.

Assume that the product collected in the preceding paragraph (0.4024 g) has a molecular weight of 392.18 g mol^{-1}. To calculate the number of moles, the mass must be divided by the molecular weight, and the result is $1.026059462 \times 10^{-3}$ moles. How many significant figures should be reported? Of the two original numbers, the mass has four significant figures, and the molecular weight has five, so the answer is properly reported as 1.026×10^{-3} (four significant figures). **When you have a combination of adding/subtracting and multiplying/dividing, both rules must be used.** If we were to combine the two previous calculations into one equation, we would have

$$\frac{(26.2391 - 25.8367)\,\text{g}}{392.18\,\text{g mol}^{-1}}$$

Note that the numerator has six significant figures and the denominator has five, so it might seem that the answer should have five. However, note that the numerator has a subtraction step in it, and the result of the subtraction has only four significant figures. Thus, the answer should have only four significant figures.

2.5 Standard Deviation

The most common method of determining the precision of a series of measurements is called the **standard deviation** s. The formal equation for standard deviation is

$$s = \sqrt{\frac{\sum (x - x_{av})^2}{N - 1}}$$

Here, the average value of the measurement (x_{av}) is obtained and subtracted from each given measurement (to keep it simple, we use x instead of x_i). The difference is then squared, and the squared differences are then summed up for each measurement. This sum is divided by $N - 1$, where N is the number of measurements. Finally, the square root is taken.

In practice, this method of taking the standard deviation is quite tedious, and a simpler method is available that is equivalent. In this approach, the equation is

$$s = \sqrt{\frac{\sum x^2 - (\sum x)^2 / N}{N - 1}}$$

Example 2.2   Suppose that we weigh a number of soil samples and obtain weights of 150, 162, 160, and 152 g. What is the standard deviation of our sample?

Answer
Using the preceding equation, we first must add up the measurements, as well as their squares:

x	x^2
150	22,500
162	26,244
160	25,600
152	23,104
$\sum x = 624$	$\sum x^2 = 97,448$

Note that $N = 4$, that is, there have been four measurements. Substituting into the foregoing equation, we write

$$s = \sqrt{\frac{97,448 - (624)^2 / 4}{3}}$$
$$s = \sqrt{34.67} = 5.88 \text{ g}$$

Standard deviation is related to probability. There is a 68 percent chance of obtaining a value within one standard deviation of the average value, a 95 percent chance of obtaining a value within two standard deviations, and a 99.7 percent chance of obtaining a value within three standard deviations. Thus, of a group of 100 weighings using this equipment, 68 of them would obtain readings within 5.88 g of 156 g; 95 of them would obtain readings within 11.8 g of 156 g, and so on.

Table 2.1 Values of t for various confidence levels[1]

Confidence levels, %	t
50	0.67
68	1.00
80	1.29
90	1.64
95	1.96
96	2.00
99.9	3.29

When the standard deviation is expressed as a percentage of the mean (percent relative standard deviation), it is called **coefficient of variation** v:

$$v = 100 \left(\frac{s}{x_{av}} \right)$$

Chemists use statistical methods to reduce the magnitude of indeterminate errors. A true mean μ of a measurement is a constant and remains unknown. However, limits, called **confidence limits**, may be set about the experimentally determined average x_{av} within which one may find the value of μ with a certain degree of confidence. The interval defined by such limits is called a **confidence interval**. For small sets of replicate data, the confidence limit for μ is given by

$$\text{Confidence limit for } \mu = x_{av} \pm \frac{ts}{(N)^{1/2}}$$

where N is the number of measurements. Values of t for various confidence levels are given in Table 2.1.

Example 2.3 In example 2.1, the average density of a substance was 1.27 g/mL. Five replicate determinations were made. The standard deviation was found to be 0.101. Calculate the 80% confidence limit and the confidence interval.

Answer

$$80\% \text{ confidence limit} = 1.27 \pm \frac{(1.29)(0.101)}{5^{1/2}} = 1.27 \pm 0.0552$$

The confidence interval is therefore 1.21 to 1.33.

If a set of data contains an outlying value that appears to be significantly different from the average, then how do we decide whether or not to reject such a value? In this regard, a method called the **Q test** is quite useful.

It is applied as follows:

1. Calculate the range or the spread of the values (see section 2.1).
2. Obtain the difference between the questionable value and its nearest neighbor.

Table 2.2 Values for rejection quotient, Q_{tab} (if $Q_{cal} > Q_{tab}$, reject the value)

No. of Readings	90% confidence	96% confidence	99% confidence
3	0.94	0.98	0.99
4	0.76	0.85	0.93
5	0.64	0.73	0.82
6	0.56	0.64	0.74
7	0.51	0.59	0.68
8	0.47	0.54	0.63

Table 2.3 Values of F at the 95% confidence level

N_2	n_1 (numerator)					
n_2 (denominator)	2	3	4	5	6	7
2	19.00	19.16	19.25	19.30	19.33	19.4
3	9.55	9.28	9.12	9.01	8.94	8.74
4	6.94	6.59	6.39	6.26	6.16	6.09
5	5.79	5.41	5.19	5.05	4.95	4.88
6	5.14	4.76	4.53	4.39	4.28	4.21
7	4.74	4.35	4.12	3.97	3.87	3.79

[handwritten annotations: N_2, N_1, degrees of freedom]

3. Calculate Q_{cal} by dividing the difference (step 2) by the spread (step 1).
4. Compare Q_{cal} with Q_{tab} (see Table 2.2). If $Q_{cal} > Q_{tab}$, then reject the questionable value; if $Q_{cal} < Q_{tab}$, then retain the value.

Example 2.4 In example 2.1, in which there are five data points, decide whether to reject the highest value of 1.432 g/mL at the 90% confidence level.

Answer
$Q_{cal} = (1.432 - 1.312)/(1.432 - 1.181) = 0.478$, which is < 0.64 at 90%. Rejection is not indicated.

To compare the precision of two different methods used for measuring the same quantity, we use the **F test**. This is designed to indicate whether there is a significant difference between two methods based on their standard deviations. F is defined as the ratio of two variances, s_1^2 and s_2^2: $F = s_1^2/s_2^2$ where $s_1^2 > s_2^2$ (variance is defined as the square of the standard deviation).

These two variances are related to two different **degrees of freedom**, N_1 and N_2, where degrees of freedom is defined as $N - 1$ for each case. If the calculated F value is greater than the tabulated value (Table 2.3) at the selected confidence level, then there is a significant difference between the variances of the two methods.

2.6 Graphing of Data

Recognizing relationships from written data is often difficult, whereas they are often easier to see in a graph. A graph illustrates the way in which one property (**the dependent variable**, y) changes when some other property (**the independent variable**, x) undergoes a controlled change. Some examples where graphing makes the relationship between variables clear include:

- How the volume (y axis) of a given amount of gas at a constant pressure changes as a function of the temperature (x axis)
- How absorbance (y axis) changes with concentration (x axis) at a fixed wavelength

Several elements make up a graph. The most common graphs use an x-y axis system (called Cartesian coordinates), with the origin being the point (0,0). Other coordinate systems are also sometimes used, such as spherical polar and cylindrical coordinates. The origin is not always included on graphs. If constructed by hand, graphs should be drawn on graph paper with a grid of at least 10 per centimeter. Various computer graphing programs are also available. Some instruments provide data directly as graphs.

In preparing a graph of some data, one must first lay out the graph coordinates. First, examine the range of x values and y values and scale each axis accordingly. The best graphs usually range from just below the smallest x and y values to just above the largest x and y values. It is often helpful to express the values for an axis all in the same power of 10 to avoid scale errors. Make sure that all your data will fit on the graph and that the data points are distributed over the entire graph. Try to avoid bunching data so that they give a vertical or horizontal line (unless you are trying to show that one of the variables is a constant, such as an equilibrium constant). The basic elements of a graph are shown in Figure 2.2.

Label the axes according to the property they represent, *including any units* (molarity, °C, and the like). When one is dealing with scales containing powers of 10, it can be very confusing to write all the exponents on the scale. A more convenient method is to multiply all of the data for that variable by some power of 10, then include that notation in the axis label. For example, suppose the x axis of a graph is of concentrations of Fe^{2+}, with values of 2×10^{-4}, 4×10^{-4}, 6×10^{-4}, and 8×10^{-4} M. It is more convenient to multiply all the values by 10^4. The scale would now have just the values 2, 4, 6, and 8 and the axis label would be "Concentration of Fe^{2+}, $M \times 10^4$" (meaning that the values on the axis are 10^4 times the actual values).

A data point is an (x, y) data pair, plotted on your graph. After plotting all data, a best-fit straight line or smooth curve should be drawn (not a jagged point-to-point connecting line). The line or curve does not necessarily have to touch all (or any) of the data points—it should go through a visual average of the points.

Straight-line graphs are the most useful in analyzing chemical data, as they provide for easy interpolation; however, raw data often do not give a straight line when the two

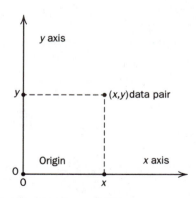

Figure 2.2   Basic graph

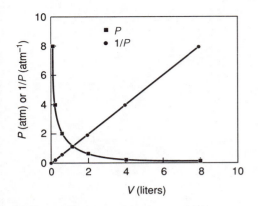

Figure 2.3 Plot of P (or $1/P$) versus V

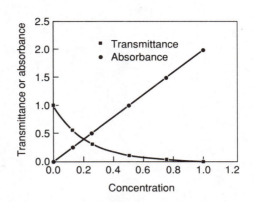

Figure 2.4 Transmittance or absorbance versus concentration

variables are graphed. Data can often be linearized, or transformed, to give a linear relationship. Transformation is a mathematical manipulation of a variable (x), such as squaring it (x^2), taking its inverse ($1/x$), its logarithm ($\ln x$) or some other mathematical function. Sometimes, both variables need to be transformed to give a linear relationship. If y is the quantity being plotted on the y axis, and x is the quantity plotted on the x axis, the equation of the resulting straight line is

$$y = mx + b$$

where m is the slope of the line ($\Delta y/\Delta x$ or rise/run), and b is the point where the line crosses the y axis (y intercept). Some examples of transforming data include the following:

1. The pressure P and volume V are recorded for a fixed amount of a gas at room temperature. When P (y axis) is plotted against V (x axis), a straight line is not obtained—a hyperbola is. This can be made linear by transforming P by taking its inverse, $1/P$, and plotting this against V. The equation of the resulting line is therefore:

$$1/P = mV + b$$

From the graph in Figure 2.3 we see that the y intercept is 0.

2. The fraction of light passing through a sample (called the fractional transmittance T) is recorded for several different concentrations of a solution. A plot of T (y axis) versus concentration (x axis) is not linear. The plot (Figure 2.4) can be linearized by calculating the inverse of the log of the transmittance (which is called the absorbance $A = 1/\log T$). The resulting equation is

$$\text{Absorbance} = m\,(\text{concentration}) + b$$

Some graphs cannot be made linear, as there is no transformation that will linearize the data. Any extrapolation must be done on the nonlinear graph. Here are two common examples:

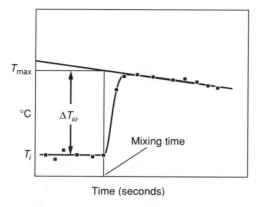

Figure 2.5 Plot of time versus temperature. Extrapolation to zero time

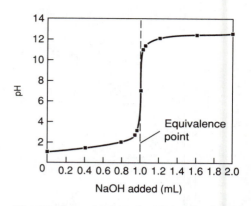

Figure 2.6 A plot of pH versus base added

1. For calculating heats of reaction (see Experiment 5), the change in temperature (ΔT) relative to the temperature before mixing is plotted versus time. A straight line is drawn through the relatively linear part of the data after the temperature change has peaked and begins to return to ambient. Generally there are a few points above or below the postmixing line that should be ignored in drawing the line, as they represent incomplete reaction or incomplete mixing. The line is extended to the time of mixing (0 minutes) to extrapolate the temperature change that would have been observed had an instantaneous reaction taken place. The temperature at which the postmixing line intersects the time of mixing is ΔT_ω for the reaction, as shown in Figure 2.5.

2. An acid is titrated with sodium hydroxide, and the titration is monitored with a pH meter. The data collected consist of pH as a function of volume of NaOH added. A typical graph looks like Figure 2.6.

   A smooth curve is drawn through the data points. The important area of the curve is where pH rises rapidly over a small volume of NaOH added. This is called the equivalence point. The exact equivalence point is taken to be the midpoint of the steep rise. From that point, drop a vertical line to the x axis to read the volume of NaOH (1.0 mL in this example). This graph is not linear, and no transformation will make it linear.

General References

1. Skoog, D.A.; West, D.M.; Holler, F.J. *Fundamentals of Analytical Chemistry*, 5th ed.; Saunders: New York, 1988.
2. There are many publications on this topic in the publication *Journal of Chemical Education*. See for example Simoes, J.A.M.; Teixeira, C.; Airoldi, C.; Chagas, A.P. *J. Chem. Educ.* **1992**, *69*, 475.
3. Szafran, Z.; Pike, R. M.; Foster, J. C. *Microscale General Chemistry Laboratory with Selected Macroscale Experiments;* Wiley: New York, 1993.

PROBLEMS—CHAPTER 2

1. The molarity of an acid solution is determined by five separate titrations as 0.1062 M, 0.1101 M, 0.0998 M, 0.1248 M and 0.1028 M.
   (*a*) Calculate the mean, median, spread, average deviation, percent relative error, standard deviation, and the coefficient of variation.

   (*b*) Calculate the confidence limit and the confidence interval for the true mean μ at a 95% confidence level.

   (*c*) Determine if any of the outlying results in this problem can be rejected at a 96% confidence level.

CHAPTER 3

Macro- and Microscale Laboratory Equipment and Experimental Techniques

Construction and Use of Microscale Glassware

Objectives
- To recognize the types of glassware and equipment commonly used in a chemistry laboratory
- To become acquainted with common experimental techniques, such as weighing, filtration, ignition, titration, preparation of solutions, and so on
- To be able to use common equipment such as balances, pipets, burets, centrifuges, and so forth
- To be able to construct special glassware for microscale work

3.1 General Facilities

Chemistry laboratories are equipped with work spaces, called **benches**. They usually contain individual drawers or lockers, wherein glassware and other equipment are stored. The benches are also equipped with water faucets (often fitted with aspirators) and gas jets. Sinks are available for washing laboratory glassware. Electrical outlets are also provided. Solid waste containers and containers for broken glass are also located inside the laboratory. For ventilation and safety purposes, hoods are available. Sometimes, small fume hoods are placed above the workstations; sometimes they are arranged in a row at one or both ends of the room.

The laboratory is also equipped with other devices such as balances, melting point and boiling point apparatus, centrifuges, and so on. Special instruments such as spectrometers, analytical balances, and other analytical tools may also be present.

Check-in Procedure

Many colleges and universities require a formal check-in procedure of their students. Some institutions may impose a glass-breakage penalty or a damage claim. Students are therefore strongly advised to follow the rules and regulations of the institution in maintaining and keeping their workplaces clean and safe. During the check-in time, you may be assigned a locker (drawer) in the laboratory. Your locker will contain part or all of the glassware (Figure 3.1) and equipment (Figure 3.2) listed shortly. You may be asked to verify the contents of your drawer against a locker inventory (see page 32).

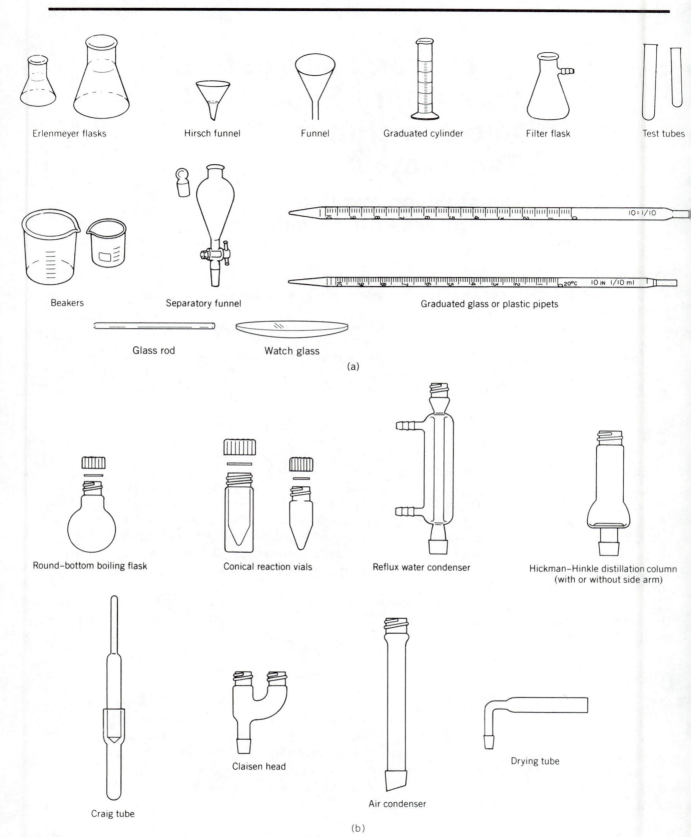

Erlenmeyer flasks   Hirsch funnel   Funnel   Graduated cylinder   Filter flask   Test tubes

Beakers   Separatory funnel   Graduated glass or plastic pipets

Glass rod   Watch glass

(a)

Round–bottom boiling flask   Conical reaction vials   Reflux water condenser   Hickman–Hinkle distillation column (with or without side arm)

Craig tube   Claisen head   Air condenser   Drying tube

(b)

Figure 3.1 (a) Common and (b) optional glassware

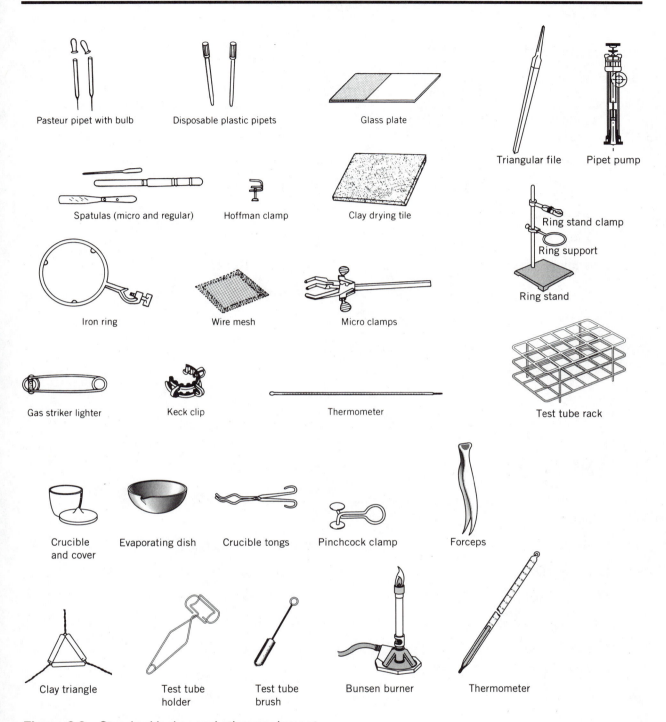

Figure 3.2 Standard locker and other equipment

3.2 Glassware

The majority of experiments presented in this textbook involve the use of microscale glassware, illustrated in Figure 3.1. Note in particular two items frequently used in a laboratory: the disposable Pasteur pipet and the glass plate which are shown in Figure 3.2. The plate is 12 × 13 in, with a half-white, half-black background (see Section 3.10). Pasteur pipets

Suggested Locker Inventory

Drawer number _____

Name _____ Lab Section _____

Mark any missing item with an X.

1.	Beakers			13.	Wire gauze	1 _____
	10 mL	2 _____		14.	Crucible tongs	1 _____
	25 mL	1 _____		15.	Tweezers	1 _____
	50/100 mL	1 _____		16.	Bunsen burner	1 _____
	250/400 mL	1 _____		17.	Rubber tubing	1 _____
2.	Erlenmeyer flasks	_____		18.	Microburner	1 _____
	10 mL	2 _____		19.	Iron ring	1 _____
	25 mL	1 _____		20.	Stirring rods	
	50/100 mL	1 _____			Regular	1 _____
3.	Graduated cylinders				Microsize	1 _____
	10 mL	1 _____		21.	Flint lighter	1 _____
	50/100 mL	1 _____		22.	Magnetic stir bar	1 _____
4.	Funnels			23.	Test-tube holder	1 _____
	Micro	1 _____		24.	Pipet bulb or pump	1 _____
	Regular	1 _____		25.	Micro metal spatula	1 _____
5.	Watch glass	1 _____		26.	Clamps	1 _____
6.	Glass plate	1 _____		27.	Rubber policeman	1 _____
7.	Ceramic plate	1 _____		28.	Triangular file	1 _____
8.	Dropper pipet	1 _____		29.	Optional micro equipment	
9.	Test tubes				Hickman still	1 _____
	10×70 mm	4 _____			Reflux condenser	1 _____
	6×50 mm	4 _____			Air condenser	1 _____
10.	Evaporating dish	1 _____			Round-bottom flask 10 mL	1 _____
11.	Microcrucibles	3 _____			Hirch funnel	1 _____
12.	Clay triangle	1 _____			Filter flask	1 _____

Check-in

_____   _____
Student's signature and date   Instructor's signature

Check-out

_____   _____
Student's signature and date   Instructor's signature

have a variety of uses in the laboratory, some of which are detailed subsequently. The use of a glass plate is strongly recommended when working at the microscale level. Since a large portion of the equipment is small and various stoppers, vials, pipets, and the like are used, placement of these items on the glass plate will assist you in keeping a neat and orderly approach to the operations at hand. The dark and white background gives a good contrast to various objects. The construction of the glass plate is described in Section 3.10.

Cleaning Glassware

No laboratory procedure should ever be done using anything but clean glassware. Even a small amount of foreign material can spell the difference between success and failure in an experiment. Absence of visible dirt on a glass surface does not necessarily mean that the glassware is clean. Very often, an invisible layer of grease may be present on the glass surface. The best way to clean glassware such as test tubes, Erlenmeyer flasks, or beakers is to use a brush and a detergent solution. Pipets, burets, and volumetric flasks may require the use of hot detergent solutions. Sometimes, to remove the oily layer from glassware, one must use an alcohol/KOH bath.

> Caution: Alcohol/KOH bath is CORROSIVE—Wear Gloves.

3.3 Handling Chemicals

Solids

Solids are generally stored in wide-mouthed bottles. A number of steps are involved when transferring solids from a reagent bottle to a reaction flask or weighing paper. If you are using weighing paper, the paper should be sharply creased down the middle so that the chemical can pour off more easily.

Read the bottle label, making sure you have the proper reagent bottle. Tilt the sealed bottle until some chemical nears the mouth of the bottle. Remove the cap and place it, **upside down**, on a clean surface.

Center the bottle over the reaction flask or weighing paper. Gently, rotate the bottle until you have transferred the desired amount of solid (Figure 3.3a). Alternatively, you can scoop the chemical from the bottle using a spatula (Figure 3.3b). Then tap the spatula until you have transferred the desired amount of solid (Figure 3.3c). To keep a solid dry, store the samples in a desiccator (Figure 3.3d).

> Never use the same spatula to obtain different chemicals unless the spatula has been carefully cleaned and dried in between.

Liquids

For microscale work, liquids and solutions are stored in small bottles (10–100 mL). Acids and aqueous ammonia are stored in glass bottles carrying glass stoppers, and bases are stored in plastic bottles (Figure 3.4a). Deionized water, acetone, alcohols, and some other reagents are dispensed from squeeze bottles. Dropper bottles fitted with eye droppers are

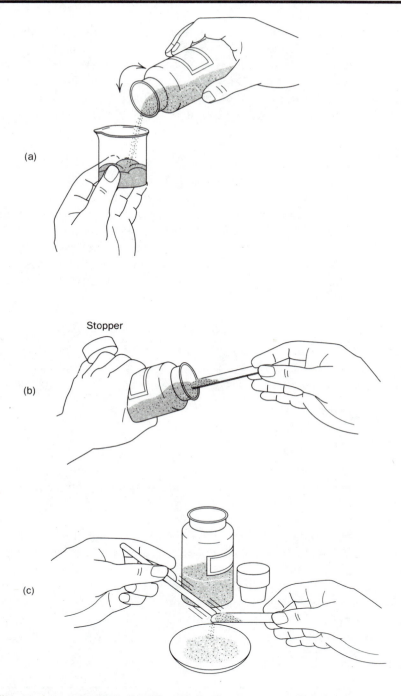

Figure 3.3 Handling and transferring of solids: *(a)* Rolling the bottle to transfer a solid; *(b)* Use of a spatula to remove a solid from a bottle; *(c)* Tapping the spatula to deliver the solid; *(d)* Desiccator

used to store indicator solutions. Read the bottle, checking for both the chemical's name *and concentration* before you transfer any liquid or solution from a bottle.

For transferring larger quantities of liquid, remove the bottle stopper. Do not set the stopper down. Instead, hold the stopper in the same hand as the bottle. Pour the liquid into the desired reaction flask or test tube, preferably down a glass rod (see Figure 3.4*b*). **Do**

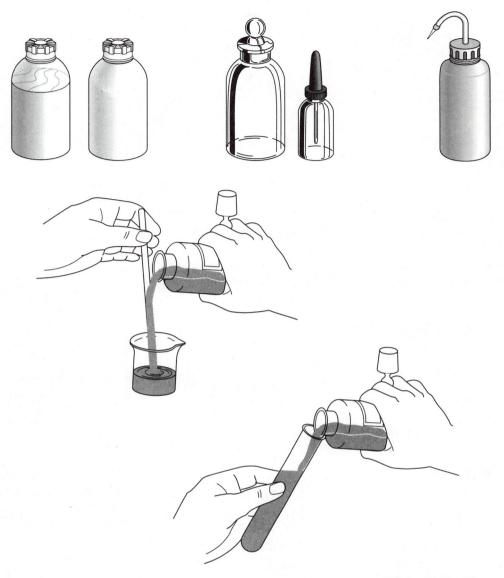

Figure 3.4 (a) Common reagent bottles; (b) transferring a liquid reagent from a bottle

not pour from a wide-mouthed bottle into a narrow-mouthed container. Doing so is a major cause of spills. Transfer smaller quantities of liquids using a Pasteur pipet.

When you are **decanting** (pouring off) a liquid from a solid, it is best to let the solid settle to one side of the container. Then pour the liquid from the beaker containing the settled solid down the side of the other beaker using a stirring rod, as shown in Figure 3.5.

Desiccators

Chemicals as well as other objects (such as weighing bottles and crucibles) tend to pick up moisture from the atmosphere. They may be kept dry by use of a desiccator, a vessel used to equilibrate these objects under a controlled atmosphere at room temperature (see Figure 3.3*d*). To maintain a controlled moisture content, a drying agent is placed in the bottom of

Figure 3.5 Decantation process

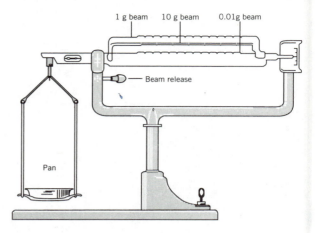

Figure 3.6 Triple-beam balance

the vessel. The efficiency of desiccants determines the equilibrium partial pressure of water in the desiccator.

3.4 Weighing

Weighing is one of the most critical measurements made at the microscale level. Various types of balances, such as the one shown in Figure 3.6, are used in laboratories. For microscale work, we recommend the use of digital balances, which weigh to the third decimal place (± 1 mg).

Digital Balances

Digital balances (Figure 3.7) come in several varieties, with sensitivities ranging from ± 10 mg to ± 0.1 mg. The balances must first be leveled. Turning on the balance automatically zeros the balance. If the balance should drift away from a zero reading, merely push the tare bar (or spot) on the front of the balance.

Place the object to be weighed on the balance pan. When the reading comes to a constant value (often, the letter *g* will appear at this point), the weight of the object can be read directly.

Figure 3.7. Mettler AJ-100 digital balance (Courtesy of Fisher Scientific, 711 Forbes Ave., Pittsburgh, PA 15219)

One great advantage of digital balances is that the mass of the weighing container can be tared out. Suppose an experiment calls for 100 mg of a particular chemical. Place the weighing container on the balance pan and press *Tare* to remove its weight. The reading should now be 0.000 g. Using a spatula, place the substance to be weighed in the weighing container. The reading on the digital readout will be the weight of the sample.

It is essential that each worker take the responsibility of keeping the balances (as well as the area around them) clean and tidy. These are expensive and delicate instruments.

Never move a balance without the laboratory instructor's approval. Some types of balances are easily thrown out of adjustment if they are improperly moved, necessitating an expensive repair.

> Chemicals should never be placed directly on the balance pan.

Solid chemicals should be weighed using aluminum or plastic weighing boats or a glass container. Weighing paper may also be used in certain situations. Liquids or solutions are generally weighed directly in the reaction flask being used or in the product flask when isolated as a liquid. In any event, a glass container is usually preferred.

Objects to be weighed should be dry and at room temperature. These requirements are particularly important on analytical balances (sensitivities of ± 0.1 mg). If the sample is not dry, its mass will decrease with time owing to evaporation of water. If it is not at room temperature, it will generate air currents as it heats or cools the air around it, and these will cause inaccurate or unsteady readings.

3.5 Measuring Liquid Volumes

Glassware such as Erlenmeyer flasks, beakers, and test tubes *are not* calibrated for use as accurate volume-measuring devices. Graduated cylinders are somewhat more accurate; burets (discussed shortly) are quite accurate if used properly; but we recommend the use of volumetric flasks or calibrated pipets for accurate measurement of small volumes of liquids at the microscale level.

Graduated Cylinders

The volume of a liquid can be approximately measured using a graduated cylinder. To measure a given volume of liquid, hold the graduated cylinder vertically at eye level and transfer the liquid into the cylinder. Look at the top curved surface of the liquid (called the **meniscus**). When making a reading (Figure 3.8), be sure to have your eye at the same level as the meniscus in order to avoid error caused by parallax. The reading is always made at the meniscus curve at the center axis of the cylinder or buret, regardless of whether the meniscus curves upward (like water) or downward (like mercury). (Since most of the liquids used in this book will have a meniscus that curves upward like that of water, the phrase "lower meniscus" will be used to mean the meniscus at the center of the vessel.)

Pipets

When a smaller volume of a reagent, solvent, or solution is required, it is convenient to use one of the following measuring devices (see Figure 3.9):

1. Glass graduated pipet with a pipet pump or bulb
2. Graduated 1 or 2 mL syringe
3. Calibrated Pasteur pipet with a pipet pump or bulb
4. Automatic delivery pipet

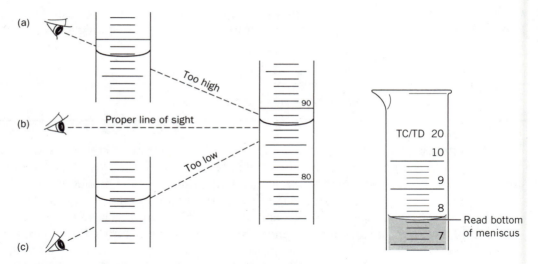

Figure 3.8   Reading a graduated cylinder or a buret

Several types of pipet pumps are commercially available. A convenient pipet pump can be made by attaching the protruding tip of a 10 or 25 mL plastic syringe (without a needle attached) to tight-fitting rubber tubing. The other end of the rubber tubing is attached to the pipet. By raising or lowering the piston of the syringe, one can fill the pipet or discharge the solution from the pipet.

The standard volumetric glass pipets (1–50 mL) come in various designs. These pipets have one calibration mark and are used to deliver a specific volume of a liquid or solution. All pipets used in a chemistry laboratory are marked TD, which stands for to deliver. A measuring pipet (called a Mohr pipet) is graduated along its length and is used to deliver different volumes of a liquid (see Figures 3. 9 and 3.10).

First rinse the pipet with distilled water. Then pump into the pipet a small amount of the liquid or solution to be transferred. Holding the pipet almost at a horizontal position, roll and tilt it back and forth so that the liquid rinses the entire interior surface of the pipet. Let the wash liquid drain through the tip. Rinse the pipet with the liquid three times.

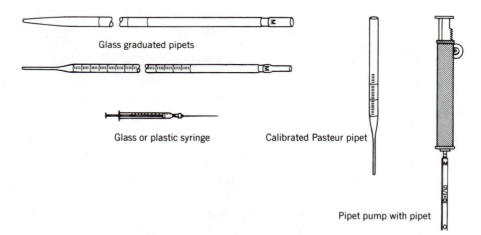

Figure 3.9   Different pipets (Courtesy of Ace Glass, Inc., Vineland, NJ; Thomas Scientific, Swedesboro, NJ).

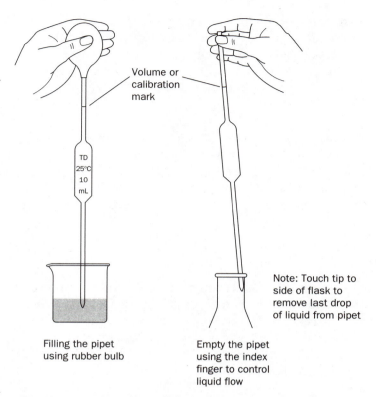

Volume or calibration mark

TD 25°C 10 mL

Note: Touch tip to side of flask to remove last drop of liquid from pipet

Filling the pipet using rubber bulb

Empty the pipet using the index finger to control liquid flow

Figure 3.10   Use of a pipet and a pipet bulb

After rinsing the pipet, draw the liquid into it until the upper level of the liquid goes above the calibration mark. Remove the pipet from the liquid and wipe the tip with tissue paper. Let the liquid level drop to the mark. Remove the hanging drop by touching the side of the tip to the liquid container. Then allow the specific volume of liquid to drain into the reaction flask. Remove the last drop by touching the tip of the pipet to the side of the container.

> Never pipet by mouth. Never try to blow out the last drop of the liquid.

A Pasteur pipet may be easily calibrated by drawing a known amount of liquid into the pipet and marking the level of the liquid on the glass with a grease pencil or permanent marker. It is recommended that you calibrate several pipets showing volumes of 0.5, 1.0, 1.5, and 2.0 mL. For permanent use, a light file mark may be scored on the pipet.

In many experiments, use of a **Pasteur filter pipet** is recommended (see Section 3.10). The Pasteur filter pipet offers two distinct advantages over the regular Pasteur pipet. First, the solution is automatically filtered each time it is taken into the pipet. This process is helpful, since it is important at the microscale level to remove dust or other material suspended in the solution. Second, when a volatile organic solvent such as acetone or methylene chloride is used, a back pressure rapidly builds up in the bulb, causing the solvent or solution to drip from the tip of the pipet. The filter helps keep this from happening.

Very small quantities of liquids are most efficiently transferred using an automatic delivery pipet. Several types are commercially available, three of which are shown in Figure 3.11. All have a control to adjust the pipet to deliver the desired volume (usually 10 μL to 1 mL). They are also designed so that the liquid being transferred comes in contact only with the plastic tip. This tip can automatically be ejected from the pipet after delivery of the liquid is complete.

There are several points to remember on the use of automatic delivery pipets:

1. They are relatively expensive, so treat them with respect.
2. Always store and use the pipets in a vertical position. Liquid running back up into the pipet mechanism can damage the controls.
3. Never immerse the pipet in a liquid—only about 5 mm of the plastic tip should come in contact with the liquid to be transferred.
4. Develop a smooth technique for depression and release of the control mechanism. Consistent results come with care and practice.

Volumetric Flasks

Volumetric flasks (Figure 3.12) should be used when the concentration of the solution being prepared must be very accurately known (three to four significant figures). Volumetric flasks are usually calibrated to contain (TC) at room temperature and must be used at the indicated temperature for greatest accuracy. There is only one graduation, which is etched on the neck of the flask. This is frequently referred to as *the mark*.

When the lower meniscus of a solution or liquid is exactly on the line, the flask contains exactly the calibrated volume. Various sizes of volumetric flasks are available. In microscale work, the 10 mL size is commonly used.

The flask should be rinsed with solvent before use. The usual procedure for preparing a solution is to fill the volumetric flask partially with solvent, add the appropriate amount

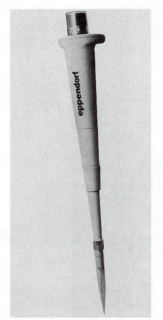 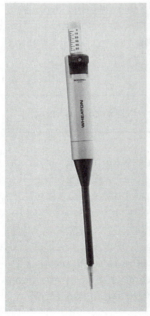 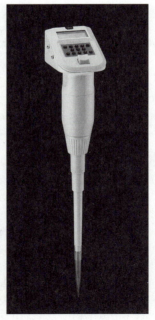

Figure 3.11 Automatic delivery pipets (Courtesy of Sargent-Welch Scientific Co., a VWR Company, Skokie, IL; Ace Glass, Inc., Vineland, NJ; Rainin Instrument Co., Woburn, MA)

Figure 3.12 Volumetric flask

of the solute, swirl the flask to dissolve and mix the solute, and then fill the flask to a little below the mark with additional solvent. The last few drops of solvent, to bring the lower meniscus to the mark, should be added with a Pasteur pipet (or medicine dropper). The stopper is replaced, and the flask is then inverted back and forth at least 10 times to ensure homogeneous mixing. When mixing the contents of the volumetric flask, be certain to hold both the flask and the stopper. Filled flasks should not be held by the neck, as they are too heavy for the glass in the flask neck to support. Broken flasks and severe cuts can be the result of improper shaking.

Burets and Microburets

Burets accurately dispense known volumes of a reagent solution (such as a base) required to react with a given amount of another chemical (such as an acid). Different sizes of burets, such as 10, 25, and 50 mL, are commercially available, but these are too large for convenient use in the microscale laboratory. Instead, the use of a microburet is recommended. A microburet can be conveniently constructed as described in Section 3.10 (see Figures 3.23 and 3.29b).

A buret is made of a long glass tube with a ground glass or Teflon™ stopcock attached at the bottom. It is graduated from top (0 mL) to bottom. An ordinary 50 mL buret (Figure 3.13) is graduated in 0.1 mL intervals and should be read to the nearest hundredth of a milliliter. A 2.00 mL microburet has graduations at 0.01 mL intervals and is read to the nearest 0.001 ml.

Before use, a buret must be cleaned using a detergent solution and buret brush. No water drops should adhere to the inside wall of a clean buret. If the stopcock is made of ground glass, it should be cleaned by dipping it in toluene to remove stopcock grease. Dry both the stopcock and the end of the buret with a tissue paper. Reapply stopcock grease and replace the stopcock.

Figure 3.13
A 50 mL buret

> CAUTION : Toluene is a flammable volatile liquid. Do not use an open flame. Carry out the step in the HOOD.

3.6 Heating Methods

Bunsen Burners and Microburners

If certain precautions are observed (no flammable solvents in use, gas turned off immediately after use), the microburner or Bunsen burner (Figure 3.14) plays a useful role in the laboratory. All laboratory burners use natural gas (methane) for combustion. A metal burner consists of a barrel attached to a base. The base contains holes that control the supply of air. A needle valve control at the bottom of the base regulates the flow of the gas. The burner is connected to a gas supply with a length of rubber tubing. With limited air supply (air inlet holes are partially or completely closed), the burner produces a cool, sooty, yellow flame called a **luminous flame** (also known as a **reducing flame**). Using more air (air inlet holes open), one obtains a nonluminous, pale blue hot flame (called an **oxidizing flame**).

Matches or flint strikers may be used to light the burner (do not use a cigarette lighter). Strikers are generally safer to use. When using a match, strike the match first, hold it near the top of the burner (but not directly above it) and then turn the gas jet on full. The gas should light within a few seconds. Holding the match directly above the burner can cause the

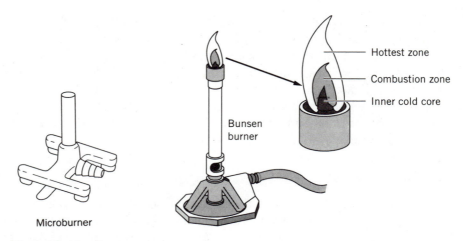

Figure 3.14 Burners

gas to blow out the match. If a striker is used, turn on the gas jet first and then squeeze the striker, sliding the flint across the friction bar so that a spark forms. Hold the striker in the fully squeezed position above the burner and adjust the burner flame if necessary. **If the flame goes out, turn the gas off and wait at least 30 seconds before relighting**. Doing this gives any uncombusted gas time to disperse before another flame is lit.

Because natural gas has no odor, manufacturers usually mix in β-mercaptoethanol for safety's sake. If the distinctive odor of this sulfur-containing compound is detected, make certain that all gas jets are tightly closed. **If the odor is still apparent, the laboratory should be evacuated until the gas leak is found**.

Sand Bath or Aluminum Block with Magnetic Stirring Hot Plate

A sand bath placed on a magnetic stirring hot plate (Figure 3.15a) makes a very efficient heating device. This arrangement allows stirring and heating to be performed simultaneously. This approach is very inexpensive and provides a nonflammable source of heat. The sand is usually contained in a glass crystallizing dish or a metal container.

Sand, being a poor conductor of heat, in effect acts as an insulator. Thus, the temperature of a container placed in the sand will vary depending upon the depth to which it is immersed. It is therefore recommended that each person be assigned his or her own equipment (sand bath and hot plate) and that a calibration of this heating source be made at the first laboratory session. The use of a sand bath is also quite effective at temperatures below 100°C, and may be used in place of steam.

An alternative to the sand bath is the aluminum block (Figure 3.15b) placed on the magnetic stirring hot plate. These blocks are now available from several supply houses (or can be easily manufactured in a metal shop). Users of this device cite two major advantages over the sand bath method: better heat transfer and no problem with breakage of a crystallization dish or spilled sand. In addition, they are relatively cheap and store easily.

> CAUTION: Hot aluminum looks like cold aluminum!

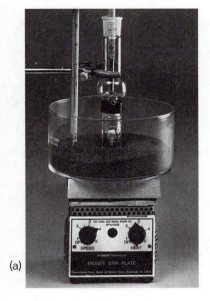

(a)

(b)

Figure 3.15 (*a*) A sand bath on a magnetic stirring hot plate; and (*b*) aluminum heating blocks (Courtesy of Ace Glass, Inc., Vineland, NJ)

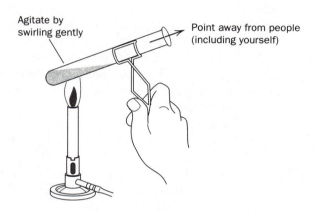

Agitate by swirling gently

Point away from people (including yourself)

Heat side of test tube gently to avoid bumping

Figure 3.16 Heating a liquid in a test tube

Heating Safety Rules

1. When heating glass containers over a burner flame, use a wire gauze square with a non-flammable ceramic center as a support.
2. Never heat thick-walled vessels, such as filter flasks or volumetric flasks.
3. When heating liquids in small test tubes, always heat from the top down to avoid explosive boiling of the liquid (Figure 3.16). Use a boiling chip to avoid bumping within the liquid. Never point the top of the tube at any other person or at yourself.
4. Never heat a flammable object over an open flame.
5. Be aware that most solvent vapors are denser than air and will sink and flow along countertops and into sinks. There they may be ignited, even by a burner some distance away.
6. When heating a solution containing solids, especially finely divided solids, be sure to stir the solution continuously to avoid bumping.
7. Never heat a closed container.

3.7 Stirring

Stirring is often required in microscale reactions and is generally carried out using a micro stirring rod, microspatula, or magnetic stir bars or vanes. A very inexpensive micro-sized magnetic stir bar may be constructed in the laboratory (see Section 3.10). Stirring is a technique whose importance is often overlooked. When the reaction gives off large quantities of heat (dilution of concentrated acids, for example), solutions should be efficiently stirred to prevent spot-heating and possibly splattering of boiling liquid. Stirring also helps achieve rapid equilibrium and is especially helpful when making pH measurements or taking kinetic data.

Heating and stirring are conveniently carried out simultaneously by using a magnetic stirring hot plate (Figure 3.17). When this type of arrangement is used, the container being stirred should be placed directly in the center of the hot plate so as to gain maximum efficiency of the magnetic flux. The spin rate of the bar or vane should be adjusted to obtain smooth mixing of the contents in the container. The insertion of a thermometer into the sand bath or aluminum block allows easy monitoring of the temperature of the container and its contents. The magnetic stir bar can be easily removed from the container using forceps or a magnetic wand.

When one wishes to mix solutions or to dissolve a solid in a given solvent and heating is not required, the use of a touch mixer (Figure 3.18) is quite convenient too. These commercially available mixers are relatively inexpensive and durable. The mixer speed is adjustable, and the mixing operation is fast and efficient. Vortex mixing is recommended over the techniques of swirling or shaking by hand, since much greater control is maintained during the mixing process.

3.8 Filtration and Gravimetric Methods

The process of separating a solid from a solution (collecting a precipitate) at the microscale level usually involves the technique of filtration. The precipitate must be quantitatively separated, washed, dried, and weighed. Two types of filtration processes are generally used.

Gravity Filtration

Gravity filtration requires a filter funnel (see Figure 3.19) with a stem to direct the filtrate (liquid) into the receiving container. Make sure that the stem of the funnel touches the side of the container. Fold a piece of filter paper in half, and then in quarters with one side slightly (by $\frac{1}{8}$ in.) smaller than the other.

Tear off a corner of the smaller quarter and open the paper into a cone so that there are three layers on one side and one on the other (see Figure 3.19a). The filter paper should be wet with solvent before filtering so that the paper adheres evenly to the top part of the funnel.

The funnel should be firmly supported using a clay triangle or cork ring. Use a glass rod to direct the liquid being filtered into the funnel (as described in Section 3.3). Do not fill the funnel more than two-thirds of the way to the top of the filter paper. A *rubber policeman* can be used to scrape solids from the decanting vessel onto the filter paper. Small loose particles can be swept into the filter paper by holding the decanting vessel in the pouring position over the funnel while using a squeeze bottle of water (or other solvent) to wash the solids out (Figure 3.19b).

Figure 3.17. Magnetic stirring hot plate (Courtesy of Ace Glass, Inc., Vineland, NJ)

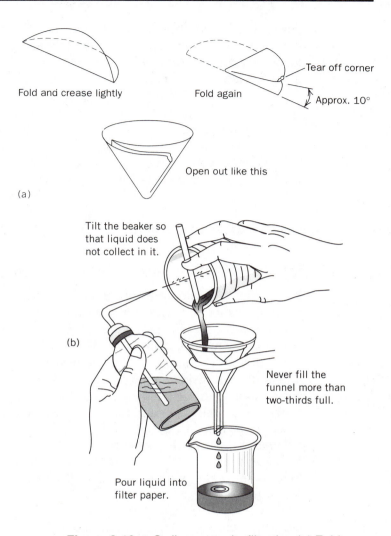

(a)

Fold and crease lightly

Fold again

Tear off corner

Approx. 10°

Open out like this

(b)

Tilt the beaker so that liquid does not collect in it.

Never fill the funnel more than two-thirds full.

Pour liquid into filter paper.

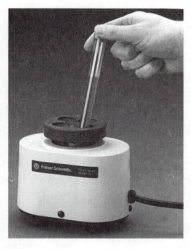

Figure 3.18. Touch mixer (Courtesy of Fisher Scientific, 711 Forbes Ave., Pittsburgh, PA 15219)

Figure 3.19. Ordinary gravity filtration (*a*) Folding of a filter paper and (*b*) technique of gravity filtration

Suction Filtration

When the amount of solid is less than 100 mg, one may use a porcelain Hirsch funnel (Figure 3.20*a*) or a *nail-filter in the funnel* (Figure 3.20*b*, see section 3.10) to collect the material. A sintered glass funnel may be substituted for a Hirsch funnel (Figure 3.20*b*). When a large quantity of a solid is to be filtered, use of a bigger filter called a Büchner funnel (Fig. 3.20*b*) is recommended. This operation can be carried out under vacuum, generally by use of a water aspirator, and is called suction or vacuum filtration. A typical arrangement is shown in Figure 3.20*a*. When using this arrangement, **always securely clamp the filter flask (filtration apparatus) to a support to prevent tipping**. Many times a valuable product has been lost as a result of not observing this simple rule.

A water trap, called vacuum trap, must *always* be placed between the filter flask and the aspirator. This is to prevent water from backing up into the filter flask and perhaps destroying the product. Be aware that when several persons are using the same water line, the water pressure can change at any time. It is recommended that the system be opened to the atmosphere by loosening the screw clamp on the trap or by removing the tubing from the filter flask before the water is turned off.

In quantitative gravimetric work, the precipitate collected on the filter is washed with a specific wash solution before it is dried and weighed. Several washings with small volumes of wash liquid are more effective in removing the contaminants from the precipitate than one washing using a large volume of wash solution.

Various types of filter paper are available. For regular work, ordinary filter paper is used. For quantitative analysis, only **ashless** filter paper is employed. Depending on the size of the precipitate, *fast*, *medium*, or *slow* porosity filter papers are selected. Filter paper of different diameters is also available. The size to be used depends on the quantity of the precipitate and the size of the funnel.

Ignition of Precipitate with Filter Paper for Gravimetric Analysis

For gravimetric work, the precipitate is ignited in a crucible to obtain a constant net weight. In many instances, during ignition a precipitate undergoes thermal decomposition, yielding a residue having a constant composition. The ignition of a precipitate is carried out according to the following procedure.

After all the liquid has drained from the funnel, carefully remove the filter paper, fold it, and place it inside a preweighed crucible. Place the crucible with its cover in a slanted position on a silica triangle supported on a ring stand. Using a burner, flame the crucible slowly to dry the paper and the precipitate. When the filter paper is dry, increase the heat and char the filter paper (Figure 3.20*c*). The paper should smolder; it must not burn off with a flame. If the paper catches fire, remove the burner immediately and cover the crucible to extinguish the fire. Under microscale conditions, one can heat more than one crucible at a time. In this case, place all the crucibles on a wire gauze positioned over a Bunsen burner (Figure 3.20*d*).

When all the paper has been charred, increase the flame until the crucible becomes red-hot. Continue to heat until all the organics and carbon deposits are removed from the crucible and the crucible cover. From time to time, rotate the crucible to expose the dark areas to strong heat. During the entire procedure, avoid direct contact between the residue and the reducing flame. When heating is complete, cool the crucible inside a desiccator to prevent condensation of atmospheric moisture and then weigh it.

3.9 Solution Preparation

Work in the fields of synthesis, quantitative analysis, and qualitative analysis depends on the chemist's ability to prepare solutions of known concentration. Virtually all aspects of life are influenced by the concentration of chemicals, including food preparation, medical diagnostics, manufacturing, and the extraction of compounds from exotic plants.

Solutions are made up of two components. The **solvent** is the material, usually present in large excess, in which the **solute** is dissolved. Solvents are usually liquids at room temperature, with water being the most common. The solvent itself may be a solution, say 0.1 M HCl. A solution such as 1.0 M $BaCl_2$ in 0.1 M HCl can also be thought of as a solvent (water) with two solutes (HCl and $BaCl_2$).

Several sets of units are commonly used to represent concentration. The most common unit used by chemists is **molarity** (*M*, defined as moles of solute per liter of solution). Others

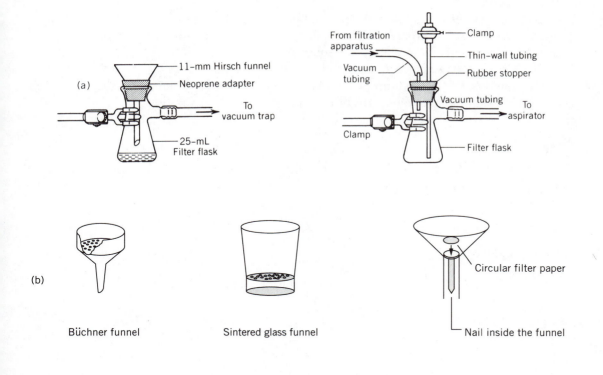

(a)
- 11–mm Hirsch funnel
- Neoprene adapter
- To vacuum trap
- 25–mL Filter flask

- From filtration apparatus
- Clamp
- Thin–wall tubing
- Vacuum tubing
- Rubber stopper
- Vacuum tubing
- To aspirator
- Clamp
- Filter flask

(b)

Büchner funnel

Sintered glass funnel

Circular filter paper

Nail inside the funnel

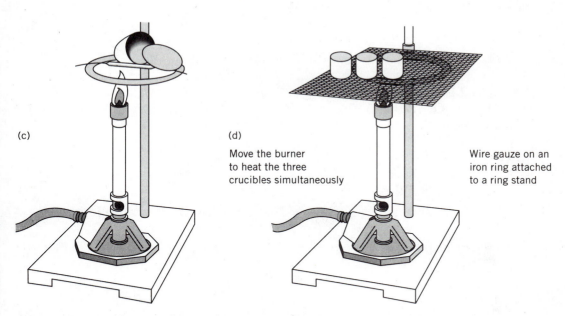

(c)

(d)

Move the burner to heat the three crucibles simultaneously

Wire gauze on an iron ring attached to a ring stand

Figure 3.20   (a) Vacuum filtration with Hirsch funnel and vacuum trap; (b) Büchner, sintered, and nail funnels; (c) ignition of a precipitate in a crucible; and (d) three microcrucibles positioned on a wire gauze

include **molality** (m, moles of solute per kilogram of solvent), normality (N, equivalents of solute per liter of solution) and, in the medical field, mg per deciliter (1 deciliter = 0.1 L). Less commonly used are percents—for solid solutes, weight:volume percent (w/v), and for liquids, volume:volume percent (v/v). For example, a 10 percent (w/v) sodium bicarbonate solution contains 10 g of sodium bicarbonate per 100 mL of solution, whereas a 5 percent (v/v) alcohol solution contains 5 mL of alcohol per 100 mL of solution.

Molarity (M)

This concentration system is based on the volume of *solution*. It is defined as the number of moles n of a solute per liter V of the solution.

$$M = n/V$$

Rearranging, we have

$$n = (M)(V)$$

The number of moles is the mass (m) divided by the molecular weight (MW)

$$n = m/MW$$

Combining these two equations, we see

$$m/MW = (M)(V)$$
$$m = (M)(V)(MW)$$

Example 3.1   You wish to make up 50.0 mL of a 0.300 M solution of oxalic acid ($C_2O_4H_2 \cdot 2H_2O$, MW=126.07 g/mol). How do you do it? How many grams of oxalic acid do you need?

Answer

$$m = (M)(V)(MW)$$
$$m = (0.300 \text{ mol/L})(0.0500 \text{ L})(126.07 \text{ g/mol}) = 1.89 \text{ g of oxalic acid}$$

You would add 1.89 g of oxalic acid to a 50 mL volumetric flask and fill to the mark with water.

Normality (N)

Normality is defined as the number of equivalents (eq) of solute per liter of solution.

$$N = n_{eq}/V$$

Just as the number of moles of a substance is the mass divided by the molecular weight, the number of equivalents (n_{eq}) of a substance is defined as the mass divided by the equivalent weight (EW).

Defining the equivalent weight is not so easy as defining a molecular weight, because the EW of a compound depends on what reaction that compound is undergoing. In an acid-base reaction, the EW is the weight of the acid (or base) that is *equivalent* to 1 mol of H^+ (or OH^-). For example, HCl gives **one** mole of H^+ (or one equivalent of H^+) per mole of HCl,

$$(36.5 \text{ g/mol HCl})(1 \text{ mol HCl}/\textbf{1} \text{ eq HCl}) = 36.5 \text{ g/eq HCl}$$

Suppose you have a 0.3 M solution of H_2SO_4. Since H_2SO_4 gives **two** moles of H^+ (or two equivalents of H^+) per mole of H_2SO_4,

$$(98.1 \text{ g/mol } H_2SO_4)(1 \text{mol } H_2SO_4/\textbf{2} \text{ eq } H_2SO_4) = 49.1 \text{ g/equiv } H_2SO_4$$

In an oxidation-reduction reaction, the EW of an oxidizing agent (or a reducing agent) is the weight *equivalent to* 1 mol of electrons (or that furnishes or reacts with 1 mol of electrons). Consider the reaction of permanganate in acid:

$$MnO_4^- + 8 H^+ + \textbf{5} e^- = Mn^{2+} + 4 H_2O$$

In this reaction, each mole of permanganate reacts with five **moles** of electrons (**five** equivalents per mol). Thus, the equivalent weight of $KMnO_4$ is

$$(158 \text{ g/mol})(1 \text{ mol}/\textbf{5} \text{ eq}) = 31.6 \text{ g/eq } KMnO_4$$

When permanganate reacts in neutral solution, the reaction is

$$MnO_4^- + 2 H_2O + \textbf{3} e^- = MnO_2 + 4 OH^-$$

In this reaction, each mole of permanganate reacts with **three** moles of electrons (three equivalents per mole). Thus, the equivalent weight of $KMnO_4$ is

$$(158 \text{ g/mol})(1 \text{ mol}/3 \text{ eq}) = 52.7 \text{ g/eq } KMnO_4$$

Thus, the normality and equivalent weight for a compound depend on the reaction.

Normality is frequently used in titrations in analytical chemistry. The reason for using normality instead of molarity is one of convenience of calculations: one equivalent of an acid will always react with one equivalent of a base; one equivalent of an oxidizing agent will always react with one equivalent of a reducing agent. At the equivalence point (end point) in any acid-base titration, since the equivalents of an acid and a base are the same, the following equalities hold:

$$eq_{acid} = eq_{base}$$

or

$$N_{acid} V_{acid} = N_{base} V_{base}$$

Just as the mass required to make up a solution of some molarity is given by

$$m = (M)(V)(MW)$$

the mass required to make up a solution of some normality is given by

$$m = (N)(V)(EW)$$

Example 3.2 How many grams of potassium permanganate ($KMnO_4$, MW = 158.03 g/mol) would be required to make 100.0 mL of 0.0500 N solution? The $KMnO_4$ solution reacts with sodium oxalate *in acidic* solution. The reaction is

$$2\ MnO_4^- + 5\ C_2O_4^{2-} + 16\ H^+ \rightarrow 2\ Mn^{2+} + 10\ CO_2 + 8\ H_2O$$

Answer

As we have already seen, *in acid*, the EW is 31.6 g/mol $KMnO_4$. Thus,

$$m = (N)(V)(EW)$$
$$m = (0.0500\ eq/L)(0.100\ L)(31.6\ g/eq) = 0.158g$$

Example 3.3 In standardizing a $KMnO_4$ solution, you titrated 1.00 mL of 0.1276 N sodium oxalate with $KMnO_4$ solution *in acid*. You needed 2.36 mL of permanganate to reach the equivalence point (end point). What is the normality of the $KMnO_4$ solution? See the reaction in Example 3.2.

Answer

$$(N_{MnO_4^-})(V_{MnO_4^-}) = (N_{oxalate})(V_{oxalate})$$
$$(N_{MnO_4^-})(2.36\ mL) = (0.1276\ N)(1.00\ mL)$$
$$(N_{MnO_4^-}) = 0.0541\ N$$

Other Concentration Terms

- Mole fraction (X). The mole fraction is defined as the number of moles of solute divided by the total number of moles (solute + solvent):

$$X_A = n_A/n_T$$

- Molality (m). Molality is defined as the number of moles of solute per kilogram of solvent.

$$m = n_A/\text{kg solvent}$$

- Weight percent (% wt). The weight percent of a solution is the weight of the solute divided by the total weight of the solution, times 100 to convert it to a percentage.

$$\%\ wt = [w_{solute}/(w_{solute} + w_{solvent})](100)$$

- Parts per million (ppm). The number of parts per million in a solution is the weight of the solute divided by the weight of the solvent, times 10^6.

$$\text{ppm} = [(w_{\text{solute}}/w_{\text{solvent}})](10^6)$$

  Since 1 L of water at room temperature weighs about 1 kg, one ppm is also defined as 1 mg of a solute in 1 kg of water. In clinical chemistry, ppm is also called milligram percent.
- Parts per billion (ppb). In environmental chemistry, even lower concentration terms are needed. The number of parts per billion in a solution is the weight of the solute divided by the weight of the solvent, times 10^9.

$$\text{ppb} = [(w_{\text{solute}}/w_{\text{solvent}})](10^9)$$

Dilution

Many materials can be purchased as solutions of a given concentration. For example, HCl is normally purchased as a 12 M solution. In many cases, we want to use some lower concentration in a reaction. The lower concentration solution is made up by **dilution** of the more concentrated solution. The equation for dilution is

$$M_1V_1 = M_2V_2$$

Here, M_1 is the molarity of the concentrated solution, V_1 is how much of that solution you will need, M_2 is the molarity of the dilute solution, and V_2 is how much of the solution you want. Note that the number of moles of solute remains the same before and after the dilution.

Example 3.4 You wish to prepare 250.0 mL of 0.100 M H_2SO_4 from a stock bottle of 6.0 M H_2SO_4. How many milliliters of the concentrated solution are required? How should the solution be made?

Answer

$$M_1V_1 = M_2V_2$$
$$(6.0 \text{ M}) \times (V_1) = (0.10 \text{ M})(250.0 \text{ mL})$$
$$V_1 = 4.17 \text{ mL}$$

Transfer 4.17 mL of 6.0 M sulfuric acid to a 250 mL volumetric flask partially filled with water. Then fill the flask to the mark.

Serial Dilution

Serial dilution is used when a series of solutions spanning several orders of magnitude of concentration must be made, or when a small quantity of very dilute solution must be prepared where it would be difficult to measure out accurately the very small quantity of solute required. In the latter case, serial dilution often avoids waste of solute and solvent by starting with a small amount of more concentrated solution and diluting it in steps until the concentration and volume desired are obtained. Serial dilution starts with a concentrated solution. A first dilution, typically 10:1 or 100:1, is made. This second solution is then used to make a third solution by diluting the second by 10:1, 100:1, or whatever is necessary and so on, until the desired volume and concentration are reached. How many steps are used in a serial dilution is up to the experimenter.

Unless very large volumes are involved, dilutions of greater than 1:100 should be avoided, as the amount of concentrated solution needed to make a dilution becomes very small and thus difficult to measure accurately. (It is also true, however, that more steps introduce greater error.)

Example 3.5   You wish to prepare 25.0 mL of 2.50×10^{-6} M potassium dichromate. How would you proceed? (MW of $K_2Cr_2O_7$ = 294.18 g/mol).

Answer
This procedure requires:

$$(0.0250 \text{ L})(2.50 \times 10^{-6} \text{ moles/L})(294.18 \text{ g/mol}) = 1.84 \times 10^{-5} \text{ g of } K_2Cr_2O_7$$

On a good analytical balance, the minimum amount that you could accurately weigh out would be about 25 mg, or 2.5×10^{-2} g, over 1000 times what you need. If you were to weigh out 25 mg and dilute it to the desired concentration directly, you would have 34 liters of solution, when you only need 25 mL! Waste solutions containing chromium cannot be discarded down the drain, as chromium is poisonous and, in some forms, carcinogenic. It is unthinkable to generate 34 liters of solution and then pay someone to haul away 33.975 L of it as waste.

Instead, weigh out 25 mg of $K_2Cr_2O_7$, and dissolve it in 100 mL of water. The resulting concentration is 8.50×10^{-4} M. Obtaining a solution that is 2.5×10^{-6} M requires a 340:1 dilution. This is conveniently done in two steps: a 100:1 dilution of the first solution (1.00 mL diluted to 100.0 mL, yielding a solution 8.50×10^{-6} M in potassium dichromate), followed by a 3.4:1 dilution as follows to get 25 mL of 2.5×10^{-6} M potassium dichromate:

$$M_1V_1 = M_2V_2$$
$$(8.50 \times 10^{-5}\text{M})(V_1) = (2.5 \times 10^{-6}\text{M})(25 \text{ mL})$$
$$V_1 = 7.4 \text{ mL}$$

When 7.4 mL of the second solution is diluted to 25.0 mL, the concentration will be 2.5×10^{-6} M. There is some waste, 99 mL of the first solution and 92.65 mL of the second solution, but the amount is less than 0.2 L as opposed to 33.975 L. (The number of moles of chromium waste is the same either way, but disposal cost is based on the volume of waste).

Multiple solute solutions

Solutions often contain more than one solute. In such cases, one calculates the amount of each solute needed, measures it in an appropriate container, mixes all the solutes together in the required volume of solvent, and stirs to dissolve and homogenize the solution.

3.10 Construction of Microscale Glassware

In the microscale laboratory one uses a variety of microglassware, some of which can be easily made by students or by instructors prior to actual laboratory work. For example, the micropycnometer used in the determination of densities of different liquids (see Experiment 1) is easy to make, inexpensive, and very easy to handle. The construction of several different pieces of equipment used in this book are described next, as are some basic techniques for the manipulation of glass. Two main types of glass tubing (rods) are used in a laboratory: flint glass (soda lime glass) and borosilicate glass. Pasteur pipets are made from the former, whereas most beakers and flasks are made from the latter. Most of our manipulations will be done using flint glass Pasteur pipets.

Cutting and Fire-Polishing Glass Tubing

Suppose you want to make a cut two inches away from the constricted part (narrow capillary end) of a Pasteur pipet. You should use the following procedure.

1. Lay the pipet on a flat surface.
2. Place the edge of the triangular file on the spot where you wish to make a cut.
3. Holding the file firmly against the spot (use the tip of your thumbnail as a guide), make a quick forward stroke with the file to make a visible deep scratch on the pipet. **Do not saw the glass**. If available, you may use a glass cutter to make the scratch.
4. Wet the scratch mark.
5. Using a folded layer of towel, grasp the glass tubing firmly with both hands placing the thumbs close to the scratch mark (the scratch mark should face away from you, and you should place your thumbs on opposite sides of the scratch).
6. Using a sharp force, pull the pipet toward you with your fingers while you push the pipet away with your thumbs against the scratch mark (see Figure 3.21).

The pipet should break at the point where the scratch was made. If the break is round and uniform, then the pieces are ready for fire polishing. If the break is jagged, repeat the process of cutting the glass until you have made a uniform cut.

> The cut end of a glass tube is very sharp and must be fire polished before use.

Warm the cut end of the pipet at the base of a Bunsen burner (or microburner) flame. When it is hot, hold the cut end in the top of a hot, nonluminous flame and rotate the glass tubing back and forth for uniform heating.

Construction and Use of a Micropycnometer

Hold a 5 in. Pasteur pipet in an oxidizing (nonluminous) Bunsen burner flame, 1 cm away from the constricted part of the pipet stem (Figure 3.22a). Rotate the body of the pipet over the flame for even heat distribution. When the glass has become sufficiently soft so that it can be pulled without much effort, remove it from the flame *and then* immediately pull

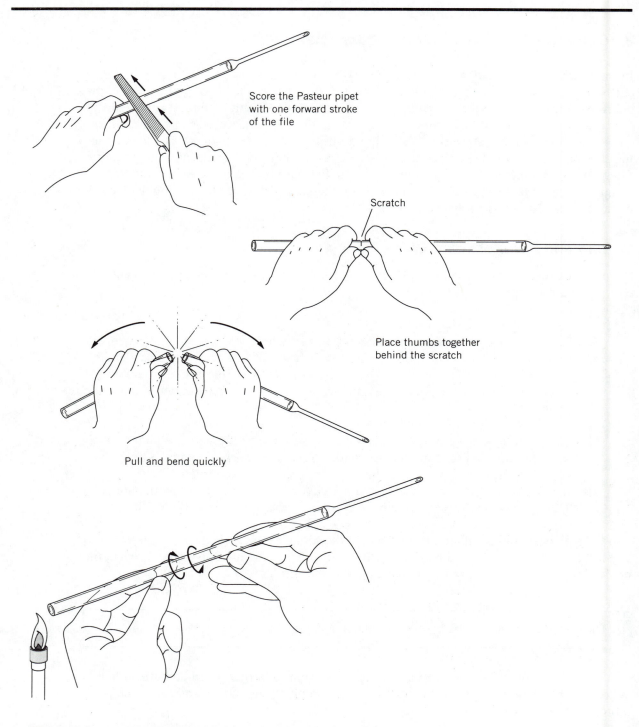

Score the Pasteur pipet with one forward stroke of the file

Scratch

Place thumbs together behind the scratch

Pull and bend quickly

Figure 3.21 (*a*) Cutting and (*b*) fire-polishing a Pasteur pipet

the two ends apart (Figure 3.22*b*) so as to form a 10 to 25 cm long thin capillary (Figure 3.22*c*). Keep the pulled pipet taut until the glass becomes rigid. At this point, you will have the stem of the pipet connected to a small bulb by a long thin capillary.

At about 0.5 cm from the rear end of the bulb, heat the capillary strongly to detach it from the main body of the pipet (Figure 3.22*c*). Save the cut-off pipet (which now has a capillary stem) for later use as a delivery pipet to transfer solutions to the micropycnometer.

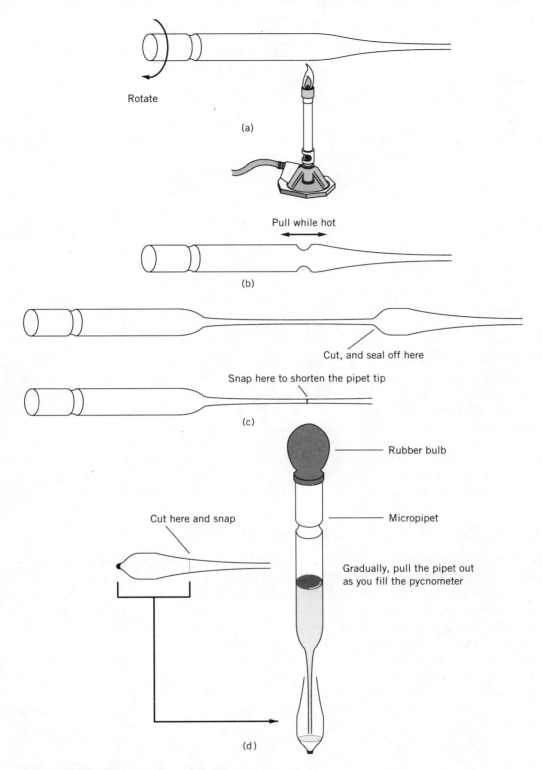

Rotate

(a)

Pull while hot

(b)

Cut, and seal off here

Snap here to shorten the pipet tip

(c)

Rubber bulb

Micropipet

Gradually, pull the pipet out
as you fill the pycnometer

Cut here and snap

(d)

Figure 3.22 Construction of a micropycnometer

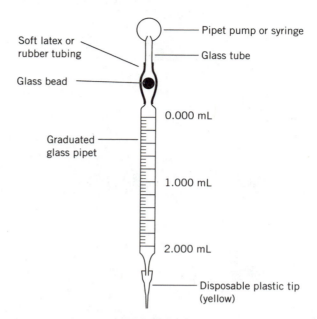

Soft latex or rubber tubing

Glass bead

Graduated glass pipet

Pipet pump or syringe

Glass tube

0.000 mL

1.000 mL

2.000 mL

Disposable plastic tip (yellow)

Figure 3.23   Construction of a microburet

Heat the just-sealed part of the pycnometer bulb strongly to form a smooth, rounded end. Cut the opposite end of the pycnometer, and fire-polish the opening. Attach a pipet bulb to the delivery pipet. Note that the width of the capillary opening must be smaller than the opening of the mouth of the pycnometer, to allow easy passage of air being displaced from inside the pycnometer when it is filled (Figure 3.22*d*).

The micropycnometer just described will have a volume of 100 to 500 μL. For smaller volumes, an **ultramicropycnometer** must be used. For an ultramicropycnometer, heat the Pasteur pipet *at the constriction* (at the end of the tapering where the tip is attached to the barrel of the pipet). The volume of the ultramicropycnometer will be 20 to 100 μL.

Construction and Use of a Microburet

The construction of the microburet is shown in Figure 3.23. The following materials are needed for its construction: a 1.00 or 2.00 mL graduated pipet, a $\sim\frac{1}{8}$ in. i.d., 6 to 8 cm long piece of rubber or *soft* latex tubing, a 6 to 7 cm long glass tube, a fine-bore plastic pipet tip (yellow), and a small smooth-surface glass bead ($\sim$4 mm) of slightly larger diameter than that of the tubing. If rubber tubing is used, clean the inside of the tubing with a triangular file, rinse the tubing with acetone, and dry.

Insert the glass bead into the latex tubing (moistening the glass bead with a drop of water or acetone will help in sliding the bead to the center of the tubing). Next, attach the tubing to the upper end of the pipet. Attach the glass tube to the free end of the latex tube. Attach the pipet tip to the lower end of the pipet. The plastic pipet tip helps in the formation of small droplets of the solution being transferred from the microburet.

Attach the buret to a buret clamp. To fill, place the tip of the microburet in the desired solution and apply suction at the other end of the buret (using a pipet bulb or pipet pump or a 5 mL syringe, see Figure 3.29*b*), *simultaneously* squeezing the glass bead. The pressure will create a passage for the solution. To deliver a solution from the microburet, press *only* the glass bead [not the tubing under the bead!] with the thumb and forefinger. You can also use a Hoffman clamp or a pinch clamp instead of the glass bead to control liquid flow from the buret. A little practice will allow you to become quite proficient at delivering small volumes of solution from the microburet.

Rinsing can be accomplished in the same fashion. Squeeze the glass bead, and fill one-third of the buret with the solution to be used as a titrant (see section 3.11). Remove the buret from the clamp. Holding the buret at a horizontal position, squeeze the glass bead to create the air passage. Tilt the buret up and down through a 30° angle. This will allow the liquid inside the buret to move back and forth.

A microburet has several advantages over larger burets. It is easy to fill (no overflow-ing), easy to control, much less expensive, and the titration requires much less time. The accuracy of such burets is $\pm$ 1 μL.

Construction of Micro Magnetic Stir Bars

Inexpensive micro magnetic stir bars may be constructed in the following manner. Take a long-tipped Pasteur pipet and seal off the bottom of the tip using a Bunsen burner or microburner. Cut and drop a 0.5 cm (or whatever length is desired) long piece of a steel paper clip down the open end of the pipet. Place the tip, just above the paper clip piece, in the flame of the burner, and with tweezers, pull off the now sealed, glass-enclosed, micro magnetic stir bar (Figure 3.24a). Several bars can be made from one pipet and paper clip.

Construction of a Pasteur Filter Pipet

A Pasteur filter pipet is used for filtering a solution as it is transferred. Insert a small cotton plug (only a tiny wisp of cotton is necessary) into the tip of a Pasteur pipet by means of a metal wire, as shown in Figure 3.24b. It is important to use the correct amount of cotton so that the plug is not so tight as to prevent easy flow of the liquid or so loose as to come out when a liquid is dispensed.

Construction of a Glass Plate Placing Mat

We recommend the use of a sandwich glass plate (12 $\times$ 13 in.) top as a placing mat for all microscale equipment used in an experiment. Obtain the following materials: two sheets of ordinary window glass (12 $\times$ 13 in.), a piece of cardboard (12 $\times$ 13 in.), a sheet of white paper (12 $\times$ 8 in.), a sheet of dark-colored paper (12 $\times$ 5 in.) and a wide (2 to 3 in.) tape (masking or plastic). Place one sheet of glass plate on the cardboard so the sides line up exactly. On the glass, position the white and dark-colored paper so that they cover the entire surface. They may overlap to some extent. Now place the second glass sheet over the paper layers to obtain a glass-paper-glass-cardboard sandwich. Carefully tape the combination all around the edges to create the finished glass plate.

Construction of a Microscale Gas Law Tube

Using a strong burner flame, rotate and heat a clean and dry 9 in. Pasteur pipet at the con-striction (Figure 3.25a). As soon as the pipet becomes soft (Figure 3.25b), remove it from the flame *and then* pull the ends apart to form a capillary (Figure 3.25c). Hold the pipet level until it cools. Snap the stem of the pipet (or just heat at the tip end of the pipet) to detach it from the capillary delivery tube, as shown in Figure 3.25d. The main body of the pipet serves as the delivery pipet (Figure 3.25e).

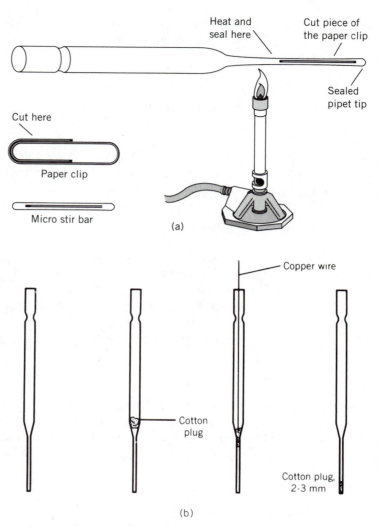

Figure 3.24 (a) Construction of a micro magnetic stir bar; (b) micro filter pipet.

The detached part of the original pipet (4 to 5 in. long) serves as the gas law tube (Figure 3.25*f*). Using the delivery capillary pipet (Figure 3.25*g*), transfer a small amount of mercury so that a 1 to 2 in. length of air column is trapped inside the tube. The length of the mercury column should be 1.5 to 2 in. Finally, use glycerin to slide the gas law tube inside a narrow bore (~3 mm i.d.) 5 to 6 in. long transparent soft Tygon® tube (Figure 3.25*h*). The Tygon® magnifies the size of the mercury column and protects the gas tube from breaking. Further, the sleeve prevents spillage of mercury.

CAUTION: Mercury is a toxic liquid. To prevent spills when transferring mercury, hold the tube inside a small plastic bowl.

Use the tube with care. A small plastic bowl with several pieces of soft tissue paper inside it should be used to hold the gas law tube during the experiment. When it is not in

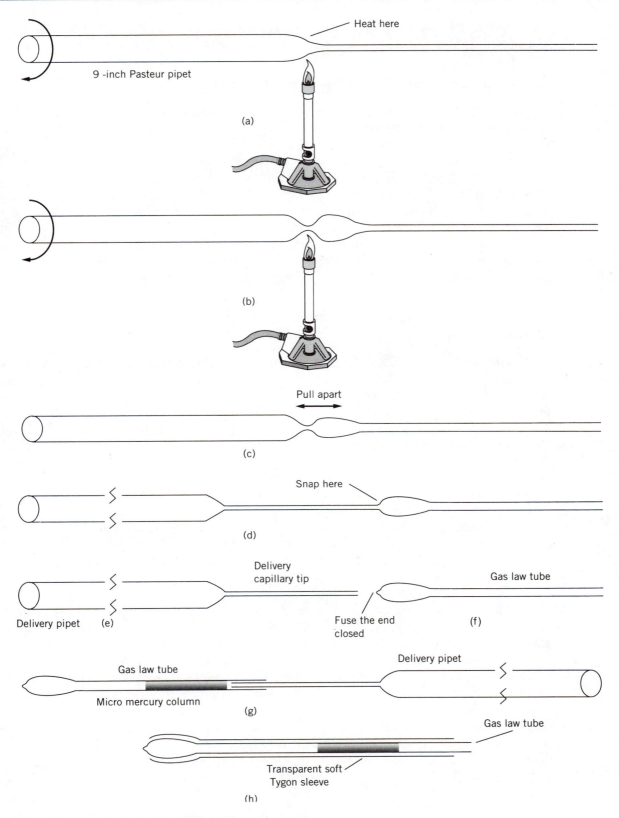

Figure 3.25 Construction of a gas law tube

use, store the tubes in a Ziploc sandwich bag. Before storing the tube, its open end may be sealed with a flame. Break the seal when the tube is to be used again.

Construction of a Bell Capillary for Boiling Point Determination

To make a bell capilary, one performs the same procedure for heating a Pasteur pipet and drawing a capillary as described earlier for the construction of a micropycnometer or a gas law tube (Figure 3.26*a* and *b*). The main body of the pipet serves as the delivery pipet (Figure 3.26*b* and *c*). The other end of the pipet, carrying a fine-bore microtip, is detached from the main body of the pipet. Fuse the end of the bell capillary and snap it from the remaining part of the original pipet tip (Figure 3.26*d* and *e*). The size of the bell capillary must be such that it can be easily inserted inside a melting point capillary tube. Using the delivery pipet, transfer a small amount of the experimental liquid into the melting point tube and then insert the bell capillary, open end down. Tap the melting point capillary on the bench to slide the capillary to the bottom (Figure 3.26*f*). You may use a centrifuge to spin the bell capillary to the bottom of the melting point capillary.

Construction of Micro and Semimicro Columns for Column Chromatography

In the construction of columns for preparative column chromatography, the glass columns are packed with the solid stationary phase. The quantity of solid required for such columns is determined by a simple rule: use 30 to 100 times the weight of packing material to the amount of the sample to be chromatographed. The size of the column should give a 1:10 ratio of diameter to height for the amount of the solid being used. In microscale work, two sizes of columns (micro and semimicro) are used:

1. For a micro column to be useful for the separation of 10 to 100 mg of the mixture, a Pasteur pipet is used. Its tip is broken at the constriction, and the pipet is packed with 0.5 to 2.0 g of solid (Figure 3.27*a*).
2. For a semimicro column, a broken buret may be used. The length of the buret should be around 10 cm above the stopcock. The column made by adding 5 to 20 g of packing material is used for separating 50 to 200 mg of mixture (Figure 3.27*b*).

Both columns are made by following a general procedure. Clamp the empty column vertically with its tip pointing downward. Use a long metal wire to insert a small plug of glass wool or cotton inside the column and place it at the bottom of the column.

Cover the plug with a 0.5 cm layer (in buret) of prewashed coarse sand. Cover the cotton plug in the Pasteur pipet with 50 mg of sand. Load the Pasteur pipet column with ~500 mg of adsorbent with gentle tapping. This procedure is called **dry packing**. Premoisten the column just before use. The buret column is packed by slurry method. At first, by adding a suitable solvent, make a slurry of the powdered solid. Next, fill the column in part with the solvent. Then open the stopcock slightly to allow the solvent to drip slowly from the bottom. Pour the slurry into the column. Gently tap the column with a rubber stopper attached to a thick glass rod. The solid packing material will settle by the action of gravity. Open the

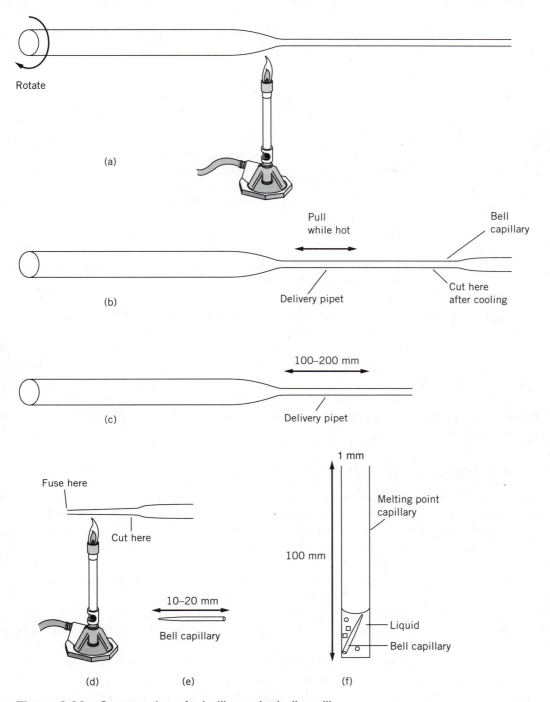

Figure 3.26 Construction of a boiling point bell capillary

stopcock to drain the solvent to the top of the column. Add a 0.5 cm layer of sand to cover the solid. The column is ready for use.

Procedure for Column Chromatography

After the column has been constructed, it is charged with a sample dissolved in a minimum amount of solvent. The pipet used for the transfer and the container in which the sample

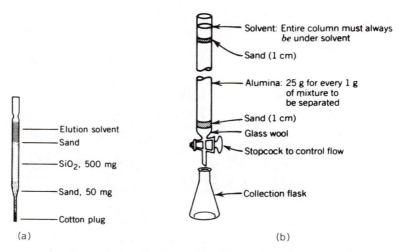

(a)

(b)

Figure 3.27 Construction of chromatographic columns: (*a*) Micro column; and (*b*) semimicro column

was prepared are also rinsed with the solvent, and the rinsings are added to the top of the column.

A suitable solvent or a mixture of solvents (eluent) is used to elute the column. The column is loaded with the eluent, and the eluate emerging from the bottom of the column is collected. The upper level of the solvent is not allowed to drop below the top of the column. Whereas the Pasteur pipet column is free flowing, the buret column flow rate can be controlled by the stopcock. In the case of a Pasteur pipet, once the column is charged, it needs constant attention.

Construction of a Viscometer

Obtain two **scrupulously** clean volumetric pipets—one 2 mL and the other 5 mL in size, a 20 to 25 cm length of soft latex tubing (5 mm i.d. × 8 mm o.d.), a stopwatch, a ring stand, three clamps, two pieces of copper wire, a 20 mL syringe with a needle or a 20 mL pipet, a thermometer, and a water bath. Clean the tubing with water followed by acetone. Dry the tubing by blowing air through it.

With an iron file make a reference mark below the bottom of the bulb of the 2 mL pipet. Insert the capillary tips of both the pipets (2 mL and 5 mL) into the openings of the latex tubing as shown in Figure 3.28. Using pieces of copper wire, secure the tubing to the pipet tips. Clamp both pipets in a vertical position in a water bath. This is the simple Ostwald-type viscometer.

Using a syringe, place ~10 to 20 mL liquid in the 5 mL pipet. Adjust the height of the 5 mL pipet arm so that the upper mark on the pipet is aligned with that scored on the 2 mL pipet (lower scratch mark). At this point the bulb of the 5 mL pipet should be half filled.

Attach a pipet pump to the top of the 2 mL pipet. Make sure that no air bubble is left in the tubing or inside the pipets. Using the pump, pull the liquid above the upper mark on the barrel of the 2 mL pipet. Remove the pipet pump. Allow the liquid in the 2 mL pipet to flow down freely. As soon as the liquid crosses the upper mark, start your stop watch. Record the time just when the liquid meniscus crosses the lower mark. Repeat the procedure until you obtain a precision of 0.5 s for the time elapsed.

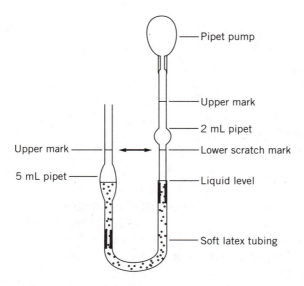

Figure 3.28 An Ostwald-type viscometer

Construction of a Nail–Filter Paper Funnel

A nail-filter funnel is a low-cost substitute for a Hirsch funnel. Cut a suitable-size soft glass rod and heat the tip of one end over a Bunsen burner flame. When it becomes soft, flatten the soft end by pressing it vertically against a cold metal surface (do not use a wooden surface). Cut the nail rod to a suitable length and place the "nail" inside the funnel stem such that the flattened head of the nail rests on the top opening of the stem of the funnel. Procure a suitable size filter paper, place it on the nail head and wet it with the solvent being used (Figure 3.20*b*). Vacuum filtration of a precipitate can be carried out easily.

3.11 Titration Procedure

Using a 50 mL Buret

In a titration, a solution of known concentration is added from a buret to a known volume of another solution (usually placed in an Erlenmeyer flask). The reaction is usually arranged to result in a color change upon completion, **the change in color indicating the end of the titration**. In most cases, a small quantity of an indicator that will change color at the desired reaction end point is added.

  If a strong acid is titrated with a strong base, for example, the pH near the equivalence point will change sharply from about 5.0 to about 9.0 upon the addition of just a few drops of base. The pH at the equivalence point is 7.0. The standard indicator solution for this type of titration is **phenolphthalein** (pronounced fee-nol-thay-leen), which undergoes a color change from colorless (in acid solution) to pink (in base solution) at a pH of 8.3. Only a drop or two of phenolphthalein is needed to see a sharp color change. Titrations of weak acids or bases have equivalence points at pH values other than 7.0, and other indicators may be in order. Some common indicators are listed in Table 3.1. Various mixtures of indicators (called **universal indicators**) are also available.

  To facilitate the reading of the buret without parallax, draw a thick ink line on a card or paper and hold this behind the buret just below the meniscus. The proper practice of using a

Table 3.1 Common Indicators: Color and pH Range

Indicator Name	Acid color	Base color	pH range
Thymol blue	Red	Yellow	1.2–2.8
Methyl orange	Red	Orange	3.2–4.4
Methyl red	Red	Yellow	4.8–6.0
Litmus	Red	Blue	4.7–8.3
Phenolphthalein	Colorless	Pink	8.2–10.0
Alizarin yellow	Yellow	Red	10.1–12.0

buret is to operate the stopcock with one hand (holding the stopcock handle with the thumb and forefinger, with the other fingers wrapped around the buret) while using the other hand to swirl the flask containing the solution being titrated (see Figure 3.29*a*).

Before filling a 50 mL or a 10 mL buret, rinse it first with deionized water and then at least three times with the solution which will be used to fill the buret. Using a funnel, add about 5 mL of the solution (0.5 mL for a microburet) to the buret with the stopcock closed. Hold the buret almost horizontally and rotate the body of the buret so that the solution rinses the inner walls. Let the solution drain through the tip. Attach the buret to a buret clamp mounted on a ring stand. Close the stopcock, put a funnel on the buret, and fill it with the solution above the zero mark (or using a pipet pump, draw the solution into the microburet). Allow the excess solution to drain. This technique will help in removing the air lock from the tip or the stopcock. **Make sure that the buret tip is filled with the solution and that there are no air bubbles**. Read the initial volume of the solution inside the buret. It is not necessary that the solution meniscus be at the 0 mL marking.

Using a pipet (see Figure 3.10), transfer a known volume of a solution (called the **analyte**) to an Erlenmeyer flask. The solution dispensed from the buret is called the **titrant**. If necessary, add a drop or two of indicator to the analyte. Place the flask under the buret.

Spread a white sheet of paper under the Erlenmeyer flask so that it will be easy to detect the end point by noting the color change of the indicator. Add the solution from the buret slowly. Toward the end of the titration, wash down the sides of the flask with deionized water to ensure that all the reactant comes in contact with the titrant.

Using a Microburet

The microscale titration procedure is basically the same as that for the macroscale titration just described. In the case of a microburet, the purpose of the stopcock as used in a 50 mL buret is served by the glass bead inside the latex tubing (see section 3.10 for details). In microscale titration, use of a fixed-volume pipet (as shown in Figure 3.10) is not recommended. Instead, a graduated microburet is used to transfer a fixed volume of an analyte. The proper operation procedure is shown in Figure 3.29*b*. Note that two variations of a microburet are illustrated. See Experiments 10 to 13 for further details.

Perform the same acid-base microscale titration several times until you learn how to handle the microburet properly. The practice will enable you to operate the glass bead–tubing arrangement properly. Improper squeezing of the glass bead will create an air bubble at the end of the micropipet tip. Before every titration, remove the air bubble from the tip by squeezing the glass bead.

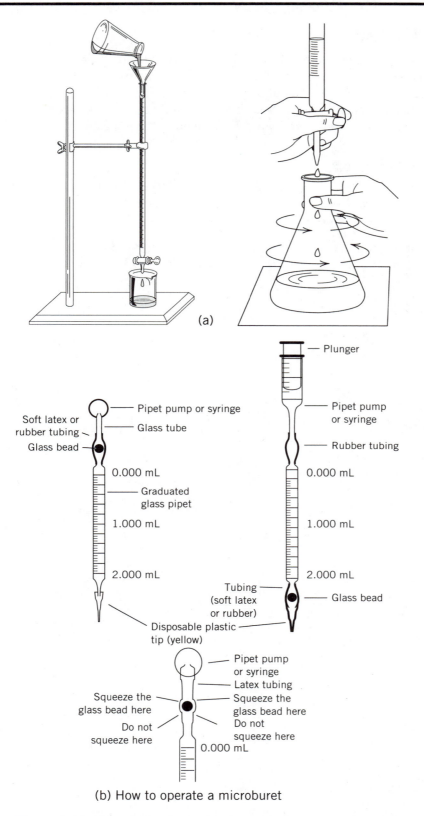

(a)

(b) How to operate a microburet

Figure 3.29 Proper titration procedure: (a) Operation of a 50 or 10 mL buret; (b) operation of a 2 mL microburet

Using a Syringe

Microscale titration can also be carried out using a graduated 1 or 5 mL syringe instead of a microburet. Replace the needle with a disposable micropipet tip. Using a syringe, transfer a known volume of a standard solution into a 10 mL Erlenmeyer flask. Add, drop by drop, the titrant from a second syringe. Near the end point, add small drops of the titrant till the reaction is complete as indicated by a color change of the indicator.

General References

1. Flash, P.; Phiri, S.; Mukherjee, G. *J. Chem. Educ.* **1994**, *71*, A5 and a reference therein.
2. Monti D. *J. Chem. Educ.* **1984**, *61*, 521 and references therein.
3. Newton, D. J. *J. Chem. Educ.* **1978**, *55*, 614.
4. Scaife, C. W. J. *J. Chem. Educ.* **1984**, *61*, 838.
5. Singh, M. M.; Szafran, Z.; Pike, R. M. *J. Chem. Educ.* **1991**, *68*, A125.
6. Singh, M. M.; Szafran, Z.; Pike, R. M. *J. Chem. Educ.* **1993**, *70*, A36.

CHAPTER 4 Report Writing and Maintaining a Laboratory Notebook

Objective

- To learn to write laboratory reports

Much scientific knowledge is obtained by carrying out experiments in a laboratory. This activity involves setting up equipment; making observations; and collecting, manipulating, and analyzing data. Scientists are often called upon to report experimental results to their peers for critical scrutiny and constructive evaluation. Without a properly maintained laboratory notebook, a chemist could not accurately remember the steps in a chemical procedure and could not disseminate the details of the findings to other chemists, who may wish to reproduce the experiment. In this chapter, the general guidelines for maintaining a laboratory notebook and writing a laboratory report are discussed.[1-6]

4.1 The Laboratory Notebook and Its Maintenance

All chemists keep a record of their laboratory work in a laboratory notebook (in this manual we also use Data Sheets: see Section 4.5). A properly maintained laboratory notebook contains the written record of the experiments as well as related results. It serves, if necessary, as a legal document to establish priority claims, such as when a new reaction was first performed. It is therefore extremely important for scientists to learn to keep detailed accounts of their work. The following points are important in connection with maintaining a laboratory notebook:

1. Use a hardbound, lined notebook.
2. All entries in the notebook should be in **ink**.
3. Leave the first few pages blank, for the writing of a table of contents when the notebook is complete. Number the blank pages in Roman numerals. The rest of the pages should be numbered in Arabic numerals. Numbers should be written on both sides of each page.
4. Write your name, the course number, the name of the college/research institute, the instructor/group leader's name, and the inclusive dates on the cover of the notebook.
5. Data (mass, volume, concentration, pH, absorbance, and the like) **including units** should be entered directly into the notebook (**never on scrap paper**), *immediately* (**not days later**) after they are obtained. The entries must be clear, unambiguous, and legible.
6. Any mistakes should be crossed out (**never erased or whited out**). A mistake should be crossed out by drawing a single line through the whole word or sentence. Never write over a mistake. Rewriting above the crossed-out mistake is permissible. If a considerable amount of material is to be removed, simply put an X mark through the material.

7. All entries should be recorded in sequence, and any calculation method should be shown. Collect large amounts of data in tables.
8. If a page is not completely filled, put an X through the empty space.
9. Sign your name on each page and show your data to your instructor/supervisor for his/her signature as a witness.
10. Entries in the notebook should be sufficiently precise so that someone else can repeat your work.

4.2 The Notebook Entry

Although the style of writing a notebook depends upon the individual, a notebook entry has several key components:[4-6]

1. The date the experimental work was conducted.
2. The title of the experiment.
3. The objective for the experiment.
4. A reaction scheme or outline of the experiment.
5. Any calculations made for the experiment.
6. A table of reagents, products, and other data.
7. The procedure that was followed, including **any precautions** that were necessary. This list will warn others who may not know about the dangers involved in performing the experiment.
8. Analytical and spectral data.
9. The signature of the person performing the experiment and of a witness.

4.3 Writing the Laboratory Report

After the data have been entered in the laboratory notebook (or data sheets), chemists must often submit a written report of their work. Most chemists write the report in the present tense, third person, and in passive voice. The past tense is used in writing the experimental section. Phrases such as "I prepared the solution" or "Prepare a solution" should be replaced by "The solution was prepared." Sentences should not begin with a number, such as "0.234 g of a compound was taken"—they should be written as "A sample of 0.234 g of a compound was taken." Typically, a full report consists of several sections. Common section choices are discussed below.

Title of the Report

The title should represent the essence of the report. After reading the title, a reader should have a good idea what the report is all about, and it should not be more than two sentences long. A title need not be a complete sentence.

Abstract

The abstract is a very brief summary (about 250 words) of the work. The experimental procedure, results, and discussion are briefly mentioned. The main finding and any new ideas developed based on the experiment should be mentioned. There should be enough information in the abstract so that readers can decide whether they are interested in reading the paper in more detail.

Introduction

This part of the report consists of a brief background to the main work carried out in the experiment. If any previous related work has been done by others, it should be referenced from the literature. The theory behind any calculation should be explained. The objective of the experiment, as well as any hypothesis that is to be proven, should be described here.

Experimental Section

This section should describe the experimental procedure that was followed. Any special equipment should be described (manufacturer, model number, and so forth). A list of chemicals (with their purities) should be given. Information about how to purchase any unusual chemical should be given. Often, an experimental procedure will follow a previously published method. In such cases, the procedure need not be repeated—citing the original literature is sufficient. If the procedure was modified in any way, the modification should be described. The method of purifying any chemical or solvent should be indicated. Details about the manner in which spectra were run should be included, for example "All visible spectra were obtained on a Beckman DU-50 at 350 nm in toluene, over a concentration range of 1–6 ppm prepared under a N_2 atmosphere."

In short, the exact method followed in the experiment must be described in sufficient detail so that the experiment can be reproduced by others.

Data Section

This section should include any data that were obtained, preferably in tabular form. The accuracy and the precision of the data should be described. Any graphs, spectra, and most figures should be included in this section. If the report includes calculations, a sample should be shown.

Discussion Section

The results should be discussed and any hypotheses should be proposed in the discussion section. If any data or figure is referred to in this section, the table or figure number should be identified for easy reference. Never try to force data to satisfy a preconceived idea of what the outcome should be. Discuss positive as well as negative results. Be critical and truthful in all evaluations. This is the part of the report where you can demonstrate how creative and imaginative you are.

Conclusion

The conclusion section should summarize the results and draw any conclusions about them. Did the data fit into a pattern? Is further work called for?

Acknowledgment

Any assistance received must be acknowledged in this section. It is customary to acknowledge financial assistance (if the work was financed by a grant, for example) in this section as well.

References

All materials gathered from outside sources for writing the report must be listed in the reference section. Reference numbers should be included in the body of the text where appropriate, using consecutive numbers. Reference numbers appear in the text either as superscripts[X] or as underlined numbers in parentheses (X). The actual cited references are listed at the end of the paper in a reference section. Cited references may consist of journals, books, conference presentations, or personal communications from others. Journal articles are cited by listing the last names of authors followed by their initials, the title of the journal (usually in abbreviated form and italicized), the year of publication (in boldface type), the volume number (in italics) and the initial page number of the article. The title of the paper is usually not mentioned in the reference unless needed for clarity. The following is an example of a journal reference:

1. Houk, K. N.; Sims, J.; Watts, C. R.; Luskus, L. J. *J. Am. Chem. Soc.* **1973**, *95*, 7301.

Books are cited by listing the name of authors (last name first), the title of the book (italics), the edition (if other than the first), the publisher (followed by a colon), city (and the state if the city is small), and the year of publication. For example,

2. Mayo, D. W.; Pike, R. M.; Butcher, S. S. *Microscale Organic Laboratory,* 2nd ed.; Wiley: New York, 1989.

4.4   Submission of a Report

The report for submission is arranged in the following order:

1. The first page contains the title of the work, your name, course number and name, college/laboratory/institute, and date. If you wish, you may include the name of the instructor/supervisor. **The prelaboratory report sheet in the manual serves this purpose.**
2. Second and subsequent pages should include the abstract, introduction, experimental procedure, data sheets (data collection and data manipulation tables), discussion, conclusion, and acknowledgment.
3. The references are given in the last page(s). It is advisable to use separate page(s) for references.
4. Tables, spectra, and graphs should be attached to the report as appendices.

4.5   Data Sheets

Prelaboratory report sheets, prelaboratory problems, data sheets (data collection and data manipulation tables), and postlaboratory problems have been included in this manual to help beginning students learn how to organize and to analyze the experimental data. Data sheets do not show how to do calculations or how to manipulate the data.

You should fill out the **prelaboratory report sheet** before coming to the laboratory, to ensure that you understand the experiment. By forcing you to think the procedure through beforehand, it will help you to carry out the experiment in a timely manner. Your instructor may collect this sheet, as well as the prelaboratory problems, before you begin your work.

Beginning students should submit laboratory reports that include these aids, plots, spectra, and the like and a short **written conclusion** of their own. More advanced students should maintain a laboratory notebook and should submit more substantial reports, in the style discussed previously (see sections 4.4 and 4.5). Attach the data sheets (data collection and data manipulation tables, prelaboratory and postlaboratory problems) to your final report. A sample of a laboratory report written by a student is reproduced below, following the references.

References

1. Ebel, H. F.; Bliefert, C.; Rusey, W. E. *The Art of Scientific Writing*; VCH: Weinheim, Federal Republic of Germany, 1987.
2. Kanare, H. M. *Writing the Laboratory Notebook*; American Chemical Society: Washington, DC, 1985.
3. Maimon, E. P.; Belcher, G. L.; Hearn, G. W.; Nodine, B. F.; O'Connor, F. W. *Writing in the Arts and Sciences*; Winthrop Publishers: Cambridge, MA, 1981.
4. Mayo, D. W.; Pike, R. M.; Trumper, K. K. *Microscale Organic Laboratory*, 3rd ed.; Wiley: New York, 1994.
5. Schoenfeld, R. *The Chemist's English*, 2nd ed.; VCH: Weinheim, Federal Republic of Germany, 1986.
6. Szafran, Z.; Pike, R. M.; Singh, M. M. *Microscale Inorganic Chemistry: A Comprehensive Laboratory Experience*; Wiley: New York, 1991.

A sample of the report written by a student is given below:

Report

Lori Murphy
CH 114, General and Advanced General Chemistry
Merrimack College
Spring 1994

Title

Complexometric titration: Determination of water hardness using a prestandardized solution of disodium ethylenediamine tetraacetate (EDTA) dihydrate

Abstract

A standardized solution of a sodium salt of EDTA was used to determine the hardness of ordinary drinking water.

Introduction

The disodium salt of H_4EDTA is extensively used in complexometric titrations. The hardness of water is due to the presence of soluble salts of calcium and magnesium in water. The sodium salt of EDTA reacts with Ca^{2+} and Mg^{2+} (M^{2+}) according to the following reaction:

$$M^{2+} + EDTA^{4-} \rightarrow M(EDTA)^{2-}$$

An indicator, Eriochrome Black T (EBT), is used. The indicator reacts with M^{2+} to form [M-EBT], which has a wine red color. When EDTA is added to the solution, it competes with EBT for the metal. Since the $[M\text{-}EDTA]^{2-}$ complex is more stable, the metal binds with EDTA, releasing free EBT. Free EBT has a sky blue color. The end point of the titration is indicated by the change of the color of the indicator from wine red to sky blue. Knowing the concentration and the volume of EDTA, it is possible to calculate the total hardness of water.

Experimental Section

The experiment was performed according to the procedure described in the laboratory textbook by Singh, Pike, and Szafran.

Materials Table

$Na_2EDTA \cdot 2H_2O$, $MgCl_2 \cdot 6H_2O$, solid NaOH, dry $CaCO_3$, HCl, ammonia buffer, Eriochrome Black T indicator. The following special glassware was obtained: two microburets and a 100 mL volumetric flask.

Data Section

See attached data sheets and sample calculations (not included in this report).

Discussion and Conclusion

For this experiment, a solid powder of Eriochrome Black T was used. At first, EDTA solution was standardized against a standard solution of $CaCl_2$ prepared from $CaCO_3$ and HCl. Next, using the standardized EDTA, the hardness of water (total Ca^{2+} and Mg^{2+}) was determined.

This experiment proved to be quite interesting in that we could find how much Ca^{2+} and Mg^{2+} were present in the water sample simply by using the titration method.

CHAPTER 5 Physical Methods in the Laboratory

EXPERIMENT **1** Determination of Density of Solids, Liquids, and Solutions

Micro- and Macroscale Experiments

Objectives

- To determine the density and specific gravity of a solid by volumetric and gravimetric methods and to calculate its thickness or radius from its density
- To determine the density of a solution or a liquid using a micropycnometer
- To learn to use a balance, buret, pipet, and burner
- To make glass capillaries and micropycnometers
- To learn to analyze data

Prior Reading

- Chapter 1   Safety precautions in the laboratory
- Section 2.2   Scientific notation
- Section 2.4   Significant figures
- Section 2.5   Standard deviation
- Section 2.6   Graphing of data
- Section 3.5   Measuring liquid volumes
- Section 3.10   Construction and use of a micropycnometer

Related Experiments

- Experiments   4 (Viscosity), 10 (Analysis of vinegar), and 34 (Polymers)

Two kinds of properties are associated with any compound: physical and chemical. Physical properties are those characteristics [state of matter (gas, liquid, or solid), color, odor, texture, melting point (mp), boiling point (bp), density (ρ), solubility, viscosity, surface tension, and refractive index] that can be studied without destroying the compound. Most chemical properties of compounds are studied by observing such chemical reactions as precipitation, combustion, oxidation-reduction, addition, and acid-base reactions. Chemical properties cannot be investigated without altering the composition of a chemical compound.

In this and subsequent experiments, a number of important physical properties of several compounds will be determined. Solids will be characterized by measuring their densities and melting points, whereas liquids will be characterized by determining their boiling points, densities, viscosities, and refractive indices.

Density

The **density** (ρ) of a substance is defined as the ratio of its mass to its volume:

$$\rho = m/V \qquad [1]$$

Table E1.1 Densities of water at various temperatures[*]

Temperature, °C	Density, g/mL	Temperature, °C	Density, g/mL
0	0.99987	24	0.99730
1	0.99993	25	0.99704
2	0.99997	26	0.99678
3	0.99999	27	0.99651
4	1.00000	28	0.99623
5	0.99999	29	0.99594
10	0.99973	30	0.99565
15	0.99910	32	0.99503
16	0.99894	35	0.99403
17	0.99877	40	0.99222
18	0.99860	50	0.98804
19	0.99841	60	0.98320
20	0.99820	70	0.97777
21	0.99800	80	0.97180
22	0.99777	90	0.96532
23	0.99754	100	0.95836

[*]*Source*: *CRC Handbook of Chemistry and Physics*, 67th ed.; CRC Press: Boca Raton, FL, 1987.

The normal units for density in SI (International System of Units) are g/mL for liquids, g/cm^3 for solids, and g/L for gases. Density is an **intensive property** of matter, that is, one that does not depend upon the size of the sample. The density of a large piece of pure copper is the same as that of a small piece. On the other hand, an **extensive property** is one that depends upon the size of the sample. Mass, volume, and length are examples of extensive properties. Note that the *ratio* of two extensive properties is independent of the sample size. Therefore, although density is calculated on the basis of the measured mass and volume of an object, it is an intensive property, because density is the ratio of two extensive properties.

The density of a pure material in the solid state is usually greater than its density in the liquid state; the density of a liquid is greater than its density in the gaseous state. There is a well-known exception to this: the density of water at 4°C is higher than the density of ice at 0°C. The density of water at different temperatures is given in Table E1.1.

Clearly, in scientific work requiring accuracy, the density of a material should be reported at a specific temperature. For rough calculations, the density of water may be taken as 1 g/mL near room temperature. To determine the density, the mass and the volume of the object must be measured.

Measurement of Mass

The mass of any liquid or solid can be easily measured on an analytical balance to a precision of one-tenth of a milligram. On normal digital balances, mass can be measured to an accuracy of ± 1 mg. Several types of balances are commonly available in a laboratory—your instructor will tell you which balance to use and will demonstrate how to operate it.

Measurement of Volume

Several methods are used to measure the volume of a liquid (mL) or a solid (cm^3). The approximate volume of a liquid can be determined by using a graduated cylinder. More

Table E1.2   Volumes of different regularly shaped solids

Solid	Volume, cm^3	Solid	Volume, cm^3
Cube of length L	L^3	Prism	LWH
Sphere of radius r	$(\frac{4}{3})(\pi r^3)$	Cone	$(1/\pi r^2)H$
Cylinder of height H	$\pi r^2 H$	Pyramid	$\frac{1}{3}$(base area)H

accurate values may be obtained by using quantitative glassware, such as graduated pipets or pycnometers.

One can determine the volume of a regularly shaped solid by measuring its dimensions. For a prism or cube, the height, width, and depth are measured with a centimeter (cm) ruler or with a micrometer and the volume calculated. The volumes of different regularly shaped solids are given in Table E1.2. Thus, the volume of a cube of length L cm is L^3 cm^3. The volume of a cube of length 2.14 cm is 2.14 cm $\times$ 2.14 cm $\times$ 2.14 cm = 9.80 cm^3.

A convenient method of determining the volume of an irregular solid is to measure the volume of a liquid it displaces. Two methods—volumetric (based on volume measurement only) and gravimetric (based on mass measurement)—are described shortly for determining the displaced volume. For this method to work, several conditions must be fulfilled. The solid must have a greater density than the liquid so that it will sink completely in the liquid. Also, the liquid must not react with the solid, and the solid must be insoluble in the liquid. Water is often selected as the liquid of choice.

Other Parameters

Once the density of an object is known, one can then determine other parameters for the object. For example, the radius of a metallic wire can be determined or the thickness of a thin metal foil can also be estimated if the density of the metal is known. The volume of an object can be calculated from its mass and density. Then the desired parameter (thickness or radius) can be obtained from the relation between the volume and the required parameter. The steps involved in the calculation of the thickness of a metal foil are shown below:

Mass of the metal foil $\quad\quad\quad = $ m, g
Density of the metal foil $\quad\quad = \rho$, g/cm^3
Length of the metal foil $\quad\quad = $ L, cm
Width of the metal foil $\quad\quad\quad = $ W, cm
Thickness of the metal foil $\quad\quad = $ T, cm
Volume of the metal foil $V \quad\quad = m/\rho = L$ cm $\times W$ cm $\times T$ cm
Thickness of the metal foil $T \quad = V/(L \times W)$
$\quad\quad\quad\quad\quad\quad\quad\quad\quad\quad T \quad = (m)/(\rho)(L \times W)$

Specific Gravity

A property closely related to density is the **specific gravity** (sp gr). It is defined as the ratio of the density of one substance to the density of a standard substance. For liquids or solids, the standard is water at 4°C, and for gases the standard is air. Specific gravity is a dimensionless quantity (it has no units). However, it changes with temperature. Thus, for a solid or liquid,

$$\text{sp gr} = \rho_{\text{substance}}/\rho_{\text{water at 4°C}} = (m/V)_{\text{substance}}/(m/V)_{\text{water}}$$

or for equal volumes,

$$\text{sp gr} = m_{\text{substance}}/m_{\text{water}}$$

Example E1.1   Several small pieces of aluminum weigh a total of 3.188 g. To measure the volume of these pieces, a student puts them in a graduated cylinder containing water. The volume of water in the cylinder before adding the aluminum was 20.0 mL. After adding the pieces, the volume of water is 21.2 mL. What is the density of aluminum?

Answer

Mass of Al pieces	$m = 3.188$ g
Volume of Al pieces	$V = 21.2$ mL $- 20.0$ mL $= 1.2$ mL
Density of Al	$\rho = m/V = 3.188$ g$/1.2$ mL $= 2.7$ g/cm^3

Note that the answer is given to two significant figures (why?).

Example E1.2   Calculate the thickness of a square piece ($L = 15.15$ cm) of aluminum foil weighing 1.769 g. Use the value for the density of Al determined in Example E1.1.

Answer

Volume of the foil	$V = m/\rho = (1.769$ g$)/(2.7$ g/cm$^3) = 0.66$ cm^3
	$V = LWT$
Thickness of the foil	$T = V/(LW)$
	$T = (0.66$ cm$^3)/(15.15$ cm $\times 15.15$ cm$) = 2.9 \times 10^{-3}$ cm

Note that the thickness is expressed to two significant figures.

Example E1.3   Calculate the specific gravity of methylene chloride (CH_2Cl_2) at 20°C. The mass of 200.00 μL of CH_2Cl_2 is 0.2650 g at 20°C. The mass of the same volume of water is 0.1996 g.

Answer

$$\text{sp gr} = m_{\text{substance}}/m_{\text{water}} \quad \text{(for equal volumes)}$$
$$\text{sp gr} = (0.2650 \text{ g})/(0.1997 \text{ g}) = 1.327$$

Note that specific gravity is dimensionless and depends upon the temperature.

Example E1.4   The densities of a series of calcium nitrate solutions were measured using a micropycnometer (pyc) at 23°C. The following data were obtained:

$$\text{Mass of empty pyc} \quad 0.649 \text{ g}$$
$$\text{Mass of pyc + water} \quad 1.049 \text{ g}$$

Mass of pyc + calcium nitrate solutions

% solution	Mass, g	% solution	Mass, g	% solution	Mass, g
10	1.068	25	1.098	40	1.124
20	1.088	30	1.106	50	1.144
Unknown	1.115				

Calculate the density of each solution. Prepare a graph of density (x axis) versus percent solution (y axis) and determine the percentage of the unknown calcium nitrate solution. What would be the specific gravity of a 40 percent solution? (Density of water at 23°C = 0.997538 g/mL.)

Answer

Mass of empty pyc $\qquad m_{\text{pyc}} = 0.649 \text{ g}$

Mass of water $\qquad m_{\text{water}} = 1.049 \text{ g} - 0.649 \text{ g} = 0.400 \text{ g}$

Volume of water (volume of pyc) $\qquad V_{\text{pyc}} = m/\rho = (0.400 \text{ g})/(0.997538 \text{ g/mL})$
$$V_{\text{pyc}} = 0.401 \text{ mL}$$

Sample calculation:

Mass of 10% solution $\qquad m_{\text{sol}} = m_{(\text{sol}+\text{pyc})} - m_{\text{pyc}}$
$$m_{10\%} = 1.068 \text{ g} - 0.649 \text{ g} = 0.419 \text{ g}$$

Density of 10% solution $\qquad \rho = 0.419 \text{ g}/0.401 \text{ mL} = 1.04 \text{ g/mL}$

Mass of known solutions

10%	20%	25%	30%	40%	50%
0.419 g	0.439 g	0.449 g	0.457 g	0.475 g	0.495 g

Density of known solutions

10%	20%	25%	30%	40%	50%
1.04 g/mL	1.10 g/mL	1.12 g/mL	1.14 g/mL	1.19 g/mL	1.23 g/mL

The graph of these data is plotted in Figure E1.1.

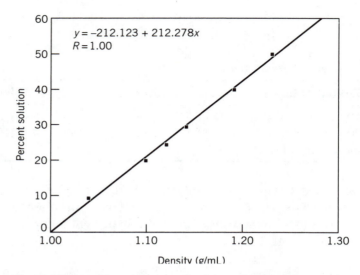

Figure E1.1. Density versus percent calcium nitrate solution

Mass of the unknown solution	$m = 1.115$ g $- 0.649$ g $= 0.466$ g
Volume of the unknown solution	$V = 0.401$ mL
Density of the unknown solution	$\rho = m/V = 0.466$ g$/0.401$ mL $= 1.16$ g/mL
% of the unknown solution	33% (from the plot)
Specific gravity of 40% solution	sp gr $= m$ of 40% solution$/m$ of same volume of water
	sp gr $= 0.475$ g$/0.400$ g $= 1.19$

General References

1. Bergendahl, T. J. *J. Chem. Educ.* **1979**, *56*, 617.
2. D'Auria, J. M.; Chesin, R. J.; Palmer, E. T. *J. Chem. Educ.* **1976**, *53*, 378.
3. McCullough, T. *J. Chem. Educ.* **1993**, *70*, 46.
4. Shearer, E. C.; Rumpel, M. L. *J. Chem. Educ.* **1974**, *51*, 140.
5. Singh, M. M.; Pike, R. M; Szafran, Z. *J. Chem. Educ.* **1993**, *70*, A36.

Experimental Section

Procedure

Part A: Density of a Regularly Shaped Solid

Macroscale experiment   Estimated time to complete the experiment: 0.5 h
Experimental Steps: The procedure involves two steps: (1) weighing and (2) obtaining
the dimensions of the object.

Obtain a regularly shaped solid from your instructor (a cube, sphere, cone, or cylinder), as well as a centimeter ruler and a vernier caliper (if necessary). Record the identification number of the object on your record sheet. Use the appropriate procedure for the solid(s) assigned to you.

Place a piece of weighing paper on the pan of the balance and weigh it. Next, place the solid object on the balance and note the total mass of the object and the weighing paper. The difference between these two weights is the mass of the object. If you are using an electronic balance, the mass of weighing paper can be tared out, and the mass of the object can be read directly.

Using a centimeter ruler, measure the appropriate dimensions of the object (length, width, and height for a block; height for a cylinder). Use the vernier caliper to determine the diameters for spheres or cylinders. A piece of fishing line can be used in place of the caliper, if desired. Measure the circumference (c) by tying the line around the widest part of the sphere. Calculate the diameter ($c = \pi d$). Tabulate your data. Calculate the volume and then the density of the object using the appropriate equations. **After you are done, return all the material to your instructor**.

Part B: Density of an Irregularly Shaped Solid by Volumetric and Gravimetric Methods

Macroscale experiment   Estimated time to complete the experiment: 2 h
Experimental Steps: Two steps are involved: (1) determining the mass and (2) the volume
of the object by the volume displacement and the gravimetric method.

Obtain a sample of metal pellets or powder from your instructor. Enter its identification number on your record sheet. Make sure that the sample is clean and dry and has no holes in it. Treat the metal with either 3 M HCl or 6 M NaOH (ask your instructor which one to use) to dissolve any surface oxide, rinse the sample with deionized water several times, and then dry the metal over some paper toweling, if necessary.

This method is based on the principle of displaced volume. Note the room temperature. Weigh an empty dry, clean graduated cylinder (W_1). Place the unknown solid inside the cylinder, and reweigh (W_2). Determine the mass of the unknown metal. Empty the cylinder by removing the metal. **Save the metal on a clean piece of paper for use as described shortly**.

Set up a clean 50.00 mL buret in a buret clamp attached to a ring stand. Fill the buret with deionized water and note the initial volume of water (V_1). Transfer 6 to 7 mL (if you are using a 10 mL graduated cylinder) or 12 to 16 mL (if you are using a 25 mL graduated cylinder) of water from the buret to the empty graduated cylinder.

> Make sure that no air bubbles are sticking to the inside wall of the cylinder and that no water drops adhere to the upper part of the cylinder.

Record the exact final buret volume (V_2) after the transfer. With a marking pencil, **mark the exact location** of the water meniscus in the graduated cylinder. Avoid parallax error. Weigh the water-filled graduated cylinder (W_3). Now, empty the graduated cylinder by draining the water into the sink. Dry the cylinder with a paper towel. Replace **quantitatively** all the metal pellets that were used previously.

Record the initial buret reading (V_3), and add water to the graduated cylinder now containing the pellets until the water level just reaches the mark you made. Record the buret reading (V_4) after the transfer. Reweigh the cylinder containing the pellets and water (W_4). Calculate the mass m of the metal pellets ($W_2 - W_1$).

Return the buret to your instructor. Also, return the clean and dry (use acetone) metal pieces to the designated container for future use.

Volumetric Calculation

The volume difference $V_2 - V_1$ gives the initial volume of the water transferred from the buret to the empty cylinder. The volume difference $V_4 - V_3$ is the amount of water necessary to fill the cylinder containing the pellets to the mark. Note that the volume V of the metal pellets is the difference between these volumes of water. Calculate the density ($\rho = m/V$) of the metal from these data. If time permits, repeat the procedure two more times and obtain the average value for the density.

Gravimetric Calculation

Calculate the mass of water ($W_3 - W_1$) in the empty cylinder and the mass of water when the graduated cylinder contained the pellets ($W_4 - W_2$). The difference of these two masses is the mass of the water displaced by the pellets. Using the density of water (see Table E1.1), calculate the volume of the displaced water (which is, of course, the volume of the pellets). Calculate the density ($\rho = m/V$) of the pellets for each of your measurement sets and take the average. Compare the two densities as determined by the volumetric method and the gravimetric method.

Part C: Determination of the Thickness of a Metal Foil

Macroscale experiment   Estimated time to complete the experiment: 0.5 h
Experimental Steps: The procedure involves one step: determining the area of the foil (length and width).

Obtain a sample of metal foil and a centimeter ruler. **The foil should be composed of the same metal for which you determined the density in Part B. Handle the foil carefully to prevent kinking or bending**. Measure the length and width of the foil with a centimeter ruler. Obtain the mass of the foil. Using the mass of the metal foil and its density (as determined previously), compute the volume of the foil. Calculate the area of the foil, and thereby obtain the thickness (thickness = volume/area). If possible, compare your result to that obtained using a micrometer. Return the foil to your instructor.

Part D: Density of Liquids and Solutions

1. Syringe/Automatic Delivery Pipet Method

> Microscale experiment   Estimated time to complete the experiment: 2 h
> Experimental Steps: The procedure involves three steps: (1) determining the mass, (2) obtaining the volume of the liquid (or the solution) using a syringe, and an optional step (3) statistical analysis.

Obtain known sample(s) or solution(s) as well as an unknown sample from the instructor. Record the identification number(s) of the materials on the data table.

Weigh an empty, **clean and dry** 5 to 10 mL weighing bottle. Using a pipet or syringe, deliver 100 to 500 μL of liquid to the weighing bottle. Record the exact volume of liquid transferred from the pipet. Weigh the container plus the liquid. Repeat these steps for other solutions or liquids. Return the weighing bottle to your instructor.

> Disposal of chemicals: Dispose of organic liquids in designated containers. Sugar or inorganic salts (ask your instructor) may be discarded into the sink.

Determine the mass of the liquids. Calculate the density of each liquid or solution from the mass and the volume. If time permits, repeat the steps several times for each liquid or solution.

Statistical Analysis (Optional)
Determine the average value of the density, the deviation from the mean, the spread, the relative deviation from the mean in parts per thousand, and the standard deviation s (for finite number of values), variance s^2, and coefficient of variation v. Calculate the 95 percent confidence interval. Use the Q test to eliminate undesirable data (outliers), if any. If two groups of students have used two different methods (pipet method or syringe method), determine the significance of any difference between the two methods using the F test ($F = s_1^2/s_2^2$, where s_1^2 and s_2^2 are the variances for two methods). For all these calculations, see Chapter 2.

2. Micropycnometer Method

Obtain the known liquid (such as organic alcohols) or solution (such as calcium chloride, sugar, potassium iodide) samples, an unknown, a micropycnometer along with a container to hold the pycnometer (see Section 3.10), a capillary delivery pipet, Pasteur pipets, Bunsen burner, acetone, and distilled water. Enter the identification number of your unknown on the record sheet.

Weigh the empty container. Place the dry pycnometer in it and record the mass. Using the capillary delivery pipet, transfer some of the known liquid or solution into the pycnometer. While filling the pycnometer, slowly pull the end of the capillary out from inside the pycnometer. This action prevents an air lock from forming inside the pycnometer. Fill the pycnometer up to the very top of the mouth. Using a wipe, clean and dry the outside of the pycnometer. Replace the filled pycnometer inside the container and weigh it once again.

Empty the pycnometer by pipeting out the liquid or solution using the same capillary pipet used to fill the pycnometer. Dispose of the liquid into a waste container. If you have more than one sample, rinse the pycnometer twice with the next liquid/solution sample. Repeat the procedure for all the liquids or solutions, including the unknown.

Finally, empty the pycnometer, rinse it twice with distilled water, and refill it with room-temperature distilled water. Reweigh the water-filled pycnometer. Clean the pycnometer twice with water, rinse it with acetone, allow it to dry, and return it to your instructor.

Calculate the mass of the empty pycnometer and the mass of added liquid. Next, from the known density of water at the given temperature (see Table E1.1), calculate the volume of water. This is also the volume of the pycnometer.

Using the masses of the other liquids and the volume of the pycnometer, calculate the density for the unknown liquid(s) or solution(s). If your knowns consist of solutions having different percentage concentrations, plot densities (y axis) versus concentrations (x axis). Using this plot, estimate the percent concentration of the unknown. If your samples consist of organic liquids, plot the number of carbon atoms (x axis) against the densities (y axis) and determine the number of carbons in your unknown. The use of a computer program for plotting the data is recommended.

Additional Independent Projects

1. Determine the density of a solid compound that is soluble in water.

   Hint: Use a solvent in which the compound is not soluble.

2. Determine the density of antifreeze (ethylene glycol) at various dilutions, using the micropycnometer method.[1]

   This experiment illustrates the concept that volumes, unlike masses, are not always additive.

3. Determine the density of single inorganic crystals.[2]

4. Determine the density of a liquid or a solid by graphing method.[3]

   Brief Outline

   For a liquid, construct six micropycnometers of different sizes. Using each pycnometer, determine the mass of the liquid and the corresponding volume of the pycnometer. Plot the mass (g) of the liquid (*y* axis) versus corresponding volume (mL) of the pycnometer (*x* axis). The slope of the straight line will give the density of the liquid.

   Instead of a micropycnometer, a 5 mL syringe without a needle may be used. Weigh the empty syringe. Draw 0.50 mL liquid in the syringe and weigh the syringe. Repeat the procedure until the total volume of the liquid inside the syringe is 4.0 mL. Plot the mass (g) versus volume (mL) and determine the density from the slope of the curve.

   Similarly for solids, determine the masses of five different portions of the same solid metal. Measure the volume of water in a 25 mL cylinder. Add the first portion of the solid. Determine the volume of water displaced. Add the second portion of the metal to the first and determine the volume of water displaced. Continue adding the metal portions successively and determine the volume of water displaced after each addition. Plot the mass (g) versus volume (mL) of the metal. The slope of the line will give the density of the metal.

   Develop the data tables (data collection and data manipulation tables) and calculate the density of the liquid and/or the solid. Write a report according to the instructions given in Chapter 4.

5. Determine the thickness of a copper penny by determining its density.

   Hint: Take 8 to 10 pennies and determine the density by volume displacement method.

6. Identify and classify plastics by density.[4,5]

   Brief Outline

   Density of plastics is determined by the buoyancy method. Several liquid mixtures of known densities are prepared. Using two miscible liquids (one having lower density than the other) prepare a series of mixtures having a range of densities in 100 mL graduated cylinders. Using the micropycnometer, determine the densities of each mixture. See reference 4 for the range of liquid mixtures that can be used. Place a small piece of plastic in each mixture. By noting whether the sample sinks or floats in a particular mixture, the identity of the plastic can established.

   Develop the data tables (data collection and data manipulation tables). Write a report according to the instructions given in Chapter 4.

7. Determine the density of plastics by the density gradient technique.[6]

References

1. Flowers, P. A. *J. Chem. Educ.* **1990**, *67*, 1068.
2. Bergendahl, T. J. *J. Chem. Educ.* **1979**, *56*, 61.
3. Richardson, W. S.; Teggins, J. E. *J. Chem. Educ.*, **1988**, *65*, 1013.
4. Kolb, K. E.; Kolb, D. K. *J. Chem Educ.* **1991**, *68*, 348.
5. Earnest, M. C. *J. Chem. Educ.* **1978**, *55*, A373. Also see *J. Chem. Educ.* **1993**, *70*, 174.
6. ANSI/ASTM Method D 1505–68.

Name	Section
Instructor	Date

PRELABORATORY REPORT SHEET—EXPERIMENT 1

Experiment title and part _____

Objective

Reactions/formulas to be used

Chemicals and solutions—their preparation

Materials and equipment table

Outline of procedure

1. A metal has a density of 2.72 g/cm^3. An empty 25.0 mL flask weighs 90.303 g. When a metal solid is placed inside the flask, it weighs 100.406 g. What volume of water will be needed to fill this flask (density of water = 1.000 g/mL)?

2. While filling the micropycnometer with liquid using a capillary pipet, the student leaves a piece of the broken pipet tip inside the pycnometer. The empty pycnometer did not contain the glass piece. Will the calculated density of the liquid be low or high? Explain your answer.

3. If some air bubbles are left in the graduated cylinder under the water level, what effect do they have on determining the density of a metal?

Name ______________________________ Section ______________

Instructor ______________________________ Date ______________

EXPERIMENT 1 DATA SHEET

Part A: Density of Regularly Shaped Solids

Sample ID number ______ Room temperature______ °C

Collection of Data

1. Mass of solid object ______ g

2. Dimensions of the object

Dimension		Run 1	Run 2	Run 3	Average
______________	cm	______	______	______	______
______________	cm	______	______	______	______
______________	cm	______	______	______	______

Manipulation of Data

3. Volume of the block ______ cm^3 (use average values of dimensions)

4. Density ______ g/cm^3

Show calculations:

Part B: Density of Irregular Objects

Solid ID number _____   Room temperature _____ °C

Collection of Data	Run 1	Run 2	Run 3
1. Mass of empty cylinder, W_1, g	_____	_____	_____
2. Mass of cylinder + pellets, W_2, g	_____	_____	_____
3. Initial buret reading, V_1, mL	_____	_____	_____
4. Final buret reading, V_2, mL	_____	_____	_____
5. Mass of cylinder + water, W_3, g	_____	_____	_____
6. Initial buret reading, V_3, mL	_____	_____	_____
7. Final buret reading, V_4, mL	_____	_____	_____
8. Mass of cylinder + water + pellets, W_4, g	_____	_____	_____

Manipulation of Data, Volumetric Method   Density of water, ρ _____ g/mL

	Run 1	Run 2	Run 3
9. Mass of pellets, g	_____	_____	_____
10. Volume of water in cylinder, mL	_____	_____	_____
11. Volume of water in cylinder containing pellets, mL	_____	_____	_____
12. Volume of water displaced by pellets, mL	_____	_____	_____
13. Density of the metal (g/cm^3)	_____	_____	_____
14. Average density (volumetric method)		_____ g/cm^3	

Manipulation of Data, Gravimetric Method   Density of water, ρ _____ g/mL

	Run 1	Run 2	Run 3
15. Mass of water in empty cylinder, g	_____	_____	_____
16. Mass of water with pellets in cylinder, g	_____	_____	_____
17. Mass of displaced water, g	_____	_____	_____
18. Volume of displaced water, g	_____	_____	_____
19. Density of the metal (g/cm^3)	_____	_____	_____
20. Average density (gravimetric method)		_____ g/cm^3	

Show calculations:

DATA SHEET, EXPERIMENT 1, PAGE 3

Part C: Thickness of a Metal Foil

Sample foil ID number _____   Room temperature _____ °C

Collection of Data

1. Density of the metal, ρ _____ g/cm^3 (as determined in Part A or B)

2. Mass of the foil, g   _____

3. Dimensions of foil

	Run 1	Run 2	Run 3	Average
Length, L cm	_____	_____	_____	_____
Width, W cm	_____	_____	_____	_____

Manipulation of data

4. Volume, cm^3 (m/ρ)   _____

5. Area, cm^2 ($L \times W$)   _____ (use average values)

6. Thickness, T cm   _____

Show calculations:

Thickness by micrometer method (cm) _____

Part D: Density of Liquids/Solutions

1. Syringe Automatic Delivery Pipet Method

Sample ID number _____   Room temperature _____ °C

Collection of Data.   Use additional sheets for other trials and statistical analysis.

Sample #	Run #	Mass of Empty container	Mass of Container + liquid	Volume of liquid
_____	(1)	_____ g	_____ g	_____ mL
	(2)	_____ g	_____ g	_____ mL
	(3)	_____ g	_____ g	_____ mL
_____	(1)	_____ g	_____ g	_____ mL
	(2)	_____ g	_____ g	_____ mL
	(3)	_____ g	_____ g	_____ mL
_____	(1)	_____ g	_____ g	_____ mL
	(2)	_____ g	_____ g	_____ mL
	(3)	_____ g	_____ g	_____ mL

Manipulation of Data

Sample #	Run #	Mass of liquid	Volume	Density	Avg. density
_____	(1)	_____	_____	_____	_____
	(2)	_____	_____	_____	
	(3)	_____	_____	_____	
_____	(1)	_____	_____	_____	_____
	(2)	_____	_____	_____	
	(3)	_____	_____	_____	
_____	(1)	_____	_____	_____	_____
	(2)	_____	_____	_____	
	(3)	_____	_____	_____	

Show calculations:

Name ______________________	Section ____________
Instructor ______________________	Date ____________

DATA SHEET, EXPERIMENT 1, PAGE 5

2. Micropycnometer Method

Sample ID number ______ Room temperature ______ °C

Density of water ______ g/mL

Collection of Data

		Run 1	Run 2	Run 3
1.	Mass of empty container, g	______	______	______
2.	Mass of container + pycnometer, g	______	______	______
3.	Mass of container + pyc + water, g	______	______	______
4.	Mass of container + pyc + liquid, g	______	______	______

If necessary, collect data for other liquids/solutions on a separate sheet.

Manipulation of Data

		Run 1	Run 2	Run 3
5.	Mass of empty pycnometer, g	______	______	______
6.	Mass of water, g	______	______	______
7.	Mass of liquid, g	______	______	______
8.	Volume of water, mL	______	______	______
9.	Volume of liquid, mL	______	______	______
10.	Density of liquid, ρ, g/mL	______	______	______
	Average density		______ g/mL	

Attach the graphs if required. From the graphs, report the following:

Concentration of the unknown: ______%

If you use an organic liquid, identify the liquid from the graph (plot of number of carbon atoms versus ρ).

1. Mercury (Hg) is the only metal that is liquid at room temperature. A sample of 10.0 mL of Hg weighs 136.0 g. What is the density of mercury?

2. The density of a lead pellet on the earth's surface is 11.43 g/cm^3. Will its density change if the same pellet is taken to the surface of the moon? Explain your answer.

3. Room temperature is recorded in all the methods described for estimating the density of a substance. Why?

4. What is the essential difference between density and specific gravity?

2 **Determination of Melting Points and Boiling Points**

Microscale Experiments

Objectives
- To characterize and to ascertain the purity of solids and liquids by determining their melting points and boiling points, respectively

Prior Reading
- Chapter 1   Safety precautions in the laboratory
- Section 2.6   Graphing of data
- Section 3.6   Heating methods
- Section 3.10   Construction of a bell capillary for boiling point determination

Related Experiments
- Experiments 19 (Synthesis of copper(II) compounds) and 32 (Synthesis of organic compounds)

Chemists use the melting point of a solid or the boiling point of a liquid to determine its purity and to characterize it. The transition of a solid to liquid at a given pressure is called **melting**. It occurs at a specific temperature known as the melting point. The reverse process is known as **freezing**.

The **melting point** (mp) of a pure solid is defined as the temperature at which the liquid and solid phases coexist in equilibrium under a particular pressure. The most common pressure at which to make this measurement is atmospheric pressure—under this condition, the melting point is called the **normal melting point**. The melting point of a pure solid is usually relatively sharp; that is, the solid melts over a relatively narrow range of temperatures ($2°C$ or so). Moderate changes in pressure do not affect the melting point significantly. However, the melting point of a solid is *very sensitive* to the presence of an impurity in the solid. Even a small amount of impurity in a solid lowers its melting point significantly and also increases the temperature range of the melt. The melting point of a solid is therefore frequently used to establish its purity.

The most familiar solid-liquid equilibrium is that between ice and water at $0°C$ and 1 atm. In the crystalline lattice (ice), water molecules are strongly held together by attractive forces called hydrogen bonds. Heating is necessary to break these attractive forces and to allow the rigid solid to make the transition to the liquid state. A heating curve as depicted in Figure E2.1 is helpful in the study of phase transitions.

When a solid is heated, its temperature increases gradually, raising the dynamic motions of the molecules until point A is reached. At this point, the solid starts to melt. The motion of the molecules in the solid becomes more vigorous and the attractions between molecules break down. During the process of melting, the compound absorbs the heat supplied and the temperature of the solid *does not increase* with time, resulting in the flat portion of the curve A–B. The heat energy required to melt one mole of a solid is called the **molar heat of fusion**, ΔH_{fus}. For ice, ΔH_{fus} is 6.00 kJ/mol. Thus, liquid water has 6.00 kJ more

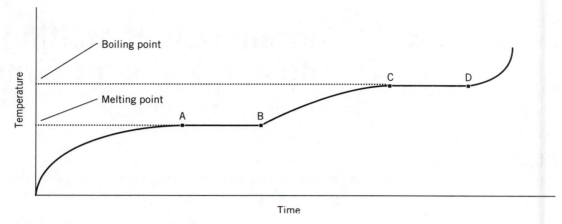

Figure E2.1 Heating curve for a solid

energy, per mole, than solid water does. Since heat must always be added in order to melt a solid, ΔH_{fus} is always a positive quantity.

Once the solid has completely melted (point B), the heat energy further increases molecular motions, and the liquid temperature increases. At any temperature, a certain fraction of the liquid molecules have sufficient energy to escape from the surface and enter the gas phase. Thus, a cup of water will evaporate well below the boiling point of water. In a closed container, an equilibrium is established between the vapor and the liquid. The pressure exerted by the vapor of the liquid on the liquid surface is called the **vapor pressure** of the liquid. The vapor pressure of a liquid increases as heat is supplied to the system, so vapor pressures get larger as the temperature increases. The temperature at which the vapor pressure is equal to the atmospheric pressure is called the **boiling point** (bp) of the liquid. This occurs at point C. Once again, the temperature remains constant during this period (line C–D) until all the liquid has evaporated.

General References

1. *CRC Handbook of Chemistry and Physics*, 68th ed.; CRC Press: Boca Raton, FL, 1988.
2. Mayo, D. W.; Pike, R. M.; Trumper, P. K., *Microscale Organic Laboratory*, 3rd ed.; Wiley: New York, 1994.

Experimental Section

Procedure

Part A: Determination of the Melting Point of a Solid

Microscale experiment This is a one-step procedure.	Estimated time to complete the experiment: 1 h

Obtain a sample of unknown solid and several melting point capillary tubes. If necessary, grind the solid to a fine powder using a mortar and pestle. Transfer about 30 mg of the sample onto a clay tile or watch glass. With the open end down, gently tap a capillary tube into the material so that some solid enters the tube (Figure E2.2). Invert the tube and tap it gently on the bench top so that the material slides down to the bottom of the tube. [A good way of driving a stubborn solid to the bottom of the tube is to drop the capillary down a

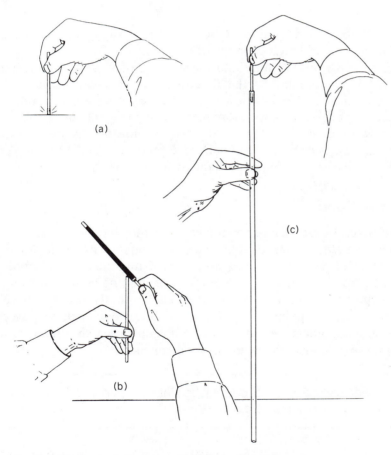

Figure E2.2   Loading a melting-point capillary tube
(a) Tap the tube into the powdered solid.
(b) Pack the solid by tapping it on the bench top or with a file.
(c) Drop it down a hollow glass tube.

Table E2.1 Known compounds and their melting points (mp, °C)

Compounds	mp range	Compounds	mp range
biphenyl	69–72	acetanilide	113–114
2,5-dimethylphenol	68–71	2-acetylbenzoic acid	114–115
4-nitrobenzylchloride	70–73	2-benzylbenzoic acid	118–119
naphthalene	80.5–81.5	benzoic acid	121–122
1-chloro-4-nitrobenzene	83–88	trans-stilbene	122–123
3,4-dibromoaniline	81–82	benzamide	131–132
4-hydroxy-3-methoxy benzaldehyde	81–83	p-toluenesulfonamide	136–138
		salicylamide	139–141
2-methoxybenzoic acid	101–103	o-chlorobenzoic acid	140–141
o-toluic acid	104–105	anthranilic acid	146–147
1,2-dihydroxybenzene	105–106	benzenesulfonamide	155–157
m-toluic acid	108–110	salicylic acid	157–159
anthralinamide	110–111		
2-aminobenzamide	110–112		
1,3-dihydroxybenzene	111–112		

long hollow glass tube (2 to 3 ft) so that it bounces repeatedly on the bench top.] Repeat the process until the tube is packed to a height of ~2 mm.

Place the loaded capillary tube into the melting point apparatus (Figure E2.3) and slowly raise the temperature. Follow the directions given by your instructor for the specific apparatus being used in the laboratory. The rate of heating must be carefully controlled, so that the thermometer records a temperature increase of no more than 2 to 3°C/min. Carefully observe the temperatures at which the melting of the sample begins (the material will soften) and ends (the material is completely liquefied). Record this as the melting point range. A pure solid will usually have a temperature range of 1 to 2°C.

Identifying the Solid

A collection of several solids with known melting points is provided in Table E2.1. Select the solids that have melting points close to that of your unknown sample. Record the melting points from the table. Make a thoroughly mixed 50/50 mixture of your sample with the first of the selected known compounds and determine the melting point of the mixture.

Repeat the procedure for the other selected known solids. If the mixture contains two different solids, the melting point of the mixture will be lower than that for the pure unknown sample. If the melting point is the same as that of the unknown, the "mixture" must actually be pure. From these data identify the unknown compound.

> Disposal of Chemicals: Put the capillary tubes into a waste glass container. Dispose of the solids in designated containers.

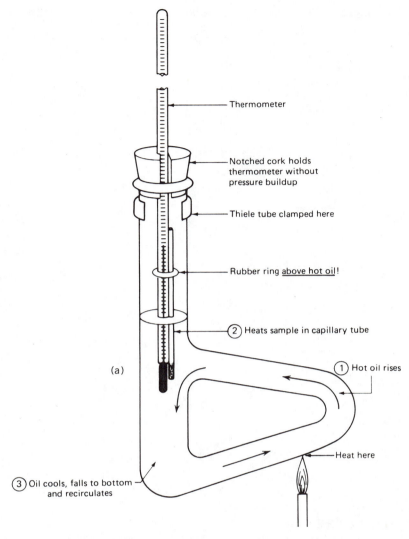

Figure E2.3   Melting point apparatus: (a) Thiele tube

Part B: Determination of the Boiling Points of Liquids

Micro and semimicroscale experiment   Estimated time to complete the experiment: 1h
Experimental Steps: Two steps are involved: (1) determining bp of liquids and (2) plotting
the data.

Obtain an unknown liquid sample from your instructor. Also acquire a known liquid sample
(an alcohol with a known number of carbon atoms in it), a Pasteur pipet, and, for method 1, a
10×100 mm test tube with a notched cork and thermometer or, for method 2, a closed-end
melting point capillary tube. Follow the appropriate instructions for determining the boiling
point of the liquids. Set up a heating bath for boiling point determinations. You may use a
Thiele tube or an electric melting point apparatus (Figure E2.3*a* and *b*).

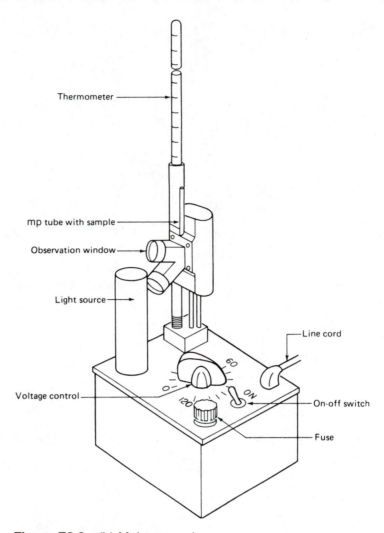

Figure E2.3 (b) Mel-temp unit

Method 1: Macroscale Method

Using a Pasteur pipet, place 0.5 to 1 mL of the unknown liquid in the test tube fitted with a notched cork and thermometer. The thermometer bulb must be held just above the liquid surface. Add a boiling stone to the test tube and clamp it to a ring stand above a wire gauze (see Figure E2.4*a*).

Using a microburner (a hot plate is recommended), heat the liquid until it vaporizes, condenses on the thermometer bulb, and drips back into the liquid. While this occurs, the temperature will remain at a constant value. Record this temperature as the boiling point of the liquid. Repeat the procedure for the known liquid.

Method 2: Semimicroscale Method

Using a twist tie, attach a 6 × 55 mm test tube to a thermometer. Place about 100 μL of the known liquid sample inside the test tube. Take a melting point capillary tube (one end sealed) and break it at a distance of 1.5 to 2.5 cm from the sealed end. This is your micro boiling point tube. Insert this, open end down, into the test tube. The open end of the capillary

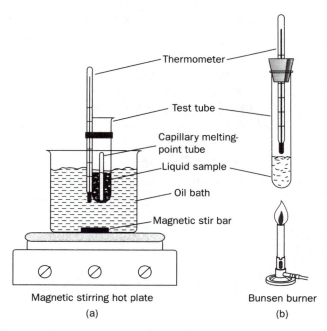

Figure E2.4 Boiling point determination: (a) Semimicroscale method; (b) Macroscale method

tube must be under the surface of the liquid. Secure the thermometer assembly to a clamp attached to a ring stand, and insert the assembly inside an oil bath held in a 10 mL beaker or a small test tube containing a stir bar (Figure E2.4*b*). Slowly heat the bath (2°C/minute) until a rapid and continuous stream of bubbles comes out of the capillary. Remove the heat and allow the bath to cool. Record the temperature at which bubbles cease and liquid begins to rise in the capillary. This is the boiling temperature of the liquid. Determine the boiling point of the unknown liquid in the same manner.

Method 3: Microscale Method

Construct (*a*) two one-end-closed micro bell capillaries that can fit inside a melting point capillary tube and (*b*) a fine microdelivery pipet from a 5 in. Pasteur pipet as described in Section 3.10 (see Figure 3.26).

Using the microdelivery pipet, transfer 5 to 10 μL of the known liquid into the melting point capillary tube. Insert an empty bell capillary (open end down) into the melting point capillary tube containing the liquid and tap it on the bench top until the bell reaches the bottom of the tube (or use a centrifuge). Put the assembly inside a melting point apparatus, as directed by your instructor. Slowly heat the bath (2°C/minute) until a rapid and continuous stream of bubbles comes out of the bell. Reduce the heat and allow the bath to cool. Record the temperature at which bubbles cease and liquid begins to rise in the bell. This is the boiling temperature of the liquid. Determine the boiling point of the unknown liquid in the same manner.

Identification of Unknown

Your instructor will post the boiling points for all the known liquids as determined by your classmates and yourself. Record all the posted values on the data sheet and make a plot of the number of carbon atoms (x axis) in the known liquids versus their boiling points (y axis). Determine the identity of your unknown sample from this plot.

> Disposal of Chemicals: Put the capillary tubes into a waste glass container.
> Dispose of the organic liquids in designated containers.

Additional Independent Project

1. Identification of unknown organic compounds by a combined method of melting point and thin layer chromatography[1] (see Experiment 26).

References

1. Levine, S. G. *J. Chem. Educ.* **1990**, *67*, 972.

Name		Section	
Instructor		Date	

PRELABORATORY REPORT SHEET—EXPERIMENT 2

Experiment title and part _____

Objective

Reactions/formulas to be used

Materials and equipment (mp, bp apparatus and others) table

Outline of procedure

PRELABORATORY PROBLEMS—EXPERIMENT 2

1. Define the melting point of a solid and the boiling point of a liquid.

2. How much heat (in kJ) is required to melt 36.04 g of ice (2.000 moles) at 0°C? (The molar heat of fusion for ice is 6.01 kJ/mole).

3. Does the melting point of ice depend upon the pressure? Explain your answer.

EXPERIMENT 2   DATA SHEET

Part A: Melting Point of Solids

Sample ID number _____ Room temperature _____ °C

Melting point range of the unknown solid _____ °C

Melting point range of pure compounds that have melting points close to that of the unknown solid:

	Compound name	**Melting point range, °C**
1.	_____	_____
2.	_____	_____
3.	_____	_____

Melting point range of the mixtures of known compounds and the unknown solid:

With Compound 1 _____ °C

With Compound 2 _____ °C

With Compound 3 _____ °C

Unknown is _____

Part B: Boiling Point of Liquids

Sample ID number: _____ Room temperature: _____ °C

Boiling point of the unknown liquid: _____ °C

Boiling point of the known liquids:

Name _____ °C _____ °C _____ °C

_____ °C _____ °C _____ °C

The unknown liquid is _____

POSTLABORATORY PROBLEMS—EXPERIMENT 2

1. In the microscale method of determining boiling points, bubbles are seen coming out of the micro bell capillary. What causes them?

2. In the microscale method of determining boiling points, one heats the liquid until bubbles are seen. The liquid is then cooled until the bubbles stop. Why is this method of determining the boiling point preferable to measuring the temperature when the bubbles start?

3. Gases can be liquefied under conditions of high pressure and low temperature. For example, nitrogen liquefies at −196°C at 1 atm pressure. Why are these conditions necessary?

Determination of the Refractive Index of a Liquid

Microscale Experiment

Objectives
- To determine the refractive indices of liquids
- To learn how to use the Abbe refractometer

Prior Reading
- Chapter 1 Safety precautions in the laboratory

One of the physical properties used to determine the purity of an organic liquid is its refractive index. When a beam of light passes from one medium to another (air to water, for example), it bends. For example, if you put a stick halfway into a pool of water, it will appear that the portion of the stick under the water surface is bent. This illusion is the outcome of the refraction of light and is caused by the different velocities of the light in the different media. The **refractive index** n is defined as the ratio of the velocity of light in vacuum to that in any medium:

$$n = \frac{v_{vac}}{v_{liq}}$$

Since the velocity of light in a vacuum is greater than in any other medium, the refractive index will always be greater than 1.

The value of the refractive index depends upon the wavelength of light. It is therefore important to use a standard wavelength of light for all the measurements. Most frequently, the wavelength of choice is the yellow sodium emission line at 589 nm (nanometer, see Appendix 1), also called the sodium D line. The Abbe-3L refractometer uses white light as the source, but compensating prisms give indices for the D line. Using a refractometer, one can determine the refractive index to up to four decimal places.

The refractive index also depends on the temperature. Since the density of any medium is temperature-dependent, this causes a change in the velocity of light. Most values for refractive index in the literature are given at 20°C, and given the symbol n_D^{20}. The value of the refractive index (when measured up to four decimal places) is also very sensitive to the purity of the medium. The refractive index is therefore quite useful for determining the purity of liquid samples.

Experimental Section

Procedure

> Microscale experiment   Estimated time to complete the experiment: 1.5 h
> Experimental Steps: Two steps are involved: (1) determining the refractive index of liquids and (2) plotting the data.

Obtain your liquid samples (a series of primary alcohols—each student pair will be assigned one liquid), including the unknown. Your instructor will provide you with specific instructions on how to use the refractometer (see Figure E3.1 for a schematic of a refractometer). Place a few drops of liquid sample between the hinged lenses of the refractometer. Adjust the refractometer so that the field of view has a well-defined light and dark split image. Record the refractive index at room temperature.

Repeat the determination at least at two other temperatures for the same liquid. The refractometer is equipped with a temperature control device. Use a low-temperature bath at 0°C (ice + water), below 0°C (ice + salt slush), and a high-temperature bath of running hot water. An aluminum or copper coil immersed in the bath is used to circulate the cold or hot water through the refractometer.

Plot the values of refractive index (y axis) versus temperature (x axis) for the liquid. If time permits, repeat the procedure for a second liquid, or for a mixture in order to determine its purity. From the data collected, identify the unknown alcohol.

> Disposal of Chemicals: Dispose of the organic liquids in designated containers.

References

1. Mayo, D. W.; Pike, R. M.; Trumper, P. K. *Microscale Organic Laboratory,* 3rd ed.; Wiley: New York, 1994.
2. Sheaver, E. C.; Rumpel, M. L. *J. Chem Educ.* **1974,** *51,* 140.

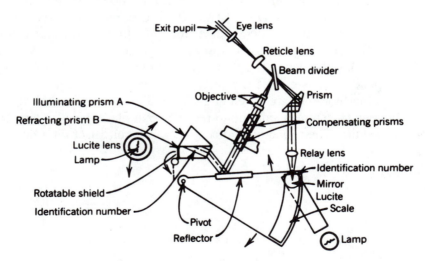

Figure E3.1 Abbe-3L Refractometer (Courtesy of Milton Roy Co., Rochester, NY)

Name ______________________ Section ______________

Instructor ______________________ Date ______________

PRELABORATORY REPORT SHEET—EXPERIMENT 3

Experiment title ______________________________________

Objective

Reactions/formulas to be used

Materials and equipment table

Outline of procedure

PRELABORATORY PROBLEMS—EXPERIMENT 3

1. On what factors does the refractive index depend?

2. What precautions must be taken to determine the correct values of the refractive index of a liquid?

3. What are the benefits of determining the refractive index of a liquid?

Name		Section	
Instructor		Date	

EXPERIMENT 3   DATA SHEET

Sample ID number _____

Known Liquids

Name of Liquid A _____

Name of Liquid B _____

Name of Liquid C _____

Temperature °C	n for Liq. A	n for Liq. B	n for Liq. C	n for Unknown
_____	_____	_____	_____	_____
_____	_____	_____	_____	_____
_____	_____	_____	_____	_____
_____	_____	_____	_____	_____

Include a graph of refractive index versus temperature for each of the liquids.

Name of the Unknown _____

POSTLABORATORY PROBLEMS—EXPERIMENT 3

1. Define refractive index. How does the refractive index of a liquid change with temperature?

2. What precautions must you take to run this experiment?

3. Why do we have to specify the wavelength of light for determining the refractive index of a liquid?

EXPERIMENT **4** **Determination of Viscosity of Liquids**

Micro- and Macroscale Experiments

Objectives
- To measure and analyze the viscosities of binary solutions (alcohol/water)
- To measure the viscosity of a given liquid at different temperatures and to calculate the activation energy to viscous flow of the liquid

Prior Reading
- Section 2.2   Scientific notation
- Section 2.5   Standard deviation
- Section 2.6   Graphing of data
- Section 3.9   Solution preparation

Related Experiments
- Experiments 1 (Determination of density) and 34 (Viscosity of polyvinyl alcohol polymer)

The **viscosity** η of a fluid is defined as its resistance to flow. This concept is familiar—thick liquids like honey move far more slowly than thin liquids like water. Thick liquids have greater viscous drag than thin liquids. The rate of flow of a liquid is determined by its viscosity. A flowing column of liquid may be considered as consisting of parallel layers of liquid. The resistance experienced by one such layer as it moves past the adjacent layer is called the viscosity of the liquid. The magnitude of this internal friction determines the viscosity of the liquid.

The **viscosity coefficient** of a liquid is expressed in units of *poise,* or P. A viscosity of 1 poise is defined as a force of 1 dyne/cm^2 that causes two parallel adjacent liquid layers 1 cm apart to move past each other with a velocity of 1 cm/s. For most common liquids at room temperature, the viscosities range from 0.002 to 0.04 P. It is therefore convenient to use the term centipoise (cP = 10^{-2} P) to report viscosities.

Viscosities are extensively used in industry to characterize lubricating oils, resins, latex paints, chocolate mousse, and so on. The stability of motor oils at higher temperatures, for example, can be determined by their viscosities. The viscosities of liquids also provide information regarding the diffusion processes in chemical reactors and thereby play an important role in chemical kinetics.

Two common methods are available for the determination of the viscosity of a liquid. For less viscous liquids, an Ostwald viscometer is used. For more viscous liquids, the falling ball or dropping needle method is preferred.

The Ostwald Viscometer Method

The Ostwald viscometer, shown in Figure E4.1, is the instrument most commonly used to determine the viscosity of liquids. Viscosity is seldom measured directly—instead, the viscosity of the unknown liquid is measured relative to that of a known liquid, thus obtaining a relative viscosity. The relative viscosity equation is

$$\eta_l = \frac{\eta_r \rho_l t_l}{\rho_r t_r}$$

Here, η is the viscosity, ρ is the density, and t is the amount of time that it takes the liquid to flow through the viscometer. The subscripts r and l refer to the reference liquid and unknown liquid, respectively. Thus, if the viscosity of the reference liquid, η_r, is known, the viscosity of the unknown liquid, η_l, is easily calculated.

Example E4.1 At 25°C in an Ostwald viscometer, pure methanol ($\rho = 0.791$ g/mL) takes 21.0 seconds to run from the upper mark to the lower one. Under the same conditions, water ($\rho = 0.997$ g/mL) takes 27.0 seconds. The viscosity of water is 0.890 cP. Calculate the viscosity of the pure methanol.

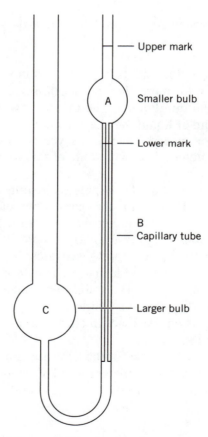

Figure E4.1 Ostwald's viscometer

Answer

$$\eta_l = \frac{\eta_r \rho_l t_l}{\rho_r t_r}$$

$$\eta_l = \frac{(0.890 \text{ cP})(0.791 \text{ g/mL})(21.0 \text{ s})}{(0.997 \text{ g/mL})(27.0 \text{ s})}$$

$$\eta_l = 0.549 \text{ cP}$$

Temperature has a profound effect on viscosity. All measurements should be made at a constant temperature. A change in temperature of 1°C results in a change of approximately 2 percent in the viscosity of a liquid. The effect of temperature on the viscosity is given by the Arrhenius equation:

$$\eta = A \exp(E_\eta / RT)$$

Here A is a constant for a given liquid. Thus, by measuring the viscosity at a variety of temperatures, the value of E_η, the **activation energy** for viscous flow, can be determined.

The Falling Ball Method

The Ostwald viscometer cannot be used for measuring the viscosity of highly viscous liquids because of the long flow times required. For viscous liquids, the falling ball method is used. The method depends on Stokes's Law, which states that the force required for a spherical object of radius r to fall through a liquid of viscosity η with a velocity v is

$$F = 6\pi r \eta v$$

As the spherical object (a solid ball) falls through the viscous liquid, it naturally experiences a viscous drag (resistance) from the liquid. If it is assumed that the ball is falling under the influence of gravity and that the velocity of the fall is constant (zero acceleration), then the force F due to the viscous drag equals the gravitational force. This can be obtained by multiplying the effective mass m of the ball by the acceleration due to gravity, g (9.806 m/s^2):

$$F = 6\pi r \eta v = mg$$

The effective mass m of the ball will depend upon the buoyancy force (Archimedes' principle) exerted by the liquid. The density of the liquid, ρ_l, accounts for the buoyancy force of the liquid. If the density of the solid ball is ρ_b, the effective mass of the ball will be $(\frac{4}{3})(\pi r^3)(\rho_b - \rho_l)$. Recall that the mass of a solid equals the product of its volume and the density and that the volume of a sphere is $(\frac{4}{3})(\pi r^3)$. Thus,

$$6\pi r \eta v = \frac{4}{3}(\pi r^3)(\rho_b - \rho_l)g$$

If the ball takes time t(s) to fall through a known distance D, then the velocity is D/t and the equation can be rearranged to obtain the viscosity of a liquid, or

$$\eta = \frac{2r^2(\rho_b - \rho_l)gt}{9D} \tag{E4.1}$$

This equation provides the absolute viscosity of the liquid, because all of the quantities in the equation can be accurately measured.

Alternatively, the falling ball method may be used to determine the relative viscosity of an unknown liquid with an apparatus that has been calibrated using another reference liquid of known viscosity. In such cases, either glycerin or castor oil makes a very satisfactory reference liquid. The formula used for the determination of relative viscosity by the falling ball method is

$$\frac{\eta_1}{\eta_2} = \frac{(\rho_b - \rho_1)t_1}{(\rho_b - \rho_2)t_2} \tag{E4.2}$$

In this equation ρ_b, ρ_1, and ρ_2 are the densities of the falling ball and the two liquids respectively; t_1 and t_2 stand for the times taken for the sphere to fall through a given distance in the two media.

As has been mentioned before, temperature has a substantial effect on viscosity. Therefore, the experiment should be carried out at constant temperature.

General References

1. Atkins, P. W. *Physical Chemistry,* 3rd ed.; W. H. Freeman: New York, 1987.
2. Crockford, H. D.; Nowell, J. W.; Baird, H. W.; Getzen, F. W. *Laboratory Manual of Physical Chemistry,* 2nd ed.; Wiley: New York, 1975.
3. Halpern, A. M.; Reeves, J. H. *Experimental Physical Chemistry: A Laboratory Textbook;* Scott, Foresman: Glenview, IL, 1988.
4. Hemmerlin, W. M.; Abel, K. B. *J. Chem. Educ.* **1991,** *68,* 417.

Experimental Section

Procedure

Part A: Determination of the Viscosity of a Liquid by Ostwald Method

Macroscale experiment   Estimated time to complete the experiment: 2.5 h
Experimental Steps: The procedure involves three steps: (1) preparing different percentage solutions (by volume), (2) determining the density by the micropycnometer method, and (3) the time of flow of each solution in a viscometer.

Disposal of Chemicals: Dispose of the organic liquids in designated containers.

Procure pure methanol, deionized water, a graduated 25 mL pipet, and five 25 mL volumetric flasks for storing the solutions. Prepare 25 mL each of 20, 40, 60, and 80 percent (by volume) methanol in water solutions by pipeting exact volumes of methanol and water. Store these solutions in stoppered containers. Determine the densities of each of these solutions by using a micropycnometer (see Experiment 1). Alternatively, your instructor may supply you with density data. Also procure an Ostwald viscometer. **You may also construct your own viscometer as described in Section 3.10.**

For accurate measurements, the viscometer must be scrupulously clean. Clean the viscometer with a cleaning solution. Then rinse it with deionized water at least five times. Using a clamp attached to a ring stand, immerse the bulb and the capillary of the viscometer in a constant-temperature water bath in an upright position. Maintain the temperature of the bath at 25°C.

Using a pipet, introduce into the viscometer a fixed volume of distilled water (5 to 10 mL) that will fill the smaller bulb A (Figure E4.1) between the upper and lower scratch marks. There should be enough liquid to fill the bottom part as well as one-third of the larger bulb C of the viscometer. In all subsequent measurements, use the same viscometer and the same volume of liquid. Attach a pipet pump to the arm of the bulb A. Using the pump, you should be able to pull the liquid level up above the upper mark of bulb A. Remove the pump and allow the level of the liquid in bulb A to fall. As soon as the lower meniscus of the liquid just crosses the upper mark, start the stopwatch.

Stop the watch exactly when the liquid meniscus passes the lower scratch mark. Record the time t_r in seconds. Repeat the process until the readings agree within 1 second.

Remove the viscometer from the bath and drain the contents. Rinse the viscometer with acetone several times and allow it to dry. Then rinse the viscometer with one of the methanol solutions. Fill the viscometer with the methanol solution as before and determine the flow time t_l in seconds. Repeat until the readings agree within 1 second. Perform the experiment with the other methanol solutions and, finally, with pure methanol. After you are done, return the clean viscometer to your instructor.

Obtain the density ρ_l for each of the methanol solutions from your instructor or determine the density by micropycnometer method (Experiment 1). Using the viscosity of water as a reference, calculate the viscosity of each solution.

The viscosity of water at 25°C is 0.8904 cP and the density is 0.997 g/mL. Determine the standard deviation for the values. Perform the Q test (90 percent level) on flow times to determine if any of the values are outlying.

Part B: Factors Affecting Viscosity of Alcohols

Macroscale experiment	Estimated time to complete the experiment: 2.5 h

Disposal of Chemicals: Dispose of the organic liquids in designated containers.

Obtain 10 mL each of a 1 mol/vol percent solution of methanol (CH_3OH, 32 g/mol, $\rho = 0.791$ g/mL), ethanol (C_2H_5OH, 46 g/mol, $\rho = 0.789$ g/mL), n-propanol (C_3H_7OH, 60 g/mol, $\rho = 0.804$ g/mL), n-butanol (C_4H_9OH, 74 g/mol, $\rho = 0.810$ g/mL), propylene glycol [$C_3H_6(OH)_2$, 76 g/mol, $\rho = 1.03$ g/mL] and glycerol [$C_3H_5(OH)_3$, 92 g/mol, $\rho = 1.26$ g/mL] in water. Store them in labeled bottles in a 25°C water bath.

To prepare stock 1 mol/vol percent solutions, take 1 mole of each of the alcohols and add enough water to make a total volume of 100 mL. Using a micropycnometer (see Experiment 1), determine the density of each of these solutions. This solution is enough for 8 to 10 students.

Place a well-cleaned viscometer in the same temperature bath as the solutions. Measure the flow time for each of the solutions (including that for water) as described in part A. Calculate the viscosity of each alcohol solution. Plot the viscosities of the solutions against (a) the molecular weights of the solutes, (b) the number of carbon atoms of the monohydric alcohols (methanol, ethanol, n-propanol, and n-butanol) and (c) the number of hydroxyl groups (OH) of n-propanol, propylene glycol, and glycerol. Which of these solute properties has the greatest effect upon the viscosities of the solutions?

Part C: Measurement of Viscosity—Falling Ball Method

Macroscale experiment	Estimated time to complete the experiment: 2.5 h

Disposal of Chemicals: Dispose of the organic liquids in designated containers.

The apparatus (Figure E4.2) consists of four parts: a 100 mL graduated glass cylinder that holds the liquid, a rubber stopper with a hole carrying a glass guide tube, and a metal ball. The bore of the glass tube must be slightly larger than the diameter of the ball. The end of the tube must extend below the surface of the test liquid by about 2 to 4 centimeters.

The purpose of the guide tube is to allow the metal ball to fall vertically through the center of the test tube. It also reduces the speed of the descending ball before it reaches the top fiducial (reference) mark. In addition, the guiding tube prevents the formation of any air bubbles on the surface of the ball as it glides through the tube.

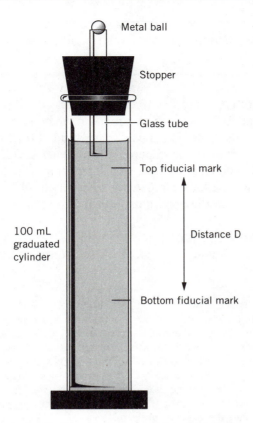

Metal ball

Stopper

Glass tube

Top fiducial mark

Distance D

100 mL
graduated
cylinder

Bottom fiducial mark

Figure E4.2   Construction of the falling
ball apparatus

Using a glass marking pencil, make two fiducial (reference) marks on the graduated cylinder as shown in the figure (or use two existing markings on the cylinder). The separation between the two fiducial marks is the distance D. During the experiment, the time (in seconds) taken by the ball to travel this distance D is measured.

Wash the glass guiding tube with cleaning solution, followed by deionized water. Rinse the tube with acetone and dry it. Construct a falling ball apparatus as just described, and obtain three steel balls.

Do not handle the steel balls with your hands—use tweezers. Set up a water bath in a 1 L beaker on a hot plate and clamp the graduated cylinder in it.

Fill the graduated cylinder to near the top with glycerin, and insert the stopper carrying the guiding tube. The end of the tube should extend below the surface of the liquid by about 3 cm. Let the system equilibrate at room temperature for 30 minutes. Record the bath temperature.

While the liquid equilibrates, weigh the three steel balls separately, measure the diameter of each, and calculate their volumes and average density (see Experiment 1). Using a micropycnometer, determine the density of glycerin at room temperature. Calculate the density of glycerin at the other temperatures using the following equation:

$$\rho = [1.2727 + aT10^{-3} + bT^2 10^2 + cT^3 10^{-9}] \text{ g/mL}$$

where T is the temperature (°C) at which ρ is calculated and $a = -0.5506$, $b = -1.016$, and $c = +1.270$.

Let one of the balls fall through the guiding tube into the liquid. Measure the time in seconds required for the ball to fall from the top mark to the bottom one. Repeat the measurement using the other two balls. Calculate the average of the three times, the standard deviation, and the confidence interval at the 95 percent confidence level. Use the Q test at the 90 percent confidence level for determining the suspected drop time for the spheres. Measure the distance between the two marks. Calculate the viscosity of glycerin using Equation E4.1 ($g = 9.806$ m/s^2). After rinsing the apparatus, return it to your instructor.

Additional Independent Project

Determine the viscosities of toluene in p-xylene. Also determine the viscosities of p-xylene at different temperatures.

Brief Outline

1. Determine the densities and flow times of 0, 20, 40, 60, 80, and 100 percent solutions of toluene in *para*-xylene. Calculate the viscosities of these solutions. Further, determine the flow times and the viscosities of pure p-xylene at the following temperatures: 25°C, 35°C, 45°C, 55°C, and 65°C. From these data you should be able to calculate the effect of temperature on viscosity. Calculate the activation energy E_η by plotting $\ln \eta$ versus $1/T$. The slope of the line gives the value of E_η/R.

2. To determine the relative viscosity of a heavy oil, perform the preceding experiment first using glycerin and then the oil. Determine the average time of fall of the spheres for both the liquids. Measure the densities of the spheres, glycerin, and the oil. The absolute viscosity of glycerin η_1 is 9.54 poise at 25°C. Using Equation 4.2, calculate the viscosity of the oil, η_2.

Develop your own data collection and manipulation tables. Write a report according to the instructions given in Chapter 4.

Name ______________________________ Section ______________

Instructor ______________________________ Date ______________

PRELABORATORY REPORT SHEET—EXPERIMENT 4

Experiment Title and part ______________________________

Objective

Reactions/formulas to be used

Materials (different solutions) and equipment (viscometer, glass tube, etc.) table

Outline of procedure

1. Define viscosity, viscosity coefficient, and mole fraction.

2. Calculate the viscosity of a liquid from the following set of data: density of water at 25°C = 0.997 g/mL; viscosity of water = 0.8904 cP; density of the liquid = 0.821 g/mL; flow time for water = 32 s; and flow time for the liquid = 24 s.

EXPERIMENT 4   DATA SHEET

Part A: Ostwald Method

Collection and Manipulation of data

Temperature _____ °C   Density of water _____ g/mL

	Trial 1	**Trial 2**	**Trial 3**	**Avg.**
Mass of the empty pycnometer	_____ g	_____ g	_____ g	
Mass of the pycnometer + water	_____ g	_____ g	_____ g	
Volume of the pycnometer	_____ mL	_____ mL	_____ mL	_____

% methanol	Mass of soln + pyc., g	Mass of soln, g	Density of soln, g/mL	Flow time, s	Viscosity η_l, centipoise
0	_____	_____	_____	_____	_____
20	_____	_____	_____	_____	_____
40	_____	_____	_____	_____	_____
60	_____	_____	_____	_____	_____
80	_____	_____	_____	_____	_____
100	_____	_____	_____	_____	_____

Part B:

Make your own data sheet.

Part C: Falling Ball Method

Average weight of steel balls   _____

Radius of steel balls   _____

Volume of steel balls   _____

Density of steel balls   _____

Fall distance D _____

Run #	Temperature	Fall time, t	Density of glycerin
1	_____	_____	_____
2	_____	_____	_____
3	_____	_____	_____
4	_____	_____	_____

(You may use the table in Part A to collect the data.)

Viscosity of glycerin _____

For statistical analysis attach a separate sheet.

POST LABORATORY PROBLEMS—EXPERIMENT 4

1. Why is it important to keep the temperature of the solutions constant during viscosity measurements?

2. If a solution contains 50.0 g of methanol (CH_3OH) and 50.0 g of water, calculate the mole fractions for methanol and water.

3. Suggest why the viscosity of an organic liquid alcohol rises as the number of OH groups increases (molecular weight remaining constant).

5 Thermochemistry and Calorimetric Measurements

Part A: Specific Heat and Atomic Weight of a Metal
Part B: Heat of Solution for the Dissolution of a Salt
Part C: Heat of Reaction Using Hess's Law

Semimicro and Macroscale Experiments

Objectives
- To learn the basic concepts of thermochemistry
- To determine the specific heat and atomic weight of a metal
- To measure the heat of reaction of chemical processes
- To apply Hess's law to obtain the enthalpy of a reaction

Prior Readings
- Chapter 1 Safety precautions in the laboratory
- Section 3.6 Heating methods

When a physical or chemical change occurs, it is usually accompanied by the transfer of heat: heat is either given off or absorbed. Consider the heating of liquid water from room temperature to its boiling point. The product (water at 100°C) has a higher energy content than the reactant (water at 25°C). The difference in these energies must be supplied by the heat source. **Thermochemistry** is the study of heat changes in chemical processes and is a part of the more general study of heat changes known as **thermodynamics** (*thermo*—heat, *dynamics*—in motion). The heating process is summarized by the equation

$$\Delta H = H_{\text{final}} - H_{\text{initial}}$$

where the symbol Δ means *the change in* and H is the symbol for **enthalpy** (heat), defined as the heat content at constant pressure. Here, ΔH would represent the amount of heat that was needed to accomplish the heating, H_{final} would be the heat content of the hot water, and H_{initial} would be the heat content of the room-temperature water. The unit of heat (energy) is the **Joule** (J), or **calorie** (cal = 4.184 J).

Since enthalpy is a **state function**, the same value of ΔH is always obtained for the same initial and final conditions, no matter how the change was accomplished. The heat change for a chemical reaction is called the **heat of reaction**, ΔH_{rxn}. Reactions for which ΔH_{rxn} is negative are called **exothermic** reactions—heat is given off to the surroundings. Reactions for which ΔH_{rxn} is positive are called **endothermic** reactions—heat is absorbed from the surroundings.

The device most commonly used to measure heat changes is called a **calorimeter**, of which the most familiar type is a thermos bottle. If the calorimeter is perfectly insulated, any change in the temperature of the calorimeter interior must be due to a chemical or physical change taking place. It should be noted that the calorimeter itself can absorb heat. Thus, the heat capacity of the calorimeter (or calorimeter constant C_c) must be determined.

In this experiment, calorimetry is used to determine the specific heat of a metal, to find the heat change for dissolving a salt in water, and to measure the enthalpy of a chemical reaction.

Specific Heat

The **heat capacity** C of a substance is the amount of heat, q, required to raise the temperature of a given quantity of any substance by 1 K (or 1°C). The amount of heat, q, necessary to raise the temperature of 1 g of a substance by 1 K (or by 1°C) is called the **specific heat capacity** s, or simply the **specific heat**.

$$q = (m)(s)(\Delta T)$$
$$C = (m)(s) \tag{E5.1}$$

As seen in equation (E5.1), the quantity of heat required to change the temperature of any substance is proportional to its mass m and to the temperature change ΔT. The specific heat is the proportionality constant in the equation. In fact, specific heat is not exactly constant—it varies somewhat with temperature. For example, the specific heats of water at 15°C (288 K) and 100°C (373 K) are 4.184 J/g K and 4.217 J/g K, respectively.

Specific heats are usually measured in water calorimeters. A given amount of a metal is heated to a known temperature and placed in a known amount of water at a known temperature in a calorimeter. Heat flows from the metal to the water and the calorimeter. The mixture reaches an equilibrium temperature, which is measured. According to the law of conservation of energy,

Heat lost by metal = heat gained by water + heat gained by calorimeter

$$-q_{met} = q_w + q_c = q_{tot}$$
$$-[(m)(s)(\Delta T)]_{met} = [(m)(s)(\Delta T)]_w + [(m)(s)(\Delta T)]_c = q_{tot} \tag{E5.2}$$
$$-[(m)(s)(\Delta T)]_{met} = [(m)(s)(\Delta T)]_w + (C_c)(\Delta T_c) = q_{tot}$$

where C_c is the calorimeter heat capacity constant or, simply, the calorimeter constant.

In these expressions, $\Delta T = T_{final} - T_{initial}$. Therefore, in equation (E5.2), ΔT for the metal is negative (the metal is cooling off) and ΔT for the water and the calorimeter is positive (they are heating up). By knowing the specific heat of water (4.184 J/g K) and the calorimeter constant, the specific heat of the metal can be calculated. Note that the specific heats of water and of the metal have different values. In 1819, Dulong and Petit proposed that 1 mole of any metal absorbs approximately 25 J/mol K. Thus, as an approximation,

$$s \text{ (J/g K)} \times \text{AW (g/mol)} \approx 25 \text{ (J/mol K)} \qquad \text{where AW = atomic weight}$$
$$\text{AW} \approx \frac{25 \text{ (J/mol K)}}{s \text{ (J/g K)}} \tag{E5.3}$$

Heats of Solution

Athletes use "instant" hot and cold packs to treat injuries. A typical pack consists of a plastic bag that contains a salt and a pouch full of water. The water is released by breaking the pouch, and the salt dissolves. Dissolving a solute (salt) in a solvent (water) often results in a change in the heat content of the system. The **heat of solution** ΔH_{sol} is defined as the heat generated (or absorbed) when a certain quantity of solute dissolves in a certain quantity of solvent. Thus, the temperature of the pack will either rise (hot pack) or drop (cold pack) depending on whether the heat of solution is exothermic or endothermic.

The heat of solution is the sum of two related quantities: the **lattice energy** U and the **heat of hydration** ΔH_h. The lattice energy of a salt is the energy required to separate one mole of a salt completely into ions in the gas phase. The enthalpy change when these gaseous ions are hydrated (surrounded by water molecules) is called the heat of hydration. Lattice energies are always positive quantities (endothermic), whereas heats of hydration are always negative (exothermic).

$$\Delta H_{sol} = U + \Delta H_h$$

As shown in Figure E5.1, the heat of solution of NaCl in water is 4 kJ/mol, which is slightly endothermic. The heat of solution is determined experimentally as follows:

$$\Delta H_{sol} = q \text{ (water)} + q \text{ (salt)} + q \text{ (calorimeter)}$$

For dilute solutions, the heat term for the salt can be neglected, giving

$$\Delta H_{sol} = [(m)(s)(\Delta T)]_w + (C_c)(\Delta T)_c \tag{E5.4}$$

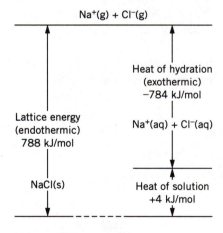

Figure E5.1 Heat of solution of NaCl

Heats of Reaction and Hess's Law

When H_2SO_4 is added to NaOH in a test tube, the test tube becomes hot. If tap water is added to solid ammonium chloride in a beaker, the beaker becomes cold. Why do such heat changes take place?

When a chemical reaction occurs, heat energy is either released or absorbed. In a chemical reaction, bonds are broken in the reactants (requiring energy) and new ones are formed in the products (releasing energy). The enthalpy change required to break a bond is called the **bond dissociation energy** or, more simply, the **bond energy**. Lists of energies for various bonds can be found in standard chemistry texts. The heat of reaction can be estimated by taking the difference in the bond energies between the products and reactants (the superscript $°$ indicates "at standard pressure"):

$$\Delta H° = \text{energy required to break bonds} - \text{energy released from new bonds}$$

or

$$\Delta H° = \sum(\text{bonds in reactants}) - \sum(\text{bonds in products})$$

For the reaction	$2H_2(g)$	$+$	$O_2(g)$	$\rightarrow$	$2H_2O(g)$
bonds broken	2 H–H		1 O=O		4 O–H
bond energy	(2)(436)		499		(4)(460) kJ/mol

$$\Delta H° = (872 + 499 - 1840)\,\text{kJ} = -469\,\text{kJ}$$

Since $\Delta H°$ is negative, the reaction is accompanied by an increase in temperature due to the release of energy.

In many cases, it is difficult to measure the heat (enthalpy) of reaction. The reaction of interest may occur along with other side reactions or may be extremely slow. In such cases, it is necessary to calculate the heat of reaction indirectly, using **Hess's Law of Heat Summation**. This law states that if a chemical reaction can be written as the sum of two or more individual reactions, the ΔH for that reaction can be obtained by adding the ΔH's of the individual reactions.

In this experiment, the heats of formation of Mg^{2+} and of MgO will be determined, using Hess's law. The three reactions sum up to give the reaction corresponding to the formation of MgO from its elements in their standard states:

$$Mg(s) + 2H^+(aq) \rightarrow Mg^{2+}(aq) + H_2(g) \qquad \Delta H_{rxn1} \qquad \text{(E5.5)}$$

$$H_2(g) + \tfrac{1}{2}O_2(g) \rightarrow H_2O(l) \qquad \Delta H_{rxn2} \qquad \text{(E5.6)}$$

$$Mg^{2+}(aq) + H_2O(l) \rightarrow MgO(s) + 2H^+(aq) \qquad \Delta H_{rxn3} \qquad \text{(E5.7)}$$

Net: $\quad Mg(s) + \tfrac{1}{2}O_2(g) \rightarrow MgO(s) \qquad \Delta H_f = \Delta H_{rxn1} + \Delta H_{rxn2} + \Delta H_{rxn3}$ (E5.8)

In the first reaction where ΔH_f is the heat of formation (E5.5), magnesium metal is dissolved in acid, forming $Mg^{2+}(aq)$ ions. By definition, the heats of formation of Mg and H_2 are zero (since they are elements in their standard states), as is the heat of formation of H^+ (since it is the ion upon which other ion heats are based). The ΔH_{rxn} for any reaction is equal to the heats of the products minus that of the reactants:

$$\Delta H_{rxn} = \sum \Delta H_f(\text{products}) - \sum \Delta H_f(\text{reactants})$$

Thus,

$$\begin{aligned}
\Delta H_{rxn1} &= \Delta H_f(Mg^{2+}) + \Delta H_f(H_2) - \Delta H_f(Mg) - 2\,\Delta H_f(H^+) \\
&= \Delta H_f(Mg^{2+}) + 0 - 0 - 2(0) \\
&= \Delta H_f(Mg^{2+})
\end{aligned} \qquad (E5.9)$$

Thus, the heat of the first reaction is also the heat of formation of Mg^{2+}. Since the reaction is exothermic, ΔH_{rxn1} is a negative quantity.

The heat for the second reaction (E5.6), ΔH_{rxn2}, is the heat of formation of H_2O. This can be obtained experimentally by combusting a quantity of hydrogen and measuring the heat given off. In practice, this reaction is very dangerous, and we will accept the value of ΔH_{rxn2} from thermodynamics tables (-285.8 kJ/mol).

The heat of the third reaction (E5.7), ΔH_{rxn3}, can be obtained by dissolving magnesium oxide (MgO) in acid, yielding magnesium ions:

$$MgO + 2\,H^+ \rightarrow Mg^{2+} + H_2O$$

Note that this reaction is the reverse of the third reaction, and its heat is therefore the negative of ΔH_{rxn3}. Thus,

$$\Delta H_{rxn} = -\Delta H_{rxn3} = \Delta H_f(Mg^{2+}) + \Delta H_f(H_2O) - \Delta H_f(MgO) - 2\,\Delta H_f(H^+)$$

The heat of formation for H^+ is zero, and the heats of formation of Mg^{2+} and H_2O are ΔH_{rxn1} and ΔH_{rxn2}, respectively. Thus,

$$\Delta H_{rxn} = -\Delta H_{rxn3} = \Delta H_{rxn1} + \Delta H_{rxn2} - \Delta H_f(MgO) - 2(0)$$

or rearranging,

$$\Delta H_f(MgO) = \Delta H_{rxn1} + \Delta H_{rxn2} - \Delta H_{rxn} = \Delta H_{rxn1} + \Delta H_{rxn2} + \Delta H_{rxn3} \quad (E5.10)$$

as was seen using Hess's Law.

General Reference

1. Gislason, E. A.; Craig, N. C. *J. Chem. Educ.* **1987,** *64,* 660.

Experimental Section

Procedure

Part A: Specific Heat and Atomic Weight of a Metal

Macroscale experiment   Estimated time to complete the experiment: 2.5 h
Experimental Steps: Three steps are involved: (1) construction and (2) calibration of a calorimeter and (3) determination of specific heat.

Construction and Calibration of a Calorimeter

Obtain two dry Styrofoam™ coffee cups, a lid with two holes (one for the stirrer and the other for the thermometer) for the cups, a glass stirrer (or small magnetic stir bar), two thermometers (±0.1°C), and a 100 mL beaker.

Construct the calorimeter by placing one cup inside the other. Insert the stirrer and a thermometer-stopper assembly through the holes in the lid and place the lid on the top cup (see Figure E5.2). Determine the mass of the assembly (cups, lid, stirrer, and thermometer together) to the nearest 0.1 g. Add 40 mL of **room-temperature** deionized water to the calorimeter, and determine the temperature of the water to ±0.1°C. Reweigh the assembly plus water to the nearest ±0.1 g.

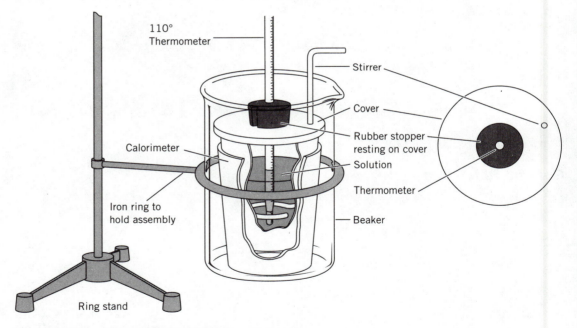

Figure E5.2 Assembly of coffee cup calorimeter

In lieu of coffee cups, other insulating materials, such as blue Styrofoam™ and Thermos™ brand "Snack Jar," may be used to construct calorimeters.[1,2] The Thermos-type plastic insulating mug with a wide base (as Mobile brand mugs) is an excellent substitute for coffee cups.

To stabilize the calorimeter setup, place the assembly inside a 250 or 400 mL beaker atop a magnetic stirrer. The beaker is held in position with the help of an iron ring. Clamp the thermometer so that its bulb is suspended just above the bottom of the calorimeter.

Weigh the 100 mL beaker to the nearest 0.1 g. Add 40 mL of deionized water and reweigh the beaker. Clamp the second thermometer to a ring stand so that its bulb is dipped completely into the water. **It must not touch the sides or the bottom of the beaker.** Heat the water on a *separate* sand bath to ~ 60°C.

With constant and gentle stirring, begin recording the temperature of the water inside the calorimeter (±0.1°C) every 30 seconds for exactly **3 minutes.** At the end of the **third minute,** record the temperature (T_{hot}) of the hot water in the beaker (±0.1°C). Remove the lid from the calorimeter and **quickly** add the hot water from the beaker all at once. Avoid splashing the water. Replace the lid immediately. Continue to stir the water inside the calorimeter, and begin recording the temperature at the **fourth minute.** Record the temperature of the contents of the calorimeter every 30 seconds for an additional 6 to 8 minutes. The temperature will rise at first, and then decrease.

Calculate the total mass of water (m_c) in the calorimeter. Remove the lid and empty the calorimeter. Clean and **dry** the calorimeter (do not use acetone to rinse the plastic cup). Reassemble the calorimeter as before for the next step.

Specific Heat Measurement

Weigh the calorimeter assembly (along with lid, thermometer, and stirrer) to the nearest ±0.1 g. Add 40 mL of water to the calorimeter and reweigh it. Measure the temperature of the water.

Obtain 6 to 10 g of an unknown metal from your instructor, and record the number of the unknown on the data sheet. Weigh a large test tube (15×85 mm) to the nearest ±0.1 g. Place 5 to 8 g of the metal in the test tube and reweigh it. Close the test tube with a loose stopper.

Place a 400 mL beaker two-thirds full of water on a hot plate and heat it to almost boiling. Clamp the metal-containing test tube inside the hot water bath as shown in Figure E5.3. The metal sample must remain well below water level. Heat the water to boiling. No water should enter the test tube. Keep the test tube inside the water bath for at least 15 minutes to attain thermal equilibrium.

Record the temperature of the calorimeter every 30 seconds for **3 minutes.** At the end of the **third minute,** quickly record the temperature (±0.1°C) of the hot water bath in the 400 mL beaker. Remove the test tube with the metal sample. Using a tissue paper, quickly dry the outside of the test tube. Remove the lid from the calorimeter momentarily and immediately transfer the metal pieces into the calorimeter. Avoid splashing water from the calorimeter. Replace the lid and stir gently. Record the temperature of the calorimeter every 30 seconds for **6 to 8 minutes.** Repeat the experiment once more if time permits.

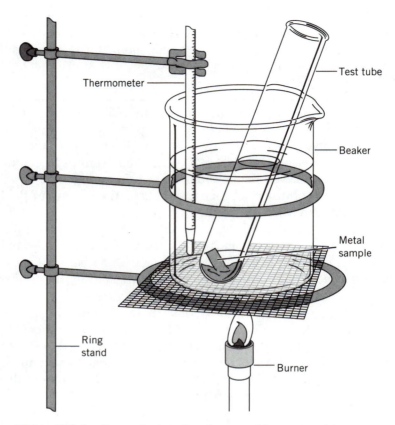

Figure E5.3   Setup for heating the metal in a test tube

Return the calorimeter assembly including the thermometer to your instructor. Metal pieces must be rinsed with deionized water and dried on a paper towel. Return the clean and dry metal pieces to the instructor. Other disposable items must be deposited in a proper container.

Calculations: Calibration of the Calorimeter

Plot time (x axis) versus calorimeter temperature (y axis). Ideally, the plot should resemble Figure E5.4.

As seen from the plot, when the hot water is added, the temperature of the mixture goes up rapidly. After the temperature has peaked, it starts decreasing and begins to return to room temperature. However, before the actual peak temperature is reached, some heat is lost to the walls of the calorimeter. To determine the maximum T_{max}, the researcher draws a straight line through the points (which are relatively linear), extrapolating it to intersect the y axis at the time $= 0$ (see Figure E5.4). Determine ΔT_c, the temperature change of the calorimeter, by subtracting the initial temperature T_i of the calorimeter (before mixing with the hot water) from T_{max}:

$$\Delta T_c = T_{max} - T_i$$

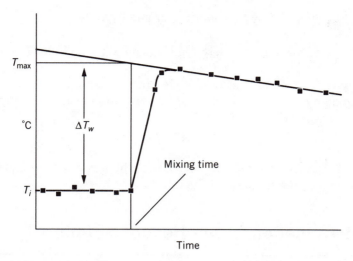

Figure E5.4 Plot of time versus temperature, with extrapolation to zero time

Now, calculate ΔT_w, the temperature change of the water, by subtracting the initial temperature T_{hot} of the hot water from the final temperature T_{max}:

$$\Delta T_w = T_{max} - T_{hot}$$

Note that ΔT_w is a negative quantity.

The amount of heat lost by hot water is

$$q_w = (4.18 \text{ J/g K})(m_w)(\Delta T_w)$$

The amount of heat gained by the calorimeter assembly (cups, water, thermometer, and stirrer), q_c, is given by

$$q_c = (4.18 \text{ J/g K})(m_c)(\Delta T_c) + C_c(\Delta T_c)$$

Therefore,

$$-q_w = q_c = (4.18 \text{ J/g K})(m_c)(\Delta T_c) + C_c(\Delta T_c)$$

All of the quantities in this equation are known except C_c, which can be calculated.

Calculations: Determination of Specific Heat and Atomic Weight

Make a plot of calorimeter temperature (y axis) versus time (x axis) as before. Obtain T_{max}, T_i, and ΔT_c, the temperature change in the calorimeter. Determine the temperature change for the metal (ΔT_{met}) by subtracting the initial temperature of the metal in the hot water, T_{hot}, from the final temperature, T_{max}:

$$\Delta T_{met} = T_{max} - T_{hot}$$

Calculate the total heat gained by the calorimeter and by water:

$$q_{tot} = q_w + q_c = (4.18 \text{ J/g K})(m_w)(\Delta T_c) + (C_c)(\Delta T_c)$$

Use the calorimeter constant that was determined in the calibration.

The heat gained by the calorimeter must be equal to the heat lost by the metal, or

$$-q_{met} = q_{tot}$$

Using equation (E5.2), calculate the specific heat of the metal. Using equation (E5.3), determine the approximate atomic mass of the metal.

Part B: Determination of the Heat of Solution

Macroscale experiment   Estimated time to complete the experiment: 1.5 h
Experimental Steps: Three steps are involved: (1) construction and (2) calibration of a calorimeter, and (3) determination of heat of solution.

Disposal of Chemicals: Dispose of the solutions in designated containers.

Set up a coffee cup calorimeter and determine the calorimeter constant C_c as described in Part A. Empty and dry the calorimeter (do not use acetone). Obtain the mass of the empty calorimeter. Add 30.0 mL of deionized water and reweigh it. Replace the lid, thermometer, and stirrer. Allow the calorimeter assembly to sit at least for 4 minutes for thermal equilibration. Now, record the temperature of the calorimeter every 30 seconds for **3 minutes.**

Obtain a solid salt from your instructor, and enter the number of the unknown on the data sheet. Weigh about 5 g (± 0.001 g) of the salt, and transfer it to the calorimeter. Replace the lid and stir the solution.

Record the temperature of the mixture every 30 seconds for **10 minutes**. Depending on the salt used, the temperature may increase or decrease. If time permits, perform a second run.

Plot the temperature (y axis) versus time (x axis). Extrapolate the final temperature T_{max} as described before. Calculate $\Delta T_w = T_{max} - T_i$. Since the specific heat of water and the calorimeter constant are known, the heat of solution can be easily calculated from equation (E5.4).

Part C: Determination of ΔH by Hess's Law

Semimicroscale experiment   Estimated time to complete the experiment: 2 h
Experimental Steps: Three steps are involved: (1) construction and (2) calibration of a calorimeter, and (3) determination of heat of reaction using Mg/MgO and HCl.

Disposal of Chemicals: Dispose of all solutions in designated containers.

Determine the calorimeter constant, C_c, as described in Part A. Clean, dry (do not use acetone), and weigh the calorimeter. Remove the lid, add 25.00 mL (± 0.01 mL) of standardized 1.00 M HCl to the calorimeter, and reweigh it.

Replace the lid, thermometer, and stirrer. Secure the calorimeter by placing the assembly on a sand bath. Allow 4 minutes for thermal equilibration, and record the temperature of the solution every 30 seconds for the next **3 minutes**.

Reaction of Magnesium (Reaction 1, ΔH_{rxn1})

Weigh 100 mg (± 1 mg) of *sanded* magnesium ribbon, coil it, and add this to the acid solution. Immediately replace the lid. Magnesium reacts with the acid to form Mg^{2+} and H_2 gas. Swirling the mixture, take temperature readings every 30 seconds until the temperature begins to fall. Continue to record the temperature at 30 second intervals for at least **12 minutes**. Empty the calorimeter, collecting the waste in a designated waste bottle. Perform a duplicate run if time permits.

Reaction of MgO (Reaction 3, ΔH_{rxn3})

Clean, dry (do not use acetone), and weigh the calorimeter. Add 25.00 mL (± 0.01 mL) of standardized 1.0 M HCl, and reweigh the calorimeter. After waiting 4 minutes for thermal equilibration, measure the temperature of the calorimeter every 30 seconds for **3 minutes**.

Weigh about 200 mg (± 1 mg) of magnesium oxide on weighing paper, and transfer it to the acid solution. **Save the weighing paper.** Immediately replace the lid. **Make sure that all the magnesium oxide is in the acid solution. You may need to tilt the cups so that the acid rinses the walls of the cup.**

While stirring the mixture, take temperature readings every 30 seconds until the temperature begins to fall. Continue to record the temperature for **12 minutes**. **Reweigh the weighing paper.** Calculate the mass of MgO actually transferred. Perform a duplicate run if time permits. Collect the solutions in a waste bottle.

Calculations

For all runs, determine the maximum temperature T_{max}, as before. Determine ΔT_c by subtracting the initial temperature of the calorimeter from T_{max}:

$$\Delta T_c = T_{max} - T_i$$

Calculate C_c.

The heat of reaction is determined using these equations:

$$\text{For Mg} \qquad -\Delta H_{rxn1} = (m_{acid} + m_{Mg})(4.18 \text{ J/g K})(\Delta T_c) + (C_c)(\Delta T_c)$$
$$\text{For MgO} \qquad -\Delta H_{rxn3} = (m_{acid} + m_{MgO})(4.18 \text{ J/g K})(\Delta T_c) + (C_c)(\Delta T_c)$$

The molar heat of reaction is given by

$$\Delta H_{rxn}^{\circ} = -\Delta H_{rxn}/n \text{ where } n = \text{\# mole of Mg/MgO used}$$

Using the literature value for $\Delta H_f^{\circ}(H_2O)$ (-285.8 kJ/mol), calculate the heat of formation of MgO, $\Delta H_f^{\circ}(MgO)$, using Equation (E5.10). Use $\Delta H_f^{\circ}(H_2O) = \Delta H_{rxn2}^{\circ}$.

Additional Independent Projects

1. Using Job's method (see Experiment 24 for details) of titrations of HCl and NaOH and of H_2SO_4 and NaOH, determine the enthalpy of neutralization.[3]

   Brief Procedure

   Determine the calorimeter constant. Mix various amounts of the acid and the base (1.00 M solutions) to a constant volume in the calorimeter and measure the change of temperature, ΔT, for each mixture. Plot ΔT (y axis) versus the volume of acid/base added (x axis). The maximum temperature change will occur when the reactants are in the stoichiometric ratio. From this information, determine the enthalpy (molar heat of reaction) of neutralization.

   Develop the data sheets, calculate the ΔH_f°, and write a short report according to the instructions given in Chapter 4.

2. Using Job's method of continuous variation, calculate ΔH_f° for the oxidation of sodium sulfite or iodide solution by aqueous sodium hypochlorite (bleach).[4]

3. Use other materials to construct a calorimeter. Measure the heat of a reaction and compare your data with those obtained by a coffee cup calorimeter.[1,2]

References

1. Brouwer, H. *J. Chem. Educ.* **1991,** *68,* A178.
2. Ruekberg, B. *J. Chem. Educ.* **1994,** *71,* 333.
3. Szafran, Z.; Pike, R. M.; Foster, J. C. *Microscale General Chemistry Laboratory with Selected Macroscale Experiments;* Wiley: New York, 1993.
4. Bigelow, M. J. *J. Chem. Educ.* **1969,** *46,* 378.

Name ______________________ Section ______________

Instructor ______________________ Date ______________

PRELABORATORY REPORT SHEET—EXPERIMENT 5

Experiment Title and part ______________________________

Objective

Reactions/formulas to be used

Materials and equipment table

Outline of procedure

PRELABORATORY PROBLEMS—EXPERIMENT 5

1. Define the following:

    (*a*) Specific heat

    (*b*) Heat capacity

    (*c*) Heat of solution

    (*d*) Heat of formation

    (*e*) Molar enthalpy

2. A student uses 100.0 mg of Mg and 300.0 mg of MgO in this experiment. She adds 50.0 mL of 1.50 M HCl to each sample separately. Write the balanced equations for the reactions. Calculate the number of moles of Mg and MgO and the limiting reagent for each reaction.

Name ______________________	Section ______________
Instructor ______________________	Date ______________

EXPERIMENT 5 DATA SHEET

Part A: Specific Heat of a Metal and Its Atomic Weight

Collection of Data

Calibration of the Calorimeter

1. Mass of the empty calorimeter ______ g

2. Mass of calorimeter + water ______ g

3. Mass of dry beaker ______ g

4. Mass of beaker + water ______ g

5. Temperature of cold water in calorimeter, T_i ______ °C ______ K

6. Temperature of hot water in beaker, T_{hot} ______ °C ______ K

7. Temperature, °C, of water in calorimeter versus time (minutes)

Time	Temp.	Time	Temp.
0	______	5.5	______
0.5	______	6.0	______
1.0	______	6.5	______
1.5	______	7.0	______
2.0	______	7.5	______
2.5	______	8.0	______
3.0	______	8.5	______
3.5	______	9.0	______
4.0	______	9.5	______
4.5	______	10.0	______
5.0	______	10.5	______

Show calculations:

Manipulation of Data

1. Mass of water in calorimeter   _____ g

2. Mass of water in beaker   _____ g

3. Total mass of water in calorimeter, m_c _____ g

4. T_{max} (extrapolated from graph)   _____ °C   _____ K

5. ΔT_c $(= T_{max} - T_i)$ _____ °C   _____ K

6. ΔT_w $(= T_{max} - T_{hot})$ _____ °C   _____ K

7. Heat loss by hot water, q_w _____ J

8. Calorimeter constant, C_c _____ J/K

Show calculations:

DATA SHEET, EXPERIMENT 5, PAGE 3

Collection of Data

Specific heat and atomic weight determination

Sample ID number _____

	Run 1	**Run 2**
1. Mass of empty calorimeter	_____ g	_____ g
2. Mass of calorimeter + water	_____ g	_____ g
3. Mass of empty dry test tube	_____ g	_____ g
4. Mass of test tube + metal	_____ g	_____ g
5. Temp. of cold water in calorimeter, T_i	_____ °C _____ K	_____ °C _____ K
6. Temp. of hot water in beaker, T_{hot}	_____ °C _____ K	_____ °C _____ K

7. Temperature, °C, of water in calorimeter versus time (minutes)

Run 1				**Run 2**			
Time min	**Temp. °C**	**Time min**	**Temp. °C**	**Time min**	**Temp. °C**	**Time min**	**Temp. °C**
0	_____	5.5	_____	0	_____	5.5	_____
0.5	_____	6.0	_____	0.5	_____	6.0	_____
1.0	_____	6.5	_____	1.0	_____	6.5	_____
1.5	_____	7.0	_____	1.5	_____	7.0	_____
2.0	_____	7.5	_____	2.0	_____	7.5	_____
2.5	_____	8.0	_____	2.5	_____	8.0	_____
3.0	_____	8.5	_____	3.0	_____	8.5	_____
3.5	_____	9.0	_____	3.5	_____	9.0	_____
4.0	_____	9.5	_____	4.0	_____	9.5	_____
4.5	_____	10.0	_____	4.5	_____	10.0	_____
5.0	_____	10.5	_____	5.0	_____	10.5	_____

Show calculations:

Manipulation of Data

		Run 1	**Run 2**
1.	Mass of water in calorimeter, m_w	_____ g	_____ g
2.	Mass of metal, m_{met}	_____ g	_____ g
3.	T_{max} (extrapolated from graph)	_____ °C _____ K	_____ °C _____ K
4.	ΔT_c ($= T_{max} - T_i$)	_____ °C _____ K	_____ °C _____ K
5.	ΔT_{met} ($= T_{max} - T_{hot}$)	_____ °C _____ K	_____ °C _____ K
6.	Calorimeter constant, C_c	_____ J/K	_____ J/K (from part A)
7.	Heat lost by metal, $-q_{met}$	_____ J/K	_____ J/K
8.	Heat gained by the calorimeter, q_{tot}	_____ J/K	_____ J/K
9.	Specific heat of the metal, s_{met}	_____ J/g K	_____ J/g K
10.	Atomic weight of the metal	_____ g/mol	_____ g/mol

Show calculations using the following equation:

$$q_{tot} = q_w + q_c = (4.18 \text{ J/g K})(m_w)(\Delta T_c) + (C_c)(\Delta T_c) = -q_{met} = -[m_{met} \times s_{met} \times \Delta T_{met}]$$

Name		Section	
Instructor		Date	

Part B: Determination of Heat of Solution

Sample ID number _____

Collection of Data

		Run 1	Run 2
1.	Mass of empty calorimeter	_____ g	_____ g
2.	Mass of calorimeter + water	_____ g	_____ g
3.	Mass of weighing paper	_____ g	_____ g
4.	Mass of weighing paper + salt	_____ g	_____ g
5.	Initial temperature of water	_____ °C	_____ °C

6. Temperature, °C, of water in calorimeter versus time (minutes)

	Run 1				Run 2		
Time min	Temp. °C	Time min	Temp. °C	Time min	Temp. °C	Time min	Temp. °C
0	_____	5.5	_____	0	_____	5.5	_____
0.5	_____	6.0	_____	0.5	_____	6.0	_____
1.0	_____	6.5	_____	1.0	_____	6.5	_____
1.5	_____	7.0	_____	1.5	_____	7.0	_____
2.0	_____	7.5	_____	2.0	_____	7.5	_____
2.5	_____	8.0	_____	2.5	_____	8.0	_____
3.0	_____	8.5	_____	3.0	_____	8.5	_____
3.5	_____	9.0	_____	3.5	_____	9.0	_____
4.0	_____	9.5	_____	4.0	_____	9.5	_____
4.5	_____	10.0	_____	4.5	_____	10.0	_____
5.0	_____	10.5	_____	5.0	_____	10.5	_____

Manipulation of Data

		Run 1	Run 2
1.	Mass of water in calorimeter, m_w	_____ g	_____ g
2.	Mass of salt	_____ g	_____ g
3.	T_{max} (extrapolated from graph)	_____ °C _____ K	_____ °C _____ K
4.	ΔT_{mix} ($= T_{max} - T_i$)	_____ °C _____ K	_____ °C _____ K
5.	Calorimeter constant, C_c	_____ J/K	_____ J/K (from part A)
6.	Heat absorbed or evolved	_____ J	_____ J
7.	Heat of solution, ΔH_{sol}	_____ J/g	_____ J/g

Part C: Heat of Reaction by Hess's Law

Collection of Data

Heat of Formation of Mg^{2+} (Reaction 1)

Molarity of HCl _____ mol/L

Volume of HCl in the calorimeter _____ mL

		Run 1	Run 2
1.	Mass of empty calorimeter	_____ g	_____ g
2.	Mass of calorimeter + acid	_____ g	_____ g
3.	Mass of weighing paper	_____ g	_____ g
4.	Mass of weighing paper + Mg ribbon	_____ g	_____ g
5.	Initial temperature of 1.0 M HCl	_____ °C	_____ °C

6. Temperature, °C, of solution in calorimeter versus time (minutes)

	Run 1				Run 2		
Time min	Temp. °C	Time min	Temp. °C	Time min	Temp. °C	Time min	Temp. °C
0	_____	6.0	_____	0	_____	6.0	_____
0.5	_____	6.5	_____	0.5	_____	6.5	_____
1.0	_____	7.0	_____	1.0	_____	7.0	_____
1.5	_____	7.5	_____	1.5	_____	7.5	_____
2.0	_____	8.0	_____	2.0	_____	8.0	_____
2.5	_____	8.5	_____	2.5	_____	8.5	_____
3.0	_____	9.0	_____	3.0	_____	9.0	_____
3.5	_____	9.5	_____	3.5	_____	9.5	_____
4.0	_____	10.0	_____	4.0	_____	10.0	_____
4.5	_____	10.5	_____	4.5	_____	10.5	_____
5.0	_____	11.0	_____	5.0	_____	11.0	_____
5.5	_____	11.5	_____	5.5	_____	11.5	_____

Show calculations:

DATA SHEET, EXPERIMENT 5, PAGE 7

Manipulation of Data

		Run 1		Run 2	
1.	Mass of the acid, m_{acid}	_____ g		_____ g	
2.	Mass of Mg, m_{Mg}	_____ g		_____ g	
3.	T_{max} (extrapolated from the graph)	_____ °C	_____ K	_____ °C	_____ K
4.	ΔT_c ($= T_{max} - T_i$)	_____ °C	_____ K	_____ °C	_____ K
5.	Calorimeter constant, C_c	_____ J/K		_____ J/K	
6.	Heat of reaction, $-\Delta H_{rxn1}$	_____ J		_____ J	
7.	Limiting reagent	_____		_____	
8.	Moles of limiting reagent	_____ mol		_____ mol	
9.	Molar heat of reaction 1, $-\Delta H^{\circ}_{rxn1}$	_____ J/mol		_____ J/mol	

Show calculations:

Collection of Data

Heat of Reaction of MgO with Acid (Reaction 3)

Molarity of HCl _____ mol/L

Volume of HCl in the calorimeter _____ mL

Collection of Data

		Run 1	Run 2
1.	Mass of empty calorimeter	_____ g	_____ g
2.	Mass of calorimeter + acid	_____ g	_____ g
3.	Mass of empty weighing paper	_____ g	_____ g
4.	Mass of weighing paper + MgO	_____ g	_____ g
5.	Initial temperature of 1.0 M HCl, T_i	_____ °C _____ K	_____ °C _____ K

6. Temperature, °C, of solution in calorimeter versus time (minutes)

Run 1				Run 2			
Time min	Temp. °C	Time min	Temp. °C	Time min	Temp. °C	Time min	Temp. °C
0	_____	6.0	_____	0	_____	6.0	_____
0.5	_____	6.5	_____	0.5	_____	6.5	_____
1.0	_____	7.0	_____	1.0	_____	7.0	_____
1.5	_____	7.5	_____	1.5	_____	7.5	_____
2.0	_____	8.0	_____	2.0	_____	8.0	_____
2.5	_____	8.5	_____	2.5	_____	8.5	_____
3.0	_____	9.0	_____	3.0	_____	9.0	_____
3.5	_____	9.5	_____	3.5	_____	9.5	_____
4.0	_____	10.0	_____	4.0	_____	10.0	_____
4.5	_____	10.5	_____	4.5	_____	10.5	_____
5.0	_____	11.0	_____	5.0	_____	11.0	_____
5.5	_____	11.5	_____	5.5	_____	11.5	_____

Show calculations:

Manipulation of Data

		Run 1		**Run 2**	
1.	Mass of the acid, m_{acid}	_____ g		_____ g	
2.	Mass of MgO, m_{MgO}	_____ g		_____ g	
3.	T_{max} (extrapolated from the graph)	_____ °C	_____ K	_____ °C	_____ K
4.	$\Delta T_c \; (= T_{max} - T_i)$	_____ °C	_____ K	_____ °C	_____ K
5.	Calorimeter constant, C_c	_____ J/K		_____ J/K	
6.	Heat of reaction, $-\Delta H_{rxn3}$	_____ J		_____ J	
7.	Limiting reagent	_____		_____	
8.	Moles of limiting reagent	_____ mol		_____ mol	
9.	Molar heat of reaction 3, $-\Delta H^{\circ}_{rxn3}$	_____ J/mol		_____ J/mol	
10.	Heat of formation of MgO	_____ J/mol		_____ J/mol	

$(\Delta H^{\circ}_{f} \, (H_2O) = -285.8 \text{ kJ/mol}, \Delta H^{\circ}_{f}(H_2O) = \Delta H^{\circ}_{rxn2}$;
for the value of ΔH°_{rxn1}; see Reaction 1 part C)

$$\Delta H^{\circ}_{f} \, (MgO) = \Delta H^{\circ}_{rxn1} + \Delta H^{\circ}_{f} \, (H_2O) + \Delta H^{\circ}_{rxn3}$$

Show calculations:

POSTLABORATORY PROBLEMS—EXPERIMENT 5

1. If a metal container is used instead of a Styrofoam cup as a calorimeter, will C_c be larger or smaller? Explain your answer.

2. Using the standard heat of formation of the species involved in the reaction,

$$Mg(s) + 2\ HCl(aq) \rightarrow MgCl_2(aq) + H_2(g)$$

Calculate the change in heat of formation (ΔH°_{rxn}) for the reaction. $\Delta H^\circ_f(HCl, aq) = 167.2$ kJ/mol and $\Delta H^\circ_f(MgCl_2) = 796$ kJ/mol.

3. Given that

$$H_2O(l) \rightarrow H_2O(g) \qquad \Delta H = 44 \text{ kJ/mol}$$
$$C_2H_5OH(l) \rightarrow C_2H_5OH(g) \qquad \Delta H = 42.2 \text{ kJ/mol}$$

explain why you feel a cooling effect when

(a) you come out of the shower

(b) you rub ethanol on your hands

Properties of Gases and Derivation of Gas Laws

Microscale Experiment

Objectives
- To develop the ideal gas laws
- To measure the atmospheric pressure using a barometer
- To calculate the value of the universal gas constant and the number of moles of a gas

Prior Reading
- Section 2.6   Graphing of data
- Section 3.10   Construction of a microscale gas law tube

Related Experiments
- Experiments 7 (Molar volume) and 9 (Electrolysis and Faraday's law)

Under certain conditions of temperature and pressure, all substances can exist as solids, liquids, or gases. Gases have remarkably simple physical properties under normal conditions. The gaseous state of any matter can be defined by four quantities: the volume V, the temperature T, the pressure P, and the quantity (mole) n.

Nitrogen and oxygen are the main constituents of the atmosphere and together account for 99 percent of its volume. All other gases (Table E6.1) make up the remaining 1 percent. Some of the rarer gases accumulate in the atmosphere as a result of biological processes, such as the decomposition of vegetation, or as a result of the pollution caused by human activities. The variation of the concentration of ozone or nitric oxide in the atmosphere is an outcome of such pollution.

Table E6.1   Average composition of dry air

Gas	Mole percent	Gas	Mole percent
Nitrogen (N_2)	78.08	Krypton (Kr)	1.10×10^{-4}
Oxygen (O_2)	20.94	Hydrogen (H_2)	5.00×10^{-5}
Argon (Ar)	0.934	Xenon (Xe)	9.00×10^{-6}
Carbon dioxide (CO_2)	0.032	Carbon monoxide (CO)	10^{-5}–10^{-6}
Neon (Ne)	1.8×10^{-3}	Nitric oxide (NO)	10^{-4}–10^{-6}
Helium (He)	5.2×10^{-4}	Ozone (O_3)	0–5×10^{-5}

Variables Used to Describe Gas Properties

A **gas** is defined as a substance that, when placed in a container, expands to occupy the total volume of the container. The SI unit for **volume** is the cubic decimeter, dm^3, also known as a liter (L, a non-SI unit still in common use). For smaller volumes, the SI unit is cubic centimeter, cm^3, which is equivalent to the milliliter (mL).

The **pressure** P of a gas is defined as the force per unit area. The SI unit is the pascal (Pa), which is defined as a force of one newton per square meter. Because the Pa is a very small pressure unit, the kilopascal (kPa) is commonly used (1 kPa = 1000 Pa). Several non-SI units are also commonly used in laboratory work. One of these is the **standard atmosphere** (atm), defined as the pressure that supports a column of mercury exactly 760 mm high at 0°C at sea level. Another non-SI unit still in common use is the **torr**, identical to 1 mm Hg. In English units, the unit of pressure is the **psi** (pound per square inch). The relationships among these pressure units are given below:

$$1 \text{ atm} = 760 \text{ mm Hg} = 760 \text{ torr} = 1.01325 \times 10^5 \text{ Pa} = 101.325 \text{ kPa} = 14.7 \text{ psi}$$

Absolute pressures are measured using a **barometer**. Pressures are also measured using various gauges (such as a tire pressure gauge). In such cases, the gauge reads the pressure *in excess* of the atmospheric pressure:

$$P_{abs} = P_{gauge} + P_{atm}$$

If your tire gauge reads 20 psi, the absolute pressure in the tire is 20 + 14.7 = 34.7 psi.

The SI unit for absolute **temperature** is the Kelvin (K, written with no degree sign). The Kelvin scale has two fixed temperatures:

1. The zero point (−273.15°C), called **absolute zero**—the lowest temperature theoretically obtainable, where the volume of a gas would be reduced to zero
2. The triple point of water—the temperature where solid, liquid, and gaseous water all coexist

One can easily convert from the Celsius scale to the Kelvin scale by adding 273.15:

$$T, \text{K} = T, \text{°C} + 273.15$$

Units in Kelvin and Celsius scales are of the same size. Thus, if the temperature of a reaction rose by 15°C, it also rose by 15 K.

The Ideal Gas Laws

It is well known that the volume of a given quantity of a gas depends upon the pressure and the temperature of the gas. When the air in a hot air balloon is heated, the balloon expands (and rises). If a tied-off balloon is placed inside a bell jar and the air is pumped out (reducing the pressure), the balloon expands. These examples illustrate the effects of the pressure and the temperature on a given volume of a gas. Note that in both the cases, the amount of gas inside the balloon does not change.

Four variables, namely, volume V, pressure P, temperature T, and the amount of gas (mole, n) determine the state of a gas. In this experiment, the effect of different pressures

(Part A) and of different temperatures (Part B) on a fixed quantity of air will be studied. In order to study the relationship between any two variables (V and P, for example), it is necessary to keep the other two variables (T and n) constant. The Combined Gas Law (the simultaneous relationship between volume, pressure, and temperature) and the zero point on the absolute temperature scale (−273.15°C) will also be determined. The following relationships will be determined from the experimental data:

1. Volume versus pressure relationship at constant T and n
2. Volume versus temperature relationship at constant P and n
3. Volume versus (pressure) × (temperature) at constant n
4. The absolute temperature

The first relationship was studied by Robert Boyle in the seventeenth century and is known as Boyle's law. The second relationship, known as Charles' Law, was first proposed by Jacques Charles in 1800. The Kelvin scale was first recognized by Lord Kelvin in 1848. More about properties of gases will be discussed in other experiments (see, for example, Experiments 7 and 9).

Example E6.1 The following data are obtained using a known volume of a gas at a constant temperature:

P, atm	0.125	0.25	0.5	1	2	4	8
V, L	8	4	2	1	0.5	0.25	0.125

Make the plot of the data as shown in Figure E6.1, and determine which is linear: (P versus V) or (P versus $1/V$).

Answer
Clearly, the graph of $1/P$ versus V in Figure E6.1 is linear.

Example E6.2 The following data were obtained for a given quantity of a gas at 1 atm pressure:

T, °C	−9	1	21	51	60	99
V, mL	6.37	6.80	7.4	7.85	8.15	9.30

(a) What relationship exists between the volume and the temperature of the gas?
(b) Plot a graph of the data to determine the value of absolute zero.

Answer

(a) As the temperature increases the volume of the gas increases. Thus, the volume is directly proportional to the temperature. In equation form,

$$V \propto T \quad \text{(where } n \text{ and } P \text{ are constant)}$$
$$V = kT \quad \text{(where } k \text{ is a constant)}$$
$$V/T = k$$

(b)   Since the absolute temperature is the temperature at which the volume of a gas will go to zero, make a plot of volume (y axis) versus temperature (x axis), and extrapolate to zero volume.

The extrapolated value is about −275°C as shown in Figure E6.2.

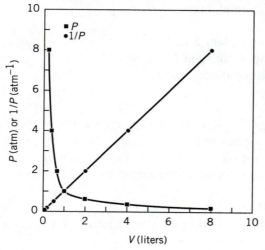

Figure E6.1   Variation of pressure with volume of a gas at a constant temperature

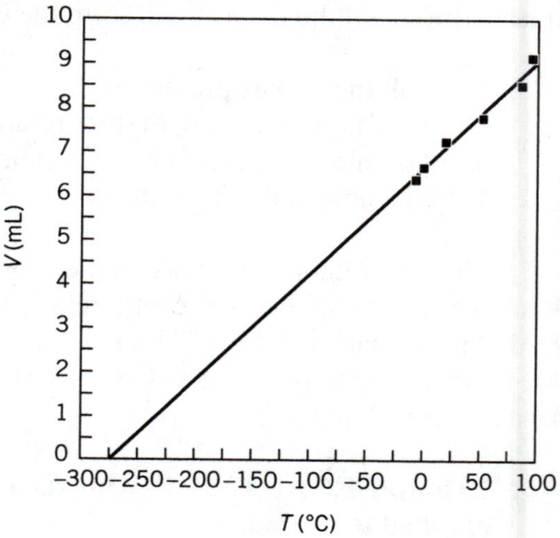

Figure E6.2   Variation of volume of gas with temperature at constant pressure

General References

1.   Davenport, D. *J. Chem. Educ.* **1962,** *39,* 252.
2.   Hermens, R.A. *J. Chem. Educ.* **1983,** *60,* 764.
3.   Sawyer, A.K. *J. Chem. Educ.* **1970,** *47,* 573.

Experimental Section

Procedure

Part A: Construction of a Micro Gas Capillary Tube

Microscale experiment   Estimated time to complete the experiment: 2.5 h
Experimental Steps: Three steps are involved: (1) constructing a gas law tube, (2) determining the effect of pressure changes, and (3) observing temperature changes on a given volume of a gas.

Obtain a clean and dry 9 in. Pasteur pipet. Construct a microscale gas law capillary tube (micro tube) containing mercury and a delivery capillary pipet as described in section 3.10. While handling the micro tube, hold it at the open end. Your instructor may provide you with the gas law tube.

HANDLE THE MICROSCALE GAS LAW TUBE WITH CARE. Do not touch the tube near the trapped air column. Follow all the precautionary measures suggested in section 3.10. The soft Tygon tubing covering the gas tube is effective in preventing mercury spills. However, if mercury does spill, report to your instructor immediately. Mercury vapors are toxic. After the experiment, return the tube to your instructor for storage.

Part B: Effect of Pressure Changes on Gas Volume

Perform this part of the experiment in a plastic bowl to prevent accidental mercury spills. As the pressure is increased or decreased on the gas sample trapped inside the micro gas capillary tube, the length L of the entrapped gas column will change. Since the area A of the tube is constant (and $V = L \times A$), the volume of the gas is therefore proportional to the length of the gas column. The pressure exerted on the trapped gas sample will depend upon the position of the micro tube, which will determine the effect of the mercury column on the trapped gas column. If the tube is held horizontally (Figure E6.3a), the mercury will exert no pressure on the gas sample and the pressure experienced by the air column will be equal to the atmospheric pressure (P_{atm}).

If the tube is held vertically upright, open end up (Figure E6.3b), the pressure on the gas sample will be equal to the pressure exerted by the Hg column (L_{Hg}) plus that of the atmosphere (P_{atm}). When the micro tube is inverted (Figure E6.3c), the pressure on the gas column is obtained by subtracting L_{Hg} from P_{atm}.

Obtain a thermometer ($-20°C$ to $110°C$), a twist tie, a 150 mm ruler (15 cm), a 400 mL beaker, a protractor, a 15 cm $\times$ 15 cm piece of cardboard, and the micro tube constructed in part A. Record the room temperature and the barometric pressure (P_{atm}) in mm Hg.

Using the twist tie, attach the micro tube to the millimeter ruler so that the zero on the ruler is just aligned with the bottom of the micro tube. **Once the adjustment is made, do not disturb the position of the ruler or the micro tube.** Handle the assembly by holding the upper end of the ruler. Carry out the operations shown in Figure E6.3a to c inside the

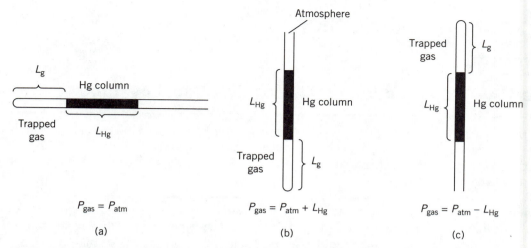

Figure E6.3 Position of the micro tube and the pressure on the gas column

plastic bowl. In each case, measure the length L_g of the *entrapped air column*. Repeat each measurement twice, and determine the average value.

Now, draw a circle of radius 5 to 6 cm on the cardboard and divide it into four quadrants. Using a protractor, mark angles of 30°, 45°, and 60° in the top right and bottom left quadrants. Fasten the cardboard vertically upright to a ring stand. Align the micro tube with each of the angles (open end up), and read the length of the entrapped gas column. Repeat this with the micro tube held open end down. Repeat each measurement twice, and determine the average value.

Calculate the effective pressure of the Hg column on the air column for all angles. The effective pressure P_{Hg} exerted by the mercury slug is given by

$$P_{Hg} = (L_{Hg})(\sin a)$$

where a is the angle between the mercury slug and the horizontal. For all open end up measurements, the effective pressure P_{Hg} should be added to the atmospheric pressure P_{atm} to calculate the total pressure P_{gas} exerted on the gas column. For the open end down measurements, P_{Hg} should be subtracted from P_{atm}. For horizontal position, P_{gas} is the same as P_{atm}.

Make two plots: one the plot of the volume V of the gas (proportional to L_g, x axis) versus the total pressure P_{gas} (y axis), and the other V (x axis) versus $1/P_{gas}$ (y axis). What conclusions can be drawn from these data? Calculate $P_{gas} \times V$ and express the result in correct units.

Part C: Effect of Temperature Changes on Gas Volume

Attach the micro tube (open end up) to a thermometer using a twist tie so that the bottom of the micro tube is aligned with the bottom of the thermometer bulb, as shown in Figure E6.4. Record the room temperature.

Hold the assembly vertically up. Using a centimeter ruler, determine the initial length of the air column from the sealed end of the tube to the bottom of the Hg slug and that of the Hg slug to the nearest ±1 mm. Using a felt tip pen (the marking ink must be insoluble in water), mark the location of the bottom of the mercury slug on the Tygon tubing.

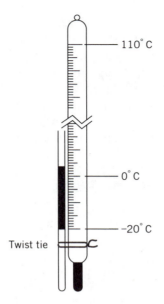

Figure E6.4 Thermometer–gas tube attachment

Place a 250 mL beaker on a wire gauze, supported on a ring stand. Fill three-fourths of the beaker with room-temperature deionized water and place the thermometer–tube assembly inside the beaker, making sure that the entrapped air and the thermometer bulb are immersed in the water and the bottom of the mercury slug remains above it. Allow 60 seconds for the temperature to stabilize. Read the temperature and mark the location of the bottom of the mercury slug on the Tygon tubing. [**Note:** Keep track of which mark is which!]

Remove the thermometer–tube assembly from the beaker, add crushed ice to the water in the beaker, and stir the mixture with a stirrer (not with the thermometer assembly). Replace the thermometer assembly in the beaker in the same way as before. After waiting for 60 seconds, record the temperature and mark the location of the bottom of the mercury slug.

Replace the ice/water bath with an ice/salt bath, in the same beaker. Following the same procedure as described before, record the temperature and mark the location of the bottom of the mercury slug. Remove the thermometer–tube setup from the bath, discard the ice/salt mixture, and rinse the beaker with water.

NOTE: The following measurements must be carried out as quickly as possible.

Fill the beaker with room-temperature deionized water. Heat the water to about 60°C. Replace the assembly in the water as before, record the temperature, and mark the location of the bottom of the mercury slug. Remove the thermometer assembly. Raise the temperature of the bath to ~100°C. Repeat the procedure of marking the bottom of the Hg column.

Remove the thermometer and the tube assembly from the beaker. Detach the micro tube. Using a centimeter ruler, measure the distances between each of the marks and the bottom of the tube to the nearest ±1 mm.

At lower temperatures (cold water, ice water, ice/salt water), the air column height decreases as a result of the contraction of the entrapped air. At higher temperatures (hot water at ~60°C and ~100°C) the gas expands, making the air column height larger.

Record these values against corresponding temperatures in °C on your data sheet. Convert Celsius degrees (°C) to Kelvin (K). Repeat the procedure once more.

Plot the volume (V) of the air (or the length of the air column, which is proportional) on the y axis versus the Kelvin temperature (T) on the x axis. Extrapolate the line drawn to zero volume. Convert the value of absolute zero to Celsius.

Part D: Combined Effect of *P* and *T* on a Given Volume of a Gas

The ideal gas law equation is

$$PV = nRT \qquad \text{(E6.1)}$$

where P is the pressure in atmospheres, V is the volume of the gas in liters, n is the number of moles of the gas under the given conditions of temperature and pressure, T is the temperature in Kelvin, and R is a constant called the **universal** or **ideal gas constant** ($R = 0.0821$ L atm/K mol). Rearranging this equation one obtains $PV/T = nR$. In the experiment, since all the measurements have been made using the same micro tube, the number of moles of air inside the tube remains constant. Therefore, the last equation may be rewritten as

$$PV/T = k \qquad \text{(E6.2)}$$

where k is a combined constant ($k = nR$).

Using the appropriate data collected in parts B and C, construct a table showing different pressures, P_{gas} (mm Hg), temperatures, T (Kelvin), and corresponding volumes, V, of the gas. Then calculate the value of k for each set using equation (E6.2). Express k values in correct units. Finally, using the relationship $n = k/R$, calculate the appropriate value of n, the number of moles of air inside the tube.

Name ______________________________	Section ______________
Instructor ___________________________	Date ________________

PRELABORATORY REPORT SHEET—EXPERIMENT 6

Experiment title ____

Objective

Reactions/formulas to be used

Materials and equipment table

Outline of procedure

PRELABORATORY PROBLEMS—EXPERIMENT 6

quantity

The following data are obtained using a fixed volume of a gas at a constant temperature:

P, mm Hg	725	760	795	820	836
V, arbitrary units	1.41	1.34	1.29	1.26	1.24

1. What variables are kept constant? What is the relationship between P and V?

2. Is the product PV constant under the experimental conditions?

3. Using graph paper, make plots of the data (P versus V) and (P versus $1/V$) and attach the plots with your prelaboratory problem sheet.

Name	______________________	Section	__________
Instructor	______________________	Date	__________

EXPERIMENT 6 DATA SHEET

Room temperature ______ °C

Atmospheric pressure ______ mm Hg (provided by the instructor)

Collection of Data

Length of air column, L_g, ______ mm

Length of Hg slug, L_{Hg}, ______ mm

Part B: Effect of Pressure Changes on Gas Volume

Tube position angle, *a*		Length of the air column, mm ($\propto V$) Trial 1	Trial 2	Trial 3	Average
0	(horizontal)	______	______	______	______
90°	(upright)	______	______	______	______
90°	(inverted)	______	______	______	______
30°	(open end up)	______	______	______	______
	(open end down)	______	______	______	______
45°	(open end up)	______	______	______	______
	(open end down)	______	______	______	______
60°	(open end up)	______	______	______	______
	(open end down)	______	______	______	______

Manipulation of Data

Angle, *a*		P_{Hg}, mm Hg	P_{gas}, mm Hg	Length V, mm	$1/P_{gas}$	$P_{gas} \times V$
0	(horizontal)	______	______	______	______	______
90°	(upright)	______	______	______	______	______
90°	(inverted)	______	______	______	______	______
45°	(open end up)	______	______	______	______	______
	(open end down)	______	______	______	______	______
60°	(open end up)	______	______	______	______	______
	(open end down)	______	______	______	______	______
30°	(open end up)	______	______	______	______	______
	(open end down)	______	______	______	______	______

Note: For open end up, $P_{gas} = P_{atm} + P_{Hg}$; for open end down, $P_{gas} = P_{atm} - P_{Hg}$

Using graph paper, plot two graphs (V versus P and V versus $1/P_{gas}$) and submit them along with your data. Show calculations. For V, use x axis, and for P_{gas} or $1/P_{gas}$, use y axis.

Answer the following questions.

1. From the nature of the plots, determine the relationship between the volume V of the gas and the pressure P at constant temperature, T. State the relation in a mathematical form.

2. In view of your answer in question 1, how do you expect that the values for the product of P times V would change? See your data table.

DATA SHEET, EXPERIMENT 6, PAGE 3

Part C: Effect of Temperature Changes on Gas Volume

Room temperature _____ °C   Atmospheric pressure _____ mm Hg

Collection of Data

1. Initial (room-temperature) length of air column, mm   _____

2. Initial (room-temperature) length of Hg slug, mm   _____

Run #		Bath temperature, °C	Length of the air column ($\propto V$) from the sealed end of the tube, mm
1	Cold water	_____	_____
	Water + ice	_____	_____
	Water + ice + salt	_____	_____
	Warm water	_____	_____
	Hot water	_____	_____
	Other water temps.	_____	_____
		_____	_____
2	Cold water	_____	_____
	Water + ice	_____	_____
	Water + ice + salt	_____	_____
	Warm water	_____	_____
	Hot water	_____	_____
	Other water temps.	_____	_____
		_____	_____

Manipulation of Data

Run #		Temperature (T), K	Volume (V) of air column	V/T (mm/K)
1	Cold water	_____	_____	_____
	Water + ice	_____	_____	_____
	Water + ice + salt	_____	_____	_____
	Warm water	_____	_____	_____
	Hot water	_____	_____	_____
	Other water temps.	_____	_____	_____
		_____	_____	_____
2	Cold water	_____	_____	_____
	Water + ice	_____	_____	_____
	Water + ice + salt	_____	_____	_____
	Warm water	_____	_____	_____
	Hot water	_____	_____	_____
	Other water temps.	_____	_____	_____
		_____	_____	_____

Using graph paper, plot your graph (V versus T) and determine the absolute zero in °C. Show calculations.

Answer the following questions.

1. From your graph, determine the effect of temperature on a given volume of a gas. State the relationship in a mathematical form.

2. What is the absolute temperature obtained from your plot? Is your value a correct one? Explain why.

Part D: Combined Effect of *P* and *T* on a Given Volume of a Gas

Data Collection and Manipulation

Use the relevant data from part B and part C to complete the following table.

From part B

Room temperature, expressed as K _____ (remains constant)

	Pressure P_{gas}, mm Hg	Volume V (L_g), mm	$PV/T = k$
1.	_____	_____	_____
2.	_____	_____	_____
3.	_____	_____	_____
4.	_____	_____	_____
5.	_____	_____	_____
6.	_____	_____	_____
7.	_____	_____	_____

From part C

Atmospheric pressure _____ mm Hg (remains constant)

	T, K	P_{gas}, mm Hg	V (L_g), mm	$PV/T = k$
1.	_____	_____	_____	_____
2.	_____	_____	_____	_____
3.	_____	_____	_____	_____
4.	_____	_____	_____	_____
5.	_____	_____	_____	_____
6.	_____	_____	_____	_____

Show calculations:

Answer the following question:

1. What is the relationship between *P*, *V*, and *T*?

POSTLABORATORY PROBLEMS—EXPERIMENT 6

1. What precautions must you take while working with a microscale gas law tube containing Hg?

2. Calculate the volume of a gas at 10.0°C under 1.20 atm pressure. The volume of the gas was 1.56 L at 25.0°C under 1.00 atm pressure.

3. What is the value of R when the pressure is expressed in kPa?

4. A 130.5 mL sample of a gas is collected at 28.5°C under 380 torr. What is the volume of the gas at STP (Standard Temperature, 0°C, and Pressure, 1.00 atm)?

Determination of the Molar Volume of a Gas and Atomic Weight of a Metal

Macroscale Experiment

Objectives
- To determine the molar volume of a gas or the atomic weight of a metal

Related Experiments
- Experiments 6 (Gas laws) and 9 (Electrolysis and Faraday's Law)

As shown in Experiment 6, the variables temperature T, pressure P, volume V, and quantity (number of moles, n) characterize the gaseous state of matter. Simple relationships, known as the **ideal gas laws**, have been established relating these variables:

Boyle's law	$P_1 V_1$	$= P_2 V_2$	T and n constant
Charles's law	V_1/T_1	$= V_2/T_2$	P and n constant
Avogadro's law	V_1/n_1	$= V_2/n_2$	P and T constant
Combined Law	$\dfrac{P_1 V_1}{T_1}$	$= \dfrac{P_2 V_2}{T_2}$	n constant

These individual relationships have been shown to be special cases of a more general relationship:

$$PV = nRT \tag{E7.1}$$

Here, R is a constant known as the **universal gas constant**. The numerical value of R depends on the units chosen for n, V, P, and T. Commonly used values are 0.0821 L atm$/$K$\cdot$ mol, 1.987 cal$/$K $\cdot$ mol and 8.315 J$/$K $\cdot$ mol.

The volume of one mole of a gas at standard temperature and pressure (STP, 273 K and 1 atm) is known as the **molar volume**. The molar volume of any gas at STP is very close to 22.4 L mol^{-1}. Rearranging equation (E7.1) to solve for the molar volume, we find

$$\frac{V}{n} = \frac{RT}{P} \tag{E7.2}$$

Since R is a constant, we see that the molar volume of an ideal gas is solely dependent on its temperature and pressure at constant n.

The purpose of this experiment is to determine either the molar volume of a gas or the atomic weight of the metal used to generate the gas.

Molar Volume of a Gas

The reaction of a metal with hydrochloric acid will be used in this experiment to generate hydrogen gas:

$$M(s) + 2\,HCl(aq) \rightarrow MCl_2(aq) + H_2(g) \tag{E7.3}$$

The hydrogen gas will be collected by the downward displacement of water from a eudiometer (gas-collecting tube). Since the collected gas will contain some water vapor, the total pressure of the gas, according to **Dalton's Law of Partial Pressures**, is

$$P_T = P_{H_2} + P_{H_2O} \tag{E7.4}$$

Here, P_{H_2O} is the vapor pressure of water at the temperature at which the gas is collected and P_{H_2} is the pressure of dry H_2 gas. In this experiment the total pressure P_T of the gas is equal to the barometric pressure. The pressure of the hydrogen gas is

$$P_{H_2} = P_T - P_{H_2O} \tag{E7.5}$$

The vapor pressure of water at various temperatures is given in Table E7.1.

The volume of dry hydrogen gas collected under the experimental conditions is converted to STP using the Combined Gas Law. This value, divided by the number of moles of hydrogen gas formed from the metal used, gives the molar volume.

Example E7.1 A volume of 35.4 mL of H_2 gas is collected at 26°C in a eudiometer tube by reacting 0.350 g Mg with excess HCl. The barometric pressure is 760 mm Hg. Calculate the molar volume of the gas at STP (1 atm = 760 mm Hg = 760 torr).

Answer

The vapor pressure of water at 26°C = 25.21 mm Hg. Thus,

$$P_{H_2} = 760.0 - 25.21 = 734.79 \text{ mm Hg}$$

Table E7.1. Vapor pressure of water at different temperatures

T, °C	P, mm Hg	T, °C	P, mm Hg	T, °C	P, mm Hg
0	4.58	16	13.63	26	25.21
5	6.54	18	15.48	28	28.35
10	9.21	20	17.54	30	31.82
12	10.52	22	19.83	40	55.3
14	11.99	24	22.38	50	92.5

Using the Combined Gas Law, we calculate the volume of the gas at STP as

$(734.79 \text{ mm Hg})(35.4 \text{ mL})/(299 \text{ K}) = (760 \text{ mm Hg})(V_2)/(273\text{K})$

$V_2 = 31.3 \text{ mL}$

Moles of Mg $= (0.0350 \text{ g})/(24.3 \text{ g/mol}) = 1.44 \times 10^{-3} \text{ mol}$

Molar volume $= V/n = (31.3 \text{ mL})/(1.44 \times 10^{-3} \text{ mol}) = 21,700 \text{ mL/mol} = 21.7 \text{ L/mol}$

The value obtained is reasonably close to the theoretical value of 22.4 L/mol H_2 at STP.

Atomic Weight of a Metal

Using the ideal gas laws, we calculate the number of moles of gas using equation (E7.1). By using the stoichiometry of the reaction, the number of moles of metal used to produce the gas can be calculated. The weight of the metal divided by the moles gives the atomic weight of the metal (1 atm $= 760$ mm Hg $= 760$ torr).

Example E7.2   A sample of 40 mg of a metal reacts with HCl as shown in the following equation to generate 24.40 mL of H_2 gas at 20°C. The barometric pressure is 775.0 mm Hg. Calculate the atomic weight of the metal.

$$M + 2 \text{ HCl} \rightarrow MCl_2 + H_2$$

Answer

$P_{H_2} = 775.0 - 17.54 = 757.5 \text{ mm Hg} = 0.997 \text{ atm}$

Volume of H_2 gas $= 0.02440 \text{ L}$

Temperature of the gas $= 293 \text{ K}$

$n = PV/RT = (0.997 \text{ atm})(0.02440 \text{ L})/(0.0821 \text{ L atm/mol K})(293 \text{ K})$

$n = 1.01 \times 10^{-3} \text{ mol } H_2 \text{ gas}$

$(1.01 \times 10^{-3} \text{ mol } H_2)(1 \text{ mol metal})/(1 \text{ mol } H_2) = 1.01 \times 10^{-3} \text{ mol metal}$

Atomic weight $= 0.040 \text{ g}/(1.01 \times 10^{-3} \text{ mol metal}) = 39.6 \text{ g/mol}$

The answer is very close to the atomic mass of calcium, 40.08 g/mol.

Experimental Section

Procedure

Obtain a 50 mL eudiometer tube, a stopper assembly (see Figure E7.1), and a widemouth 1 L beaker. The eudiometer tube must be clean and free of grease. Wrap the stopper of the eudiometer with a copper wire that ends in a small loop to hold the metal strip or the packet containing the metal powder. Obtain a metal strip or sample of powder from your instructor. If it is a ribbon, rub it with fine sandpaper to remove any oxide film. Place the strip (with forceps) in a *tared* 10 mL beaker and weigh it to the nearest milligram. If it is a powder, weigh it using weighing paper. Put the powder in an empty tea bag (to empty a tea bag, tear a corner of the bag and remove the tea leaves). Record the weight on the data sheet.

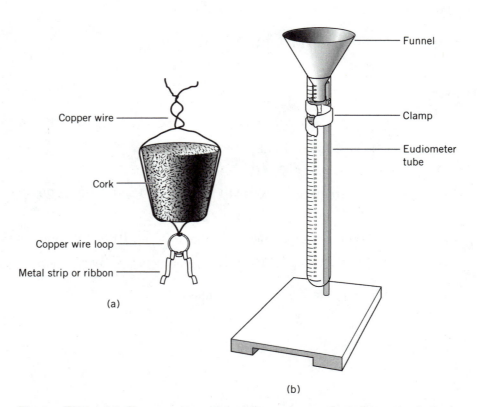

Figure E7.1   (*a*) Stopper assembly (shown larger than the actual size); (*b*) Eudiometer tube

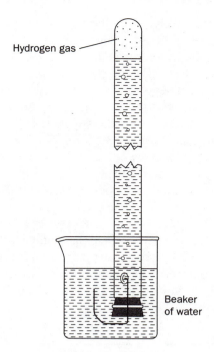

Figure E7.2 Eudiometer tube assembly inside water

Clamp a eudiometer tube to a ring stand. Using a funnel, pour 9 to 10 mL of 6 M HCl into the eudiometer. (If the metal is a powder, add 10 mL of 12 M HCl.) Fill the rest of the eudiometer with room-temperature water, adding the water slowly to avoid undue mixing of the acid. (For a powder, heat water to 50 to 60°C and add the hot water to the eudiometer tube.) Note that "adding water to an acid" is contrary to what the safety rule says: "Add acid to water." In this case, the addition of water to HCl does not generate much heat. (CAUTION: HCl is corrosive).

Insert the stopper assembly holding the metal sample about 3 to 4 cm into the eudiometer. *Loosely* press the stopper into the eudiometer—make certain that no air bubbles are trapped under the stopper.

Place 1.5 L of water in a 2 L beaker. Remove the eudiometer tube (it may be hot; use a paper towel) from the clamp, put your finger over the stopper, and immediately invert the tube and immerse it in the beaker (Figure E7.2). Make sure that no water is lost during this step.

Remove your finger, and clamp the eudiometer so that the end is 3 to 4 cm from the bottom of the beaker. Bubbles of hydrogen gas will be seen collecting in the eudiometer.

If the sample of metal is not inserted far enough into the eudiometer, small bubbles of gas may escape from the bottom of the eudiometer. If this occurs, the experiment must be repeated.

When all the metal has reacted and no more bubbles appear, allow the equipment to cool to room temperature (heat is generated in the reaction). Tap the side of the eudiometer to free any adhering bubbles of gas. Without removing the eudiometer from the beaker, remove the stopper from the eudiometer by pulling on the copper wire. Close the end of the eudiometer tube with your finger and transfer the tube to a water-filled 2 L graduated cylinder or a soda bottle (from which the neck part has been cut off). Raise or lower the eudiometer to make the water levels inside the eudiometer and outside equal. This step equilibrates the pressure of the collected hydrogen gas to atmospheric pressure.

If you cannot make the levels equal, measure the difference in the two levels in millimeters with a meter stick. To convert this reading from millimeters of H_2O to millimeters of Hg, multiply the value by the ratio of the density of water to that of mercury (density of water = 1.00 g mL^{-1} and that of mercury = 13.6 g mL^{-1}). This step corrects the barometric pressure to give the total pressure of the wet gas.

Record the volume of gas collected in the eudiometer. Place a thermometer in the water in the soda bottle, and record this value as the gas temperature. Look up the vapor pressure of water at this temperature and enter this value into the data table, along with the barometric pressure.

Repeat the experiment if time permits. Calculate either the molar volume or the atomic weight of the metal (as suggested by your instructor). For the calculation of atomic weight of the metal, your instructor will tell you what the stoichiometry of the reaction is. Return the **clean** apparatus assembly to your instructor including the stopper and the eudiometer tube.

Disposal: Ask your instructor about the disposal procedure. Solutions containing small amounts of HCl (after neutralization with sodium carbonate) may be discarded in the sink.

Additional Independent Projects
1. Demonstration of Raoult's Law and vapor pressure.[1]

Reference

1. Koubek, E.; Paulson, D. R. *J. Chem. Educ.* **1983**, *60*, 1069.

Name ______________________ Section ____________

Instructor ______________________ Date ____________

PRELABORATORY REPORT SHEET—EXPERIMENT 7

Experiment Title and part ______________________

Objective

Reactions/formulas to be used

Materials and equipment table

Outline of procedure

1. What precautions must be taken in collecting H_2 gas in a eudiometer?

2. What is the value of R if P is expressed in units of kPa?

3. You wish to make a barometer for use in the laboratory. You do not have a supply of mercury, so you think, "Why not use water?" How high a column of water must be used to measure a pressure of 1 atm?

4. A volume of 43.0 mL gas is collected at 25.0°C at a pressure of 763.0 torr by the downward displacement of water. How many moles of dry gas are present?

Name ______________________ Section ____________

Instructor ______________________ Date ____________

EXPERIMENT 7 DATA SHEET

Collection of data

Room temperature ______ °C

		Trial 1	Trial 2	
1.	Mass of metal sample, g	______	______	
2.	Total volume of hydrogen gas, mL	______	______	
3.	Temperature of the gas, °C	______	______	
4.	Barometric pressure, mm Hg	______	______	(provided by the instructor)
5.	Difference in water levels (if necessary)	______	______	(Table E7.1)
6.	Vapor pressure of water at experimental temperature, mm Hg	______	______	

If you use a container to measure the mass of the metal sample, calculate its mass and insert the value in step 1.

Manipulation of Data

1. Pressure of dry hydrogen gas, P_{H_2}, atm _____ _____

2. Temperature, K _____ _____

3. Volume of wet hydrogen gas, V_1, L _____ _____

4. Volume of dry gas at STP, V_2, L _____ _____

Molar Volume

5. Theoretical moles of hydrogen gas generated from the reaction _____ _____

6. Molar volume for hydrogen gas _____ _____

7. % deviation from literature value _____ _____

8. Average value (if two trials made) _____

or

Atomic Weight

9. Number of moles of the gas, n _____ _____

10. Atomic weight of metal (if unknown is used) (Your instructor will provide you with the stoichiometry of the reaction.) _____

Show calculations:

POSTLABORATORY PROBLEMS—EXPERIMENT 7

1. If the metal strip had not been cleaned and an oxide film was left on the ribbon, would the volume of hydrogen gas generated be higher or lower than the true value? Explain.

2. If some air enters the eudiometer tube, what effect will it have on the atomic weight determination?

3. What is the value of R in SI units?

Solvent Extraction

Part A: Determination of the Distribution Ratio of Benzoic Acid
Part B: Base Extraction of Benzoic Acid

Microscale Experiments

Objectives
- To learn about solvent extraction
- To determine the solubility characteristics of a compound partitioned between an immiscible solvent pair
- To investigate the use of a chemical reagent in altering the solubility of a compound

Prior Reading
- Chapter 1   Safety precautions in the laboratory
- Section 3.4   Weighing
- Section 3.5   Measuring liquid volumes: use of automatic delivery pipet
- Section 3.3   Handling chemicals
- Section 3.8   Filtration using a Hirsch funnel
- Section 3.9   Solution Preparation
- Section 3.10   Preparation of a Pasteur filter pipet
- Section 3.11   Titration (optional)

Related Experiment
- Experiment 25 (Extraction of Caffeine)

Substances tend to dissolve in solvents with which they are chemically similar. A common way of saying this is that **like dissolves like.** The most frequently used solvent is water, so chemists are interested in compounds that interact strongly with water. Species that dissolve well in water usually have several characteristics in common, such as being ionic, having polar bonds, or having the ability to form hydrogen bonds. Hydrogen bonds are unusually strong intermolecular attractions, formed between a partially positively charged hydrogen atom on one molecule and a negatively charged atom on a second molecule (usually N, O, or F).

The hydroxyl group (–OH) is one of the best hydrogen bonding groups found in compounds because it is polar. Oxygen is more electronegative than hydrogen and withdraws electrons from it in their bond (that is, the electrons are not shared equally). Thus, the hydrogen atom takes on a partial positive charge, and the oxygen atom takes on a partial negative one.

Hydrogen bonds are about 5 to 10 percent as strong as covalent bonds. Examples of hydrogen bonding systems include the following:

$$H_2O \cdots HOH \qquad H_3N \cdots HOH \qquad H_2O \cdots HOCH_2CH_3$$

Water–water   Ammonia–water   Water–ethanol

The hydroxyl end of the ethanol molecule can hydrogen bond strongly to water. Therefore, ethanol and water are miscible in all proportions. The attractive forces set up between the two molecules are nearly as strong as in water itself; however, the attraction is somewhat weakened by the presence of the nonpolar hydrocarbon unit, CH_3CH_2-.

A similar situation occurs in the carboxylic acids. Acetic acid (vinegar), having an even more polar –OH group than ethanol (due to the electron-withdrawing C=O nearby), hydrogen bonds strongly to water. Since the nonpolar hydrocarbon unit is small, this acid is also miscible in all proportions with water. One would expect as the nonpolar group becomes larger, the solubility of the acid would decrease as the nonpolar group becomes the dominant structural feature. We would further expect an enhanced solubility of the molecule in nonpolar solvents. Experimentation has demonstrated that this is true. Table E8.1 summarizes the solubilities of a number of straight-chain carboxylic acids in water. Note that as the length of the carbon chain increases, the solubility in water decreases. Benzoic acid is not a straight-chain carboxylic acid but instead contains a cyclic hydrocarbon group, the phenyl unit C_6H_5-.

Similar to the straight-chain acids, benzoic acid contains a polar group (the –COOH group) and a nonpolar group (the phenyl unit).

The Partition Coefficient

A given substance, if placed in a mixture of two immiscible solvents, will **partition** (distribute) itself in a manner that is a function of its solubility in each of the two solvents.

Table E8.1   Solubility of Carboxylic Acids

Name of acid	Formula of acid	Solubility, g/100 mL H_2O at 20°C
Acetic	CH_3COOH	Infinite
Propanoic	CH_3CH_2COOH	Infinite
Pentanoic	$CH_3(CH_2)_3COOH$	3.7
Hexanoic	$CH_3(CH_2)_4COOH$	1.0
Decanoic	$CH_3(CH_2)_8COOH$	0.003 (at 15°C)
Benzoic	C_6H_5COOH	0.30

Suppose, for example, some solute X is more soluble in water (a polar solvent) than it is in methylene chloride (a less polar solvent). Water and methylene chloride are not soluble in each other and thus form an immiscible solvent pair (and form two layers). Now, some X is added to the solvent pair and the system is shaken thoroughly. After equilibrium is reestablished, most of the X will be dissolved in the water layer, and only a little will be in the methylene chloride layer. The **partition coefficient K** is defined as the ratio of the amount of X found in each layer:

$$K = \frac{X_{\text{methylene chloride}}}{X_{\text{water}}}$$

The partition coefficient is constant for a given solvent pair at a given temperature. Determination of the partition coefficient for a particular compound in various solvent pairs often gives valuable information to aid in the isolation and purification of the species using extraction techniques.

The preceding equation does not apply to a system where the solute undergoes dissociation or association in either phase. In practice, chemists are more interested in determining the fraction of a solute in one phase or the other, regardless of its dissociation or association. It is useful to introduce another term, the **distribution ratio D**:

$$D = \frac{c_1}{c_2}$$

where c_1 and c_2 are the concentrations in solvent 1 and solvent 2 of the solute in all its forms: dissociated, undissociated, and associated.

Part A

The procedure for determining the distribution ratio is demonstrated in part A of this experiment, in which benzoic acid is partitioned in a water–methylene chloride mixture. The distribution equation is therefore

$$D = \frac{(\text{g}/100 \text{ mL})_{\text{organic layer}}}{(\text{g}/100 \text{ mL})_{\text{water layer}}}$$

This expression uses grams per 100 mL, but other concentration units such as grams per liter (g/L), parts per million (ppm), or molarity (M) would be equally valid. Since the distribution ratio is dimensionless, any concentration units may be used, provided the units are the same for both phases. If *equal volumes* of both solvents are used, the equation reduces to the ratio of the masses (*m*) of the given species in the two solvents, or

$$D = \frac{m_{\text{organic layer}}}{m_{\text{water layer}}}$$

If the mass of the benzoic acid that is in the methylene chloride layer (organic layer) and in the water layer can be determined, the distribution ratio for benzoic acid can be calculated.

Single Extraction or Multiple Extractions?

A problem often encountered in the laboratory is to determine the best method of extracting a species from a solution. If V mL of an aqueous solution containing x_o millimoles (mmol) of a solute is extracted n times using v mL of a solvent, then the millimoles of the solute left in the aqueous solution x_n is given by

$$x_n = x_o \left(\frac{V}{vD + V} \right)^n$$

It is therefore useful to extract a solute using a small volume of the liquid several times rather than extracting the solute once using a large volume of the liquid.

The percent extraction (%E) of a solute from solvent 1 into solvent 2 can be calculated. The fraction of the solute extracted is equal to the millimoles (or grams) of the solute in solvent 1 divided by the total number of millimoles (or grams) of the solute in both the solvents. The millimoles are given by the molarity times the volume of the solvents used. Thus, the percent extraction is given by

$$\%E = 100 \left(\frac{M_1 V_1}{M_1 V_1 + M_2 V_2} \right)$$

which can be expressed in terms of the distribution ratio D as follows:

$$\%E = \frac{D \times 100}{D + (V_2/V_1)}$$

where M_1 and M_2 are the molarities of the solute in the two solvents and V_1 and V_2 are the volumes.

Example E8.1 Iodine, I_2, is highly soluble in carbon tetrachloride and slightly soluble in water. A sample of 1.00 mL of a 1.00×10^{-3} M aqueous solution of I_2 is extracted with 1.00 mL of carbon tetrachloride (CCl_4). After the first extraction, the concentration of I_2 in the aqueous layer is found to be 1.17×10^{-5} M. Calculate the distribution ratio D for the distribution of I_2 between CCl_4 and water. What is the percentage of I_2 extracted?

Answer

mg of I_2 in 1.00 mL of water
$$= (1.00 \text{ mL})(1.00 \times 10^{-3} \text{ mmol/mL})(254.0 \text{ mg } I_2/\text{mmol})$$
$$= 2.54 \times 10^{-1} \text{ mg } I_2$$

mg of I_2 left in water after the first extraction
$$= (1.00 \text{ mL})(1.17 \times 10^{-5} \text{ mmol/mL})(254.0 \text{ mg } I_2/\text{mmol})$$
$$= 2.97 \times 10^{-3} \text{ mg } I_2$$

mg of I_2 extracted by CCl_4
$$= 254.0 \times 10^{-3} - 2.97 \times 10^{-3} = 2.51 \times 10^{-1} \text{ mg } I_2$$

$D = \dfrac{m_{\text{organic layer}}}{m_{\text{water layer}}} \qquad D = \dfrac{2.51 \times 10^{-1}}{2.97 \times 10^{-3}} = 84.5$

$\%E = \dfrac{D \times 100}{D + V_2/V_1}$

$\%E = 100\,[84.5/(84.5 + 1)] = 98.8\%$

Part B

Separation of Organic Acids and Bases

The separation of organic acids and bases constitutes an important application of solvent extraction. In Table E8.1, the solubility of benzoic acid was given as only 0.30 g/100 g water, so the compound is only barely soluble. The solubility of benzoic acid can be increased in the following way.

Benzoic acid reacts with sodium hydroxide in aqueous solution to form a salt. The salt, being ionic, dissolves in the more polar aqueous phase.

(water-insoluble)   (water-soluble)

Sodium bicarbonate, being a base, behaves in a similar manner toward benzoic acid:

$$C_6H_5 \overset{O}{\underset{\|}{C}} - OH + Na^+HCO_3^- \rightleftharpoons C_6H_5 \overset{O}{\underset{\|}{C}} - O^-Na^+ + H_2CO_3$$

Benzoic acid   Sodium bicarbonate   Sodium benzoate   Carbonic acid

$$H_2CO_3 \rightleftharpoons CO_2 + H_2O$$

These reactions change the solubility characteristics of the water-insoluble acid. The water phase containing the salt may then be extracted with an immiscible organic solvent to remove any impurities, leaving the acid and salt in the water phase. The salt is then reacted with hydrochloric acid, causing the precipitation of the original insoluble organic acid in a relatively pure state.

In a similar fashion, organic bases such as amines can be rendered more water-soluble by treating them with dilute hydrochloric acid to form hydrochloride salts:

(slightly water-soluble)   (water-soluble)

(slightly water-soluble)   (water-soluble)

Test for a carboxylic acid

Carbon dioxide gas is generated in the reaction of benzoic acid and sodium bicarbonate. This reaction is also observed with acetic acid, and in fact, all carboxylic acids:

$$CH_3\overset{\displaystyle O}{\overset{\|}{C}}—OH + HCO_3^- \rightleftharpoons CH_3\overset{\displaystyle O}{\overset{\|}{C}}—O^- + H_2CO_3 \rightleftharpoons CO_2 \cdot$$

Once saturation of the solution occurs, bubbles of CO_2 gas form. The observed effervescence can be used as a qualitative test for the presence of carboxylic acids and the –COOH group.

References

1. *CRC Handbook of Chemistry and Physics,* 68th ed.; 1 CRC Press: Boca Raton, FL, 1988.
2. Mayo, D. W.; Pike, R. M.; Trumper, P. K. *Microscale Organic Laboratory,* 3rd ed.; Wiley: New York, 1994.

Experimental Section

Procedure

Part A: Determination of the Distribution Coefficient of Benzoic Acid

> Microscale experiment   Estimated time to complete the experiment: 2 h
> Experimental Steps: Three steps are involved: (1) partitioning benzoic acid between water and an organic solvent, (2) determining the mass of benzoic acid recovered, and (3) performing the titration of benzoic acid left in the aqueous solution.

Obtain three 5.0 mL vials and number them. Weigh vial #1, and accurately add about 50 mg of benzoic acid. Using an automatic delivery pipet (a different one for each solvent), add 700 μL of methylene chloride followed by 700 μL of water. The methylene chloride layer, having a greater density than water, will settle to the bottom. Construct a Pasteur filter pipet (see Section 3.10). Obtain a sample of anhydrous sodium sulfate.

> The methylene chloride addition should be carried out in the HOOD. Disposal: Dispose of the chemicals in designated containers. Halogenated organic compounds (which contain F, Cl, Br, I) must be handled carefully. Recover these halogenated species by distillation. Leftover compounds must be disposed of in a special container marked "Halogenated organic waste."

Cap the vial and shake the mixture (use a vortex mixer if available) until the benzoic acid dissolves. After thorough mixing, allow the two layers to separate (remember to vent the vial). Draw the lower (methylene chloride) layer *quantitatively* into a Pasteur filter pipet and carefully transfer it to vial #2, containing 100 mg of anhydrous, granular sodium sulfate (see Figure E8.1). Rinse the Pasteur filter pipet with two small portions (total volume: 1 mL) of fresh methylene chloride and transfer the washings into vial #2. Recap the vial.

> Be careful not to overfill the pipet to ensure that solvent does not come in contact with the rubber bulb. More than one transfer may be necessary.

Allow the methylene chloride solution to dry for 10 to 15 minutes over the sodium sulfate. Using a Pasteur filter pipet, transfer the solution to the previously weighed vial #3, containing a small boiling stone. Rinse the sodium sulfate with an additional 700 μL of methylene chloride and combine the rinse with the solution in vial #3. Place vial #3 in a warm sand bath [HOOD] and carefully evaporate the solvent to obtain a solid residue of benzoic acid.

> The boiling stone will help to avoid explosive boil-up of the solvent system when the vial is placed in the sand bath.

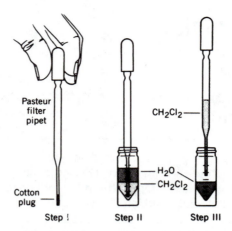

Figure E8.1 Extraction process with a Pasteur filter pipet. The bottom CH_2Cl_2 layer contains the product.

If available, evaporate the solvent under a gentle stream of nitrogen gas in a warm sand bath in the **HOOD**. Do not heat too strongly, as the benzoic acid may sublime. It is important to warm the solution while evaporating the solvent; otherwise the heat of vaporization will rapidly cool the solution. In this latter case, as the cold solid acid precipitates from the saturated solution, moisture will condense from the air entrained in the evaporation process and contaminate the surface of the recovered material.

Weigh the vial and determine the weight of solid benzoic acid. Break up the filter cake with a microspatula, and reheat the vial and contents in a sand bath to remove the last traces of solvent and any water that remains in the system. Cool and reweigh. Repeat this operation until a constant weight is obtained for the benzoic acid that dissolved in the methylene chloride layer.

The original mass of benzoic acid used minus the amount of benzoic acid recovered in the methylene chloride layer equals the mass of the benzoic acid that dissolved and remained in the water layer. Since equal volumes of both solvents were used, the distribution ratio may be determined from the ratio of the mass of benzoic acid in the methylene chloride solvent to the mass of benzoic acid in the water layer. Calculate the distribution ratio D for benzoic acid in the solvent pair used in this experiment, and the percentage of benzoic acid that was extracted in the methylene chloride layer.

Optional Experiment

Quantitatively transfer the leftover benzoic acid in the aqueous layer in vial #1 into a 25 mL Erlenmeyer flask. Wash the vial with two small portions of water (total volume must not exceed 1 mL) and transfer the washings to the same flask. Titrate the benzoic acid in the aqueous layer with standardized 0.1 M NaOH, using phenolphthalein as an indicator (see Section 3.11). Calculate the mass of benzoic acid left in the aqueous layer. This amount of benzoic acid should be close to that calculated by the difference in the first part of the experiment.

Part B: Base Extraction of Benzoic Acid

Experimental Steps: Two steps are involved: (1) extracting benzoic acid with a basic solution and (2) testing for a carboxylic acid.

Add 100 mg of benzoic acid to a centrifuge tube, followed by 3 mL of diethyl ether.

Dispense the ether in the HOOD.

Stir the mixture with a glass rod to dissolve the solid (use a vortex mixer if one is available). Add 2 mL of 3 M NaOH, and mix as before. After mixing, allow the two layers to separate.

Using a Pasteur filter pipet, separate the lower (aqueous) layer quantitatively and transfer it to a 10 mL Erlenmeyer flask. Add a second 2 mL portion of 3 M NaOH to the remaining ether layer. Remove the aqueous layer as before, and transfer it to the same Erlenmeyer flask. Stopper the flask.

Add 6 M HCl dropwise to the aqueous solution in the Erlenmeyer flask until the solution is distinctly acidic to litmus paper. Cool the flask in an ice bath for 10 minutes. If the acid does not precipitate, add 200 μL of saturated NaCl solution.

Collect the white precipitate of benzoic acid by suction filtration using a Hirsch funnel (or nail–filter paper funnel). Carefully wash the solid acid (dropwise) with 1 mL of ice cold water. Air-dry the benzoic acid on a clay plate or on preweighed filter paper. Weigh the product, and calculate the percent recovery. Take the melting point and compare your result to the literature value.

Test for Carboxylic Acid

Place 1 mL of 10 percent sodium bicarbonate on a small watch glass. Add the pure acid sample to the bicarbonate solution. Evolution of bubbles of carbon dioxide indicates the presence of an acid. Perform this test for carboxylic acids on your recovered benzoic acid and on any unknowns that are provided.

Disposal of Chemicals: Dispose of the organic chemicals in designated containers.

Additional Independent Project
1. Rapid semimicro extraction of caffeine.[1]

References
1. Neuzil, E. F. *J. Chem Educ.* **1990**, *67,* A262.

Name ______________________ Section ____________

Instructor ______________________ Date ____________

PRELABORATORY REPORT SHEET—EXPERIMENT 8

Experiment Title and part ______________________

Objective

Reactions/formulas to be used

Materials (chemicals and solutions) and equipment table

Outline of procedure

Table for the optional experiment (attach the table to the data sheet)

1. What is the difference between a partition coefficient and a distribution ratio?

2. In an experiment, 48.0 mg of benzoic acid is recovered from a solution of 50 mg of benzoic acid in ether. What is the percent recovery?

Name ______________________	Section ____________
Instructor ______________________	Date ____________

EXPERIMENT 8 DATA SHEET

Part A Determination of the Distribution Ratio of Benzoic Acid

Room temperature ______ °C

Collection of Data	Trial 1	Trial 2
1. Mass of vial #1, g	______	______
2. Mass of benzoic acid and vial #1, g	______	______
3. Volume of water, mL	______	______
4. Volume of methylene chloride, mL	______	______
5. Mass of vial #3 + stone, g	______	______
6. Mass of benzoic acid + vial #3 + stone, g	______	______

Manipulation of Data	Trial 1	Trial 2
1. Mass of benzoic acid taken, g	______	______
2. Mass of benzoic acid in CH_2Cl_2 in vial #3, g	______	______
3. Mass of benzoic acid left in H_2O in vial #1, g	______	______
4. Distribution ratio *D*	______	______
5. Percent extracted	______	______

Show calculations:

Part B   Base Extraction of Benzoic Acid

Collection and Manipulation of Data	Trial 1	Trial 2
1. Mass of benzoic acid taken, g	_____	_____
2. Mass of the acid recovered, g	_____	_____
3. Percent yield of benzoic acid	_____	_____
4. Melting point of benzoic acid	_____	_____

5. Observations on the reactions of
    unknowns with bicarbonate

Unknown A _____

Unknown B _____

Unknown C _____

Unknown D _____

POSTLABORATORY PROBLEMS—EXPERIMENT 8

1. A sample of 2.00 mL of a 0.10 M aqueous solution of an organic acid is extracted with 2.00 mL of an organic solvent. After the extraction, 0.050 mmol of the acid is left in the aqueous layer.

   (*a*) What is the concentration of the acid in the organic layer?

   (*b*) Calculate the partition coefficient, K, for the distribution.

   (*c*) What percentage is extracted by the organic layer?

2. In Part B, if instead of using sodium bicarbonate, you used ammonium hydroxide, would benzoic acid have been extracted in the same manner? Explain.

EXPERIMENT **9** # Electrolysis of an Aqueous KI Solution: Determination of Faraday's Constant and Avogadro's Number

Microscale Experiment

Objectives
- To determine the value of Faraday's constant and of Avogadro's number experimentally
- To construct an electrolytic cell for an electrolysis of a salt solution
- To collect and measure the gas evolved in a reaction

Prior Reading
- Section 3.3   Handling chemicals
- Section 3.4   Weighing
- Section 3.7   Stirring
- Section 3.8   Filtration

Related Experiments
- Experiments 6 (Gas properties), 7 (Molar volume), and 32 (Preparation of iodoform: Part B)

Electrolysis is the process in which electrical energy is used to cause a nonspontaneous chemical reaction to occur. Michael Faraday, an English physicist and chemist, carried out extensive investigations on the electrolysis of aqueous solutions. His results were summarized into what are now referred to as **Faraday's Laws of Electrolysis**. These laws accurately describe the relationship between the amount of electric current passed through a cell and the chemical reactions that result. Faraday's Laws of Electrolysis can be stated as follows:

1. The amount of product formed (or reactant consumed) at each electrode is proportional to the quantity of electricity passing through the electrolytic cell.

2. For a given quantity of electric charge, the amount of any metal that is deposited is directly proportional to its equivalent weight (atomic weight/charge on the metal).

The electrolysis of molten salts and aqueous solutions of salts has achieved industrial importance for the production of a number of basic chemicals. These include aluminum and sodium metals, chlorine gas, and sodium hydroxide. The electroplating technique is also based on electrolytic methods.

The amount of a chemical reaction that occurs at an electrode is directly proportional to the number of electrons transferred at the electrode. A **faraday** is defined as the total charge carried by Avogadro's number of electrons. Thus, one faraday represents the charge on one mole of electrons.

In this experiment, the value of the faraday will be determined by measuring the charge required to reduce one mole of H^+ ions. The electrical units used in electrolysis calculations are the coulomb C and the ampere A. These units are related in the following way:

1 faraday $= (1.602 \times 10^{-19} C/\text{electron})(6.022 \times 10^{23} \text{electrons/mol}) = 9.650 \times 10^4 C/\text{mol}$

1 coulomb $= 1$ ampere second (quantity of electricity when a current of 1 ampere flows for 1 second)

1 ampere $= 1$ coulomb per second

The charge on an electron is 1.602×10^{-19} coulombs.

Calculations involving passage of electric current through a cell can be treated in the same way as any other stoichiometric problem. The current in amperes is converted to coulombs, then to faradays, and then to moles of electrons. Moles of electrons may be treated in the same manner as moles of any chemical species. Using Faraday's laws, one can calculate the amount of substance produced if the current and the time are known.

(Current, A)(time, s) $\leftrightarrow$ Coulombs, C $\leftrightarrow$ Faraday, F $\leftrightarrow$ Moles of electrons

$\leftrightarrow$ Moles of chemical species

Example E9.1   A current of 0.120 A is passed through a dilute solution of sulfuric acid for exactly 5 minutes. How many coulombs are passed through the solution? How many electrons?

Answer
    The reaction involved is

$$2 H^+(aq) + 2 e^- \rightarrow H_2(g)$$
$$\#Coul = (\#Amps)(\#secs) = (0.120 \text{ A})(5 \text{ min})(60 \text{ s/min}) = 36.0 \text{ C}$$
$$\#e^- = 36.0 \text{ C}(1 \text{ e}^-/1.60 \times 10^{-19} \text{ C}) = 2.25 \times 10^{20} e^-$$

The foregoing information can be used to calculate Avogadro's number in the following manner.

Example E9.2   Suppose that 5.00 mL of hydrogen gas is collected over water at 21°C and 725 torr, under the conditions described in Example E9.1. How many moles of hydrogen are obtained? What value is calculated for Avogadro's number? What value is obtained for Faraday's constant? The vapor pressure of H_2O gas at 21°C is 18.7 torr.

Answer
Using Dalton's law, we calculate

$$P_{total} = P_{H_2O} + P_{H_2}$$
$$P_{H_2} = 725 \text{ torr} - 18.7 \text{ torr} = 706.3 \text{ torr}$$

Using the ideal gas law, we then calculate

$$PV = nRT$$

$$n = \frac{706.3 \text{ torr}(1 \text{ atm}/760 \text{ torr})(5.00 \times 10^{-3} \text{ L})}{(0.0821 \text{ L atm}/\text{K mol})(294 \text{ K})} = 1.93 \times 10^{-4} \text{ mol H}_2$$

From the reaction, we see that two moles of electrons are required for each mole of hydrogen that is produced. Thus,

$$(1.93 \times 10^{-4} \text{ mol H}_2)(2 \text{ mol e}^-/\text{mol H}_2) = 3.86 \times 10^{-4} \text{ mol e}^-$$

We can now calculate the approximate value of Avogadro's number N by taking the ratio of the two results (see example E9.1):

$$N = 2.25 \times 10^{20} \text{ e}^-/3.86 \times 10^{-4} \text{ mol e}^- = 5.83 \times 10^{23} \text{ e}^-/\text{mol}$$

Faraday's constant is the number of coulombs per mol of electrons:

$$F = (36.0 \text{ C})/(3.86 \times 10^{-4} \text{ mol e}^-) = 9.33 \times 10^4 \text{ C/mol e}^-$$

Example E9.3   An aqueous solution of KI is electrolyzed in a cell using Pt electrodes. The half reactions are

$$2 \text{ I}^-(\text{aq}) \rightarrow \text{I}_2(\text{aq}) + 2\text{e}^-$$
$$2 \text{ H}_2\text{O}(\text{l}) + 2 \text{ e}^- \rightarrow \text{H}_2(\text{g}) + 2 \text{ OH}^-(\text{aq})$$

If a current of 9.00 mA is passed through the cell for 12.0 minutes, how many grams of I_2 can be produced? (Use 1 faraday $= 9.65 \times 10^4$ C/mol e$^-$.)

Answer

$$\#\text{Coul} = (9.00 \times 10^{-3}\text{A})(12.0 \text{ min})(60 \text{ s/min}) = 6.48 \text{ C}$$

The reaction shows that two moles of electrons are produced with each mole of I_2, so

mol e$^-$ = $(6.48\text{C})(1 \text{ mol e}^-/96, 500 \text{ C}) = 6.72 \times 10^{-5}$ mol e$^-$

g I$_2$ = $(6.72 \times 10^{-5}$ mol e$^-)(1$ mol I$_2/2$ mol e$^-)(254$ g I$_2/$ mol I$_2) = 8.53 \times 10^{-3}$ g I$_2$

Electrolysis of an Aqueous Solution of KI in the Presence of Acetone

At room temperature and pressure, water does not dissociate spontaneously to form hydrogen and oxygen. In order to get this reaction to occur, energy must be added. The current passed in an electrolysis *could* be the source of that energy, except for the fact that pure water is a poor conductor of electricity. Salt water, however, is a much better conductor of electricity. In this experiment, the salt KI is added to water to generate the ions K^+ and I^-. When an aqueous solution of KI is electrolyzed, the following reduction half reactions are possible at the cathode:

$$K^+(aq) + e^- \rightarrow K(s) \qquad\qquad E° = -2.92 \text{ volt}$$
$$2\,H_2O(l) + 2e^- \rightarrow H_2(g) + 2\,OH^-(aq) \qquad E° = -0.83 \text{ volt}$$

A higher potential indicates a more favorable reaction. Since the potential for H_2O reduction is higher, water will be reduced in preference to K^+ ion to generate H_2 gas and OH^-.

At the anode, the following oxidation half reactions are possible:

$$2\,I^-(aq) \rightarrow I_2(s) + 2\,e^- \qquad\qquad E° = -0.54 \text{ volt}$$
$$2\,H_2O(l) \rightarrow O_2(g) + 4\,H^+(aq) + 4\,e^- \qquad E° = -1.23 \text{ volt}$$

Here, the potential for the formation of I_2 is more positive. This suggests the formation of solid iodine.

The half reactions and the overall cell reaction are therefore as follows:

At cathode	$2\,H_2O(l) + 2\,e^- \rightarrow H_2(g) + 2\,OH^-(aq)$
At anode	$2\,I^-(aq) \rightarrow I_2(s) + 2\,e^-$
Overall	$2\,I^-(aq) + 2\,H_2O(l) \rightarrow H_2(g) + I_2(s) + 2\,OH^-$

Thus, we can see that hydrogen gas and iodine solid are produced by the electrolysis of KI solution, and the solution will become basic due to the formation of OH^-. Iodine generated reacts with the excess KI to form KI_3, which is soluble in water.

$$I_2(s) + KI(aq) \rightarrow KI_3(aq)$$

In the presence of OH^-, I_2 (KI_3) reacts with acetone to form iodoform, CHI_3, which, being insoluble, is collected in the cell (see Experiment 32, part B: preparation of iodoform) as a yellow powder.[1,2] This is the well-known iodoform test for methyl ketones.[3]

A diagram of an electrolytic cell describing the electrolysis of an aqueous solution of KI is shown in Figure E9.1a and b. The cell consists of a pair of nichrome (preferably platinum, Pt) wire electrodes (enclosed in Beral pipet tips) immersed in an aqueous KI solution containing acetone. The electrodes are connected to a DC source (such as a 12 V battery).

Note: The experiment may be carried out without using an ammeter. The number of moles of I_2 produced by the electrolysis can be found from the moles of H_2 gas formed.

References

1. Helle, K.; Rijks, J. A.; Janssen, L. J. J.; Schuyl, J. W. *J. Chem. Educ.* **1969,** *46,* 518.
2. Singh, M. M.; Pike, R. M., Szafran, Z.; Davis, J. D.; Leone, S. A. *J. Chem. Educ.* 1994.
3. Mayo, D. W.; Pike, R. M.; Trumper, P. K. *Microscale Organic Laboratory,* 3d ed.; Wiley: New York, 1994.

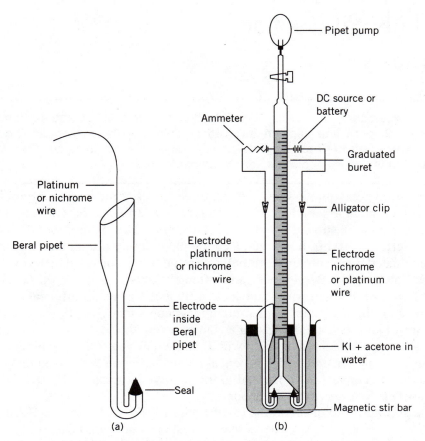

Figure E9.1 (*a*) Nichrome (or platinum) wire sealed inside a Beral pipet; (*b*) electrolytic cell

Experimental Section

Procedure

Microscale experiment   Estimated time to complete the experiment: 2 h
Experimental Steps: Three steps are involved: (1) setting up the apparatus, (2) collecting H_2 gas by electrolysis of KI solution in an aqueous solution of acetone, and (3) calculating either the Faraday constant or Avogadro's number.

Procure some solid potassium iodide, acetone, three pieces of nichrome (or preferably platinum, Pt) wire of suitable length, a 50 mL beaker or a large test tube fitted with a three-hole stopper, a stir bar, a stirring motor, a DC source or a 12V battery (or two 6V batteries connected in series), a 10 mL buret with a cotton plug loosely fitted in the mouth or a gas-collecting graduated tube (eudiometer tube) with a cotton plug inserted in the mouth, a pipet pump, a thermometer, two Beral pipets, and a funnel that will fit inside the beaker.

Place 30 mL of distilled water and about 1.66 g of KI in a 50 mL beaker (or a large-size test tube carrying a stopper) containing a magnetic stirring bar. Stir the solution on a magnetic stirrer to dissolve the KI. Using a clamp, secure the beaker to a ring stand.

Insert one end of a nichrome (or Pt) wire (enclosed in a cut-off Beral pipet except for a 1 cm section at the bottom end, which is bent as shown in Figure E9.1a) through the stopper into the beaker and connect the other end to the positive terminal of the DC source via an alligator clip. This wire serves as the anode.

Place a second piece of nichrome (or Pt) wire (again enclosed in a Beral pipet, see Figure E9.1a) in the other hole of the stopper so that one end is in the beaker and the other end is connected to a terminal of the ammeter via an alligator clip. This is the cathode of the cell. Position both electrodes under the lip of an inverted funnel as shown in the figure. Make sure the bare portion of the wire of each electrode is completely inside the funnel lip **but do not allow them to touch**. If they are not positioned correctly, some of the generated hydrogen gas may escape.

Lower the 10 mL buret (open end down with stopcock open) with the cotton plug over the stem of the funnel as shown in the figure. **If a battery is used as the DC source in the electrolytic cell, be sure *not* to complete the circuit until you are ready to begin the experiment.**

Add 0.50 mL of acetone to the reaction solution and with stirring, thoroughly mix the solution. Place a pipet pump on the buret tip (or a rubber bulb) and draw this solution into the buret so as to fill it up past the graduation marks. Close the stopcock. Remove the pump (or the bulb), and open the stopcock slowly to let the liquid level fall to the point where graduations begin on the buret.

Connect a **third wire** to the negative side of the DC source and to the other terminal of the ammeter. This completes the electrolytic cell (see Figure E9.1a and b). As soon as you turn on the DC source or complete the circuit using two 6V batteries in series or one 12V battery, record the time. Also, record the ammeter reading (a 9V battery may be used, but it will slow down the reaction). At this point, the evolution of hydrogen gas begins.

During the electrolysis, be careful not to move the electrodes since doing so may cause the current to fluctuate. If the current is not steady, record an average value. Continue the electrolysis until 6 to 9 mL of hydrogen gas has collected (approximately 15 to 30 minutes depending upon voltage of the battery used). Record the time when you stop the reaction.

Measure and record the volume of hydrogen gas collected. During electrolysis a yellow solid (iodoform) is also formed. The cotton plug will not allow the solid to enter the buret.

Measure the height of the water column in the buret above the surface of the solution in the beaker. Measure the temperature of the solution, and obtain the barometric pressure and the vapor pressure of water at the solution temperature. Record these values as well. If time is available, repeat the experiment. Rinse and clean the apparatus and return to your instructor.

Use the following relationships to find the pressure due to the dry H_2 gas:

$$P_{H_2} = P_{barometric} - P_{water\ column} - P_{water\ vapor}$$
$$P_{water\ column} = (\text{height of water, mm})/(13.6\ \text{mm water/torr})$$

The same setup can be used to synthesize iodoform (see Experiment 32). If instructed to collect the yellow product (iodoform), disconnect the cell, filter the residue, and collect the product on a Hirsch funnel. Dry the product and determine its melting point.

Disposal of Chemicals: Dispose of the solids in designated containers. Collect the liquid from the cell in a separate container.

Additional Independent Projects

1. Electrolysis using other organic compounds such as ethanol, 2-propanol, and methyl ethyl ketone.[1–3]

   Brief Outline

   Instead of acetone, other organic compounds such as ethanol and 2-propanol may be used in this reaction. Perform the electrolysis using these compounds. Determine the percent yield of the product. Develop the data tables and submit a written report according to the instructions given in Chapter 4.

2. Electrolysis (*a*) at various pH values and (*b*) using other metallic wires as electrodes.

3. Determination of pH in the electrolytic cell.[4]

   Brief Outline

   One of the byproducts produced in this experiment is acetic acid, which may react with OH⁻ produced. Determine the pH of the solution at different time intervals. Using a standardized solution of thiosulfate, one can also determine the amount of iodine left unreacted.

   Develop the data tables and submit a written report according to the instructions given in Chapter 4.

References

1. Helle, K.; Rijks, J. A.; Janssen, L. J. J.; Schuyl, J. W. *J. Chem. Educ.* **1969**, *46*, 518.
2. Singh, M. M.; Pike, R. M.; Szafran, Z. 13th Biennial Conference on Chemical Education, Bucknell University, Lewisburg, PA, 1994.
3. Singh, M. M.; Pike, R. M., Szafran, Z.; Davis, J. D.; Leone, S. A. *J. Chem. Educ.* (in press).
4. See Experiment 28 for details.

Name ______________________ Section ______________

Instructor ______________________ Date ______________

PRELABORATORY REPORT SHEET—EXPERIMENT 9

Experiment Title ____

____

Objective

Reactions/formulas to be used

Materials and equipment table

Outline of procedure

1. What gas is being generated at the anode? _____

   What process is occurring at the anode? _____

   What process is occurring at the cathode? _____

2. How many moles of electrons flow through a light bulb that draws a current of 3.0 A in a period of 1.5 minutes?

3. How many faradays of electricity are involved in each of the following electro-chemical reactions?

   (a) 0.500 mol of Br_2 are converted to Br^-

   (b) 2.0 L of O_2 at STP are converted to H_2O in acid solution

EXPERIMENT 9   DATA SHEET

Room temperature _____ °C

Collection of Data

Amount of KI added _____ g

Volume of acetone added _____ mL

		Trial 1	**Trial 2**
1.	Initial volume of water in buret, mL	_____	_____
2.	Final volume of water in buret, mL	_____	_____
3.	Volume of H_2 gas collected, mL	_____	_____
4.	Temperature of reactant solution, °C	_____	_____
5.	Height of water column in buret, mm	_____	_____
6.	Barometric pressure, torr	_____	_____
7.	Vapor pressure of water, torr	_____	_____
8.	Time reaction was started	_____	_____
9.	Time reaction was stopped	_____	_____
10.	Elapsed time for reaction, s	_____	_____
11.	Current, A	_____	_____

Determination of Faraday's Constant and Avogadro's Number

Manipulation of Data

1.   Pressure of H_2 gas, atm   _____

2.   Moles of H_2 gas produced, mol   _____

3.   Number of coulombs, C   _____

4.   Faraday's constant, C/mol e^- _____

5.   Avogadro's number, e^-/mol e^- _____

POSTLABORATORY PROBLEMS—EXPERIMENT 9

1. Calculate the percent error in the Faraday constant and in Avogadro's number as calculated in this experiment versus the accepted values.

2. If the amperage in your electrolysis cell were doubled, what effect would this have on the time required to produce the same amount of hydrogen gas that you collected in the experiment?

3. A current of 2.5 amperes was passed through a solution containing gold ions for 20 minutes. Gold metal deposited at the cathode. How many grams of gold metal were collected?

$$Au^+ + e^- \rightarrow Au$$

4. A current of 0.57 A is passed through an electrolytic cell containing molten $CaCl_2$ for 2.0 hours. Calcium metal is deposited at the cathode. Write the electrode reactions and calculate the amount of calcium metal (in grams) formed at the cathode.

10 Acid-Base Titration:
Part A: Analysis of Vinegar
Part B: Analysis of a Carbonate/Bicarbonate Mixture
Micro- and Macroscale Experiments

Objectives
- To learn how to prepare primary and secondary standard solutions
- To become acquainted with the titrimetric methods
- To standardize solutions
- To determine the concentration of an acid that is an active ingredient in a household commercial product
- To analyze a mixture using acid-base titration

Prior Reading
- Chapter 2   Mathematical methods and manipulation of data
- Section 3.4   Weighing
- Section 3.5   Measuring liquid volumes: use of quantitative glassware
- Section 3.9   Solution preparation
- Section 3.10   Construction and use of a microburet
- Section 3.11   Titration procedure

Related Experiments
- Experiments 11 (Equivalent weight), 12 (Redox titration), 13 (K_{sp} by titration), 14 (Complexometric titration), 25 (Ion exchange), and 28 (pH titration)

Activities a Week Before
- Dry a sample of sodium carbonate and store it in a desiccator

A chemist often has to determine the amount of a chemical (**analyte**) in a solution or a mixture. This can be done either by instrumental analysis (such as by using an atomic absorption or visible spectrophotometer) or by volumetric analysis (such as by titration). In **volumetric analysis** the volume of a standard reagent needed to react completely with the analyte is measured. This reagent, called the **titrant**, is a solution of known concentration. It is added in small increments from a buret, to the analyte taken in an Erlenmeyer flask, in a process called **titration**. This method of analysis is based on a quantitative chemical reaction such as

$$t\text{T} + a\text{A} \rightarrow \text{products}$$

where t moles of the titrant T react with a moles of the analyte A. The end of the titration is called the **equivalence point**.

Consider, for example, a reaction where the titrant, permanganate (MnO_4^-) solution (purple), is added to a hot acidic solution of the analyte, oxalate ($C_2O_4^{2-}$), which is colorless:

$$5\,C_2O_4^{2-}(aq) + 2\,MnO_4^-(aq) + 16\,H^+(aq) \rightarrow 2\,Mn^{2+}(aq) + 10\,CO_2(g) + 8\,H_2O(l)$$

Suppose that the analyte solution contains 10 millimoles of the oxalate anion. Since the balanced equation indicates that MnO_4^- and $C_2O_4^{2-}$ react in a 2:5 mole ratio, the equivalence point is reached when exactly 4 millimoles of MnO_4^- have been added.

Ideally, in any titration we want to measure the equivalence point. In actuality, however, what we measure is the **end point** of the titration. The end point is indicated by a sudden change in a physical or chemical property of the solution. In many titrations, an **indicator** is added to the solution being titrated. The indicator changes color at the end point, thereby visually indicating the end of the titration. In some other titrations, the color change of one of the reactants indicates the end point. In still other instances, a change in pH, conductivity, or optical absorbance may mark the end point.

In the foregoing example, as the purple permanganate solution is added in increments to the colorless oxalate solution, the solution remains colorless, indicating that the oxalate reacts with permanganate to form the colorless product (Mn^{2+}). When no more oxalate is left to react with the permanganate, the addition of one additional drop of permanganate will impart a faint purple color to the solution. This indicates the end point of the titration. In this case, since an additional drop of permanganate solution is required to see the color change, the end point is slightly different from the equivalence point (by that one drop). The volume of the titrant used is then recorded. Since the volume of the titrant and its concentration are known, the amount of oxalate consumed can be easily calculated.

The difference between the end point and the equivalence point results in an unavoidable error, called **titration error**. In many instances, one can correct for this titration error by performing a **blank titration**.

In the previous example, a blank titration under identical conditions can be carried out by adding permanganate to a solution containing no oxalate. The volume of permanganate solution needed to generate the same faint purple color of the solution is subtracted from the total volume of permanganate required to titrate the oxalate solution.

There are several requirements for a titration to be successful. The reaction between the analyte and the titrant must be stoichiometric, quantitative, and rapid. There must be no other side reaction, and the reaction must be specific.

Reactions Suitable for Titrations

Four different types of chemical reactions are suitable for the titrimetric method of analysis.

Acid-Base Titrations
Simple acids and bases react with each other, forming a salt and water. This is called a **neutralization reaction**.

$$H^+(aq) + OH^-(aq) \rightarrow H_2O(l)$$

The end point can be easily detected by adding an indicator or by noting the sudden change in pH with a pH meter.

Oxidation-Reduction (Redox) Titrations
Chemical reactions involving oxidation-reduction reactions are widely used in titrimetric methods. Potassium permanganate or dichromate solutions are extensively used as titrants in such reactions (Experiments 12 and 13).

Complexometric Titrations
Certain organic compounds called **chelating agents**, such as ethylenediaminetetraacetic acid (EDTA) or its sodium salt, can form complexes with a number of metal ions and are widely used in titrations (Experiment 14).

Precipitation Titrations
In precipitation titrations, the titrant forms an insoluble product with the analyte. Chloride, for example, can be measured by titrating it with silver nitrate, forming the insoluble silver chloride. An indicator such as chromate anion (Mohr's method) or iron(III)ammonium sulfate (Volhard method, a back titration of Ag^+ with SCN^-) may be used to indicate the end point.

Primary Standards

A standard solution can be prepared by dissolving an accurately weighed sample of a solid to make a known volume of solution. Such standard solutions can be used to **standardize** other **secondary standard** solutions by titration. The solid used to prepare a standard solution is called a **primary standard**. A primary standard should have the following characteristics:

1. It must be available in a state of high purity. Impurities cannot exceed 0.01 to 0.02 percent.
2. The compound must be stable in air and nonhygroscopic.
3. It must have a known composition and must not undergo any change on heating.
4. It should have a reasonably high molecular weight or equivalent weight.

General References
1. Bombi, G. G.; Macca, C. *J. Chem. Educ.,* **1990**, *67*, 1072 and references therein.
2. Day, Jr., R. A.; Underwood, A. L. *Quantitative Analysis,* 5th ed.; Prentice-Hall: Englewood Cliffs, NJ, 1986.
3. Harris, D. C. *Quantitative Chemical Analysis,* 2nd ed.; W. H. Freeman: New York, 1987.
4. Pietrzyk, D. J.; Frank, C. W. *Analytical Chemistry,* 2nd ed.; Academic Press: New York, 1979.

Part A: Determination of Acetic Acid, the Active Ingredient in Household Vinegar

Many food items contain a variety of added chemicals; for example:

"Cola," "Soda," or "Pop": Carbonated water (water in which CO_2 is dissolved), high-fructose corn syrup, phosphoric acid, caffeine, and citric acid

Roast Chicken Breast: Sodium lactate, salt, carrageenan, sodium nitrite, and sodium phosphate

These chemicals have been added to food for some purpose. The companies that produce these items must ensure that the proper quantities of chemical ingredients are used. To do this, the manufacturers have quality control laboratories where analyses are done by analytical chemists. These chemists maintain strict quality control of the products, which must satisfy certain rigid standard specifications. Preparations that do not meet these specifications are either discarded or recycled.

The active ingredient in household vinegar is acetic acid ($C_2H_4O_2$, MW = 60.05 g mol^{-1}). The concentration of acetic acid in household vinegar cannot exceed 4 to 5 percent by weight. In this experiment, you will act as an analytical chemist and determine the amount of acetic acid in a sample of household vinegar.

The percentage of acetic acid in vinegar can be found in the following way:

1. An exact volume of vinegar is taken from the bottle. Convert the volume into mass by using the density ρ of the solution [determined using a micropycnometer (see experiment 1) or given by your instructor].

$$\text{g vinegar} = (V_{\text{vinegar}})(\rho_{\text{vinegar}})$$

2. Using the volume of a prestandardized ~0.10 M NaOH solution added from the buret to react completely with the acid, calculate the number of moles of NaOH used:

$$\text{mol base} = (V_{\text{base}})(M_{\text{base}})$$

3. The stoichiometry of the reaction,

$$C_2H_4O_2(aq) + Na^+(aq) + OH^-(aq) \rightarrow Na^+(aq) + C_2H_3O_2^-(aq) + H_2O(l)$$

tells us that 1 mole of sodium hydroxide reacts with 1 mole of acetic acid, or

$$\text{mol acid} = \text{mol base}$$

4. Convert the moles of acetic acid into mass units, or

$$\text{g acid} = (\text{mol acid})(60.05 \text{ g mol}^{-1})$$

5. Finally, compute the percentage of acetic acid in the vinegar,

$$\% \text{ acid} = 100\left(\frac{\text{g acid}}{\text{g vinegar}}\right)$$

Example E10.1 The density of vinegar was determined using a pycnometer. The empty pycnometer weighed 0.495 g at 23°C. Filled with vinegar, it weighed 0.899 g. Filled with water, it weighed 0.896 g. The density of water at 23°C is 0.9975 g mL^{-1}. A 0.825 mL sample of NaOH was standardized using 0.815 mL of 0.0969 M potassium hydrogen phthalate (KHP). The standardized solution was then used to titrate 0.250 mL of the vinegar. The volume of NaOH needed was 2.205 mL.

(*a*) Calculate the density of the vinegar.
(*b*) Find the concentration of acetic acid in the vinegar in moles per liter, grams per liter, and percent by weight.

Answer

Mass of vinegar = 0.404 g   Mass of water = 0.401 g

$$V_{pyc} = \frac{0.401\,g}{0.9975\,g/mL} = 0.402\,mL \qquad \rho_{vinegar} = \frac{0.404\,g}{0.402\,mL} = 1.005\,g/mL$$

g vinegar = (0.250 mL)(1.005 g/mL) = 0.251 g = 251 mg

mmol KHP = (0.815 mL)(0.0969 mmol/mL) = 7.90×10^{-2} mmol KHP = mmol NaOH

[from the reaction shown below in Eq. (E10.1)]

M (NaOH) = (7.90×10^{-2} mmol NaOH)/(0.825 mL NaOH) = 0.0958(mmol/mL)

mmol of acetic acid = mmol NaOH = (2.205 mL NaOH)(0.0958 mmol/mL) = 0.211 mmol

M (acetic acid) = (0.211 mmol acid)/(0.250 mL vinegar) = 0.844 M

g/L acid = (0.844 mol/L)(60.05 g/mol) = 50.7 g/L acetic acid

g acid = (0.211 mmol acid)(60.05 mg/mmol) = 12.7 mg

$$wt\ \% = 100\left(\frac{12.67\,mg}{251\,mg}\right) = 5.06\%$$

Experimental Section

Procedure

Microscale experiment   Estimated time to complete the experiment: 3 h
Experimental Steps: The procedure involves four steps: (1) preparation of a ~0.1 M NaOH solution, (2) standardization of the base, (3) titration of the unknown vinegar solution with NaOH, and (4) determination of the density of vinegar.

Part A: Analysis of Vinegar

Procure a 250 mL plastic bottle, a pair of 2 mL burets, a pipet pump, a 10 mL Erlenmeyer flask, a 25 mL volumetric flask, a 10 mL graduated cylinder, a funnel, micro stir bar, a micropycnometer (see Section 3.10), a stirring motor, 1 percent and 0.25 percent phenolphthalein solutions, micro-tipped Pasteur pipet, potassium hydrogen phthalate (KHP), NaOH, and samples of two different brands of vinegar. Enter the brand name(s) on the data sheet, page 2.

Preparation of ~0.1 M Sodium Hydroxide Solution

Prepare a ~0.1 M solution of NaOH by adding 1.0 g of solid NaOH to 250 mL of deionized water in a labeled *plastic* bottle (alternatively, dilute ~2 mL of 50 percent supplied NaOH solution to 250 mL). Stopper the bottle and shake it well. The bottle should be kept stoppered, as sodium hydroxide absorbs CO_2 from the atmosphere.

CAUTION: NaOH is corrosive—handle the chemical with care. In case of skin contact, wash the area with copious amounts of water and report to your instructor.

Standardization of Sodium Hydroxide Solution with KHP

The reaction between NaOH and potassium hydrogen phthalate (KHP), [(HOOC)C_6H_4(COOK), MW = 204.23 g/mol] is

$$NaOH + HOOC-C_6H_4-COOK \rightarrow NaOOC-C_6H_4-COOK + H_2O \qquad (E10.1)$$

Using 2 mL pipets, construct (see Section 3.10) a pair of 2.00 mL graduated microburets and clamp onto a buret clamp. Label one microburet "acid" and the other "base." Make sure that the microburets do not leak. Before use, rinse a 25 mL volumetric flask and a 10 mL Erlenmeyer flask with deionized water. Place the micro stir bar in the Erlenmeyer flask.

Accurately weigh ~525 mg of KHP (a primary standard acid) and using a funnel, transfer it quantitatively to a 25 mL volumetric flask. Half-fill the flask with deionized water, washing down the sides of the funnel. The preparation of this solution must be very accurate and precise.

Shake the volumetric flask well to dissolve all the KHP. Remove the funnel, and add four drops of 1 percent phenolphthalein to the solution. Fill the flask to the mark with deionized water and mix the solution. This is your KHP primary standard solution. Label the flask. Calculate the exact molarity of the primary standard solution and record the value on your data sheet.

Rinse the base buret with the ~0.1 M NaOH three times and refill it with the base. Record the initial volume of NaOH on the data sheet. Similarly, rinse and fill the acid buret with the KHP solution. Record the initial volume of the KHP solution on the data sheet.

For details of the titration procedure, see Section 3.11. From the acid buret, add 1.00 mL (or a little more, but never less than 1.00 mL) of the KHP solution to the 10 mL Erlenmeyer flask containing a micro magnetic stir bar. Record the exact volume on your data sheet. Using a 10 mL graduated cylinder, add 1 mL of water to this solution and place it on a magnetic stirring hot plate (if not available, mix the solution manually). *Place a white sheet of paper under the flask for color contrast.* Dropwise, add the NaOH solution from the buret to the KHP solution. Stir the solution or swirl the flask after each drop is added. Continue to titrate the acid until the end point is reached. This is indicated by the appearance of a *faint* pink color of phenolphthalein that persists for 30 seconds or more. A dark pink/red color indicates that you have passed the end point. In that event, repeat the titration. Record the final buret reading on the report sheet. Calculate the molarity of the NaOH solution, and record it on the data sheet. Repeat the procedure twice (or more if time permits), and calculate the average molarity of the base.

Titration of Vinegar with NaOH

Rinse the 10 mL Erlenmeyer flask thoroughly with deionized water. Rinse the acid buret with deionized water at least two times. Then, rinse it twice with one of the vinegar solutions.

Fill the acid buret with the vinegar and record the initial volume on the data sheet. Fill the base buret with the standardized NaOH. Record the initial volume of NaOH on your data sheet.

Transfer no more than ~0.250 mL of the acid from the buret to the flask. Record the exact volume of the acid transferred on the data sheet. Using a graduated cylinder, add 2 mL of deionized water to the acid solution. Add one drop of 0.25 percent phenolphthalein indicator (use a *fine-tip* Pasteur pipet to minimize the size of the drop) and add a micro stir bar.

Following the same procedure just described, titrate the vinegar solution in the flask by adding the standardized NaOH solution from the base buret. The end point is indicated by the persistent faint pink color of the solution in the flask. Record the final volume of NaOH added from the buret on the data sheet. Repeat the titration once more. Record the volume of NaOH for the second titration.

Repeat the titration with the second brand of the vinegar, entering all data on the data sheet. **Return the rinsed and cleaned burets and other equipment/glassware to your instructor before you leave the laboratory.**

Determination of Density of Vinegar

Using a micropycnometer (see Experiment 1), determine the density of the vinegar. Enter your data in the data table.

Calculate the concentration of acetic acid in both brands of vinegar in terms of moles per liter, grams per liter, and finally weight percent. From the cost of the bottle, determine which brand is the better buy.

Alternative Procedure

Macroscale method	Estimated time to complete the experiment: 4 h

The general titration procedure for this method is the same as that for the microscale method. Weigh accurately 204 ± 1 mg KHP; transfer it quantitatively to a 125 mL Erlenmeyer flask; add 20 mL of water and a drop of 1 percent phenolphthalein; and titrate the standard solution with NaOH solution delivered from a 50 mL buret as described before (see Section 3.11).

Repeat the titration at least three times. Using the stoichiometry of reaction (E10.1), calculate and record the average molarity of NaOH solution on your data sheet.

Obtain samples of two different brands of vinegar. Using a graduated pipet, transfer 1.00 mL of the original vinegar into the Erlenmeyer flask, add 20 mL deionized water and one drop of 1 percent phenolphthalein indicator. Titrate the acid with the standardized NaOH solution according to the procedure described above. Record the final volume of NaOH. Repeat the titration once more. Now, titrate the other brand of the acid in a similar manner. Calculate the concentration of acetic acid in both brands of vinegar in terms of moles per liter, grams per liter, and finally weight percent. From the cost of the bottle, determine which brand is the better buy.

Part B: Analysis of a Mixture of Soluble Carbonate and Bicarbonate

In the first part of this experiment, a solution of hydrochloric acid is standardized using a primary standard solution of sodium carbonate. The neutralization reactions that occur are as follows:

$$1. \qquad CO_3^{2-}(aq) + H^+(aq) \rightarrow HCO_3^-(aq) \qquad (E10.2)$$

$$2. \qquad HCO_3^-(aq) + H^+(aq) \rightarrow H_2CO_3(aq) \qquad (E10.3)$$

$$3. \qquad H_2CO_3(aq) \rightarrow H_2O(l) + CO_2(g) \qquad (E10.4)$$

Adding 1 and 2: 4. $\qquad CO_3^{2-}(aq) + 2\,H^+(aq) \rightarrow H_2CO_3(aq)$	(E10.5)

Since two protons are neutralized by sodium carbonate, two end points are detected in this titration. One end point occurs at pH ~ 8.0 (step 1, detected by phenolphthalein indicator) and the other appears at pH 3.8 to 5.4 (step 2, detected by bromcresol green indicator). For bromcresol indicator to work, the solution is boiled to drive off any dissolved CO_2 as shown in step 3.

In the second part of the experiment, an unknown carbonate is titrated with a standardized HCl solution. Total alkalinity (bicarbonate + carbonate) is determined by titrating the mixture with HCl (steps 5 and 6).

5. $HCO_3^-(aq) + H^+(aq) \rightarrow H_2CO_3(aq)$ (E10.6)

6. $CO_3^{2-}(aq) + 2H^+(aq) \rightarrow H_2CO_3(aq)$ (E10.7)

A second portion of the unknown is then treated with excess standard NaOH to convert any bicarbonate to carbonate (step 7). The carbonate is then precipitated by adding a solution of $CaCl_2$ (step 8), and the excess NaOH is then back-titrated with standard HCl.

7. $OH^-(aq) + HCO_3^-(aq) \rightarrow CO_3^{2-}(aq) + H_2O(l)$ (E10.8)

8. $Ca^{2+}(aq) + CO_3^{2-}(aq) \rightarrow CaCO_3(s)$ (E10.9)

This step will determine the amount of bicarbonate present in the mixture. From the total alkalinity and the bicarbonate concentration, one can find the amount of the carbonate present in the unknown. The indicator selected will change color before HCl reacts with $CaCO_3$.

Example E10.2   Suppose 110 mg of sodium carbonate, Na_2CO_3 (MW = 106 g/mol), is present in 10 mL of solution. A 2.00 mL sample of this solution is titrated with HCl to the end point, requiring 4.00 mL of the acid. Calculate the molarity and the normality of HCl solution.

Answer

$mmol\ Na_2CO_3 = (110\ mg)/(106\ mg/mmol) = 1.04\ mmol$

$M\ (Na_2CO_3) = (1.04\ mmol)/(10\ mL) = 0.104\ M$

$mmol\ Na_2CO_3\ consumed = (2.00\ mL)(0.104\ mmol/mL) = 0.208\ mmol\ Na_2CO_3$

Since every millimole of Na_2CO_3 neutralizes two millimoles of H^+ [reaction (E10.5)],

$mmol\ HCl\ used = 2(mmol\ Na_2CO_3) = 2(0.208\ mmol) = 0.416\ mmol\ HCl$

$M\ (HCl) = (0.416\ mmol)/(4.00\ mL\ HCl) = 0.104\ M\ HCl$

$N\ (HCl) = (number\ of\ eq\ of\ H^+/mol\ HCl)(M, HCl)$

$N\ (HCl) = (1\ eq\ H^+/mol\ HCl)(0.104\ mol/L, HCl) = 0.104\ N$

Alternative method (see Section 3.9)

Since Na_2CO_3 reacts with two equivalents of HCl,

$EW\ (Na_2CO_3) = (106\ g/mol)/(2\ eq\ H^+/mol) = 53.0\ g/eq = 53.0\ mg/meq$

Number of meq (Na_2CO_3) = (110 mg)$/$(53.0 mg/meq) = 2.08 meq

$$N (Na_2CO_3) = (2.08 \text{ meq})/(10 \text{ mL}) = 0.208 \text{ meq/mL} = 0.208 \text{ N } Na_2CO_3$$
$$(N, Na_2CO_3)(V, Na_2CO_3) = (N, HCl)(V, HCl)$$
$$(0.208 \text{ N})(2.00 \text{ mL}) = (N, HCl)(4.00 \text{ mL})$$

N (HCl) = 0.104 N

Example E10.3 A 1.00 g sample of a mixture of Na_2CO_3, $NaHCO_3$, *and some impurity* is dissolved in a 50 mL volumetric flask. One mL of this solution is titrated with 0.113 M HCl. The volume of HCl needed to reach the bromcresol green end point is 1.25 mL. A second 1-mL aliquot of this solution is then treated with 2.00 mL of 0.100 M NaOH (an excess). The carbonate is removed by adding $CaCl_2$. The excess NaOH is then titrated with 0.113 M HCl. The volume of HCl needed is 0.850 mL. Calculate the weight percent of Na_2CO_3, $NaHCO_3$, and impurity in the sample.

Answer

Total mmol HCl used = (1.25 mL)(0.113 mmol/mL) = 0.141 mmol HCl

mmol HCl to titrate excess NaOH = (0.850 mL)(0.113 mmol/mL)
$$= 0.0961 \text{ mmol HCl}$$

So, there were 0.0961 mmoles of excess NaOH.

Total mmol NaOH added = (2.00 mL)(0.100 mmol/mL) = 0.200 mmol NaOH

mmol NaOH used in converting $NaHCO_3$ to Na_2CO_3 = (0.200 − 0.0961) mmol
$$= 0.104 \text{ mmol NaOH}$$

Since NaOH reacts with $NaHCO_3$ in a 1:1 ratio to form Na_2CO_3, there must have been 0.104 mmol of $NaHCO_3$ (step 7) [reaction (E10.8)].

Total mmol HCl used = (mmol to titrate $NaHCO_3$) + (mmol to titrate Na_2CO_3)

0.141 mmol HCl = (0.104 mmol to titrate $NaHCO_3$) + (mmol to titrate Na_2CO_3)

mmol HCl used to titrate Na_2CO_3 = (0.141 − 0.104) mmol = 0.037 mmol HCl

Since Na_2CO_3 reacts with HCl in a 1:2 ratio [reaction (E10.5)],

mmol Na_2CO_3 = mmol HCl$/2$ = 0.037 mmol$/2$ = 0.019 mmol Na_2CO_3

Mass of $NaHCO_3$ in 1 mL = (0.104 mmol)(84.0 mg/mmol) = 8.74 mg/mL

Mass of $NaHCO_3$ in original sample (in 50 mL) = (50 mL)(8.74 mg/mL) = 437 mg

% NaHCO$_3$ = 100(437 mg/1000 mg sample) = 43.7%

Mass of Na$_2$CO$_3$ in 1 mL = (0.019 mmol)(106 mg/mmol) = 2.0 mg/mL

Mass of Na$_2$CO$_3$ in original sample (in 50 mL) = (50 mL)(2.0 mg/mL) = 100 mg

% Na$_2$CO$_3$ = 100(100 mg/1000 mg sample) = 10%

% impurity = 100% − (43.7 + 10)% = 46.3%

General References

1. Day, Jr., R. A.; Underwood, A. L. *Quantitative Analysis,* 5th ed.; Prentice-Hall: Englewood Cliffs, NJ, 1986, pp. 685–690.
2. Pietrzyk, D. J.; Frank, C. W. *Analytical Chemistry,* 2nd ed.; Academic Press: New York, 1979.
3. Skoog, D. A.; West, D. M. *Analytical Chemistry: An Introduction,* 4th ed.; Saunders College Publishing, New York, 1986.
4. Szafran, Z.; Pike, R. M.; Foster, J. C. *Microscale General Chemistry Laboratory with Selected Macroscale Experiments,* Wiley: New York, 1993, p. 243.

Experimental Section
Procedure

> Microscale method   Estimated time to complete the experiment: two labs of 2.5 h
> Experimental Steps: This procedure involves four steps: (1) standardizing $\sim$0.1 M HCl solution with sodium carbonate, (2) standardizing 0.1 M NaOH solution, (3) titrating an unknown solution containing carbonate and bicarbonate with standardized HCl solution, and (4) determining bicarbonate by back-titrating NaOH after separating carbonate as $CaCO_3$.

Part B: Analysis of a Carbonate/Bicarbonate Mixture

A week before, dry a sample of Na_2CO_3 at 160°C for 1 h.

Standardization of Hydrochloric Acid Solution with Sodium Carbonate

Obtain a 25 mL volumetric flask, a pair of 2 mL burets, a 2 mL graduated pipet, a pipet pump, a funnel, $\sim$0.1 M HCl solution, a dry sample of Na_2CO_3, two 10 mL Erlenmeyer flasks, a micro stir bar (see Section 3.10), a stirring motor, 1 percent phenolphthalein, and 1 percent bromcresol green. Obtain a sample of unknown and record its number.

Place an accurately weighed sample ($\sim$265 mg) of dried Na_2CO_3 in a labeled 25 mL volumetric flask, using a funnel for transferring the solid. Boil 100 mL of deionized water and then cool it to room temperature. This step removes dissolved CO_2. Half-fill the flask with this water and dissolve the solid. Add 2 drops of 1 percent phenolphthalein and fill the flask to the mark with deionized water and shake it well. The solution will be pink. Calculate the molarity and enter the value on the data sheet.

For details about the titration procedure, see Section 3.11. Set up two 2 mL microburets. Label one of them as the "acid" buret and the other as the "base" buret. Rinse and fill the acid buret with the HCl solution. Similarly, fill the base buret with the standard Na_2CO_3 solution. Record the initial volume of the acid and the base in each buret. Transfer 1.00 mL of the Na_2CO_3 solution to a 10 mL Erlenmeyer flask containing the stir bar. While stirring, slowly titrate the solution with the HCl solution to the end point (pink solution turns colorless). At this point, about half (actually a slight excess) of the total amount of HCl necessary to titrate the base solution has been added [this can be viewed as having converted the Na_2CO_3 to $NaHCO_3$, reaction (E10.2)]. Estimate the volume of HCl needed to reach the final end point (twice as much). You would actually need slightly less than the estimated volume of the acid.

Add 2 drops (fine-tip pipet) of bromcresol green indicator to the solution. Continue to titrate the solution adding the acid dropwise until a blue-green color is obtained.

> CAUTION: Avoid adding too many drops of the acid. See below.

Stop the titration at this point, and boil the sample (under constant stirring) for 2 to 3 minutes to expel any dissolved CO_2 gas. If necessary, add enough freshly boiled deionized water to make up the loss of water. After boiling, the solution should turn blue again (if it does not, you have added too much acid; in that event, repeat the procedure). Cool the solution to room temperature. Continue to add HCl from the buret dropwise until the blue solution turns *faint* yellow-green.

> CAUTION: Only a drop or two of HCl are needed to see the indicator color change.

Record the volume of HCl on your data sheet. Repeat the procedure at least two more times. Calculate the molarity of the acid in each case and determine the average. Also find the normality of the acid.

Titration of a Mixture of Carbonate and Bicarbonate

The NaOH and HCl solutions prepared and standardized in parts A and B respectively will be used here. Set up two microburets (labeled "acid" and "base"). Obtain two 25 mL Erlenmeyer flasks, a 10 percent solution of $CaCl_2$, a 25 mL volumetric flask, a graduated pipet, a 10 mL graduated cylinder, a micro stir bar, and a magnetic stirring hot plate. Record your unknown number in the report sheet. Rinse and fill the acid buret with 0.100 M HCl. Similarly, rinse and fill the base buret with the 0.100 M NaOH solution (see part A).

If solid, the unknown must be dried and stored in a desiccator before the laboratory begins. Weigh 200 to 300 mg of the unknown (your instructor will tell you how much) and transfer the sample directly into the 25 mL volumetric flask, using a funnel. Add freshly **boiled and cooled** (to remove dissolved CO_2) water through the funnel to the flask. Gently swirl the flask to dissolve the sample. Rinse down the sides of the funnel and the neck of the flask. When the sample has dissolved completely, add enough water to fill to the mark. Stopper the flask and mix well.

If the unknown is a solution, transfer it *quantitatively* into the 25 mL volumetric flask (using a funnel), rinse the sample container with 2 mL of water at least twice, and transfer the washings into the flask. Fill the volumetric flask up to the mark and shake well.

Titrations for Total Alkalinity

Rinse the pipet with the unknown solution and transfer 2 mL of this solution into a 25 mL Erlenmeyer flask. Add 1 mL of deionized water (using a graduated cylinder) to the flask. Add two drops of bromcresol green indicator to the Erlenmeyer flask, and titrate the solution with the standardized HCl solution from the acid buret as described before (standardization of hydrochloric acid solution with sodium carbonate). You may have to refill the acid buret to complete the titration. (If one buretful of acid is not enough to complete the titration, a better alternative is to take less than 2 mL of the unknown solution and repeat the titration.) Repeat this procedure two more times. Wash the Erlenmeyer flask with water before the next titration.

Titrations for NaHCO$_3$

Pipet 2 mL (or a smaller volume) of the unknown solution into a clean 25 mL Erlenmeyer flask. Add accurately 5.00 mL of the standardized NaOH solution to the flask from the base buret. Swirl to mix the contents completely and add 0.5 mL of 10 percent $CaCl_2$ solution to the flask, using a graduated cylinder. Stir the solution thoroughly to precipitate solid $CaCO_3$.

Add one drop of 0.25 percent phenolphthalein solution (from a fine-tip Pasteur pipet) and titrate immediately with the standardized HCl solution from the acid buret. Repeat the procedure two more times.

Record all results on the data sheet. Calculate the total alkalinity from the first set of titrations. From the second set of titrations, calculate the excess amount of NaOH left after the reaction. From this, obtain the amount of $NaHCO_3$ converted into Na_2CO_3. Using this value and the total alkalinity, compute the amount of Na_2CO_3 that has reacted with the acid. Find the percentage of $NaHCO_3$ and Na_2CO_3 in the original unknown. Example E10.3 gives a sample calculation.

Average the three determinations, and find the standard deviation. Using the rejection quotient (Table 2.2), determine if any of the individual data should be rejected.

Alternative Procedure

Macroscale method	Estimated time to complete the experiment: two labs of 3 h

Dispose of all the titrated solutions according to the directions given by your instructor. Leftover chemicals and solutions must be returned to your instructor.

The general procedure is the same as that for the microscale method. Obtain two 50 mL burets, label them "acid" and "base," and clamp them to a buret stand. Also obtain a 10 mL pipet, a 10 percent solution of $CaCl_2$, 1 percent phenolphthalein indicator, a 100 mL volumetric flask, and a 125 mL Erlenmeyer flask, as well as an unknown (liquid or solid sample). Procure 250 mL of ~0.1 M NaOH solution. Standardize the 0.1 M NaOH solution (Part A). Also, prepare and standardize 250 mL of 0.1 M HCl solution (see the following paragraph). Rinse and fill the base buret with the NaOH solution and the acid buret with the HCl solution. Make sure that there is no air lock in the stopcock and that the tip of the buret is full of the solution. If the unknown is a solid, dry it at 110°C for two hours.

Weigh ~0.106 ± 0.001 g of Na_2CO_3 (previously dried at 160°C for 1 h) directly into a 125 mL Erlenmeyer flask. Dissolve it in 10 mL of freshly **boiled and cooled** water (to remove CO_2). Titrate the solution with the HCl following the procedure just described (see part B). Add two drops of 1 percent phenolphthalein indicator for the first end point and two drops of bromcresol green indicator for the second end point.

Repeat the procedure two more times. Calculate the molarity and the normality of the HCl solution.

If the unknown is a solid, weigh 1.0 to 1.5 g (±0.001 g) of the unknown and transfer it quantitatively into a 100 mL volumetric flask. Dissolve the solid in freshly **boiled and cooled** deionized water, then fill the flask to the mark with deionized water and mix well. If a liquid sample is given as the unknown, transfer the liquid into a 100 mL volumetric flask and rinse the original container at least twice, adding the washings to the volumetric flask. Fill to the mark with deionized water, and mix well.

Titration for Total Alkalinity

Pipet 10 mL of the unknown solution into a 100 mL Erlenmeyer flask and titrate with standardized HCl, using bromcresol indicator (two drops) as described in Part B. Record the initial and final volumes of HCl in the acid buret on the data sheet. Repeat this procedure with two more samples of the unknown.

Titration for NaHCO₃

Pipet a 10 mL aliquot of the unknown and 25 mL of standardized NaOH solution into a 125 mL Erlenmeyer flask. Mix the solution well. Add 4 mL of 10 percent (wt/wt) CaCl₂ solution, and stir the solution to precipitate CaCO₃. Add one drop of phenolphthalein indicator (the solution turns pink), and titrate immediately with a standardized HCl solution to a colorless end point. Repeat the procedure with two more samples of the unknown.

Record all results on the data sheet. Calculate the total alkalinity of the unknown sample from the first HCl titration. Calculate the bicarbonate concentration from the second HCl titration. Estimate the concentration of the carbonate. See Example E10.3 for a sample calculation.

Calculate the average of the three determinations and the standard deviation. Using the rejection quotient, determine if any of the individual data should be rejected.

Additional Independent Projects

1. Nonaqueous titration of aspirin in crude aspirin or in an analgesic tablet

   Brief Outline

   Methanol is used as a solvent in this titration. The titrant is 0.01 M sodium methoxide solution prepared by dissolving a known amount of metallic sodium (**CAUTION:** Sodium is a toxic and highly reactive metal. It reacts with water violently, causing fire) in methanol. Reagent grade acetylsalicylic acid is used as the primary standard (0.01 M solution). Bromcresol purple is used as an indicator. Excedrin™ tablets may be used as analgesic tablets, or you may use the aspirin that you prepared in Experiment 32. The titration must be performed under microscale conditions.

   Develop your own data table and submit the report as described in Chapter 4.

2. Writing chemical equations from titration data[1]
3. Determination of bicarbonate in Alka-Seltzer[2]
4. Acid-base titration of a product of photochemical bromination of an alkane[3]

   Brief Outline   **Time required to complete the experiment is 1 h**

   An alkane is a hydrocarbon having the general formula C_nH_{2n+2}. Heptane is a hydrocarbon having the formula C_7H_{16}. In the presence of sunlight (under photolytic conditions), heptane reacts with bromine (Br_2) to form HBr, which can be extracted into an aqueous layer and titrated with standardized 0.1 M NaOH using phenolphthalein indicator.

   $$C_7H_{16} + Br_2 \rightarrow C_7H_{15}Br + HBr$$

   Using a graduated pipet, transfer 4 mL heptane (density = 0.684 g/mL) to a small 15 to 20 mL weighing bottle (or a test tube) carrying a micro stir bar and fitted loosely with a stopper. Add 1.5 to 2.0 mL deionized water followed by ~0.100 ± 0.001 g bromine (d = 3.102 g/mL; MW = 159.8 g/mol; **ask the instructor to add the bromine.** Bromine vapors are corrosive, use a **HOOD**). Once inside the test tube, bromine, being heavy, will settle to the bottom. Replace the stopper and stir the mixture gently. Allow the mixture to stand in a sunny spot (or use a tungsten lamp). Swirl the mixture every 2 to 3 minutes until all the added bromine has reacted. Add one drop of 0.25 percent phenolphthalein, and titrate the HBr collected in the aqueous layer with a standardized 0.1 M NaOH solution. How many theoretical moles of Br_2 have been taken? How many moles of HBr have formed? What is the percent error between the moles of Br_2 taken and the moles of HBr formed?

Develop the data collection and manipulation tables and submit a written report according to the instructions given in Chapter 4.

References

1. State, H. M. *J. Chem. Educ.* **1962,** *39,* 297.
2. Peck, L.; Irgolic, K.; O'Connor, R. *J. Chem. Educ.* **1980,** *57,* 517.
3. Deck, E.; Deck, C. *J. Chem. Educ.* **1989,** *66,* 75.

Name ______________________ Section ______________

Instructor ______________________ Date ______________

PRELABORATORY REPORT SHEET—EXPERIMENT 10

Experiment title ______________________________

and part ______________________________

Objective

Reactions/formulas to be used

Materials and equipment table

Outline of procedure

1. You want to prepare 50 mL of 0.10 M NaOH from exactly 50 percent (wt/wt) NaOH solution. How many milliliters of 50 percent solution are required?

2. What is a primary standard? Can KOH be used as a primary standard? Explain.

ACID-BASE TITRATION

3. An aqueous sample is analyzed and found to contain 4 percent NaOH (4 g in 96 mL H_2O).

   (*a*) Calculate the concentration of the solution in molarity (assume the density of water is 1.00 g/mL and the volume of the solution is 100 mL).

   (*b*) If this solution is neutralized with 1 M sulfuric acid, how many mL of sulfuric acid will be needed?

   (*c*) What is the normality of the NaOH solution?

Name ______________________	Section ______________
Instructor ______________________	Date ______________

EXPERIMENT 10 DATA SHEET

Part A: Analysis of Vinegar

Data Collection and Manipulation Table

1. Preparation of 0.1 M Sodium Hydroxide Solution

1. Mass of NaOH or volume of 50% NaOH taken ______ g or mL

Volume of water added ______ mL

2. Preparation of KHP and Standardization of Sodium Hydroxide Solution

Preparation of KHP Standard Solution (~0.1 M) in 25 mL volumetric flask

2. Initial mass of KHP + container ______ g   Show calculations

3. Final mass of KHP + container after transfer ______ g

4. Mass of KHP taken ______ g

5. Concentration of KHP ______ mol/L

Standardization of 0.1 M NaOH Solution

	Trial 1	**Trial 2**	**Trial 3**
6. Initial volume of KHP	______ mL	______ mL	______ mL
7. Final volume of KHP	______ mL	______ mL	______ mL
8. Volume of KHP taken	______ mL	______ mL	______ mL
9. Initial volume NaOH	______ mL	______ mL	______ mL
10. Final volume NaOH	______ mL	______ mL	______ mL
11. Volume of NaOH used	______ mL	______ mL	______ mL
12. Concentration of NaOH	______ M	______ M	______ M
		Average molarity	______ M
		Average normality	______ N
		Standard dev.	______ M
		Rel. standard dev.	______ %

Show calculations (use a separate sheet if necessary):

DATA SHEET, EXPERIMENT 10 PART A, P. 2

Data Collection Table

Brand 1 name ________

Brand 2 name ________

Molarity of NaOH ______ M

Molar mass of acetic acid: 60.05 g/mol

Density Data

1. Room temperature ______ °C
2. Mass of the pycnometer ______ g
3. Mass of pycnometer + vinegar ______ g
4. Mass of pycnometer + water ______ g

Titration Data

	Brand 1		Brand 2	
	Trial 1	**Trial 2**	**Trial 1**	**Trial 2**
5. Initial acid volume	______ mL	______ mL	______ mL	______ mL
6. Final acid volume	______ mL	______ mL	______ mL	______ mL
7. Initial base volume	______ mL	______ mL	______ mL	______ mL
8. Final base volume	______ mL	______ mL	______ mL	______ mL

Show calculations:

Name	______________________	Section	______________
Instructor	______________________	Date	______________

DATA SHEET, EXPERIMENT 10 PART A, P. 3

Data Manipulation Table

Density Calculation

9.	Mass of vinegar	______	g
10.	Mass of water	______	g
11.	Density of water	______	g/mL (provided)
12.	Volume of water	______	mL
13.	Volume of pycnometer	______	mL
14.	Density of vinegar	______	g/mL

Calculations involving the titration

		Brand 1		**Brand 2**	
		______________		______________	
15.	Acid volume	______ mL	______ mL	______ mL	______ mL
16.	Mass of acid taken	______ g	______ g	______ g	______ g
17.	Base volume	______ mL	______ mL	______ mL	______ mL
18.	Base moles	______ mol	______ mol	______ mol	______ mol
19.	Acid moles	______ mol	______ mol	______ mol	______ mol
20.	Average acid moles		______ mol		______ mol
21.	Acid molarity		______ mol/L		______ mol/L
22.	Acid mass in the aliquot taken		______ g		______ g
23.	g of acid/L		______ g/L		______ g/L

Show calculations:

24. % acetic acid _____ % _____ %

25. Cost of the acid $ _____ $ _____

26. Which is the better buy? Brand name _____

Show all the calculations:

EXPERIMENT 10   DATA SHEET

Part B: Analysis of a Carbonate/Bicarbonate Mixture

Data Collection and Manipulation Table

1. Preparation of 0.1 M Hydrochloric Acid Solution

    1.   Volume of concentrated HCl taken _____ mL   Diluted to a volume of _____ mL

2. Standardization of Hydrochloric Acid with Sodium Carbonate

Preparation of Na_2CO_3 Standard Solution ($\sim$0.1 M) in 25 mL flask

    2.   Initial mass of Na_2CO_3 + container _____ g

    3.   Final mass of Na_2CO_3 + container _____ g

    4.   Mass of Na_2CO_3 taken _____ g

    5.   Concentration of Na_2CO_3 solution _____ mol/L

Show calculations:

Standardization of 0.1 M HCl solution using the standard Na_2CO_3 solution above

Volume of HCl from phenolphthalein step _____ mL _____ mL _____ mL

Estimated total volume of HCl needed _____ mL _____ mL _____ mL

		Trial 1	Trial 2	Trial 3
6.	Volume of Na_2CO_3	_____ mL	_____ mL	_____ mL
7.	Initial HCl volume	_____ mL	_____ mL	_____ mL
8.	Final HCl volume	_____ mL	_____ mL	_____ mL
9.	Volume of HCl used	_____ mL	_____ mL	_____ mL
10.	Concentration of HCl	_____ M	_____ M	_____ M
	Mean molarity	_____ M		
	Mean normality	_____ N		

Show calculations:

3. Titration of a Mixture of Carbonate and Bicarbonate

Data Collection Table

Molarity of HCl _____ mol/L   Sample Number _____

Molarity of NaOH _____ mol/L

If the solid unknown is taken, record the mass taken.

1. Mass of unknown before transfer _____ g

2. Mass of unknown after transfer _____ g

3. Mass of unknown transfer _____ g

First set of Titrations for Total Alkalinity (titration with standardized HCl solution)

4. Volume of unknown taken for each of three trials

    _____ mL _____ mL _____ mL

(Note: the numbers in ( ) indicate the number of times the buret has been filled to complete one acid titration. Preferably, adjust the volume of the unknown.)

5. Trial 1	**(1)**	**(2)**	**(3)**	and so on
Initial buret volume	_____ mL	_____ mL	_____ mL	
Final buret volume	_____ mL	_____ mL	_____ mL	
Volume of HCl used	_____ mL	_____ mL	_____ mL	
Total volume of HCl used			_____ mL	

6. Trial 2	**(1)**	**(2)**	**(3)**	and so on
Initial buret volume	_____ mL	_____ mL	_____ mL	
Final buret volume	_____ mL	_____ mL	_____ mL	
Volume of HCl used	_____ mL	_____ mL	_____ mL	
Total volume of HCl used			_____ mL	

7. Trial 3	**(1)**	**(2)**	**(3)**	and so on
Initial buret volume	_____ mL	_____ mL	_____ mL	
Final buret volume	_____ mL	_____ mL	_____ mL	
Volume of HCl used	_____ mL	_____ mL	_____ mL	
Total volume of HCl used			_____ mL	

DATA SHEET, EXPERIMENT 10 PART B, P. 3

Second Set of Titrations (after adding a known volume of ~0.1 M NaOH)

		Trial 1	Trial 2	Trial 3
8.	Volume of the unknown	_____ mL	_____ mL	_____ mL
9.	Volume of 0.1 M NaOH	_____ mL	_____ mL	_____ mL
10.	Volume of HCl used in titration	_____ mL	_____ mL	_____ mL

(If the micro buret is refilled, record the total volume of HCl in step 10.)

Calculation. Show only for one trial.

Concentration of HCl _____ mol/L   Concentration of NaOH _____ mol/L

First Set of Titrations for Total Alkalinity

1. Total mmol HCl used _____ mmol HCl

Second Set of Titrations for NaHCO$_3$

2.	Total mmol NaOH added	_____ mmol NaOH
3.	mmol HCl needed to titrate excess NaOH	_____ mmol HCl
4.	mmol NaOH used in converting NaHCO$_3$ to Na$_2$CO$_3$	_____ mmol NaOH
5.	mmol of NaHCO$_3$ converted to Na$_2$CO$_3$ (= mmol NaOH)	_____ mmol NaHCO$_3$
6.	mmol of HCl consumed by NaHCO$_3$(= mmol NaHCO$_3$)	_____ mmol HCl
7.	mmol HCl consumed by Na$_2$CO$_3$	_____ mmol HCl
8.	mmol Na$_2$CO$_3$ (half that of HCl)	_____ mmol Na$_2$CO$_3$
9.	Mass of NaHCO$_3$ in the aliquot taken	_____ g NaHCO$_3$
10.	Mass of Na$_2$CO$_3$ in the aliquot taken	_____ g Na$_2$CO$_3$
11.	% NaHCO$_3$ in the mixture	_____ %
12.	% Na$_2$CO$_3$ in the mixture	_____ %

Show your calculation for each step (See Example E10.3): Use a separate sheet.

1. What precautions did you take in setting up a micro or a macro buret?

2. What amount of oxalic acid ($H_2C_2O_4 \cdot 2H_2O$, MW $= 126.0$ g/mol) must be used to prepare 10 mL of 0.130 M oxalic acid solution? This acid has two replaceable acidic protons. What would be the normality of the acid?

3. If you take 50 mL of 5 percent acetic acid, how many millimoles of acetic acid have you taken?

Acid-Base Titration: Determination of the Equivalent Weight of an Acid

Micro- and Macroscale Experiment

Objectives
- To find the equivalent weight of an acid using the acid-base titrimetric method

Prior Reading
- Chapter 2   Mathematical methods and manipulation of data
- Section 3.4   Weighing
- Section 3.5   Measuring liquid volumes: use of quantitative glassware
- Section 3.9   Solution preparation
- Section 3.10   Construction and use of a microburet
- Section 3.11   Titration procedure
- Experiment 10   Acid-base titrations

Activities a Week Before
- Standardization of NaOH solution is suggested.

Related Experiments
- Experiments 10 (Acid-base titration), 12 (Redox titration), 13 (K_{sp} by titration), 14 (Complexometric titration), 25 (Ion exchange), and 28 (pH titration)

We have seen that the molarity of a solution is defined as the number of moles of solute per liter of solution. A second way of stating the concentration is called the **normality**. Normality is defined as the number of equivalents (eq) of solute per liter of a solution, or

$$N = \frac{eq}{L} \tag{E11.1}$$

Just what is an equivalent? Consider the reaction of sulfuric acid with sodium hydroxide:

$$H_2SO_4 + 2\,NaOH \rightarrow Na_2SO_4 + 2\,H_2O$$

One mole of sulfuric acid reacts with *two moles* of sodium hydroxide. Another way of saying this is that one mole of sulfuric acid *is equivalent to* (completely reacts with) two moles of sodium hydroxide. In an acid-base reaction, an **equivalent** is defined as the amount of acid providing one mole of H^+, or the amount of base providing one mole of OH^-. Thus, H_2SO_4 has 2 eq/mol, since one mole of H_2SO_4 provides 2 eq of H^+.

Similarly, NaOH has 1 eq/mol, since it provides one mole of OH^- per mole of NaOH. In oxidation-reduction reactions, an equivalent is defined as the amount of material providing (or requiring) one mole of electrons. Thus, in the half reaction

$$Fe \rightarrow Fe^{3+} + 3e^-$$

iron (Fe) has 3 eq/mol, since one mole of iron provides three moles of electrons.

Just as the molecular weight of a substance is the number of grams in a mole, the **equivalent weight (EW)** of a substance is defined as the number of grams in an equivalent. Thus,

$$EW = \frac{g}{eq} = \left(\frac{g}{mol}\right)\left(\frac{mol}{eq}\right) = \frac{MW}{eq} \tag{E11.2}$$

Combining equations (E11.1) and (E11.2) and rearranging, we have

$$g = N\left(\frac{eq}{L}\right) \times V(L) \times EW\left(\frac{g}{eq}\right)$$

The relationship between molarity and normality is that

$$N = nM$$

where n is the number of H^+, OH^-, or electrons produced by each mole of an acid, base, or an oxidation-reduction reagent, respectively.

Normality units are frequently used in titrations due to convenience of calculations: one equivalent of an acid *will always* react with one equivalent of a base; one equivalent of an oxidizing agent will always react with one equivalent of a reducing agent. Thus, at the equivalence point in any titration,

$$eq_{acid} = eq_{base}$$

and

$$N_{acid}V_{acid} = N_{base}V_{base}$$

The equivalent weight depends upon the reaction that occurs. Consider, for example, the following reactions of the permanganate ion, MnO_4^-:

1. $MnO_4^-(aq) + 8 H^+(aq) + 5 e^- \rightarrow Mn^{2+}(aq) + 4 H_2O(l)$
2. $MnO_4^-(aq) + 4 H^+(aq) + 3 e^- \rightarrow MnO_2(s) + 2 H_2O(l)$
3. $MnO_4^-(aq) + 8 H^+(aq) + 4 e^- \rightarrow Mn^{3+}(aq) + 4 H_2O(l)$

In reaction (1), the MnO_4^- has 5 eq/mol and an EW = $158.03/5$ = 31.606 g/eq.
In reaction (2), the MnO_4^- has 3 eq/mol and an EW = $158.03/3$ = 52.677 g/eq.
In reaction (3), the MnO_4^- has 4 eq/mol and an EW = $158.03/4$ = 39.508 g/eq.

Experimental Section

Procedure

Microscale method   Estimated time to complete the experiment: 2.5 h
Experimental Steps: Three steps are involved in this procedure: (1) preparing ~0.1 M
NaOH, (2) standardizing it using KHP (potassium hydrogen phthalate), and (3) determin-
ing the equivalent weight of an acid by acid-base titration.

**For preparation of a 0.1 M NaOH solution and its standardization with KHP, see
Experiment 10**. Obtain a microburet, two 10 mL Erlenmeyer flasks, a magnetic stirring
hot plate, a micro stir bar (see Chapter 3, section 10), a 10 mL graduated cylinder, and an
unknown acid from your instructor (record its identification number on the data sheet).

Accurately weigh 20 to 30 mg of the *dry* acid and transfer it to the 10 mL Erlenmeyer
flask. Add 1 to 2 mL of water to the acid, and gently swirl the contents to dissolve the solid.
**NOTE: In some cases, the acid may not be completely soluble—it will go into solution
when NaOH is added in the next step**. Add one drop of 0.25% phenolphthalein solution,
using a fine-tipped Pasteur pipet, to the mixture. Rinse and fill the buret with standardized
0.1 M NaOH solution. Record the initial buret reading.

Titrate the acid sample according to the procedure described in Experiment 10, part A.
The end point is indicated by the appearance of a persistent faint pink color of the solution
in the flask. Record the final buret reading. If you need more than one buret full of NaOH,
repeat the titration using a smaller amount of the unknown. Alternatively, you may have to
refill the buret to complete the titration. Every time the buret is refilled, record the initial
and final buret readings. Determine the *total* volume of the base used.

Using fresh samples of the unknown acid, repeat the titration three more times. Calculate the equivalent weight of the acid in each case and compute the average value, standard
deviation and the relative standard deviation. **Rinse and clean all the pieces of glassware
and return them to your instructor**.

Disposal: Dispose of all the solutions according to the directions given by your instructor.

Alternative Procedure

Macroscale method	Estimated time to complete the experiment: 3 h

For preparation of a 0.1 M NaOH solution and its standardization with KHP, see Experiment 10. Obtain a 50 mL buret, a 10 mL pipet, a 125 mL Erlenmeyer flask, and a dropper. Also obtain a sample of the unknown acid from your instructor. Record the sample number on the report sheet.

Accurately weigh 500 mg of the unknown acid and transfer it to the 125 mL Erlenmeyer flask. Add 20 mL of water to the acid, and gently swirl the contents to dissolve the solid. Add one drop of phenolphthalein solution to the mixture. Rinse and fill the buret with standardized 0.1 M NaOH solution. Record the initial buret reading. **NOTE: In some cases, the acid may not be completely soluble—it will go into solution when NaOH is added in the next step.**

Titrate the acid sample according to the procedure described in Experiment 10. The end point is indicated by the appearance of a persistent *slight* pink color of the solution in the flask. Record the final buret reading.

Using fresh samples of the unknown acid, repeat the titration twice more. Calculate the equivalent weight of the acid in each case and compute the average value, standard deviation, and the relative standard deviation.

Additional Independent Project

1. Determine the equivalent weight of potassium permanganate by titrating with standardized oxalic acid solution in the presence of H_2SO_4.[1]

Note

1. See Experiment 12.

Name ______________________________ Section ______________

Instructor ______________________________ Date ______________

PRELABORATORY REPORT SHEET—EXPERIMENT 11

Experiment title and part ______________________________

Objective

Reactions/formulas to be used

Materials and equipment table

Outline of procedure

1. Why do chemists often use normality as the concentration unit in analytical chemistry?

2. What is the equivalent weight of the following species?
   (a) H_3PO_4 in the reaction

   $$2\,NaOH(aq) + H_3PO_4(aq) \rightarrow Na_2HPO_4(aq) + 2\,H_2O$$

   (b) $KMnO_4$ in the reaction

   $$MnO_4^- + 8\,H^+ + 5\,e^- \rightarrow Mn^{2+} + 4\,H_2O$$

3. What is the equivalent weight of an unknown acid, if 0.0200 g of the acid requires 4.44 mL of 0.100 N NaOH to reach the end point?

EXPERIMENT 11   DATA SHEET

Determination of Equivalent Weight of an Acid: Microscale and Macroscale Methods

Data Collection and Manipulation Tables

1. Preparation of 0.1 M Sodium Hydroxide Solution

1.   Mass of NaOH taken _____ g or   Volume of NaOH _____ mL

   volume of 50 percent NaOH taken _____ mL

2. Standardization of Sodium Hydroxide Solution with KHP

Preparation of KHP standard solution (~0.1 M) in 10 mL flask

2.   Initial mass of KHP + container   _____ g

3.   Final mass of KHP + container after transfer   _____ g

4.   Mass of KHP taken   _____ g

5.   Concentration of KHP   _____ mol/L

Show calculations:

Standardization of 0.1 M NaOH Solution using the standard KHP solution above

		Trial 1	Trial 2	Trial 3
6.	Volume of KHP	_____ mL	_____ mL	_____ mL
7.	Initial volume NaOH	_____ mL	_____ mL	_____ mL
8.	Final volume NaOH	_____ mL	_____ mL	_____ mL
9.	Volume of NaOH used	_____ mL	_____ mL	_____ mL
10.	Concentration of NaOH	_____ M	_____ M	_____ M

Average molarity   _____ M

Average normality   _____ N

Standard dev.   _____ M

Rel. standard dev.   _____ M

Show calculations: use a separate sheet

Standardization of 0.1 M NaOH using three different mass samples of KHP

		Trial 1	Trial 2	Trial 3
1.	Initial mass of KHP + container	_____ g	_____ g	_____ g
2.	Final mass of KHP + container	_____ g	_____ g	_____ g
3.	Mass of KHP taken	_____ g	_____ g	_____ g
4.	Moles of KHP taken	_____ mol	_____ mol	_____ mol

Show calculations:

		Trial 1	Trial 2	Trial 3
5.	Initial NaOH volume	_____ mL	_____ mL	_____ mL
6.	Final NaOH volume	_____ mL	_____ mL	_____ mL
7.	Volume of NaOH used	_____ mL	_____ mL	_____ mL
8.	Concentration of NaOH	_____ M	_____ M	_____ M

Average molarity _____ M

Average normality _____ N

Standard dev. _____ M

Rel. standard dev. _____ M

Show calculations:

3. Titration of an Unknown Acid and Determination of its Equivalent Weight.

Concentration of NaOH _____ N (meq/mL, see data sheet page 2) Unknown no. _____

		Trial 1	Trial 2	Trial 3
1.	Acid mass before transfer	_____ g	_____ g	_____ g
2.	Acid mass after transfer	_____ g	_____ g	_____ g
3.	Acid mass transferred	_____ g	_____ g	_____ g

Titration with NaOH [Note: The numbers in () indicate the number of times the base buret has been filled to complete one acid titration.]

4. Trial 1

	(1)	(2)	(3)	and so on
Initial buret volume	_____ mL	_____ mL	_____ mL	
Final buret volume	_____ mL	_____ mL	_____ mL	
Volume of NaOH used	_____ mL	_____ mL	_____ mL	
Total volume of NaOH used			_____ mL	

5. Trial 2

	(1)	(2)	(3)	and so on
Initial buret volume	_____ mL	_____ mL	_____ mL	
Final buret volume	_____ mL	_____ mL	_____ mL	
Volume of NaOH used	_____ mL	_____ mL	_____ mL	
Total volume of NaOH used			_____ mL	

6. Trial 3

	(1)	(2)	(3)	and so on
Initial buret volume	_____ mL	_____ mL	_____ mL	
Final buret volume	_____ mL	_____ mL	_____ mL	
Volume of NaOH used	_____ mL	_____ mL	_____ mL	
Total volume of NaOH used			_____ mL	

	Trial 1	Trial 2	Trial 3
7. Equivalent weight of the acid	_____ g/eq	_____ g/eq	_____ g/eq
8. Average equivalent weight			_____ g/eq
		Standard deviation	_____
		Rel. standard dev	_____

Calculations: Show at least one full calculation including all the setups and units.

Name ______________________________	Section ______________
Instructor ______________________________	Date ______________

POSTLABORATORY PROBLEMS—EXPERIMENT 11

1. Define the equivalent weight of an acid. How does it differ from the equivalent weight of an oxidant?

2. Describe how one would prepare a sample of 0.012 N $KMnO_4$. Assume the following reaction:

$$MnO_4^-(aq) + 5\,Fe_2^+(aq) + 8\,H^+(aq) \rightarrow Mn^{2+}(aq) + 5\,Fe^{3+} + 4\,H_2O(l)$$

3. Describe how to prepare 5.0 mL of 0.100 M H_2SO_4 from concentrated sulfuric acid (36.0 M).

Redox Titrations: Titration of Sodium Hypochlorite in Bleach

Microscale Experiment

Objective
- To carry out an oxidation-reduction (redox) titration

Prior Reading
- Section 3.10   Construction of a microburet
- Section 3.11   Titration procedure

Related Experiments
- Experiments 10 (Acid-base titration), 13 (K_{sp} by titration), 14 (Complexometric titration), 25 (Ion exchange), and 28 (pH titration)

Activities a Week Before
- Preparation and standardization of sodium thiosulfate solution is recommended

Bleach is commonly used to "remove" stains from clothing in the laundry. In reality, the stain is not actually removed—instead, it is made colorless. The active ingredient in chlorine bleaches is the salt sodium hypochlorite, $NaOCl$, or more specifically, the hypochlorite ion, OCl^-. The label on most bleaches reads "active ingredient, sodium hypochlorite, 5.25%, and inert ingredients 94.75%." The inert ingredients consist of water, fragrances, and so forth. Nonchlorine bleaches are also available.

The strength of a chlorine bleach is proportional to the amount of hypochlorite in it. There is no convenient direct way to measure the amount of hypochlorite. Instead, an indirect method, where the hypochlorite ion reacts with the iodide anion to produce iodine, is used. The amount of iodine produced is measured by the titration.

When KI is added to hypochlorite bleach two reactions take place simultaneously.

$$OCl^-(aq) + H_2O + 2\,e^- \rightarrow Cl^-(aq) + 2\,OH^-(aq) \qquad (E12.1)$$

$$2\,I^-(aq) \rightarrow I_2(aq) + 2\,e^- \qquad (E12.2)$$

Reactions (E12.1) and (E12.2) in which electrons are transferred are called **oxidation-reduction** or **redox** half reactions. In (E12.1), the chlorine oxidation state changes from $(+I)$ to $(-I)$ through the gain of two electrons.

The element that gains the electrons is said to be **reduced**. Whenever a reduction occurs, an oxidation must occur with it to supply the necessary electrons. In this case, the added iodide ion is oxidized, producing iodine (I_2) as shown in reaction (E12.2). The oxidation state of I^- changes from $(-I)$ to (0). Each iodide gives up one electron, for a total of two electrons from two iodide anions. The element that loses the electrons is said to be **oxidized**. Without iodide ion or some other material present to be oxidized, the hypochlorite reduction could not occur. The overall reaction is obtained by combining reactions (E12.1) and (E12.2):

$$OCl^-(aq) + H_2O + 2\,I^-(aq) \rightarrow Cl^-(aq) + 2\,OH^-(aq) + I_2 \qquad (E12.3)$$

Solutions of iodine in water ($KI + I_2$) are brown when concentrated, and yellow when dilute. The iodine present can be titrated using thiosulfate ion, $S_2O_3^{2-}$, according to the equation

$$I_2(aq) + 2\,S_2O_3^{2-}(aq) \rightarrow 2\,I^-(aq) + S_4O_6^{2-}(aq) \text{ (tetrathionate anion)} \qquad (E12.4)$$

When most of the iodine has reacted, the solution will turn from brown to a lighter yellow. At this point, the chemist adds starch. Even small quantities of iodine will form a highly colored blue complex with starch. Thiosulfate ion is then added one drop at a time. When the last trace of iodine has reacted, the blue color of the solution disappears, providing a clear solution.

Working backward, the sequence is as follows:

1. We know how much thiosulfate was added to reach the end point. We therefore know how much iodine was present [reaction (E12.4)].
2. Since we now know how much iodine was present, we know how much iodide was oxidized to form iodine [reaction (E12.2)] by the hypochlorite ion and thus how much original hypochlorite ion was present [reaction (E12.1)].

For every hypochlorite ion that is reduced, one iodine molecule is produced. The two materials are therefore equivalent to each other—the number of moles of one equals the number of moles of the other if sufficient iodide ion is present. Note that two thiosulfate ions are required to titrate one iodine molecule and that one iodine molecule results from one hypochlorite ion.

The sodium thiosulfate solution used in this titration must be standardized by titrating iodine liberated in a reaction between a standard solution of KIO_3 and KI. When excess potassium iodide is added to an acidic solution of KIO_3, iodide is oxidized to free iodine according to the following reaction:

$$IO_3{}^-(aq) + 5\,I^-(aq) + 6\,H^+(aq) \rightarrow 3\,I_2(aq) + 3\,H_2O(l) \qquad (E12.5)$$

The free iodine thus generated is titrated with a solution of sodium thiosulfate according to reaction (E12.4). Starch is used as an indicator in this titration. By combining equations (E12.4) [after multiplying (E12.4) by 3] and (E12.5), the following overall reaction is obtained:

$$6\,S_2O_3^{2-}(aq) + IO_3^-(aq) + 6\,H^+(aq) \rightarrow 3\,S_4O_6^{2-}(aq) + I^-(aq) + 3\,H_2O(l) \qquad (E12.6).$$

Example 12.1   A 1.00 mL aliquot of 0.120 M KIO_3 solution containing five drops of 2 M sulfuric acid is treated with 5 mL of 10 percent KI solution in water. In the presence of a starch indicator solution, the liberated iodine is titrated with 5.25 mL of sodium thiosulfate. What is the molarity of the thiosulfate solution? (Equation E12.6).

Answer

$$(0.120 \text{ mmol } KIO_3/\text{mL})(1.00 \text{ mL}) = 0.120 \text{ mmol } KIO_3$$

$$\frac{(0.120 \text{ mmol } KIO_3)(6 \text{ mmol thiosulfate}/1 \text{ mmol } KIO_3)}{5.25 \text{ mL thiosulfate}} = 0.137 \text{ M}$$

Example 12.2   A 0.300 mL sample of bleach weighing 0.310 g was allowed to react with KI. The iodine liberated required 4.40 mL of 0.137 M thiosulfate (thio) solution to titrate it. What is the weight of the hypochlorite (OCl^-) and what is the percent hypochlorite in the bleach? (Equations E12.3 and E12.4).

Answer

$(4.40 \text{ mL thio})(0.137 \text{ mmol/mL}) = 0.603 \text{ mmol thio}$

$(0.603 \text{ mmol thio})(1 \text{ mmol } I_2/2 \text{ mmol thio})(1 \text{ mmol } OCl^-/1 \text{ mmol } I_2) =$

$(0.302 \text{ mmol } OCl^-)(51.45 \text{ mg/mmol}) = 15.5 \text{ mg } OCl^-$

$\% \ OCl^- \text{ in bleach } = 100(15.5 \text{ mg}/310 \text{ mg}) = 5.00\%$

General References

1.   *Vogel's Textbook of Inorganic Analysis,* 4th ed.; revised by Bassett, J.; Denney, R. C.; Jeffery, G. H.; Mendham, J. Longman Scientific & Technical: London, England, 1986.
2.   Silberman, R. G.; Zipp, A. P. *J. Chem. Educ.* **1986**, *63*, 1098.
3.   Ludeman, S. M.; Brandt, J. A.; Zon, G. *J. Chem. Educ.* **1976**, *53*, 377.

Experimental Section

Procedure

Obtain a 25 mL volumetric flask, an amber-colored bottle for storing the thiosulfate solution, one 25 mL and one 10 mL Erlenmeyer flask, two microburets (2 mL preferred), soluble starch or a cornstarch-based Eco-Foam[TM] packaging pellet, solid KI (or 10 percent KI solution), sodium thiosulfate, 2 M H_2SO_4, potassium iodate, and an automatic delivery pipet (or 2 mL graduated pipet).

Preparation of a Starch Indicator Solution

[This solution may be provided by your instructor.] Dissolve one cornstarch-based Eco-Foam packaging pellet[1] in 100 mL deionized water. Alternatively, add a paste of ~1.0 g soluble starch to 100 mL of boiling water. Let the starch solution cool. Add ~0.1 g KI to stabilize the solution.

Preparation of 0.2 M Sodium Thiosulfate Solution

Dissolve ~2.5 g of $Na_2S_2O_3 \cdot 5 H_2O$ in 50 mL of deionized water in a bottle. If the solution is to be saved for more than a week, add 5 mg of sodium carbonate to the solution. Label the bottle, including the date of preparation. After proper dilution, this solution may be used for Experiment 13.

Preparation of 0.0500 M Potassium Iodate Solution

Accurately weigh 0.268 ± 0.001 g of potassium iodate and, using a funnel, transfer the solid to a 25 mL volumetric flask. Add water, rinsing down the funnel. Swirl the flask to dissolve the solid. Fill the flask to the mark, and calculate the concentration of KIO_3 solution to the third decimal place.

Standardization of Sodium Thiosulfate Solution

Set up two microburets, marking one "thiosulfate" and the other "iodate." Rinse and fill the thiosulfate buret with the thiosulfate solution. Similarly, rinse and fill the iodate buret with the iodate solution. Record the initial volumes of the solutions in each buret.

Transfer accurately $~1.000 \pm 0.001$ mL of the solution from the iodate buret to a 25 mL Erlenmeyer flask. Add 5 mL of water followed by six drops of 2 M H_2SO_4. Add ~200 mg of solid KI (or 2 mL of 10 percent KI solution) to the flask and mix the contents. The solution should become reddish brown as a result of the liberation of iodine. Titrate the liberated iodine immediately with the sodium thiosulfate solution. When the reddish brown color of iodine diminishes to a pale yellow, add 2 mL of water and 15 to 20 drops (use a Pasteur pipet) of ~1 percent starch solution. The solution will assume a violet color. Continue to titrate the solution dropwise until the solution becomes colorless. Record the final thiosulfate buret reading. Perform the titration two more times.

Titration of the Bleach Solution

Weigh 100 µL of the bleach to be analyzed in a tared 10 mL Erlenmeyer flask. Record the mass on your data sheet. Add 2 mL of water to the flask, washing down any droplets of bleach on the sides, and add 1 mL of 6 M acetic acid. Add ~200 mg of potassium iodide (or 2 mL of 10 percent KI solution) to this solution and swirl the contents of the flask to dissolve the KI. The solution should turn red-brown, indicating the formation of iodine.

Using a microburet, titrate the iodine solution with the standardized (~0.200 M) sodium thiosulfate, $Na_2S_2O_3$. Record the initial microburet reading on the data sheet. Add the thiosulfate slowly, with swirling. The color will change from red-brown to light yellow. At this point, add 15 to 20 drops (use a Pasteur pipet) of ~1 percent starch solution, which will cause the solution to turn dark violet.

> NOTE: As there is no harm in adding the starch too early, it should be added as soon as the solution color lightens and shows any hint of yellow color.

Continue titrating, one drop at a time, swirling the flask between drops, until the solution becomes colorless. Record the final microburet reading on the data sheet.

Repeat the procedure twice with two more samples of the same bleach, and average the results. If time permits, perform the same analysis using a sample of a *different* brand of bleach. Determine the amount of hypochlorite (OCl^-) in *each* sample of the bleach taken, convert the mass of the hypochlorite anion to that of sodium hypochlorite (NaOCl) and, finally, calculate (in terms of weight) the percent sodium hypochlorite in the bleach (see examples E12.1 and E12.2). **Rinse, clean, and return all the glassware and other equipment.**

> Disposal: Dispose of all the solutions in designated containers or ask your instructor about the disposal procedure.

Additional Independent Projects

1. Determination of the vitamin C (ascorbic acid) content of a vitamin tablet or orange juice[2,3]

   Brief Outline

   This experiment is based upon the microscale iodometric method[3] described in the literature. Vitamin C, or ascorbic acid, has the composition $C_6H_8O_6$. It is oxidized by I_2 to dehydroascorbic acid. Prepare a 0.01 M KIO_3 solution in a 100 mL volumetric flask. Weigh ~ 100 ± 1 mg pulverized (powdered) vitamin C tablet in a 10 mL beaker; add 0.3 M H_2SO_4 and stir. To remove insoluble solids, filter the solution directly to a 100 mL volumetric flask and dilute to the mark with 0.3 M H_2SO_4. Pipet 1.00 mL of the vitamin solution to a 10 mL Erlenmeyer flask, add 0.100 to 0.200 mL of KIO_3 solution followed by a few crystals of KI. Stir the solution to dissolve KI. In the presence of H^+, KI reacts with KIO_3 to generate I_2, which then oxidizes ascorbic acid to dehydroascorbic acid ($C_6H_6O_6$). Titrate the excess I_2 with a standardized 0.0200 M sodium thiosulfate solution. Calculate the percent vitamin C in the tablet. Instead of a vitamin C tablet, one can use orange juice.

2. Determination of the stoichiometry of the oxidation-reduction reaction[4] of Fe(III) with hydroxyl ammonium ion, NH_3OH^+, supplied as $(NH_3OH)_2SO_4$.

   Note: In this titration, a standardized solution of potassium permanganate is used. We recommend that the standardization of permanganate solution be carried out using a standard solution of sodium oxalate under microscale conditions. For this titration see General Reference 1.

3. Oxidation of bromcresol green.[5]

4. Convert the above experiment 12 to a macroscale one.

References

1. De Moura, J. M. *J. Chem. Educ.* **1992**, *69*, 860.
2. Haddad, P. *J. Chem . Educ.* **1977**, *54*, 192.
3. Kumar, V.; Courie, P.; Haley, S. *J. Chem. Educ.* **1992**, *69*, A213.
4. Child, W. C.; Ramette, R. W. *J. Chem. Educ.* **1967**, *44*, 109.
5. Pickering, M.; Heller, D. *J. Chem. Educ.* **1987**, *64*, 81.

Name ________________________________ Section ______________

Instructor ______________________________ Date ______________

PRELABORATORY REPORT SHEET—EXPERIMENT 12

Experiment title____

Objective

Reactions/formulas to be used

Materials and equipment table

Outline of procedure

1. Balance each of the following redox reactions in acid solution:

   (a) $MnO_4^-(aq) + HSO_3^-(aq) \rightarrow Mn^{2+}(aq) + HSO_4^-(aq)$

   (b) $OCl^-(aq) + I_2(s) \rightarrow Cl^-(aq) + IO_3^-(aq)$

2. Balance each of the following redox reactions in alkaline solution:

   (a) $Al(s) + OH^-(aq) \rightarrow Al(OH)_4^-(aq) + H_2(g)$

   (b) $Pb(OH)_3^-(aq) + OCl^-(aq) \rightarrow PbO_2(s) + Cl^-(aq)$

3. Sodium bismuthate ($NaBiO_3$) oxidizes $MnCl_2$ to MnO_4^- and is consequently reduced to $Bi(OH)_3$. How many grams of $MnCl_2$ will be oxidized by 5.0 g of $NaBiO_3$?

EXPERIMENT 12   DATA SHEET

1. Mass of KIO_3 + container _____ g

2. Mass of KIO_3 + container after transfer _____ g

3. Mass of KIO_3 taken _____ g

4. Concentration of KIO_3 solution _____ mol/L

Concentration of Sodium Thiosulfate ($Na_2S_2O_3$)

5. Mass of thiosulfate _____ g

		Trial 1	Trial 2	Trial 3
6.	Initial buret reading (KIO_3)	_____ mL	_____ mL	_____ mL
7.	Final buret reading (KIO_3)	_____ mL	_____ mL	_____ mL
8.	Volume of KIO_3 taken	_____ mL	_____ mL	_____ mL
9.	Initial buret reading (thiosulfate)	_____ mL	_____ mL	_____ mL
10.	Final buret reading (thiosulfate)	_____ mL	_____ mL	_____ mL
11.	Volume of thiosulfate used	_____ mL	_____ mL	_____ mL
12.	Conc. of thiosulfate solution	_____ mol/L	_____ mol/L	_____ mol/L
13.	Average concentration of thiosulfate			_____ mol/L

Show calculations:

DATA SHEET, EXPERIMENT 12, P. 2

Brand name of Bleach A ____________________

Conc. of thiosulfate ______ mol/L (from the previous page)

Data Collection

		Trial 1	Trial 2	Trial 3
1.	Mass of bleach	______ g	______ g	______ g
2.	Initial buret reading (thio)	______ mL	______ mL	______ mL
3.	Final buret reading (thio)	______ mL	______ mL	______ mL
4.	Vol. of thiosulfate added	______ mL	______ mL	______ mL

Data Manipulation

5. Determine the number of millimoles of thiosulfate that were added.

Show calculations. ______ mmol ______ mmol ______ mmol

6. Determine the number of millimoles of iodine that were titrated. The reaction is

$$I_2 + 2\,S_2O_3^{2-} \rightarrow 2\,I^- + S_4O_6^{2-}$$

Show calculations. ______ mmol ______ mmol ______ mmol

7. Determine the number of millimoles of hypochlorite (OCl^-) in the bleach. The reactions occurring are these:

$$OCl^- + H_2O + 2\,e^- \rightarrow Cl^- + 2\,OH^-$$
$$2\,I^- \rightarrow I_2 + 2\,e^-$$

Show calculations. ______ mmol ______ mmol ______ mmol

Average OCl^- ______ mmol

Mass OCl^- ______ mg

Mass NaOCl ______ mg

DATA SHEET, EXPERIMENT 12, P. 3

8. Determine the average weight and weight percent of NaOCl in the bleach.

Name of Bleach B _____

Data Collection

		Trial 1	Trial 2	Trial 3
1.	Mass of bleach	_____ g	_____ g	_____ g
2.	Initial buret reading (thio)	_____ mL	_____ mL	_____ mL
3.	Final buret reading (thio)	_____ mL	_____ mL	_____ mL
4.	Vol. of thiosulfate added	_____ mL	_____ mL	_____ mL

Data Manipulation

5. Determine the number of millimoles of thiosulfate that were added.

Show calculations.   _____ mmol   _____ mmol   _____ mmol

6. Determine the number of millimoles of iodine that were titrated. The reaction is

$$I_2 + 2\,S_2O_3{}^{2-} \rightarrow 2\,I^- + S_4O_6{}^{2-}$$

Show calculations.   _____ mmol   _____ mmol   _____ mmol

7. Determine the number of millimoles of hypochlorite (ClO^-) in the bleach. The reactions occurring are these:

$$OCl^- + H_2O + 2\,e^- \rightarrow Cl^- + 2\,OH^-$$
$$2\,I^- \rightarrow I_2 + 2\,e^-$$

Show calculations.   _____ mmol   _____ mmol   _____ mmol

Average OCl^- _____ mmol

Mass OCl^- _____ mg

Mass NaOCl   _____ mg

Experiment 12   Redox Titrations: Titration of Sodium Hypochlorite in Bleach   259

8. Determine the average weight and weight percent of NaOCl in the bleach.

9. Which brand is a better buy?

POSTLABORATORY PROBLEMS—EXPERIMENT 12

1. A commercial bleach is titrated with 0.50 M sodium thiosulfate. If 2.0 g of the bleach requires 1.1 mL of the thiosulfate solution, what is the percentage of NaOCl in the bleach?

2. What is the purpose of adding starch solution to an iodine solution being titrated with a solution of sodium thiosulfate?

13 Determination of the Solubility Product Constant K_{sp}, and the Common Ion Effect

Microscale Experiment

Objectives

- To determine the solubility product constant (K_{sp}) of a sparingly soluble compound by using an oxidation-reduction titration method and to study the common ion effect

Prior reading

- Section 3.4   Weighing
- Section 3.5   Measuring liquid volumes
- Section 3.8   Filtration and gravimetric methods
- Section 3.9   Solution preparation
- Section 3.10   Construction of a microscale buret, stirring bar, and so forth
- Experiment 12 Redox titrations

Related Experiments

- Experiments 10 (Acid-base titration), 11 (Equivalent weight), 12 (Redox titration), 14 (Complexometric titration), 25 (Ion exchange), and 28 (pH titration)

Activities a Week Before

- Prepare and standardize a sodium thiosulfate solution. Also, prepare solid calcium iodate.

Many ionic solids exhibit a very limited solubility in water at room temperature. Such salts are called **sparingly soluble**, **slightly soluble**, or even **insoluble salts**. No salt is truly insoluble—all have a finite solubility in water, even if this solubility is extremely small. The silver halides, AgX (X = Cl$^-$, Br$^-$, and I$^-$) are examples of such slightly soluble salts. In water, the following equilibrium occurs:

$$AgX(s) \rightleftharpoons Ag^+(aq) + X^-(aq)$$

The equilibrium expression for this reaction is written (assuming X = Cl$^-$) as

$$K_{eq} = \frac{[Ag^+][Cl^-]}{[AgCl]}$$

The concentration of any solid (such as AgX) is constant at a given temperature. Its concentration can therefore be included in the constant, and the preceding expression may be rewritten as

$$K_{sp} = K_{eq}[AgCl] = [Ag^+][Cl^-]$$

where K_{sp} is called the **solubility product constant** (see Appendix 5).

Example E13.1 The K_{sp} of AgI in water is 8.51×10^{-17}. Calculate the solubility of AgI in water. What would be the solubility of AgI if 10.0 mL of a 0.01 M solution of NaI is added to 990 mL of the preceding solution?

Answer

For the solution of AgI, since each mole of AgI dissociating produces one mole of Ag^+ and one mole of I^-, $[Ag^+] = [I^-]$. Therefore,

$$K_{sp} = [Ag^+][I^-] = [Ag^+]^2 = 8.51 \times 10^{-17}\ mol^2/L^2$$
$$[Ag^+] = 9.22 \times 10^{-9}\ mol/L$$

Thus, the molar solubility of AgI is 9.22×10^{-9} mol/L.

The added salt, NaI, is very soluble and dissociates completely. The quantity of iodide ion produced by the dissociation of NaI is *much larger* than that produced by the dissociation of the slightly soluble AgI. As an excellent approximation, we may assume that all of the iodide present in solution comes from NaI. Thus,

$$mol\ I^- = (1.0 \times 10^{-2}\ L)(0.01\ mol/L) = 1.0 \times 10^{-4}\ mol$$
$$Total\ volume = 990\ mL + 10\ mL = 1000\ mL = 1.00\ L$$
$$[I^-] = (1 \times 10^{-4}\ mol)/(1.00\ L) = 1 \times 10^{-4}\ mol/L$$
$$K_{sp} = [Ag^+][I^-]$$
$$8.51 \times 10^{-17} = [Ag^+][1 \times 10^{-4}]$$
$$[Ag^+] = 8.51 \times 10^{-13}\ mol/L$$

Thus, the solubility of AgI in the solution containing NaI is 8.51×10^{-13} M, a much lower solubility than in pure water. This decrease in solubility in a solution containing a second salt with an ion in common with the first salt is termed the **common ion effect**.

In this experiment, the K_{sp} of the slightly soluble $Ca(IO_3)_2$ will be determined. It dissociates in aqueous solution according to the following equilibrium:

$$Ca(IO_3)_2(s) \rightleftharpoons Ca^{2+}(aq) + 2\ IO_3^-(aq) \tag{E13.1}$$
$$K_{sp} = [Ca^{2+}][IO_3^-]^2 \tag{E13.2}$$

In a saturated solution of $Ca(IO_3)_2$, the dissociation reaction (E13.1) shows that

$$[Ca^{2+}] = (1/2)[IO_3^-]$$

Thus,

$$K_{sp} = (1/2)[IO_3^-][IO_3^-]^2 = (1/2)[IO_3^-]^3 \tag{E13.3}$$

The iodate concentration can be experimentally determined through an oxidation-reduction titration with a standardized solution of $Na_2S_2O_3$ in the presence of KI, using a starch indicator solution, as done previously in Experiment 12. When KI solution is added to iodate [released from $Ca(IO_3)_2$] in the presence of acid, the I^- ion (from KI) reduces the iodate ion to iodide and itself is oxidized to iodine according to the following reaction:

$$IO_3^-(aq) + 6\,H^+(aq) + 5\,I^-(aq) \rightarrow 3\,I_2(aq) + 3\,H_2O(l) \tag{E13.4}$$

The amount of reddish brown iodine formed in reaction (E13.4) may be determined by titration with sodium thiosulfate solution using a starch indicator solution:

$$I_2(aq) + 2\,S_2O_3^{2-} \rightarrow 2\,I^-(aq) + S_4O_6^{2-}(aq) \tag{E13.5}$$

The overall oxidation-reduction reaction, combining (E13.4) and $3\times$(E13.5), is

$$6\,Na_2S_2O_3 + KIO_3 + 3\,H_2SO_4 \rightarrow 3\,Na_2S_4O_6 + KI + 3\,Na_2SO_4 + 3\,H_2O \tag{E13.6}$$

Using equation (E13.6) and the concentration of $Na_2S_2O_3$, one can calculate the concentration of IO_3^- produced according to the reaction (E13.1). From this, the K_{sp} for $Ca(IO_3)_2$ can be determined.

General References

1. Gotlib, L. J. *J. Chem. Educ.* **1990**, *67*, 937.
2. Lagier, C.; Olivieri, A. *J. Chem. Educ.* **1990**, *67*, 934.

Experimental Section

Procedure

> Microscale method   Estimated time to complete the experiment: 2.5 h
> Experimental Steps: Five steps are involved in this procedure: (1) preparation and % yield determination of calcium iodate, (2) preparation of 5.00×10^{-3} M standard solution of potassium iodate and a ~0.02 M sodium thiosulfate solution, (3) standardization of the sodium thiosulfate solution, (4) preparation of saturated solutions of calcium iodate in pure water and in $CaCl_2$ or $Ca(NO_3)_2$ solution, and (5) titration of the solution saturated with calcium iodate.

Obtain two 100 mL volumetric flasks, a 10 mL graduated cylinder, three 25 mL beakers, three micro stir bars, a magnetic-stirring hot plate, a Hirsch funnel, filter paper, a suction filtration apparatus, a sample of $Ca(NO_3)_2$ or $CaCl_2$ solution of known concentration (assigned by your instructor), 0.2 M KIO_3 solution, 0.1 M $Ca(NO_3)_2$ or $CaCl_2$ solution, cornstarch-based Eco-Foam pellets or soluble starch[1], four 25 mL Erlenmeyer flasks, and two microburets (preferably 2 mL capacity).

Preparation of a Starch Indicator Solution

See Experiment 12.

Preparation of 5.00×10^{-3} M Potassium Iodate Solution

Accurately weigh about 0.107 ± 0.001 g of KIO_3 and dissolve it in deionized water in a 100 mL volumetric flask to provide a $\sim5.00 \times 10^{-3}$ M solution.

Preparation and Standardization of 0.02 M Sodium Thiosulfate Solution

Prepare a 0.02 M sodium thiosulfate solution by dissolving 0.50 g of $Na_2S_2O_3$ in 100 mL of deionized water. Standardize the thiosulfate solution with the $\sim5.00 \times 10^{-3}$ M KIO_3 solution following the general titration procedure described in Experiment 12.

Preparation of Solid $Ca(IO_3)_2$ and Calculation of its Percent Yield

Using a 10 mL graduated cylinder, measure and mix 9.0 mL of 0.2 M KIO_3 solution with 10.0 mL of 0.1 M $Ca(NO_3)_2$ or $CaCl_2$ solution in a 25 mL beaker. Stir the mixture with a stirring rod. A white precipitate of calcium iodate, $Ca(IO_3)_2$, forms. Allow the mixture to stand for 5 to10 minutes.

Collect the crystals of calcium iodate on a filter by suction filtration using a Hirsch funnel. Transfer all the solids from the beaker to the funnel using three 1 mL portions of deionized water. Wash the solid with two additional 1 mL portions of cold water.

Air-dry the solid on the funnel under suction, and then place the solid on a small watch glass or a ceramic plate for further drying. Weigh the product and determine the percent yield (see Experiment 20 or 32). Using a spatula, divide the dry solid into two equal portions. **The solid must be dry** for the next part of the experiment.

Preparation and Titration of Solutions Saturated with IO_3^-

Place one portion of the solid calcium iodate in 10 mL of deionized water in a 25 mL beaker marked #1, and stir the mixture thoroughly. Transfer the second portion of the solid to another 25 mL beaker marked #2, and add 10 mL of $CaCl_2$ or $Ca(NO_3)_2$ solution (concentration assigned by your instructor). Let these mixtures stand for at least 15 minutes with occasional stirring, so that the supernatant liquid becomes saturated with IO_3^-.

Take two **clean and dry** 25 mL Erlenmeyer flasks and two filter funnels (if they are wet, rinse them with acetone and place in an oven for 15 minutes to dry). Place a **dry** piece of filter paper in the funnel. **Do not wet the filter paper.** Filter the mixture from beaker #1, collecting the filtrate directly in the dry Erlenmeyer flask. Label this flask "A." Similarly, filter the second solution using a separate **dry** filtration unit and collect the filtrate in a different flask, labeled "B."

Titration of the First Filtrate in Flask A

Rinse and fill a clean 2 mL microburet with the filtered solution in flask A (which is saturated with IO_3^-). Rinse and fill a second buret with the standardized 0.02 M thiosulfate solution. Record the initial volumes of the two burets to ± 0.001 mL.

Transfer 1.000 mL of the IO_3^- solution from the buret into a clean 25 mL Erlenmeyer flask, and add 5 mL of deionized water followed by 1 mL of 10 percent KI solution. Add 5 to 6 drops of 2 M sulfuric acid. Swirl the flask.

Titrate the liberated iodine with thiosulfate solution. When the solution assumes a light yellow color, add 10 to 15 drops of starch indicator (use a Pasteur pipet) and complete the titration as described in Experiment 12. Note the final volume of the thiosulfate solution. Repeat the procedure twice more using different volumes of the IO_3^- solutions.

Titration of the Second Filtrate

Rinse and fill the iodate microburet with the filtrate from flask B. Transfer 1.000 mL of this solution from the buret. Complete the titration with the thiosulfate solution as described before for the filtrate from flask A. Record the initial and final volumes in the burets. Repeat the procedure at least two more times. **This titration will require less thiosulfate solution if the common ion effect (Ca^{2+} in this case) is real, as there will be less dissolved IO_3^-.**

Rinse, clean, and return all the glassware and other equipment to your instructor.

Disposal: Dispose of the solutions to designated containers or ask your instructor for the proper disposal procedure.

Calculations

For each run, calculate $[IO_3^-]$ in the first filtrate (filtrate A) and second filtrate (filtrate B). Determine the average concentration for each. Report these values to your instructor, who will post the values against respective $[Ca^{2+}]$ on the chalk board. Calculate the K_{sp} for the first filtrate. Summarize the class's results by preparing a plot of $[IO_3^-]$ on the y axis versus $[Ca^{2+}]$ (used for preparing the second filtrate) on the x axis including your average $[IO_3^-]$ in deionized water where $[Ca^{2+}] = 0$.

Additional Independent Project

1. Study the oxidation-reduction chemistry of tungsten.[2]
2. Determine the solubility product of copper(II) tartrate.[3]

   Brief Outline

   This method uses the spectrophotometric method (Bausch & Lomb Spectronic 20) to determine the K_{sp} of copper(II) tartrate, $CuC_4H_4O_6$ (abbreviated as CuT). It dissociates according to the equation

   $$CuT(s) \rightleftharpoons Cu^{2+}(aq) + T^{2-}(aq)$$

   Prepare a stock (5.0×10^{-2} M) solution of $CuSO_4 \cdot 5H_2O$ (0.1248 g) in a 100 mL volumetric flask. Starting from this stock solution, prepare four more standard solutions (calculate the concentrations) of copper sulfate by diluting 4.00, 6.00, 8.00, and 12.00 mL (use a graduated pipet) to 25 mL (use a 25 mL volumetric flask). Measure the absorbances of each of these standard Cu^{2+} solutions at 675 nm wavelength (see Chapter 8 and Experiments 21 and 24) and construct a calibration curve. Stir 100 to 150 mg of copper(II) tartrate in ~10 mL of deionized water at 45°C. Filter the solution and measure its absorbance. From the calibration curve, obtain the concentration of copper(II), $[Cu^{2+}]$. Calculate the K_{sp} for copper(II) tartrate.

3. Determine the solubility product of calcium sulfate by ion exchange–complexo-metric titration.[4]

References

1. De Moura, J. M. *J. Chem. Educ.* **1992**, *69,* 860.
2. Pickering, M.; Monts, D. L. *J. Chem. Educ.* **1982**, *59,* 316.
3. Thomsen, M. W. *J. Chem. Educ.* **1992**, *69,* 328.
4. Koubek, E. *J. Chem. Educ.* **1976**, *53,* 254.

Name ______________________ Section ______________

Instructor ______________________ Date ______________

PRELABORATORY REPORT SHEET—EXPERIMENT 13
USE EXTRA SHEET IF NECESSARY

Experiment Title ______________________

Objective

Reactions/formulas to be used

Chemicals and solutions—their preparation

Materials and equipment table

Outline of procedure

1. For measuring the K_{sp} of $Ca(IO_3)_2$ we did not weigh out the iodate salt or water. Why?

2. The K_{sp} of Hg_2Cl_2 is 1.48×10^{-18} mol^2/L^2. Calculate the solubility of Hg_2Cl_2 in mg/mL and in parts per million (ppm).

3. A mixture is made of 1.00 mL of 0.1 M $Pb(NO_3)_2$ and 2.00 mL of 0.087 M Na_2SO_4. Write an equation for this reaction. Which species is expected to precipitate? Will it precipitate under the given conditions? Support your answer with calculations.

EXPERIMENT 13   DATA SHEET

Data Collection and Manipulation Table

1. Mass of KIO_3 + container _____ g

2. Mass of KIO_3 + container after transfer _____ g

3. Mass of KIO_3 taken _____ g

    Mass of $Na_2S_2O_3$ taken _____ g

    Concentration of KIO_3 solution _____ mol/L

Calculations:

Concentration of Thiosulfate Solution

		Trial 1	Trial 2	Trial 3
4.	Initial buret volume (KIO_3)	_____ mL	_____ mL	_____ mL
5.	Final buret volume (KIO_3)	_____ mL	_____ mL	_____ mL
6.	Volume of KIO_3 taken	_____ mL	_____ mL	_____ mL
7.	Initial buret volume (thiosulfate)	_____ mL	_____ mL	_____ mL
8.	Final buret volume (thiosulfate)	_____ mL	_____ mL	_____ mL
9.	Volume used (thiosulfate)	_____ mL	_____ mL	_____ mL
10.	Concentration of thiosulfate solution	_____ mol/L	_____ mol/L	_____ mol/L

Average concentration of thiosulfate _____ mol/L

Show sample calculations:

11.   Yield of $Ca(IO_3)_2$ _____ g   % yield _____

Show calculations:

Solubility Product Constant (first filtrate, flask A)

1. Volume of IO_3^- taken   _____ mL   _____ mL   _____ mL

2. Initial buret volume (thiosulfate)   _____ mL   _____ mL   _____ mL

3. Final buret volume (thiosulfate)   _____ mL   _____ mL   _____ mL

4. Volume used (thiosulfate)   _____ mL   _____ mL   _____ mL

5. Moles of thiosulfate   _____ mol   _____ mol   _____ mol

6. Moles of IO_3^- _____ mol   _____ mol   _____ mol

7. Molarity of IO_3^- (mol/L)   _____   _____   _____

8. Molarity of Ca^{2+} (mol/L)   _____   _____   _____

9. Calculate the value of K_{sp} for each trial, and determine the average value.

                                        _____   _____   _____

                                     Average   _____

Show calculations:

DATA SHEET, EXPERIMENT 13, P. 3

Common Ion Effect and K_{sp} (second filtrate, flask B)

Concentration of Ca^{2+} solution assigned by the instructor _____ mol/L

Concentration of thiosulfate _____ mol/L

		Trial 1	Trial 2	Trial 3
1.	Volume of IO_3^- taken	_____ mL	_____ mL	_____ mL
2.	Initial buret volume (thiosulfate)	_____ mL	_____ mL	_____ mL
3.	Final buret volume (thiosulfate)	_____ mL	_____ mL	_____ mL
4.	Volume used (thiosulfate)	_____ mL	_____ mL	_____ mL
5.	Moles of thiosulfate	_____ mol	_____ mol	_____ mol
6.	Moles of IO_3^-	_____ mol	_____ mol	_____ mol
7.	Molarity of IO_3^- (mol/L)	_____	_____	_____
8.	Molarity of Ca^{2+} (mol/L)	_____	_____	_____

9. Calculate the average solubility of $Ca(IO_3)_2$ in the presence of the common ion Ca^{2+}. Obtain values from your classmates for their solutions, and construct a graph of $[IO_3^-]$ (y axis) versus $[Ca^{2+}]$ (x axis). What does this graph show?

Show calculations:

1. Write a balanced reaction for the following change:

$$BrO_3^- + H^+ + Br^- \rightarrow Br_2 +$$

2. What happens when concentrated HCl is added to a solution of $BaCl_2$?

3. The solubility of Ag_2CrO_4 in water at 25°C is 7.80×10^{-5} mol/L. Calculate K_{sp}.

14 Complexometric Titration: Determination of Water Hardness or Calcium in a Calcium Supplement

Micro- and Macroscale Experiments

Objectives
- To become acquainted with the complexometric titration
- To determine the water hardness or calcium content in a calcium supplement

Prior Reading
- Section 3.4   Weighing
- Section 3.5   Measuring liquid volumes: use of quantitative glassware
- Section 3.9   Solution preparation
- Section 3.10   Construction and use of a microburet
- Section 3.11   Titration procedure

Related Experiments
- Experiments 10 (Acid-base titration), 11 (Equivalent weight), 12 (Redox titration), 13 (K_{sp}), 25 (Ion exchange), and 28 (pH titration)

Activities a Week Before
Dry a sample of $CaCO_3$ in an oven at 100°C for one hour and store it in a desiccator

Collect a sample of water for analysis

Untreated well water contains dissolved mineral ions, such as calcium (Ca^{2+}), magnesium (Mg^{2+}), iron (Fe^{2+}), bicarbonate (HCO_3^-), chloride (Cl^-), and sulfate (SO_4^{2-}). Water samples containing major quantities of these ions are known as **hard water**. For example, if water containing dissolved CO_2 passes through underground limestone ($CaCO_3$), the following reaction takes place:

$$CaCO_3(s) + H_2O(l) + CO_2(g) \rightarrow Ca^{2+}(aq) + 2\,HCO_3^-(aq)$$

The soluble $Ca(HCO_3)_2$ salt is what makes the water hard. This is a serious problem for water heaters, tea kettles, and the like. In boilers, when hard water is heated and evaporated, a solid layer of **scale**, consisting of $CaCO_3$ and $MgCO_3$, is gradually deposited on the walls, drastically reducing the boiler efficiency. Hard water also reacts with soaps and detergents. Soap contains soluble Na^+ or K^+ salts of long-chain fatty acids (palmitic, stearic, and oleic acids). These acid anions form insoluble salts with Ca^{2+} and Mg^{2+} cations, and these salts prevent the formation of suds. This reduces the cleaning capacity of the soap and leaves a sticky residue on clothes.

The removal of Ca^{2+} and Mg^{2+} ions from hard water is called **water softening**. **Temporarily hard water** contains bicarbonates of Ca^{2+} and Mg^{2+}, which can be softened simply by boiling the water to decompose HCO_3^- to CO_2 and H_2O. This decomposition produces $CaCO_3$ and $MgCO_3$, which precipitate. **Permanently hard water** cannot be softened simply by boiling.

The concentrations of Ca^{2+}, Mg^{2+}, and Fe^{2+} in water determine its hardness. The concentration of many metal cations can be determined in solution by the use of **complexometric titration**. In this technique, a complexing agent called a **chelating ligand** is used to bind and subsequently remove the metal ions from water. One of the most frequently used complexing agents is ethylenediaminetetraacetic acid (H_4EDTA). H_4EDTA and its disodium salt (abbreviated as $\overline{EDTA}$) are shown below.

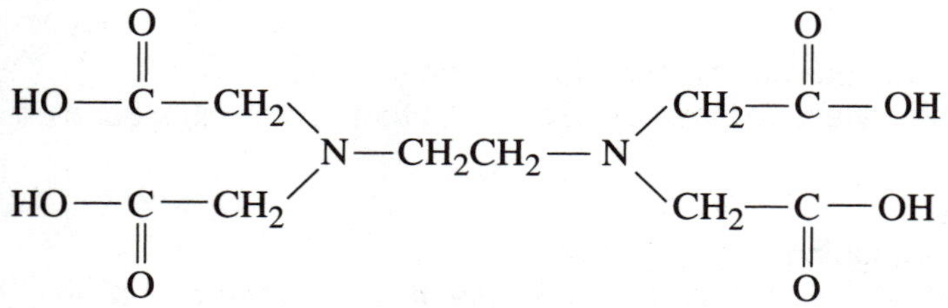

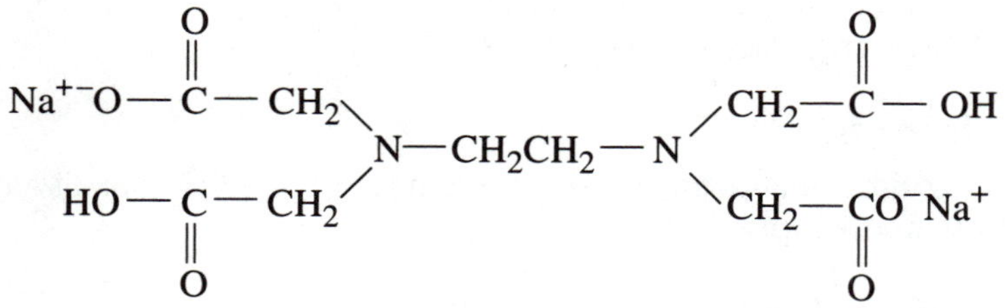

H_4EDTA (frequently abbreviated as H_4Y) or its disodium salt, Na_2H_2Y, dissociates in water to establish a series of equilibria as shown below (pK values at 20°C are also given):

$$\begin{aligned} H_4Y &\rightleftharpoons H_3Y^- + H^+ & pK_1 &= 2.0 \\ H_3Y^- &\rightleftharpoons H_2Y^{2-} + H^+ & pK_2 &= 2.7 \\ H_2Y^{2-} &\rightleftharpoons HY^{3-} + H^+ & pK_3 &= 6.2 \\ HY^{3-} &\rightleftharpoons Y^{4-} + H^+ & pK_4 &= 10.3 \end{aligned}$$

The given dissociation constants show that two of the protons are strongly acidic and two are much more weakly acidic. In this analysis, we will use the disodium salt, EDTA. The exact position of equilibrium is determined by controlling the pH of the solution. At pH = 10, the predominant species is Y^{4-}. At very low pH, H_4Y predominates. It is important that $[Y^{4-}]$ be maximized because the metal cations M^{2+} react with Y^{4-} quantitatively. This condition is achieved only at a pH of ~10 or higher. For this purpose, an ammonia buffer solution is used.

EDTA reacts with a metal cation (for example, Ca^{2+} or Mg^{2+}) to form a complex ion, M(EDTA).

$$EDTA^{4-}(aq) + M^{2+}(aq) \rightarrow M(EDTA)^{2-}(aq)$$

The end point of this reaction can be determined by titration using an indicator such as Eriochrome Black T (EBT) at pH 10. The free indicator is sky blue. When added to a solution containing M^{2+} cations, it forms a wine red complex:

$$\text{Excess } M^{2+}(aq) + \underset{\text{blue}}{EBT^{2-}} \rightarrow \underset{\text{wine red}}{[\text{M-EBT}](aq)} + M^{2+}(aq)$$

Since only a very small amount of the indicator is added, most of the metal ions remain free or uncomplexed. When a solution of EDTA is added to hard water, it forms a more stable complex with the metal cations than does EBT. When the end point in the EDTA titration is reached, all of the free M^{2+} ions have been removed as the complex ion, $M(EDTA)^{2-}$, including the ones from the [M-EBT] complex. The end point is therefore detected by the change of the color of the indicator from wine red back to sky blue. This color change is quite sharp when magnesium ions are present in solution. If the sample being titrated does not contain any Mg^{2+}, a small amount of Mg^{2+} is usually added to the EDTA solution as it is prepared to ensure a sharp end point. By knowing the concentration of EDTA used in the titration, the concentration of M^{2+} can be calculated.

$$M_{\text{EDTA}} V_{\text{EDTA}} = M_{M^{2+}} V_{M^{2+}}$$

Usually, the hardness of water is expressed in parts per million (ppm) of $CaCO_3$. From the total concentration of M^{2+} as determined by the titration, the concentration of $CaCO_3$ in grams per liter of water can be calculated. The result is finally converted into ppm of $CaCO_3$. In the optional experiment, the procedure for determining Ca^{2+} separately is described; it involves precipitating Mg^{2+} as $Mg(OH)_2$ with a strong base.

References

1. Day, Jr., R. A.; Underwood, A. L. *Quantitative Analysis,* 5th ed.; Prentice-Hall: Englewood Cliffs, NJ, 1986.
2. Harris, D. C. *Quantitative Chemical Analysis,* 2nd ed.; W. H. Freeman, New York, 1987.

Experimental Section

Procedure:

Microscale method   Estimated time to complete the experiment: 2.5 h

Experimental Steps: Four steps are involved in this procedure: (1) preparation of a 0.01 M standard solution of calcium chloride, (2) preparation of a 0.01 M disodium EDTA solution, (3) standardization of the EDTA solution, and (4) determination of hardness of water or of calcium in a tablet.

Disposal procedure: All of the solutions can be disposed of in a marked container. The chemicals used in this experiment are generally not damaging to the environment.

Obtain a 25 mL beaker, 10 and 100 mL graduated cylinders, a 100 mL volumetric flask, and a 100 mL bottle. Procure a solution of Eriochrome Black T (or a solid mixture of 100 mg EBT with 10 g NaCl and 100 mg of hydroxylamine hydrochloride). Obtain samples of $Na_2EDTA \cdot 2H_2O$, $MgCl_2 \cdot 6H_2O$, $CaCO_3$, 0.1 M NaOH, 6 M HCl, NH_4Cl, and aqueous NH_3 (6 M solution).

Preparation of 0.01 M Disodium EDTA Solution

Set up two 2 mL microburets (see Section 3.10). Weigh about 0.4 g of $Na_2EDTA \cdot 2H_2O$ and about 0.01 g of $MgCl_2 \cdot 6H_2O$ and transfer them into a clean 25 mL beaker. Dissolve the solids in 20 mL of water. If the solution is turbid, add a drop or two of 0.1 M NaOH, until a clear solution is obtained. Have the instructor check the clarity of the solution. Transfer the solution to a labeled bottle and dilute it to a total volume of 100 mL.

Preparation of 0.01 M Calcium Chloride Solution

Obtain a thoroughly cleaned and rinsed 100 mL volumetric flask. Accurately weigh 0.100 g of powdered $CaCO_3$ that has been previously dried at 100°C for one hour. Transfer the solid quantitatively to the 100 mL volumetric flask using a funnel. Rinse down the solid using 10 to 20 mL of deionized water added dropwise with a Pasteur pipet.

Add 6 M HCl dropwise (Pasteur pipet) to this mixture, until effervescence of CO_2 ceases. **Do not add excess acid.** Before adding another drop of the acid, wait until all effervescence clears. Then, add water to the mark and mix the solution thoroughly. Calculate the molarity of the solution and enter the result on your data sheet. The reaction taking place is

$$CaCO_3 + 2\,HCl \rightarrow CaCl_2 + H_2O + CO_2$$

Preparation of Ammonia/Ammonium Chloride Buffer Solution

Quantities given here are sufficient for an entire class. In a **Hood**, add 27 g of NH_4Cl to 228 mL of concentrated NH_3 (**CAUTION:** Concentrated NH_3 is corrosive and toxic) in a 500 mL Erlenmeyer flask. Dilute this solution to 400 mL by adding 172 mL of water.

Transfer the solution to a bottle and label it as "NH$_4$Cl/NH$_3$ buffer." Indicate the date of preparation of the buffer. Your instructor may provide this solution.

Standardization of Disodium EDTA Solution

Set up two 2.00 mL microburets: one for CaCl$_2$ and the other for the EDTA solution. Obtain a clean 10 mL Erlenmeyer flask containing a micro stir bar, a 1 mL pipet, and a magnetic-stirring hot plate. Rinse the CaCl$_2$ buret with the CaCl$_2$ solution and the other with the EDTA solution. Fill the burets with their respective solutions. Record the initial volume of the solution in each of the burets.

Rinse the Erlenmeyer flasks with deionized water several times. Transfer 2.00 mL of the CaCl$_2$ solution into the 10 mL Erlenmeyer flask. Add 0.2 to 0.3 mL of ammonia buffer, followed by either 0.2 mL of Eriochrome Black T solution or a quarter-pea-sized portion of solid Eriochrome Black T. Stir the contents of the flask to mix the indicator.

The solution will assume a *wine red* color. With constant stirring, titrate the solution by adding EDTA solution until the wine red color changes to a distinct *blue*. No trace of red color should remain in the solution.

Record the final volume of EDTA solution on your data sheet. You may have to practice finding the end point several times. Save a solution at the end point to use as a color comparison for other titrations.

Repeat the titration using at least two more aliquots of CaCl$_2$ solution. Determine the molarity of the EDTA solution and enter it on the data sheet. The reaction of EDTA with CaCl$_2$ is in a 1:1 mole ratio.

Determination of Water Hardness

Empty the CaCl$_2$ buret and rinse it thoroughly with a sample of hard water (supplied by your instructor, or you may use tap water) several times. Refill the buret with a sample of hard water.

> NOTE: If sea or river water is used, it must be diluted as follows: pipet 2.00 mL of sea water into a 50 mL volumetric flask and add deionized water to the mark. At the end of your calculation, you must use this dilution factor to calculate the hardness of sea water.

Refill the EDTA buret. Record the initial volumes in these burets on the data sheet.

Transfer a 2.00 mL (2.000 g) hard water sample into a 10 mL Erlenmeyer flask. Add 0.2 to 0.3 mL of the ammonia buffer solution and EBT indicator as described earlier. Complete the titration as before. Record the final volume of EDTA. Repeat the procedure at least three more times. From these data and using the calculated EDTA molarity, calculate the percentage of M^{2+} in the hard water sample. Finally, calculate the total hardness of water in terms of CaCO$_3$ (grams per liter of water) and parts per million of CaCO$_3$ as indicated here. Assuming the density of hard water to be 1.00g/mL

$$\% \, Ca = 100 \frac{(V_{EDTA})(M_{EDTA})(mol \, Ca^{2+}/mol \, EDTA)(AW \, Ca^{2+}/mol \, Ca^{2+})}{2.00 \, g \, solution}$$

Calculate the moles of Ca^{2+} and grams of $CaCO_3$ per liter of water. From these, estimate the concentration of Ca^{2+} in water in parts per million (milligrams of Ca^{2+} per liter of water).

$$\frac{\text{g } CaCO_3}{\text{L solution}} = \text{mol } Ca^{2+} \left(\frac{\text{mol } CaCO_3}{\text{mol } Ca^{2+}}\right)\left(\frac{100.1 \text{ g } CaCO_3}{\text{mol } CaCO_3}\right)\left(\frac{1000 \text{ mL solution}}{2.00 \text{ mL solution}}\right)$$

If you have diluted your water sample, multiply your answer by the dilution factor. Determine the average of the three results.

Optional Procedure: Separate Determination of Calcium and Magnesium

Pipet 2.00 mL of the sample into a 10 mL Erlenmeyer flask. Using a Pasteur pipet, add 2 drops of 50 percent (weight per weight) NaOH to the solution.

> CAUTION: NaOH is highly corrosive; avoid skin contact. In the case of skin contact, wash the area with copious amount of cold water. Report to your instructor.

Swirl the mixture for 2 minutes to precipitate the $Mg(OH)_2$ completely (you may not actually see a precipitate). Add ~ 10 mg of solid hydroxynaphthol blue, disodium salt. This indicator is used because it remains blue at a higher pH than EBT does. Titrate the sample with the EDTA solution as described before to a *blue* end point. Allow the solution to stand for at least 5 minutes with occasional swirling to redissolve any $Ca(OH)_2$ precipitate. If the solution turns wine red, add additional EDTA solution to reach the blue end point. Repeat this step as necessary. Perform a blank titration using 2.00 mL of deionized water. Calculate the concentration of Ca^{2+}. From the total concentration of Ca^{2+} and Mg^{2+} as determined before (see the previous part), calculate the concentration of Mg^{2+}.

Alternative Procedure

> Macroscale method   Estimated time to complete the experiment: 3.5 h

Preparation of 0.01 M EDTA Solution

Transfer about 2.0 g of $Na_2EDTA \cdot 2H_2O$ and 0.05 g of $MgCl_2 \cdot 6H_2O$ into a clean 250 mL beaker. Dissolve the solid in 100 mL of water. If the solution is turbid, add a few drops of 0.1 M NaOH until a clear solution is obtained. Transfer the solution to a labeled 500 mL bottle and dilute with an additional 400 mL of water. This solution is enough for two students.

Preparation of 0.01 M Calcium Chloride Solution

Accurately weigh 0.250 g $CaCO_3$, and transfer it into a rinsed 250 mL volumetric flask (use a funnel). Wash the solid down the side of the funnel with water. Add 6 M HCl dropwise to this mixture until the effervescence of CO_2 ceases. Add water to the mark and mix thoroughly.

Preparation of Ammonia/Ammonium Chloride Buffer Solution

See Microscale method.

Standardization of Disodium EDTA Solution

Obtain a clean and rinsed 250 mL volumetric flask, a 50 mL buret, a 25 mL pipet, and a 250 mL Erlenmeyer flask. Rinse the pipet with $CaCl_2$ solution and transfer 50.00 mL of the $CaCl_2$ solution into a 250 mL Erlenmeyer flask. Add 5 mL of ammonia buffer followed by either five drops of indicator solution (Eriochrome Black T) or a half-pea-size portion of solid indicator. Swirl the contents of the flask to dissolve all the indicator. The solution will have a wine red color.

Rinse and fill the buret with EDTA solution. Record the initial volume of EDTA in the buret. Titrate the solution in the Erlenmeyer flask by adding EDTA solution until the wine red color changes to a distinct *blue*. No trace of red color should remain in the solution. Enter the final reading on the buret. Repeat the titration with at least two more aliquots of $CaCl_2$ solution. Calculate the molarity of the EDTA solution.

Determination of Water Hardness

Refill the EDTA buret. Record the initial volume on the data sheet. Transfer 50.00 mL of the hard water sample and add 5 mL ammonia buffer solution and EBT indicator as described before. Complete the titration as before. Record the final volume of EDTA. Repeat the procedure at least three more times. From these data and using the EDTA molarity, calculate the percentage of M^{2+} in the water sample. Finally calculate the total hardness of water in terms of $CaCO_3$ (grams per liter of water) and parts per million of $CaCO_3$ in water. For calculations, see the microscale procedure.

Additional Experiment.   Determination of Calcium in Calcium Supplements by Macroscale Method

Calcium is a biologically important metal. Of the calcium in the human body, 99 percent is found in bones and 1 percent in the teeth. Blood serum contains about 5 mmol/L of calcium. Calcium deficiency causes several health problems, such as osteoporesis.

Many calcium supplement tablets are sold commercially, such as calcium gluconate (the salt of *D*-gluconic acid, $HOOC[CH(OH)]_4CH_2OH$) and calcium lactate (the salt of lactic acid, $CH_3CHOHCOOH$, which is the weak acid found in sour milk, apples and other fruits, beer, and wine).

Calcium can be determined in any of these supplements by titrating it with a standardized EDTA solution. The following is a macroscale procedure that can be scaled down to microscale levels. Prepare the following solutions according to the procedure described before: 250 mL of 0.01 M EDTA solution, a primary standard 0.01 M $CaCl_2$ solution, and an ammonia buffer solution. Standardize the EDTA solution. Record your data on the data sheet.

Obtain a calcium sample from the instructor. Powder it in a mortar, and dry the sample in an oven at 100°C for one hour. Transfer 0.1000 g of the powder to a 250 mL Erlenmeyer flask, and add 20 to 25 mL of deionized water to dissolve the sample. Some heating may be necessary. If it does not dissolve completely, ignore the turbidity. Add a small amount of the indicator Eriochrome Black T (half the size of a pea of solid or five drops of solution) and titrate the mixture with standardized EDTA solution according to the procedure outlined before. The color will change from *wine red* to *blue*. Record the final volume of EDTA solution. Repeat the procedure two more times and calculate the percent calcium in the supplement.

$$\% \, \text{Ca} = 100 \times \frac{(V_{\text{EDTA}})(M_{\text{EDTA}})(\text{mol Ca}^{2+}/\text{mol EDTA})(\text{AW Ca}^{2+}/\text{mol Ca}^{2+})}{\text{g sample}}$$

Additional Independent Projects

1. Determination of the concentration of salt on a pretzel by titrating with a standard solution of silver nitrate[1]
2. Titration of calcium and magnesium in milk[2]
3. Determination of Na_4EDTA in household chemicals[3]

References

1. Day, Jr., R. A.; Underwood, A. L. *Quantitative Analysis,* 5th ed.; Prentice-Hall: Englewood Cliffs, NJ, 1986; pp. 685–690.
2. McCormick, P. G. *J. Chem. Educ.* **1973,** *50,* 136.
3. Kump, K. I.; Palocsay, F. A.; Gallaher, T. N. *J. Chem. Educ.* **1978,** *55,* 265.

Name ______________________________	Section ____________
Instructor ___________________________	Date ______________

PRELABORATORY REPORT SHEET—EXPERIMENT 14

Experiment title ____

and part ____

Objective

Reactions/formulas to be used

Materials and equipment table

Outline of procedure

1. Suppose that in the preparation of 0.1 M EDTA you have used contaminated deionized water, containing a trace amount of Mg^{2+}. Your standard 0.1 M calcium chloride solution was, however, prepared with a pure sample of deionized water (i.e., it did not contain any Mg^{2+}). You standardized the EDTA solution with the calcium chloride solution.

   (a) How would the presence of Mg^{2+} affect the molarity of EDTA?

   (b) If you use the standardized EDTA solution containing the impurity to determine water hardness, how would it affect the result?

EXPERIMENT 14   DATA SHEET

Determination of Hardness of Water

Data Collection and Manipulation Tables

1. Preparation of 0.01 M Disodium EDTA Solution

1. Mass of EDTA before transfer   _____ g

2. Mass of EDTA after transfer   _____ g

3. Mass of EDTA taken   _____ g

4. Mass of $MgCl_2 \cdot 6H_2O$ taken   _____ g

5. Volume of the solution   _____ mL

2. Preparation of 0.01 M Calcium Chloride Solution

6. Mass of $CaCO_3$ before transfer   _____ g

7. Mass of $CaCO_3$ after transfer   _____ g

8. Mass of $CaCO_3$ taken   _____ g   Volume of the flask _____ mL

9. Concentration of $CaCl_2$ solution   _____ M

3. Preparation of Ammonia Buffer Solution

10. Volume of concentrated ammonia   _____ mL

11. Mass of NH_4Cl added   _____ g

12. Total volume of the buffer   _____ mL

4. Standardization of EDTA Solution

		Trial 1	Trial 2	Trial 3
13.	$CaCl_2$ volume taken	_____ mL	_____ mL	_____ mL
14.	Initial EDTA volume	_____ mL	_____ mL	_____ mL
15.	Final EDTA volume	_____ mL	_____ mL	_____ mL
16.	Concentration of EDTA	_____ M	_____ M	_____ M
17.	Average concentration of EDTA solution		_____ M	

Show calculations:

Data Collection

Concentration of EDTA solution _____ M

Sample number and/or source _____

5. Determination of Hardness of Water

		Trial 1	Trial 2	Trial 3
18.	Volume of water sample	_____ mL	_____ mL	_____ mL
19.	Initial volume EDTA	_____ mL	_____ mL	_____ mL
20.	Final volume EDTA	_____ mL	_____ mL	_____ mL

Data Manipulation

21.	Dilution factor, if any	_____		
22.	Moles of EDTA used	_____ mol	_____ mol	_____ mol
23.	Average moles of EDTA		_____ mol	
24.	Average moles of Ca^{2+}		_____ mol	
25.	Percent Ca^{2+}		_____ %	
26.	Hardness of water (g $CaCO_3$/L water)		_____ g/L	
27.	ppm $CaCO_3$		_____ ppm	

Optional Procedure

6. Determination of Calcium Separately in Hard Water

28.	Volume of water sample	_____ mL
29.	Volume of EDTA used after adding NaOH	_____ mL
30.	Moles of EDTA used for Ca^{2+} in water	_____ mol
31.	Moles of Mg^{2+}	_____ mol
32.	ppm Ca^{2+}	_____ ppm
33.	ppm Mg^{2+}	_____ ppm

Show calculations: use a separate sheet of paper if necessary.

DATA SHEET, EXPERIMENT 14, P. 3

Determination of Calcium in Calcium Supplements

1. Preparation of 0.01 M Disodium EDTA Solution

1. Mass of EDTA before transfer _____ g

2. Mass of EDTA after transfer _____ g

3. Mass of EDTA taken _____ g

4. Mass of $MgCl_2 \cdot 6H_2O$ taken _____ g

5. Volume of the solution _____ mL

2. Preparation of 0.01 M Calcium Chloride Solution

6. Mass of $CaCO_3$ before transfer _____ g

7. Mass of $CaCO_3$ after transfer _____ g

8. Mass of $CaCO_3$ taken _____ g   Volume of the flask _____ mL

9. Concentration of $CaCl_2$ solution _____ M

3. Preparation of Ammonia Buffer Solution

10. Volume of concentrated ammonia _____ mL

11. Mass of NH_4Cl added _____ g

12. Total volume of the buffer _____ mL

4. Standardization of EDTA Solution

		Trial 1	Trial 2	Trial 3
13.	$CaCl_2$ volume taken	_____ mL	_____ mL	_____ mL
14.	Initial EDTA volume	_____ mL	_____ mL	_____ mL
15.	Final EDTA volume	_____ mL	_____ mL	_____ mL
16.	Concentration of EDTA	_____ M	_____ M	_____ M
17.	Average concentration of EDTA solution			_____ M

Show calculations:

		Trial 1	**Trial 2**	**Trial 3**
18.	Mass of Ca tablet before transfer	_____ g	_____ g	_____ g
19.	Mass of Ca tablet after transfer	_____ g	_____ g	_____ g
20.	Mass of tablet taken	_____ g	_____ g	_____ g
21.	Initial volume EDTA	_____ mL	_____ mL	_____ mL
22.	Final volume EDTA	_____ mL	_____ mL	_____ mL

(If the microburet needs to be refilled, report the total volume in step 22.)

23.	Moles of EDTA used	_____ mol	_____ mol	_____ mol
24.	Moles of Ca^{2+} present in the sample taken	_____ mol	_____ mol	_____ mol
25.	Percent Ca^{2+} in the tablet	_____ %	_____ %	_____ %
26.	Average percent Ca^{2+}		_____ %	

Show calculations:

POSTLABORATORY PROBLEMS—EXPERIMENT 14

1. Explain why a pH 10 ammonia buffer is required for the EDTA titration of water hardness. Look at the chart of pK_a values. Suggest which other buffers could be used.

2. What is the hardness (ppm $CaCO_3$) in a sample of water, 2.00 mL of which requires 0.560 mL of 0.0127 M EDTA solution for titration?

Identification of Chemical Compounds in Solution

Microscale Experiment

Objectives
* To investigate ionic compounds (acids, bases, and salts) and their reactions in aqueous solutions
* To know the solubility of these species in aqueous solutions

Prior Reading
* A sound knowledge of chemical reactions and solubility rules is essential.

Related Experiments
* Experiments 16 (Qualitative analysis) and 30 (Organic qualitative analysis)

Activities a Week Before:
Prelaboratory assignment must be done a week before. This laboratory requires a good deal of prelaboratory preparation.

In this experiment, a set of chemical reactions of acids, bases, and salts in aqueous solutions will be used to identify a specific compound in solution. Based on the Arrhenius concept, an **acid** is a compound that (in water) increases the concentration of H^+ or H_3O^+ in solution. A **base** is any substance that produces OH^- ions. Alternative definitions of acids and bases include the Brønsted-Lowry theory, the Lewis theory, and the solvent system definition (see your textbook).

Salts are compounds consisting of a cation (positively charged species) other than H^+ and an anion (negatively charged species) other than OH^-, oxide (O^{2-}), and peroxide (O_2^{2-}) ions. In a salt, the cation or anion may be simple ions like Ca^{2+} or Cl^-; polyatomic ions such as NH_4^+ (ammonium) or ClO_4^- (perchlorate); or complex species such as $[Ag(NH_3)_2]^+$ [diamminesilver(I)].

Acid-base Reactions

An acid turns blue litmus paper red; a base turns red litmus paper blue. Thus, it is easy to identify an acidic or a basic solution. An acid neutralizes a base, forming water and a salt. The reaction is usually exothermic (it generates heat). An example of such a reaction is

$$H_2SO_4 + 2\,NaOH \rightarrow Na_2SO_4 + 2\,H_2O + heat$$

Some acids and bases are stronger than others. A useful guide to the degree of acidity and basicity of chemical compounds can be obtained from the acid or base ionization constants (K_a and K_b). The values for these constants are given in Appendix 4. Larger K_a and K_b values indicate stronger acids and bases, i.e., more complete ionization.

Some soluble salts react with water, producing acidic or basic solutions. Ammonium chloride (NH_4Cl) in water, for example, generates an acidic solution and sodium acetate ($NaCH_3COO$) results in the formation of a basic solution:

$$NH_4Cl(aq) \rightarrow NH_4^+(aq) + Cl^-(aq) \qquad \text{(dissociation)}$$

$$NH_4^+(aq) + H_2O(l) \rightarrow NH_3(aq) + H_3O^+$$

$$Na(CH_3COO)(aq) \rightarrow Na^+(aq) + CH_3COO^-(aq) \qquad \text{(dissociation)}$$

$$CH_3COO^-(aq) + H_2O(l) \rightarrow CH_3COOH(aq) + OH^-$$

Precipitation Reactions

Some salts are soluble in water and dissociate into the corresponding cations and anions. Other salts are insoluble and do not dissociate. The mixing of two solutions may result in the formation of a **precipitate** (insoluble compound). The precipitate may be colored or white. The following general solubility rules are helpful:

Soluble Salts

1. Salts of Li^+, Na^+, K^+, and NH_4^+ are soluble.
2. NO_3^- (nitrate), NO_2^- (nitrite), ClO_4^- (perchlorate), ClO_3^- (chlorate), BrO_4^- (perbromate), BrO_3^- (bromate), IO_4^- (periodate), CH_3COO^- (acetate), and $Cr_2O_7^{2-}$ (dichromate) salts are soluble.
3. Most Cl^- (chloride), Br^- (bromide), I^- (iodide), and SCN^- (thiocyanate) salts are soluble. The major exceptions are those of Pb^{2+}, Ag^+, and Hg_2^{2+}. (Note: Lead halides are soluble in hot water.)
4. SO_4^{2-} (sulfate) salts are soluble except those of Sr^{2+}, Ba^{2+}, and Pb^{2+}. Note that Ca^{2+}, Ag^+, and Hg^{2+} sulfates are slightly soluble.

Insoluble Salts

5. BO_3^{3-} (borate), CO_3^{2-} (carbonate), HCO_3^- (bicarbonate), SiO_3^{2-} (silicate), PO_3^{3-} (phosphite), PO_4^{3-} (phosphate), AsO_3^{3-} (arsenite), AsO_4^{3-} (arsenate), S^{2-} (sulfide), SO_3^{2-} (sulfite), and F^- (fluoride) salts are usually insoluble. Exceptions are the salts of these anions with Group I metal cations and NH_4^+ (as noted in the foregoing rule 1). Silver fluoride is also soluble.
6. O^{2-} (oxide) and OH^- (hydroxide) compounds are insoluble except those of the Group I metal cations, Ca^{2+}, Sr^{2+}, Ba^{2+}, and NH_4^+.
7. CrO_4^{2-} (chromate) salts are insoluble except those of the Group I cations, NH_4^+, Mg^{2+}, Ca^{2+}, and Sr^{2+}.

The solubility of a sparingly soluble salt depends upon the solubility product constant, K_{sp}. The values of K_{sp} can be found in Appendix 5. Higher K_{sp} values indicate greater solubility.

A note of caution is needed here. Many times, a simple salt will be insoluble in water, but on addition of an excess anion will become soluble owing to formation of a complex. For example, the compound $Zn(OH)_2$ is insoluble (as are most hydroxides). Upon addition of more hydroxide, the complex $[Zn(OH)_4]^{2-}$ forms, which is soluble.

Evolution of Gases

Some reactions proceed with the evolution of gases. The identification of these gases can be helpful in identifying the parent compounds. H_2S gas has a rotten egg smell, SO_2 has a pungent smell, and NO_2 is a brown gas.

> All three gases are toxic. Avoid breathing them.

Some examples of gas-forming reactions include the following:

$$Na_2CO_3(aq) + 2\,HCl(aq) \rightarrow 2\,NaCl(aq) + CO_2(g) + H_2O(l)$$
$$Na_2S(aq) + H_2SO_4(aq) \rightarrow Na_2SO_4(aq) + H_2S(g)$$
$$Na_2SO_3(aq) + 2\,HCl(aq) \rightarrow 2\,NaCl(aq) + H_2O(l) + SO_2(g)$$

Redox Reactions and Formation of Colored Solutions

Some reactions produce colored solutions. For example, when KI is oxidized, I_2 is produced (see Experiments 9 and 12), which imparts a brown color to the solution:

$$2\,KI(aq) + NaOCl(aq) + H_2SO_4(aq) \rightarrow I_2(aq) + K_2SO_4(aq) + NaCl(aq) + H_2O(l)$$

Prelaboratory Assignment

Read the following experimental procedure. Select any nine solutions from the following list and prepare a chart of the principal reactions between any two pairs of solutions. The object is to predict what reactions are possible when any two solutions are mixed. Try to include all possible combinations for the set selected.

Silver nitrate, $AgNO_3$

Sulfuric acid, H_2SO_4

Sodium sulfate, Na_2SO_4

Sodium bromide, $NaBr$

Potassium chlorate, $KClO_3$

Sodium sulfide, Na_2S

Sodium thiosulfate, $Na_2S_2O_3$

Sodium hydroxide, $NaOH$

Barium chloride, $BaCl_2$

Potassium chromate, K_2CrO_4

Hydrochloric acid, HCl

Nitric acid, HNO_3

Sodium sulfite, Na_2SO_3

Ammonium chloride, NH_4Cl

Sodium nitrite, $NaNO_2$

Potassium iodide, KI

Potassium iodate, KIO_3

Bleach, $NaOCl$

Water, H_2O

Barium hydroxide, $Ba(OH)_2$

Reference

1. Zuehlke, R. W. *J. Chem Educ.* **1966,** *43,* 601.

Experimental Section

Procedure

Microscale method   Estimated time to complete the experiment: 3 h
Experimental Steps: This procedure involves two steps: (1) making a matrix table to record the results of the tests performed and (2) analyzing and classifying the results to identify the solutions.

Disposal procedure: Dispose of all the solutions in marked containers.

You will be given nine unknown solutions from the list of 20 given in the prelaboratory assignment. The test tubes are numbered 1 through 9. *The matrix table must be made before coming to the laboratory.* You will also be provided with a list containing the names of the nine solutions, but the identity of each of these will not be given. The task is to determine which test tube contains which solution.

CAUTION: H_2S gas, Br_2, and I_2 are all hazardous chemicals. Do not smell them. Perform these tests inside a HOOD.

Litmus paper should be used to determine the acidity or basicity of the solutions; lead acetate paper, to test for H_2S gas (if a wet lead acetate paper is held at the mouth of the test tube where H_2S gas is generated, it will turn black because of the formation of black PbS); and toluene solvent, to detect the presence of bromine and iodine. Toluene is insoluble in water and forms a separate layer when added to a reaction mixture in a test tube. Bromine and iodine are soluble in toluene, imparting a brown or violet color, respectively.

After determining which solution is acidic or basic, mix equal volumes of any two solutions (two or three drops of each) in a clean micro test tube. No other solutions except the nine solutions given may be used to identify the unknowns. After mixing each pair of solutions, observe carefully to see if any reaction has taken place in terms of the evolution of gas, precipitation, heat changes, color changes, and so on.

From your observations, you should be able to determine which test tube contains which solution. Carefully observe the reactions. After each reaction is performed, enter your observations in the chart as shown in Table E15.1.

Another chart like the one shown in Table E15.2 may also be drawn in which to write your observations for the last set of reactions.

If, at the end of the first 40 minutes, you are unable to complete the identification, ask your instructor to provide you with the set of the original 20 solutions. Select those solutions that are in your original list. **Do not remove these 20 solutions from the table; others may need the samples.**

Table E15.1 Summary of the reactions (by solution number)

	1	2	3	4	5	6	7	8	9
1	?	-	-	-	W/ppt*	Gas	-	-	-
2	-	?	-	-	W/ppt	-	-	-	-
3	-	-	?	-	-	Heat	-	-	-
4	-	-	-	?	W/ppt	Cloudy	-	-	-
5	W/ppt	W/ppt	-	W/ppt	?	W/ppt	-	-	W/ppt
6	Gas	-	Heat	Cloudy	W/ppt	?	-	-	-
7	-	-	-	-	-	-	?	-	-
8	-	-	-	-	-	-	-	?	-
9	-	-	-	-	W/ppt	-	-	-	?

*white precipitate

Table E15.2 Reaction results (by solution number)

	2	4	7	8	9
2	?	No rxn	Brown soln.	No rxn	No rxn
4	-	?	No rxn	No rxn	No rxn
7	-	-	?	No rxn	No rxn
8	-	-	-	?	No rxn
9	-	-	-	-	?

Using your selected nine solutions, perform the standard tests as already outlined, and try to match your preliminary findings with those obtained earlier. In this way, you can confirm your initial conclusions. **If any of the solution required to perform a particular test is not in the set given to you, use the solution you need from those provided by your instructor.**

Discuss your results with the instructor before writing the final report. Include the following information in your report:

1. The matrix table and the list of the nine chemicals properly identified. To identify the chemicals, put the number of the test tube containing the unknown by the side of the name of the chemical in the list.
2. A short write-up describing how you arrived at your conclusions.
3. A series of balanced reactions that helped in the identification of the unknowns.

Example of the Procedure

An example of the procedure is now illustrated. Suppose you are given the following nine solutions:

$$Na_2S \qquad Na_2S_2O_3 \qquad KI$$
$$KIO_3 \qquad BaCl_2 \qquad H_2O$$
$$NaOH \qquad H_2SO_4 \qquad Na_2SO_4$$

Table E15.1 is constructed on the basis of the observations made on the reactions taking place when two solutions are mixed. Note that the table has columns and rows numbered 1 through 9.

First, each solution is tested with litmus paper (or better, use pH paper). It is found that solution 3 is basic and that solution 6 is acidic. Solution 1 also shows slight basic character. From the given list of chemicals, we can conclude that solution 3 is NaOH and solution 6 is sulfuric acid.

From the table, it is observed that solution 5 is the most reactive because it forms white precipitates (w/ppt) with many of the solutions. Solution 1, when mixed with solution 6, produces a gas that turns moist lead acetate paper black. The only salt from the list that reacts with sulfuric acid to form H_2S gas is Na_2S. Therefore, solution 1 contains Na_2S.

$$H_2SO_4(aq) + Na_2S(aq) \rightarrow Na_2SO_4(aq) + H_2S(g)$$

Solution 6, when added to solution 5, forms a white precipitate. A look at the list shows that since solution 6 is H_2SO_4, solution 5 must be $BaCl_2$, which forms $BaSO_4$ (an insoluble sulfate) according to the following reaction:

$$BaCl_2(aq) + H_2SO_4(aq) \rightarrow BaSO_4(s) + 2\,HCl(aq)$$

Notice that solution 5 also forms white precipitates with solutions 1, 2, 4 and 9. Since solution 1 is known to be Na_2S, the other three solutions must be KIO_3, $Na_2S_2O_3$ and Na_2SO_4. To identify which is which, we will have to adopt a new strategy involving the use of other reactions.

Out of the nine chemicals given in the list, five solutions (2, 4, 7, 8, 9) are still unknown. Three of them (2, 4, and 9) are KIO_3, Na_2SO_4 and $Na_2S_2O_3$ (all of these solutions form a precipitate with solution 5). The remaining two (7 and 8) must then be H_2O and KI (no precipitate with solution 5).

It is known that KIO_3 (or $KClO_3$) quantitatively oxidizes KI in the presence of an acid to form I_2. Such a mixture will become dark brown in color as a result of the dissolved I_2 in KI (KI_3). Elemental I_2 thus formed can be quantitatively reduced back to iodide, I^-, by the thiosulfate anion, $S_2O_3^{2-}$. The resulting solution thus becomes colorless. The sequence of these oxidation and reduction reactions is

$$5\,KI(aq) + KIO_3(aq) + 3\,H_2SO_4(aq) \rightarrow 3\,K_2SO_4(aq) + 3\,I_2(s) + 3\,H_2O(l)$$
$$I_2(s) + KI(aq) \rightarrow KI_3(aq) \quad \text{(brown)}$$
$$KI_3(aq) + 2\,Na_2S_2O_3(aq) \rightarrow KI(aq) + 2\,NaI(aq) + Na_2S_4O_6(aq)$$

These reactions will establish the identity of the three solutions KI, $Na_2S_2O_3$, and KIO_3. Now, make a series of mixtures by mixing these five solutions two at a time. It can be easily seen that no more than 10 such mixtures need be made. Add to each of these mixtures one drop of H_2SO_4 (solution 6).

Table E15.2 can be constructed in a similar manner to Table E15.1. As can be seen from the table, when H_2SO_4 is added to the mixture of solutions 2 and 7, the mixture becomes brown, indicating the formation I_2 (KI_3). Thus, 2 and 7 must be KI and KIO_3. If to this brown mixture a drop of toluene is added, the toluene layer will become violet, indicating the presence of I_2. To this mixture, if several drops of solution 4 (solutions 8 and 9 do not react) are added and the mixture stirred with a glass rod, the iodine color disappears. Solution 4 is then $Na_2S_2O_3$.

Now, whereas solution 5 forms a white precipitate with solution 2 (see Table E15.1) as well as with solution 9, it does not produce any precipitate with solution 7. Solution 2 must be KIO_3 (it also produces I_2 from KI), which forms a white crystalline precipitate of $Ba(IO_3)_2$. Solution 9 is Na_2SO_4. This conclusion leaves solution 7 to be KI. The last remaining solution (solution 8), by the principle of exclusion, must be water (see the list of the chemicals given). We can further confirm our findings by performing known tests with the supplied solutions in the laboratory.

Some Important Hints

1. To detect ammonia, gently heat the mixture of a solution of an ammonium (NH_4^+) salt and NaOH over a low flame. If you hold a wet red litmus paper at the mouth of the test tube, it will turn blue.
2. To observe the formation of brown NO_2 gas when a nitrite (NO_2^-) salt solution is heated with dilute HCl in a test tube, hold the test tube against a white wall or paper. The upper part of the test tube will appear slightly brown.
3. When silver nitrate is added to a sulfide (S^{2-}) solution, a black precipitate of Ag_2S forms.
4. Addition of NaOH solution to silver nitrate solution forms a dark precipitate.
5. Note the colors of AgCl (curdy white), AgBr (light yellow), and AgI (yellow).
6. A nitrite (NO_2^-) solution yields a white crystalline precipitate of $AgNO_2$ with $AgNO_3$ solution; the crystals are, however, soluble in excess water.
7. $BaCl_2$ solution forms a precipitate with Na_2SO_3 (soluble in an acid, smell of SO_2 given off), Na_2S, and KIO_3 (not with $KClO_3$) solutions.
8. A mixture of Na_2S with KIO_3 forms a white precipitate.

Additional Independent Project

1. Analysis of an alloy[1]

Reference

1. Watson, N. V. *J. Chem. Educ.* **1971,** *48,* 324.

Name ______________________ Section ____________

Instructor ______________________ Date ____________

PRELABORATORY REPORT SHEET—EXPERIMENT 15

Experiment Title ______________________

____

Objective

Reactions/formulas to be used

Materials and equipment table

Outline of procedure

PRELABORATORY PROBLEMS—EXPERIMENT 15

1. You are given the following solutions for identification. By using only these solutions as suggested in the procedure, make a table of all the possible reactions. Also, make a second chart if necessary. From these charts, try to identify the chemicals.

1. HCl	2. NH_4Cl	3. $AgNO_3$	4. Na_2S	5. Bleach (NaOCl)
6. $NaNO_2$	7. $KClO_3$	8. Na_2SO_3	9. KI	

The numbers indicate the test tube numbers.

2. Write balanced chemical equations for all the reactions that you have used to identify the solutions. Submit the chart and your deductive reasoning as to how you arrived at your conclusion. Use separate sheets.

3. What are precipitation reactions? Write the balanced net ionic reactions for the following:

$$Na_2S(s) + HCl(aq) \rightarrow$$

$$AgNO_3(aq) + NaOH(aq) \rightarrow$$

$$BaCl_2(aq) + H_2SO_4(aq) \rightarrow$$

$$H_2SO_4(aq) + Na_2SO_3(aq) \rightarrow$$

4. Write balanced chemical equations for the following:

$$Cl_2(g) + NaOH(aq) \rightarrow$$

What is the name and type of the following reaction?

$$Na_2S(g) + HNO_3(aq, conc) \rightarrow$$

POSTLABORATORY PROBLEMS—EXPERIMENT 15

1. Which of the following compounds will form gaseous products when treated with dilute sulfuric acid (cold or hot)?

Compounds	**Reactions**
Na_2SO_4 or Na_2SO_3	
Na_2S or $NaClO_3$	
K_2SO_4 or KCl	
KIO_3 or K_2CO_3	

2. Why does NH_4Cl behave as an acid in aqueous solution?

3. Which is the stronger oxidizing agent, KIO_3 or $KClO_3$ (see Appendix 6)?

4. Write a balanced reaction to illustrate the use of $KBrO_3$ as an oxidizing agent in an acid medium.

5. The description of the procedure is based on the scientific method involving the use of such terms as "Definition of a Problem," "Prediction," "Hypothesis," "Experimentation," "Data Collection and Interpretation," and "Proof of Hypothesis." Indicate which parts of the procedure you developed belong to these processes. Use a separate sheet if necessary.

An Introduction to Inorganic Qualitative Analysis: General Methods and Techniques

(Read This Chapter in Conjunction with Experiment 16)

Objectives
- To become familiar with the methods of inorganic qualitative analysis

Prior Reading
From any general chemistry textbook:
- Chemical reactions and types of chemical reactions
- Chemical equilibria: acid-base, complex, and solubility equilibria and their role in qualitative analysis

Chemistry is the experimental science that studies the properties of matter—its composition, analysis, and reactivity. Analytical chemistry is the sub-area of chemistry that specifically deals with the analysis of matter. It is divided into two broad general areas: **qualitative** and **quantitative** analysis. Qualitative analysis is concerned with **determining the identity** of the constituents present in a sample. Quantitative analysis deals with the determination of **how much** of a particular constituent is present in a sample. This discussion pertains to qualitative analysis of inorganic salts.

6.1 Common Reactions Encountered in Inorganic Qualitative Analysis

Most soluble inorganic salts dissociate into ions (cations and anions) in aqueous solution. These salts, as well as some complexes and the strong acids HCl, HBr, HI, $HClO_4$, H_2SO_4, and HNO_3, are almost completely dissociated in solution, forming the corresponding ions. Therefore, reactions occurring in aqueous solutions are often conveniently considered to be occurring between ions (for example, between Pb^{2+} and SO_4^{2-}).

To a certain degree, all chemical reactions are reversible. Whenever a reaction takes place, products are formed. If they are not removed from the reaction zone, products react or decompose to some extent to regenerate the reactants by a reverse reaction. At some point, the rate of formation of the products becomes equal to the rate of decomposition of the products, that is, a **dynamic equilibrium** is established. Under equilibrium conditions, most chemical reactions do not go to completion. The extent to which a reaction occurs depends on the magnitude of the **equilibrium constant** (K_{eq}) for the reaction and the relative amounts of reagents present. According to **Le Chatelier's principle**, the equilibria can be shifted by adding or removing reagents and by changing physical conditions. By controlling the reaction conditions, one can force a precipitation reaction to occur, make sparingly soluble compounds dissolve, or form a complex with metal cations so that they will not interfere with tests for other ions of interest.

Several types of common reactions encountered in qualitative analysis are listed below (see also Experiment 15).

(a) Acid-Base Reactions and Changes in pH: the Brønsted-Lowry Concept

According to the Brønsted-Lowry concept, an acid is a substance that donates a proton (H^+) to any other substance. The substance that accepts a proton is called a base. The following are examples of Brønsted-Lowry acid-base reactions. In these reactions, there is a change in the concentration of H_3O^+ and OH^- in solution.

$$2\,H_3O^+(aq) + CO_3^{2-}(aq) \rightleftharpoons 3\,H_2O(l) + CO_2(g) \tag{6.1}$$

$$Al(OH)_3(s) + OH^-(aq) \rightleftharpoons Al(OH)_4^-(aq) \tag{6.2}$$

$$Al(OH)_3(s) + 3\,H_3O^+(aq) \rightleftharpoons Al^{3+}(aq) + 6\,H_2O(l) \tag{6.3}$$

$$NH_3(aq) + H_2O(l) \rightleftharpoons NH_4^+(aq) + OH^-(aq) \tag{6.4}$$

In reaction (6.1), H_3O^+, an acid, reacts with carbonate ion (CO_3^{2-}) to produce water and carbon dioxide gas. In reaction (6.2), aluminum hydroxide, $Al(OH)_3$, dissolves in basic solution to form the tetrahydroxoaluminate ion, $Al(OH)_4^-$. This lowers the pH, as hydroxide ion is consumed. In reaction (6.3), $Al(OH)_3$ dissolves in acidic solution to form the Al^{3+} ion. This consumes H_3O^+, thereby raising the pH. Species such as $Al(OH)_3$, which react with either acid or base, are called **amphoteric substances**. In reaction (6.4), water behaves as an acid. A solution of 0.1 M NH_3 in distilled water has a pH of 8.9, whereas concentrated ammonia (14.5 M) has a pH of 10.0, an increase of hydroxide concentration by a factor of 10. The pH and concentration of ammonia can determine whether an ion precipitates as a hydroxide salt or forms a soluble ammonia or hydroxide complex. Sometimes, no reaction occurs at all.

Buffers (see Experiment 28 for a detailed discussion) are solutions that resist changes in pH when small amounts of a strong acid or base are added. A solution is said to be buffered when it has (approximately) equal amounts of a weak acid (or base) and its conjugate salt. Common examples are CH_3COOH/CH_3COONa and NH_3/NH_4Cl.

(b) Precipitation Reactions

The solubility of any sparingly soluble compound is determined by its solubility product constant, K_{sp} (see Experiment 13 for detailed discussion). When the concentration of reacting ions in solution exceeds the solubility product of a particular salt, **precipitation** will occur. Tables of K_{sp} values are available (see Appendix 5). However, the following **solubility rules** or **guidelines** are valuable in qualitative analysis.

Table 6.1   Solubility Rules for Common Ions

1. All nitrates (NO_3^-) and nitrite (NO_2^-) salts of the cations mentioned in this chapter are *soluble*.

2. Salts of the Group I (periodic table) cations (Na^+, K^+, Rb^+, and Cs^+) and of ammonium (NH_4^+) are *soluble*.

3. All halides (Cl^-, Br^-, and I^-) and thiocyanates (SCN^-) are *soluble*, except for $Ag^+, Cu^+, Tl^+, Pb^{2+}$, and Hg_2^{2+}.

4. All sulfates (SO_4^{2-}) are *soluble* except for Pb^{2+}, Ba^{2+}, and Sr^{2+} (Ca^{2+}, Hg_2^{2+}, and Ag^+ are somewhat soluble).

5. Thiosulfates ($S_2O_3^{2-}$) are *soluble*, except for Pb^{2+}, Ba^{2+}, and Ag^+ ($Ag_2S_2O_3$ decomposes in excess thiosulfate with reduction of Ag^+).

6. Sulfites (SO_3^{2-}), carbonates (CO_3^{2-}), phosphates (PO_4^{3-}), oxalates ($C_2O_4^{2-}$), and chromates (CrO_4^{2-}) are all *insoluble* in neutral or basic solution, with the exception of Mg^{2+} and the ions listed in Rule 2. Similarly, oxides (O^{2-}) and hydroxides (OH^-) are *insoluble*, except for $Ca^{2+}, Sr^{2+}, Ba^{2+}$ and the ions listed in Rule 2. All are soluble in acidic solution.

7. Fluorides (F^-) are *insoluble*, except for Ag^+, Fe^{3+}, and the ions listed in Rule 2. Some transition metal fluorides are soluble as a result of complex ion formation.

The solubility of a particular species may be affected by the presence of a particular reagent that causes a competing reaction. For example, silver chloride, AgCl, is insoluble in distilled water, but it is soluble in aqueous ammonia because of the formation of the soluble silver ammine complex, $Ag(NH_3)_2^+$:

$$AgCl(s) \rightleftharpoons Ag^+(aq) + Cl^-(aq)$$
$$Ag^+(aq) + 2\,NH_3(aq) \rightleftharpoons Ag(NH_3)_2^+(aq)$$

Detecting the presence of a precipitate is not always trivial. The solution and the precipitate may or may not change color. The presence of a precipitate is difficult to detect when a solution is dark in color. In such cases, it helps to dilute a small sample of the solution, or to examine a thin layer of it against a light. A note on terminology: *clear* means not milky; *colorless* means like water. The two terms are not synonymous.

(c) Decomposition

A **decomposition** reaction is one where a compound changes into one or more different, less complex products. For example, when HgO is strongly heated, it decomposes into metallic mercury and oxygen gas:

$$2\,HgO(s) \overset{\triangle}{\rightleftharpoons} 2\,Hg(l) + O_2(g)$$

(d) Oxidation-Reduction Reactions

Oxidation-reduction, or redox, reactions are used frequently in qualitative analysis to dissolve an insoluble compound or to convert an ion to a different oxidation state. For example,

$$2\,NaBr(aq) + H_2O_2(aq) + 2\,H^+(aq) \rightarrow Br_2(l) + 2\,Na^+(aq) + 2\,H_2O(l) \qquad (6.5)$$

$$2\,Cr(OH)_3(s) + 3\,H_2O_2(aq) + 4\,OH^-(aq) \rightarrow 2\,CrO_4^{2-}(aq) + 8\,H_2O(l) \qquad (6.6)$$

Bromine produced in (6.5) can be detected by adding a drop or two of toluene solvent (in which Br_2 is soluble), making the solvent layer brown in color. The anion CrO_4^{2-} formed in (6.6) is easily detected by its distinctive yellow color.

In laboratories, nitric acid (HNO_3) and hydrogen peroxide (H_2O_2) solution are the most frequently used oxidizing agents. Ferrous (Fe^{2+}), stannous (Sn^{2+}), thiosulfate ($S_2O_3^{2-}$), oxalate ($C_2O_4^{2-}$), and iodide (I^-) ions are often used as reducing agents. Hydrogen peroxide in acidic solution also acts as a reducing agent.

(e) Complex Formation Reactions

Many common anions and neutral molecules can donate one or more lone pairs of electrons (thus acting as **Lewis bases or donors**) to a **Lewis acid** to form a bond. Metal cations are usually Lewis acids and as such accept two to six electron pairs from Lewis bases in forming complexes. In a complex ion, the Lewis bases are known as **ligands**. Water and ammonia are examples of neutral ligands. Anions that readily act as ligands include Cl^-, Br^-, I^-, SCN^-, $C_2O_4^{2-}$, OH^-, F^-, and CN^-.

Metal cations in aqueous solution tend to have a fixed number (most often six) of water molecules acting as ligands, although the water molecules are often not included when writing reactions. For example, the reaction

$$FeCl_3(\text{in water}) \rightarrow Fe^{3+}(aq) + 3\,Cl^-(aq)$$

should more properly be written as

$$FeCl_3(s) + 6\,H_2O(l) \rightleftharpoons [Fe(H_2O)_6]^{2+}(aq) + 3\,Cl^-(aq)$$

The aqueous solutions of many transition metal ions are colored because water molecules, acting as ligands, form colored aquo complex ions. Some ligands have a higher affinity for the metal ion than water molecules do. Such ligands replace the water molecules from an aquo complex, resulting in the formation of a new complex ion. Often, this reaction is accompanied by a color change that can serve as a diagnostic test for the presence of a given metal cation. In other cases, a metal can form a very stable complex. This prevents the metal cation from reacting with a reagent being used to test for another metal cation. For example, to detect Co^{2+} in the presence of Fe^{3+}, F^- ions are added. The fluoride anions form a stable complex with Fe^{3+}, $(FeF_6)^{3-}$, which is almost colorless. When an aqueous solution of KSCN (SCN^-) is added to Co^{2+} in the presence of Fe^{3+}, a blue complex, $[Co(SCN)_4]^{2-}$, forms. The iron cation, being complexed by fluoride anions, fails to react with SCN^- to form the characteristic red complex $Fe(SCN)^{2+}$.

(f) Catalytic Reactions

When a **catalyst** is added to a reaction system, the activation energy is lowered and equilibrium is reached more quickly. Many metal cations act as catalysts because they can exist in more than one oxidation state. By being oxidized or reduced, they can often catalyze a reaction by increasing the rate at which electrons are transferred.

6.2. Common Laboratory Techniques for Qualitative Analysis

A successful chemical analysis depends upon a number of factors. Some of these are described below.

(a) Knowledge of Chemical Reactions

A sound knowledge of the chemical reactions taking place in aqueous solution is an essential component of a successful qualitative analysis.

(b) Cleanliness of Glassware and Other Equipment

The use of clean glassware is an important requirement for qualitative analyses. Since the reactions will be carried out using drops of solutions, it is recommended that small test tubes (10×75 mm), small pointed stirring rods (Figure 6.1a; for obvious reasons, metal spatulas are not recommended except for transferring solids) and other small-size glassware such as 10 mL beakers, Pasteur pipets (5 inch and 9 inch), medicine droppers, and 12 mL or 15 mL centrifuge tubes (Figure 6.1b) be employed. Several small glass stirring rods can be constructed by cutting 10 cm × 2 mm pieces from an originally long glass rod. Fire-polish the ends. Make sure that all glassware has been scrupulously cleaned and dried.

(c) Heating of Solutions and Use of a Hot Water Bath

Most chemical reactions are accelerated by heating. A hot water bath is a safe method of heating. Set up a 250 mL beaker, three-quarters filled with deionized water, on a hot plate. A test tube rack (Figure 6.2) made by drilling several holes in a thick plastic or thin aluminum plate (11×11 cm) may be employed to hold several test tubes in the same water bath. The rack should be placed over the mouth of the beaker being used as a water bath.

> CAUTION: Never heat a test tube directly over an open flame; use a hot water bath.

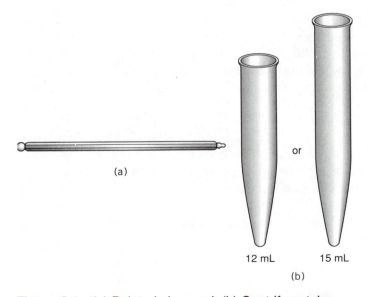

(a)

or

12 mL   15 mL

(b)

Figure 6.1   (a) Pointed glass rod; (b) Centrifuge tube

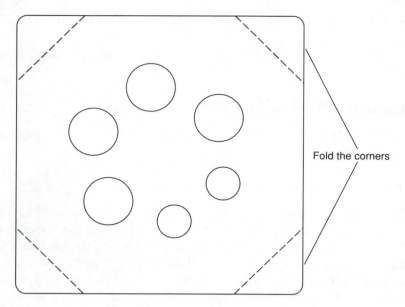

Figure. 6.2   Test tube rack with holes

(d) Use of a Medicine Dropper or a Pasteur Pipet

Always use a *clean* medicine dropper or a Pasteur pipet to add a reagent or to transfer a solution from one test tube to another. Obtain a 50 mL beaker containing deionized water. Rinse the dropper *once* with deionized water from the beaker and dispose of the wash liquid into a designated container. Rinse the pipet or the dropper a second and a third time with deionized water. Generally, all reagents will be provided in dropping bottles fitted with medicine droppers.

> To avoid contamination, always replace the screw cap carrying the dropper on the correct bottle.

Before use, it may be desirable to calibrate the droppers or Pasteur pipets. A medicine dropper (1.5 mm inside and 3.0 mm outside diameter) delivers ~20 drops of water per milliliter. For calibration, fill a Pasteur pipet with water, hold the pipet at a 45° angle to the horizontal surface, and deliver water drops to fill up to the 1 mL mark in a 10 mL graduated cylinder. In a few cases, a micro capillary pipet constructed from a Pasteur pipet is used (see Section 3.10).

(e) Mixing

The best way to mix a solution in a test tube is to stir the solution with a glass rod of appropriate size. Rinse the glass rod thoroughly with deionized water after its use.

> Never mix a solution by closing the mouth of the test tube with your fingers or a stopper.

(f) Complete Precipitation of Solids

One of the most frequently used methods in qualitative analysis is to separate ions from a solution in the form of precipitates. Use the correct amount of the precipitating reagent. Stir the mixture thoroughly and let the solution stand in a hot water bath for several minutes, if so directed. Allow the precipitate to settle. Add a drop of the reagent to the liquid above the precipitate to test for the completeness of precipitation. Incomplete precipitation of an ion may interfere with the analysis.

(g) Separation of a Precipitate

A precipitate may be separated from the solution either by filtration (see Section 3.8, Figure 3.19 for details) or by centrifuging the solid in a centrifuge tube. Most often, the precipitation is carried out in a centrifuge tube. If the precipitation is done in a test tube, transfer the mixture (the precipitate and the solution) to the centrifuge tube with a pipet. Never fill the centrifuge tube to more than 1 cm from the top. Place the centrifuge tube in the holder inside the centrifuge (Figure 6.3), and insert a second centrifuge tube containing the same volume of material in the opposite slot—this serves to balance the centrifuge. Never use a centrifuge tube with cracks or chips—it may shatter. Close the cover on the centrifuge. If no cover is available, a plastic container should be used as a replacement.

Turn on the centrifuge and let it run for 20 to 45 seconds or until it reaches its highest speed. Turn off the centrifuge and allow it to come to rest.

> CAUTION: Never slow the centrifuge with your hands!

Remove the centrifuge tubes, and make sure that the precipitate has been deposited at the bottom of the tube containing the sample. The liquid above the precipitate is called the **supernatant** liquid.

Figure 6.3 Centrifuge (Courtesy of Fisher Scientific, 711 Forbes Ave., Pittsburgh, PA 15219)

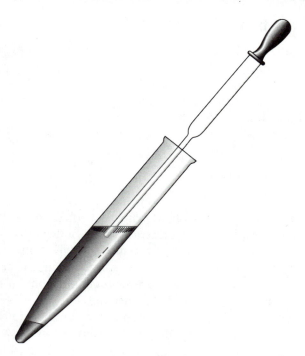

Figure 6.4   Removing the supernatant liquid from
a centrifuge tube

The supernatant liquid may be decanted by carefully tipping the centrifuge tube and pouring off the liquid into another test tube. Alternatively, the liquid may be withdrawn using a Pasteur pipet (Figure 6.4). If the supernatant liquid is not clear, a Pasteur **filter** pipet should be used (see Section 3.10, Figure 3.24).

(h) Washing of Precipitates

It is important to wash all the precipitates so that they are free from the reagents that were used. Otherwise, these reagents may interfere in the analysis. Deionized water is usually used as the wash liquid. The precipitate should be washed at least twice. Add the required amount of wash water and stir the mixture. Allow the solid to settle at the bottom of the test tube (or use a centrifuge tube and centrifuge the solution) and pipet out the wash liquid. The wash liquid should be disposed of in an appropriate waste container.

(i) Evaporation

Evaporation of a solvent is carried out either to concentrate the solution or to obtain a dry product. Where a rapid concentration of a solution is desired, the solution should be placed in a test tube carrying a micro stir bar. The test tube is placed inside a boiling water bath (use a test tube rack). To evaporate a solution to dryness, transfer the solution to a porcelain crucible. Heat the crucible on a hot sand bath or over a Bunsen burner flame.

6.3   Special Tests

Several special tests are performed for detecting micro quantities of inorganic solids. Some of these tests are very specific to a metal cation or to an anion.

(a) Spot Tests

Spot tests are often employed to identify a particular cation. A spot test may be performed on a spot plate, a ceramic plate that is half white and half black. It contains several small depressions, called spot holes, that are meant for adding and mixing solutions. Small test tubes (10×75 mm) may substitute for spot plates.

To perform a spot test, add one drop of the solution to be tested to a depression on the spot plate. Using *another* pipet, add one drop of the **spot test reagent** to the same depression. Mix the solutions using a micro stirring rod. Observe whether any color change has occurred or whether a precipitate has formed.

Other methods of performing spot tests include placing a drop of the solution to be tested onto a test strip (2×2 cm) made from a filter paper (Whatman No. 120 is best) that has been previously treated with a spot test reagent and dried. For example, lead acetate paper (test strip moistened with lead acetate solution) is used to detect the presence of sulfide anion.

> The spot plate and the delivery pipet must be scrupulously clean.

(b) Flame Test

An extremely simple test for some cations is the **flame test**. This test depends on the excitation of valence electrons to higher energy levels (due to the absorption of energy) when a metal cation is placed in a hot flame. The electrons emit radiation of a characteristic wavelength, imparting a color to the flame, when they return to lower energy levels. For example, Cu^{2+} cations impart a green color to the flame of a Bunsen burner.

Insert a nichrome or, preferably, platinum wire (6 to 8 cm long) into a cork and make a loop at the end. To conduct a flame test, clean the piece of wire by dipping the looped end in 6 M HCl and holding it in the hottest part (the outer light blue zone at the base of the flame) of a Bunsen burner flame. Repeat the procedure until no color is imparted to the flame. Next, dip the loop in the solution containing the cation, and insert the loop into the flame. Look carefully, because often the color of the flame does not persist.

The flame colorations due to different ions are given in Table 6.2. A sodium flame, which is bright yellow, will mask the violet color produced by potassium salts. To detect potassium in the presence of sodium, view the color of the flame through a cobalt blue glass. Modified flame colors through cobalt glass are also given in the table.

Table 6.2 Flame Test

Color	Flame color through cobalt blue glass	Ions
Yellow	None	Sodium cation
Violet	Crimson	Potassium cation
Brick red	Light green	Calcium cation
Crimson red	Purple	Strontium cation
Yellow-green*	Blue-green	Barium cation
Green	None	Borate anion
		Copper cation
Livid blue**	None	Lead, bismuth cations

* To observe the color of the flame, heat the wire for a longer time (more than 30 seconds).
** The wire gets corroded.

Table 6.3   Color of Borax Beads

Color of bead	Metal
Green when hot; blue when cold	Copper
Yellow or red when hot; yellow when cold	Iron
Yellow when hot; green when cold	Chromium
Violet (amethyst), hot and cold	Manganese
Blue, hot and cold	Cobalt
Brown when cold	Nickel

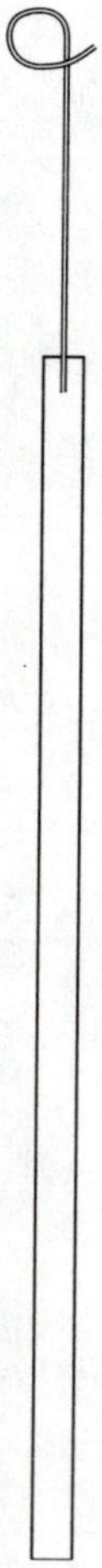

(c) Borax Bead Test

Using a 6 to 8 cm long nichrome (or preferably platinum) wire fused to a glass rod (or inserted into a cork), make a wire loop (Figure 6.5) by wrapping the wire around the tip of a lead pencil. The diameter of the loop should not exceed 0.3 cm. Heat the empty loop strongly in a Bunsen flame, and then dip it quickly into powdered borax, $Na_2B_4O_7 \cdot 10H_2O$. Some of the solid will adhere to the wire. Introduce the wire carrying the borax salt into the flame again. At first the borax will swell, and then it will collapse into a colorless, transparent glasslike bead, consisting of a mixture of sodium metaborate and boron oxide:

$$Na_2B_4O_7 \cdot 10H_2O \xrightarrow{\triangle} 2\,NaBO_2 + B_2O_3 + 10\,H_2O$$

The bead will not fall off the loop if the wire is held horizontally. A tiny amount of the test solid is then adhered to the hot bead. Reheat the bead at the hottest zone of the flame (bluish cone) until the test solid dissolves in the molten bead. Remove the bead from the flame and allow it to cool. Note the color of the bead (Table 6.3).

6.4   Strategy for the Development of an Analytical Scheme for the Separation of Cations

Qualitative chemical analysis in the introductory chemistry laboratory is one of the most important learning tools. It introduces an impressive array of chemical concepts (equilibrium, specificity of reactions, properties of ions in solution, and so on) into the first-year chemistry curriculum. Both the **investigative and directive methods of qualitative analysis** will be developed. Both these approaches are based upon **scientific methods** that involve the following steps:

Figure 6.5

Making a nichrome wire loop

(a) Experimentation

Perform the tests with known cations/anions. If necessary, perform specific special tests for known cations/anions.

(b) Observation

Carefully observe the reaction. See if any change has occurred: Has any precipitate formed, any color changed, or any gas evolved?

(c) Collection and Classification of Data

Collect all the observations, along with experiments performed and inferences drawn, in a table. Try to find a pattern in them and classify the behavior of cations/anions according to their chemical reactions.

(d) Development of an Analytical Scheme

Using the information collected in steps (a) through (c), develop the complete analytical scheme for the identification of cations/anions.

To assist you in the investigative approach, we will start with known cations/anions and their reactions. In the development of an analytical scheme, there is one major problem: given an aqueous solution, how can one identify which cations and anions are present? Many tests give similar results with several different cations/anions.

The most common way to solve this problem is to subdivide the cations into smaller groups by **selective precipitation**, in which a small subgroup of cations is *chemically* precipitated. They can then be *physically* separated from those remaining in solution by centrifuging. The precipitate (solid, abbreviated as *ppt*) settles out, and the solution (supernatant liquid, abbreviated as *snt*) is transferred into another container. In this fashion, an initially large group is separated into smaller groups until, finally, a definitive test can be performed to verify the presence or absence of each specific cation.

For the development of an analysis scheme, the concepts of acid-base equilibrium, buffer action, and solubility products of various precipitates must be understood and examined. It is also useful to consider the formation constants of various complex ions formed by the metal cations. Table 6.4 lists K_{sp} values of a variety of precipitates. Formation constants are given in Table 6.5.

6.5 The Analytical Subgroups

Group 1 (Insoluble Chloride Group)

From Table 6.4, determine which cations form **insoluble chlorides.** This group is known as **Group 1—The Insoluble Chloride Group.** Aqueous HCl is usually used to precipitate the metals in this group.

Group 2 (A+B) (Insoluble Hydroxide Group)

Many cations form hydroxides that are insoluble in aqueous ammonia but are readily soluble in alkaline (NaOH) solutions. They can therefore be subdivided based on hydroxide solubility in aqueous NH_3 and in NaOH solution (see Table 6.4). This group is known as **Group 2—The Insoluble Hydroxide Group.**

Group 2 is handled in the following way:

1. After precipitating the Group 1 cations as insoluble chlorides, NH_3/NH_4^+ buffer is added to the supernatant liquid so that a pH of about 9 to 10 is achieved. The hydroxides (or oxides) of the Group 2 cations will precipitate.

2. The precipitate containing the Group 2 cations is treated with 6 M NaOH and H_2O_2 (basic oxidation). The Subgroup 2A cations remain as precipitates, whereas the Subgroup 2B cations go into solution.

Table 6.4   Solubilities of Metal Compounds (K_{sp})

Cation	Precipitating Agents					
	Cl^-	OH^-	$C_2O_4^{2-}$	CO_3^{2-}	SO_4^{2-}	Comments
Ag^+	1.7×10^{-10}	1.5×10^{-8}	1.3×10^{-11}	8.5×10^{-12}	7.0×10^{-5}	a
Hg_2^{2+}	1.1×10^{-18}	decomp.		8.9×10^{-17}		b
Pb^{2+}	1.2×10^{-5}	1.4×10^{-20}	2.7×10^{-11}	4.0×10^{-14}	1.8×10^{-8}	c
Mn^{2+}	soluble	4.5×10^{-14}	insoluble	9.8×10^{-11}		d
Fe^{3+}	soluble	1.1×10^{-36}	soluble			d
Bi^{3+}	soluble	insoluble	soluble			d
ZrO^{2+}	soluble	i as ZrO_2				d
Al^{3+}	soluble	3.7×10^{-33}	sol. alkali			e
Cr^{3+}	soluble	6.7×10^{-31}		soluble	soluble	e
Sn^{4+}	soluble	1.0×10^{-26}	sol. alkali			e
Sb^{3+}	soluble	i as Sb_2O_3	soluble			e
Ba^{2+}	soluble	5.0×10^{-3}	1.6×10^{-7}	2.6×10^{-9}	1.1×10^{-10}	f
Ca^{2+}	soluble	7.9×10^{-6}	2.6×10^{-9}	6.2×10^{-9}	2.3×10^{-4}	f
Co^{2+}	soluble	2.5×10^{-16}	insoluble	insoluble	soluble	f
Sr^{2+}	soluble	3.2×10^{-4}	1.6×10^{-8}	1.6×10^{-9}	3.4×10^{-7}	f
Cu^{2+}	soluble	6×10^{-20}	sol. NH_3	1×10^{-10}	soluble	d
Mg^{2+}	soluble	5.6×10^{-12}	sol. NH_3			d
Ni^{2+}	soluble	1.6×10^{-16}	sol. NH_3			d
Cd^{2+}	soluble	1.2×10^{-14}	sol. NH_3			d
Zn^{2+}	soluble	6.9×10^{-17}	sol. NH_3			e
Li^+, Na^+, K^+	soluble	soluble	soluble			
Mo^{6+}	soluble	soluble	soluble			
NH_4^+	soluble	soluble	soluble			

Note: a-soluble in aqueous NH_3; b-insoluble in aqueous NH_3; c-soluble in hot water; d-insoluble in NaOH solution; e-soluble in NaOH solution; f-hydroxides soluble in aqueous NH_3, oxalates ($C_2O_4^{2-}$) insoluble in aqueous NH_3.

Group 3 (Insoluble Carbonate Group)

After removing Groups 1 and 2, one can see that some of the remaining cations form carbonates that are insoluble in NH_3 solution. Thus, the Group 3 cations are precipitated from solution by adding NH_3 to adjust the pH to about 9 to 10 and then adding ammonium carbonate solution. This group is known as **Group 3—The Insoluble Carbonate Group.**

Group 4 (Alkali-Insoluble Hydroxides/Oxides)

After removing the Group 1, 2, and 3 cations, determine which cations form hydroxides or oxides that are insoluble in basic (NaOH) solution. Since these *are* soluble in NH_3 solution, it is necessary to remove any NH_3 that may be present in the solution. Ammonia can be removed by evaporating the solution to dryness in a crucible over a low flame. Any ammonium ion is converted to NH_3 gas.

Table 6.5 Soluble Complexes of Metal Cations

Cation	Ligands or other ions			
	NH_3	OH^-	O^{2-}	Other
Ag^+	$Ag(NH_3)_2^+$			
Pb^{2+}		$Pb(OH)_4^{2-}$		
Cu^{2+}	$Cu(NH_3)_4^{2+}$	$Cu(OH)_4^{2-}$		
Co^{2+}	$Co(NH_3)_6^{2+}$			$Co(SCN)_4^{2-}$
Co^{3+}	$Co(NH_3)_6^{3+}$			$Co(NO_2)_6^{3-}$
Ni^{2+}	$Ni(NH_3)_6^{2+}$			
Zn^{2+}	$Zn(NH_3)_4^{2+}$	$Zn(OH)_4^{2-}$		
Al^{3+}		$Al(OH)_4^-$		
Cd^{2+}	$Cd(NH_3)_4^{2+}$			
Sn^{4+}		$Sn(OH)_6^{2-}$		$SnCl_6^{2-}$
Fe^{3+}				$FeCl_4^-$, $FeSCN^{2+}$, FeF_6^{3-}, $Fe(CN)_6^{3-}$
Sb^{3+}			SbO_3^{3-}	
Mn^{7+}			MnO_4^-	

A small quantity of concentrated HNO_3 is then added to the *cooled* crucible, which is heated to dryness once again. The treatment with HNO_3 converts all the oxides/hydroxides into nitrates. This residue is dissolved in HCl to yield the cations. The **Group 4 or Alkali Insoluble Hydroxides** are then precipitated by the addition of NaOH, forming insoluble hydroxides. Sodium and potassium remain in solution and are commonly determined only by flame tests.

General References

1. Szafran, S.; Pike, R. M.; Foster, J. C. *Microscale General Chemistry Laboratory with Selected Macroscale Experiments;* Wiley: New York, 1993.
2. Vogel's *Textbook of Macro and Semimicro Qualitative Inorganic Analysis,* 5th ed., revised by G. Svehla; Longman: London, 1979.

Qualitative Analysis: Separation and Identification of Ions

Part A: Separation of Group 1 Cations (Chloride Group)
Part B: Separation of Group 2 Cations (Hydroxide Group)
Part C: Separation of Group 3 Cations (Carbonate Group)
Part D: Separation of Group 4 Cations
Part E: Separation of Anions
Part F: Analytical Scheme for Identification of an Inorganic Salt

Microscale Experiments

Objectives
- To develop qualitative inorganic analytical schemes
- To become familiar with spot tests, the borax bead test, and flame tests

Prior Reading
- Chapter 6   An Introduction to Inorganic Qualitative Analysis

Related Experiments
- Experiments 15 (Identification of chemical compounds in solution), 26 (Paper chromatography), 30, 31 (Organic qualitative analysis)

This experiment describes the procedure for the qualitative analysis of several groups of cations and anions. Two different approaches have been adopted for the analysis:

1. **Investigative approach**. This approach is used for cations of Groups 1 through 3. For these cations, a series of tests will be performed to identify the known ions by wet analysis. The results of the analysis will be used to develop an analysis scheme for identifying ions present in an unknown solution.
2. **Directive approach**. This approach provides the procedure for completing the qualitative analysis scheme. This approach is used for the separation of Group 4 cations, the anions, and for the analysis of an unknown inorganic solid.

Experimental Section

Part A   Qualitative Analysis of Group 1 Cations—
Investigative Approach

> Complete the prelaboratory report sheet and the prelaboratory problem sheet before coming to the laboratory. For concentrations of common acids and bases, see Appendix 3.

The Group 1 cations include the lead (Pb^{2+}), mercurous (Hg_2^{2+}), silver (Ag^+), and tungsten (as tungstate, WO_4^{2-}) ions. The objectives of this experiment are (*a*) to perform the specific tests on the individual known cations and to enter the observations on the data sheet and (*b*) to develop a scheme for the qualitative analysis of an unknown solution containing these cations. Flame tests for Na^+ and K^+ may also be performed (if cations from other groups are absent). A special test for NH_4^+ (ammonium ion) may also be performed here.

Procedure

> Microscale experiment   Estimated time to complete the experiment: 2 h
> Important Note: The use of a centrifuge is recommended (see Chapter 6, section 2*g* and *h*) for the separation of precipitates. After removing the supernatant liquid with a Pasteur pipet, wash the residue twice with two to four drops of water, recentrifuge the mixture, and remove the supernatant liquid. Always save the supernatant liquid in a marked container for further analysis unless instructed otherwise. Collect the residue for further testing.

> Safety Note: The Group 1 cations are toxic. Avoid contact with skin. If spilled, wash the area immediately with running water. Report any accident to your instructor. Wear eye protection at all times.
> Disposal procedure: Dispose of the solutions only in marked containers.

Known solutions containing Group 1 cations are provided.

Test Reagents

The reagents used in this section include 6 M HCl (used as a Group 1 precipitating reagent), concentrated HCl, aqueous NH_3 (for detecting Hg_2^{2+} and separating Ag^+ and WO_4^{2-}), 0.25 M tin(II) chloride solution (for detecting WO_4^{2-}), 0.5 M KI solution (for detecting Pb^{2+} and Ag^+), and 6 M HNO_3 (for separating silver from tungsten). They are provided in reagent dropping bottles.

Special Test Reagents

Other reagents such as a 1 percent solution of diphenylcarbazide in ethyl alcohol and a combination of $KHSO_4$, H_2SO_4, and phenol (Defacqz test reagent) are also provided for spot tests.

Obtain five centrifuge tubes, several Pasteur pipets and bulbs, and a stirring rod. Fill a 50 mL beaker with deionized water for rinsing the Pasteur pipets during the experiment. Set up two hot water baths: one in a 100 mL beaker and the other in a 250 mL beaker. Place a perforated aluminum or plastic plate (see Figure 6.2) on the 250 mL water bath to serve as a test tube rack.

Take 2 to 3 mL of the unknown sample solution. This solution may contain from one to three cations. Record the ID number of the unknown solution on your record sheet. Also obtain samples of each of the known solutions.

Using a Pasteur pipet, place ~20 drops of the unknown solution on a clean watch glass. Place the watch glass on the boiling water bath for complete evaporation of the solution. While the evaporation is taking place, perform the wet tests in Section A (which follows) on known cations. When the solution has completely evaporated, remove the watch glass from the water bath. Collect the dry solid. Perform the tests described in Section B (which follows) on the solid.

Section A: Analysis of Known Cations

1. Wet Tests: Separation of Cations Using Reagents

> CAUTION: HCl is corrosive. Ammonia is an irritant. Nitric acid is corrosive and a strong oxidizing agent.

Using a dropper or a Pasteur pipet, place four drops of each known cation solution in a separate centrifuge tube labeled with the cation symbol. Add four drops of water to each centrifuge tube. Add two to four drops of 6 M HCl to each of the test tubes. To avoid confusion, always mark the centrifuge tube with the symbol of the cation being tested and the reagent added.

Stir each mixture with a stirring rod, and then let the solutions stand for 10 to 20 seconds. Observe the mixture closely to detect if any change has taken place. You may see the formation of a precipitate, a color change, or the like. Record all observations on the data sheet.

If precipitation has occurred, record "ppt" in the appropriate data box, followed by its color code (w for white, y for yellow, r for red, b for black, br for brown, and so on). **Save all centrifuge tubes that contain a precipitate. They will be used to perform additional tests, described shortly.**

Any precipitate may be separated from the supernatant liquid by placing the centrifuge tube inside a centrifuge and centrifuging for 45 to 60 seconds. Remove the centrifuge tube and decant the supernatant liquid, leaving behind the solid. Discard the supernatant liquid (or save it if it is needed for separating the Group 2–4 cations). Wash the precipitate with two to four drops of water, recentrifuge, discard the supernatant liquid, and collect the residue. Similarly, separate the solid in the other centrifuge tubes. **Use these residues for the specific tests described below.**

Empty any centrifuge tubes that do not contain any precipitate into a designated waste container and rinse them with deionized water. Using fresh samples of the known cations, repeat the foregoing procedure with the next reagent ($NH_3(aq)$; HNO_3), recording all observations. If time permits, test each cation solution with each reagent.

---

2. Additional Tests and Treatment of the Precipitates

Lead(II), Pb^{2+}

To the residue in the centrifuge tube that contained the lead(II) known solution and 6 M HCl, add 0.5 to 1 mL of cold water and heat the resulting mixture in a hot water bath for several minutes. What happens to the $PbCl_2$ residue? Record your observation on the record sheet.

Now, add two drops of KI solution to the Pb^{2+} solution. Scratch the side of the centrifuge tube with a glass rod and allow the mixture to cool. Do you see a precipitate? What is the color of the precipitate? Record your observations on the data sheet.

Mercury(I), Hg_2^{2+}

Add five drops of aqueous NH_3 to the solid Hg_2Cl_2 residue present in the centrifuge tube that contained Hg_2^{2+} solution and 6 M HCl. Stir the mixture, being sure to break up the solid. Enter your observation on the data sheet.

Silver(I), Ag^+

Add an excess of aqueous NH_3 (five to six drops) to the AgCl solid left in the centrifuge tube that contained Ag^+ and 6 M HCl. Using a stirring rod, stir the mixture thoroughly. Touch the end of the stirring rod to a piece of red litmus paper. If the solution is not basic (red litmus did not turn blue), add more NH_3 solution.

Stir the solution while heating it in the hot water bath (use the test tube rack). Do you obtain a clear solution? Record your observations on the data sheet. While stirring with a glass rod, add 6 M HNO_3 dropwise to the foregoing solution until the solution is distinctly acidic (blue litmus turns red). Record your observations on the data sheet.

For treatments of solutions and precipitates containing silver, see the section "Silver Recovery and Reuse" at the end of this part.

Tungsten(VI), WO_4^{2-}

To four drops of the known tungstate solution, add four drops of 6 M HCl. Gently boil the mixture over a flame. Is there any change in the solid residue (tungstic acid, H_2WO_4)? Enter your observations on the data sheet. Remove and discard the supernatant liquid from the mixture. Add an excess of aqueous NH_3 (five to six drops) to the residue, and stir the mixture. Is the residue soluble in ammonia solution? Acidify the resulting solution with 6 M HCl until it is just neutral (avoid adding excess HCl). Add 6 drops of tin(II) chloride solution, followed by 6 to 10 drops of concentrated HCl and heat the solution to boiling over a flame. Does the color of the solution change? Record your observations.

3. Special Tests

Flame Test

Obtain a small amount ($\sim$10 mg, or $\frac{1}{8}$ the size of a pea) of the solid known samples from your instructor, as well as a nichrome or platinum wire, and perform the flame test described in Chapter 6, section 3*b*. Observe the color of the flame. Record your observations on the data sheet. Lead(II) salt will produce a light blue color.

Spot Test for Lead(II) Cation

Following the general procedure described in Chapter 6, section 3a, perform the spot tests for Pb^{2+} using a spot plate. Place a drop of Pb^{2+} solution in the depression of the spot plate and add a drop of KI solution. Observe the color change. Avoid adding too much KI solution—excess KI forms a soluble and colorless species, $[PbI_4]^{2-}$.

Spot Test for Mercury(I) and Mercury(II) Cations

Spot a piece of filter paper with a freshly prepared 1 percent solution of diphenylcarbazide in ethyl alcohol. Add one drop of 0.5 M HNO_3 to the same spot. Allow the filter paper to dry for a few minutes. Place a drop of the test solution containing the mercury ions on the same spot on the filter paper. The appearance of a violet color confirms the presence of mercury. In the presence of 0.5 M HNO_3, this test is very specific for mercury ions.

Spot Test for Tungsten

> Note: Concentrated sulfuric acid is corrosive and an oxidizer. Phenol is toxic and corrosive.

Collect four or five drops of a tungstate salt solution in a microcrucible and evaporate it to dryness on a slow flame. Add 10 to 20 mg of potassium hydrogen sulfate, $KHSO_4$, to the residue. Heat the crucible until the solid fuses and a clear melt is obtained. Using a pair of tongs, remove the crucible from the flame and **allow it to cool to room temperature**. Add two drops of concentrated (18 M) H_2SO_4, and stir the mixture with a glass rod. Add a few crystals of solid phenol. The appearance of red color is the confirming test for tungsten.

Sodium Hydroxide Test for Ammonium Ion

Add three drops of 3 M NaOH to six drops of NH_4^+ solution in a test tube (10×75 mm). Heat the test tube in a hot water bath. Using a pair of forceps, hold a piece of wet red litmus paper on the mouth of the test tube; be careful not to touch the mouth of the test tube with the litmus. Evolution of ammonia gas will turn the red litmus paper blue, confirming the presence of an ammonium salt.

Section B: Development of a Qualitative Analysis Scheme for Group 1 Cations

Using the observations made during the tests performed with the known cations, develop a scheme for the qualitative analysis of a mixture of cations of this group. Draw a flow chart for your scheme. Show the flow chart to your instructors for their approval. Using 10 to 15 drops of the unknown mixture, perform the analysis following your scheme.

Since you are using a significant amount (10 to 20 drops) of the unknown mixture, you will have to use a larger amount of group reagents for the analysis. This requirement is reflected in the procedure described next.

Guideline for the Development of a Qualitative Analysis Scheme

1. Place 10 to 15 drops of your unknown solution in a centrifuge tube. First, separate the cations of Group 1 from other groups by acidifying (test with litmus paper) the solution with four drops of 6 M HCl. If Group 1 cations are present in your unknown solution, they will precipitate from the solution as insoluble chlorides.

2. Centrifuge the solution and decant the supernatant liquid (save the supernatant liquid if other cations may be present), leaving the solid in the centrifuge tube. Add four drops of deionized water to the solid and stir. Decant and discard the wash liquid. **Use the solid residue for the analysis of the Group 1 cations according to your qualitative analysis scheme.**

Special Tests on the Dry Residue of the Unknown Mixture

Collect the solid residue **from the watch glass** (see preceding) and perform the flame tests for Na^+ and K^+ (that is, for those cations that do not form a precipitate with HCl). Also carry out the sodium hydroxide test for NH_4^+ as well as the flame test and the spot test for Pb^{2+}. Assume that the solution does not contain any cations from other groups.

Submit your laboratory report in one of the following formats:

1. Prelaboratory report sheet, data sheets, prelaboratory problems, and postlaboratory problems.
2. A short written report of your experiment (see Chapter 4). Attach the prelaboratory report sheet and the data sheets along with your written report including the prelaboratory and the postlaboratory problems.

Example of a Flow Chart

Suppose an unknown solution contains the cations A^+, B^+, and C^+. When tested with an aqueous solution of HCl, the cations A^+ and B^+ form chloride precipitates, whereas C^+ does not form any precipitate. The residue ACl is soluble in boiling water, but BCl is not soluble. In the presence of aqueous NH_3, BCl becomes black or dark colored. A solution of ACl in hot water forms a yellow precipitate with KI solution. A flow diagram for the separation of A^+, B^+, and C^+ is shown below.

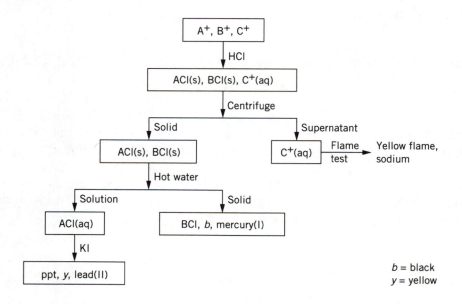

Silver Recovery and Reuse[1]

> Handle all silver salts with caution. Use a pair of gloves to protect your hand from being stained.

Collect all the silver waste in a 25 mL beaker. Add dilute (6 M) hydrochloric acid to the waste (if it is a solution) to convert all the unreacted silver salt solutions into silver chloride. Filter the precipitate of silver chloride under suction. Using 10 mL portions of deionized water, wash the residue at least 5 to 6 times. Transfer the solid silver chloride residue to a 25 mL beaker. Add, with stirring, an excess amount of concentrated ammonia (use **HOOD**) solution to dissolve solid silver chloride. Warm the solution on a sand bath to accelerate the rate of dissolution. If any solid remains, filter the mixture. Add an excess of 1 M solution of ascorbic acid (vitamin C) in water to the filtrate. Ascorbic acid will reduce silver (I) to silver metal. Metallic silver will slowly precipitate as a gray powder. Allow the metal to settle at the bottom of the beaker. Decant off the the supernatant liquid, or suction filter the suspension. Wash the solid several times with deionized water followed by two 5 mL portions of acetone. Allow the solid to dry on the filter under suction. Transfer the solid to the same beaker in which the reduction was carried out. Add sufficient dilute (6 M) HNO_3 to dissolve the metal, forming silver nitrate solution, which can be reused. For other methods of recovery of silver from silver waste, see references 2 and 3.

References

1. Hill, J. W.; Bellows, L. *J. Chem. Educ.* **1986,** *63,* 357.
2. Renganathan, S.; Mehta, B. J. *J. Chem. Educ.* **1976,** *53,* 347.
3. Rawat, J. P.; Kamoonpuri, S. I. M. *J. Chem. Educ.* **1986,** *63,* 537.

Name ______________________________ Section ______________

Instructor ______________________________ Date ______________

PRELABORATORY REPORT SHEET—EXPERIMENT 16, PART A

Experiment title ______________________________

Objective

Reagents, reactions/formulas to be used

Materials and equipment table

Outline of procedure

1. Complete and balance the following reactions. If a reaction does not occur, write "no rxn."

$Pb(NO_3)_2(aq) + HCl(aq) \rightarrow PbCl_2(s) +$

$PbCl_2(s) + H_2O(l) + heat \rightarrow$

$AgNO_3(aq) + HCl(aq) \rightarrow$

$AgNO_3(aq) + KI(aq) \rightarrow$

$AgCl(s) + NH_3(aq) \rightarrow Ag(NH_3)_2{}^+(aq) +$

$Hg_2{}^{2+}(aq) + HNO_3 \rightarrow$

$Hg_2Cl_2(s) + NH_3 \rightarrow HgNH_2Cl(s) + Hg(l) +$
    white   white   black

2. Suggest a scheme (flow diagram) for the separation of a mixture containing NH_4^+ and Pb^{2+} cations and tungstate anion. Mention the confirmatory tests (the tests that are conclusive for a particular cation) for each.

Name ______________________	Section ______________
Instructor ______________________	Date ______________

EXPERIMENT 16, PART A DATA SHEET

Data Collection Table

Unknown ID number ______

1. Wet analysis of known individual cations

Write balanced chemical reactions for each of the positive tests under the "Inference drawn" column.

Experiment performed	Observation made	Inference drawn
1.		
2.		

Use a separate sheet if necessary.

Summary Table

Reagents Used	WO_4^{2-}	NH_4^+	Ag^+	Hg_2^{2+}	Pb^{2+}
Aq. HCl					
Aq. NH_3					
Aq. HNO_3					

Use a separate sheet if necessary.

2. Additional tests and treatment of the precipitates

Use a separate sheet if necessary.

3. Special tests

Experiment performed	Observation made	Inference drawn
Flame tests		
Hg_2^{2+}		
Ag^+		
WO_4^{2-}		
Pb^{2+}		
NH_4^+		
Spot tests		
Other tests		

Analysis Scheme for the Unknown Sample

Make a table similar to that on Data Sheet page 1 or develop a scheme for analyzing the unknown solution, suggesting the reaction sequence, and draw the flow diagram as shown in the example. Show it to your instructor and obtain his or her approval before you perform the unknown analysis.

POSTLABORATORY PROBLEMS—EXPERIMENT 16, PART A

1. What did you observe when you performed the following tests?

   Flame test for Pb^{2+} _____

   NaOH test for NH_4^+ _____

   Flame test for Ag^+ _____

   Flame test for tungsten _____

2. List the metal cations of this group that formed precipitates with HCl solution.

3. What happens when
   (a) ammonia is added to a mixture of AgCl and $PbCl_2$ solids?

   (b) ammonia is added to a mixture of Hg_2Cl_2 and $PbCl_2$?

   (c) nitric acid is added to a solution of AgCl in aqueous NH_3?

Part B: Qualitative Analysis of Group 2 Cations— Investigative Approach

> Complete the prelaboratory report sheet and the prelaboratory problem sheet before coming to the laboratory. For concentrations of common acids and bases, see Appendix 3.

The Group 2 cations are divided into two subgroups. **Subgroup 2A** consists of bismuth(III), Bi^{3+}; iron(II) or (III), Fe^{2+} or Fe^{3+}; manganese(II), Mn^{2+}; titanium(IV), Ti^{4+}; and zirconium(IV), ZrO^{2+}. **Subgroup 2B** consists of aluminum(III), Al^{3+}, and tin(II) or (IV), Sn^{2+} or Sn^{4+}.

The objectives of this experiment are (*a*) to perform the tests on the individual known cations and to enter observations on the data sheet and (*b*) to develop a scheme for the qualitative analysis of an unknown solution. Known solutions containing the cations of this group are provided by your laboratory instructor.

Test Reagents

The reagents are as follows: 6 M HCl, 6 M aqueous NH_3, 6 M NaOH, 3 percent H_2O_2 (hydrogen peroxide), solid $NaHCO_3$ (sodium bicarbonate), and 6 M HNO_3.

Special Test Reagents

These reagents include Alizarin Red S; solid sodium bismuthate, $NaBiO_3$; solid $SnCl_2$ [tin(II) chloride]; 0.2 M KSCN (potassium thiocyanate solution); 0.2 M $K_4Fe(CN)_6$ [potassium hexacyanoferrate(II)]; 0.5 M cobalt(II) chloride solution; Al powder; 0.1 M $HgCl_2$ solution; and aluminon.

Procedure

> Microscale experiment   Estimated time to complete the experiment: 3–5 h
> Important note: The use of a centrifuge is recommended for separating precipitates from supernatant liquids [see Chapter 6.2(*g*) and (*h*)]. After removing the supernatant liquid with a Pasteur pipet, wash the residue twice with two to four drops of water and recentrifuge the mixture. Discard the supernatant liquid and collect the residue for further testing. Always save the supernatant liquid in a marked container for further analysis unless instructed otherwise.

> Disposal procedure: Dispose of the solutions only in marked containers.

Obtain seven 10×75 mm test tubes (or use a well plate), seven centrifuge tubes, several Pasteur pipets, a rubber bulb, a small stirring rod, and one 50 mL beaker to serve as a source of water for rinsing Pasteur pipets. Set up a water bath in a 250 mL beaker three-quarters full of deionized water. Obtain 2 to 3 mL of the unknown solution, which may contain **one, two, or three cations**.

Section A: Analysis of Known Cations

1. Wet Tests: Separation of Cations Using Reagents

> CAUTION: HCl is corrosive; NH₃ is an irritant. HNO₃ is corrosive and a strong oxidizing agent. Hydrogen peroxide is a strong bleaching and oxidizing agent.

Using a Pasteur pipet, place four drops of each of the known cation solutions in separate centrifuge tubes labeled with the cation symbols. Add four drops of water to each test tube, followed by one drop of 6 M HCl solution. Stir the mixture, and then allow it to stand for 30 seconds. If no precipitate is formed, the Group 1 cations are absent. (In this case, since pure solutions of Group 2 cations are provided, one would not expect to see any Group 1 cations forming chloride precipitates.)

Add 6 to 10 drops of 6 M NH₃ solution (pH ~ 10, test with pH paper) to each of the centrifuge tubes. Stir the mixture. **The solution must be distinctly basic.**

Observe the mixture closely to detect if any change has taken place. You may see the formation of precipitates, some of which may be colored. Record all the observations on the data sheet by writing "ppt" followed by its color code (*w* for white, *y* for yellow, *r* for red, *b* for black, *br* for brown, and so on). The reactions involved in this part are as follows:

$$Bi^{3+}(aq) + 3\,NH_4OH(aq) \rightarrow Bi(OH)_3(s) + 3\,NH_4^+(aq)$$

$$ZrO^{2+}(aq) + 2\,NH_4OH \rightarrow ZrO(OH)_2(s) + 2\,NH_4^+(aq)$$

$$Ti^{4+}(aq) + 4\,NH_4OH(aq) \rightarrow Ti(OH)_4(s) + 4\,NH_4^+(aq)$$

$$Mn^{2+}(aq) + 2\,NH_4OH(aq) \rightarrow Mn(OH)_2(s) + 2\,NH_4^+$$

$$Al^{3+}(aq) + 3\,NH_4OH(aq) \rightarrow Al(OH)_3(s) + 3\,NH_4^+(aq)$$

$$Sn^{4+}(aq) + 4\,NH_4OH \rightarrow Sn(OH)_4(s) + 4\,NH_4^+(aq)$$

Save all centrifuge tubes that contain precipitates, and proceed to step 2 for additional tests. Empty the contents of the centrifuge tubes that do not contain any precipitate into a designated waste container. Rinse them with deionized water.

Place fresh samples of each of the known cations in one each of the seven test tubes, and follow a similar procedure as before except add four drops of 3 percent H₂O₂ followed by 6 to 10 drops of 6 M NaOH solution. Stir the mixtures with a stirring rod and heat in a hot water bath. Determine which cations form hydroxide or oxide precipitates that are insoluble in excess NaOH, and which precipitates redissolve in excess base. Record your observations.

> Hint: Initially all the cations will form precipitates; however, with the addition of excess NaOH, some of the solids will redissolve.

2. Additional Tests

Centrifuge the contents of the centrifuge tubes containing precipitates (from the first wet test). Discard the supernatant liquid. Wash each residue with four drops of water, and stir to break up the precipitate. Recentrifuge, and discard the wash.

Add four drops of 3 percent H_2O_2, followed by four drops of 6 M NaOH solution to each residue. Stir the mixture and allow it to stand for two minutes. **The Group 2B cations will dissolve, forming clear solutions** [of $Al(OH)_4^-$ and $Sn(OH)_6^{2-}$] **and the Group 2A cations will not dissolve. The color of the solid residues may change.** The insoluble residues are $Bi(OH)_3$ (white), $Fe(OH)_3$ (brown), Ti^{4+} (white gelatinous), $ZrO(OH)_2$ (white), $Mn(OH)_2$ (brown), and MnO_2 (black). Record your observations on the data sheet.

$$Mn^{2+}(aq) + 2\,OH^-(aq) \rightarrow Mn(OH)_2(s)$$
$$Mn(OH)_2(s) + H_2O_2(aq) \rightarrow MnO_2(s) + H_2O(l)$$
$$ZrO^{2+}(aq) + 2\,OH^-(aq) \rightarrow ZrO(OH)_2(s)$$

Add six to eight drops of 6 M HCl to the tubes that contain the solids. These solids will dissolve in HCl, producing clear solutions containing $Bi^{3+}(aq)$, $Ti^{4+}(aq)$, ZrO^{2+}, Fe^{3+}, and Mn^{4+} (some may be reduced back to Mn^{2+}). Use the solutions just generated (both Group 2A and 2B) for the spot tests described next.

3. Special Tests

Spot Tests or Other Specific Tests for Different Cations
Following the general procedure described in Chapter 6, section 3*a*, perform the following spot tests for the solutions just given. (Note: Flame tests are not performed because the cations of this group do not produce characteristic colors in a flame.)

Zirconium(IV) Cation (ZrO^{2+})
Place four drops of the ZrO^{2+} solution in a test tube, and add a drop of 6 M HCl and two drops of 3 percent H_2O_2. Stir the mixture, heat it in a water bath for two minutes, and cool it to room temperature. Add one drop of Alizarin Red S reagent (HARS) to the cooled solution. Using pH paper, check the pH of the solution. It should be ~ 5. Heat the mixture in a water bath. Observe the color of the mixture and enter the observation on the data sheet.

$$Zr(OH)_4(aq) + HARS(aq) \rightarrow Zr(OH)_3ARS(aq) + H_2O$$

Manganese(II) Cation (Mn^{2+})
Place five drops of the Mn(II) solution in a test tube, add a drop of 6 M HCl, and several small portions (one-quarter the size of a pea) of solid sodium bismuthate, $NaBiO_3$ (**Caution: $NaBiO_3$ is an oxidizer**). If no color develops, add two drops of 6 M HNO_3. Allow the mixture to settle. Record the change in color of the supernatant liquid.

$$2\,Mn^{2+}(aq) + 14\,H^+(aq) + 5\,BiO_3^-(aq) \rightarrow 2\,MnO_4^-(aq) + 5\,Bi^{3+}(aq) + 7\,H_2O(l)$$

Bismuth(III) Cation (Bi^{3+})
Transfer five drops of Bi^{3+} solution to a test tube. Add a drop of 6 M HCl, several drops of 6 M NaOH, and five drops of $SnCl_2$ solution. If no precipitate appears, add a few more drops of 6 M NaOH. Observe the color of the mixture and enter your observation on the data sheet. Note that on dilution $Bi^{3+}(aq)$ may form an insoluble white precipitate, BiOCl.

$$Bi^{3+}(aq) + OH^-(aq) \rightarrow Bi(OH)_3(s)$$

$$Sn^{2+}(aq) + 3\,OH^-(aq) \rightarrow Sn(OH)_3^-(aq)$$

$$2\,Bi(OH)_3(s) + 3\,Sn(OH)_3^-(aq) + 3\,OH^-(aq) \rightarrow 2\,Bi^0 + 3\,Sn(OH)_6^{2-}(aq)$$

Iron(II) or (III) Cation (Fe^{2+} or Fe^{3+})

Two spot tests for iron are performed. In both the tests iron must be in the 3+ oxidation state.

1. Place four drops of the iron solution on a spot plate and add a drop of 6 M HNO_3 followed by a drop of $K_4Fe(CN)_6$. Record the intense color change of the solution.

$$4\,Fe^{3+}(aq) + 3\,Fe(CN)_6^{4-}(aq) \rightarrow Fe_4[Fe(CN)_6]_3(s)$$

2. Take four drops of the iron solution, and add four drops of 6 M HNO_3 followed by a drop of KSCN. Record the intense color change on the data sheet.

$$Fe^{3+}(aq) + SCN^-(aq) \rightarrow Fe(SCN)^{2+}(aq)$$

Titanium(IV) (Ti^{4+})

Place five drops of the Ti(IV) solution on a spot plate or in a small test tube. Add one drop of 6 M H_2SO_4 followed by one drop of 3 percent H_2O_2. The solution turns orange-yellow, confirming the presence of titanium.

Aluminum(III) and Tin(II) Cations (Al^{3+} and Sn^{4+})

Place four drops each of the desired solution in a centrifuge tube. Add 10 drops of 6 M NH_3(aq), and stir the solution. Does a precipitate appear in each case? Record any observations on the data sheet. Decant the supernatant liquid, and add four drops of 3 percent H_2O_2 followed by 10 drops of 6 M NaOH to the residue. Stir the mixture and heat it for two minutes in a water bath. Allow the mixture to stand for two minutes and centrifuge the solution, if necessary. Acidify the basic supernatant solution with 6 M HCl (blue litmus paper turns red).

Neutralize the excess acid by adding 1 M $NaHCO_3$ solution. Tin(IV) will form a precipitate, while aluminum(III) remains in solution. Centrifuge the mixture, then remove and collect the supernatant liquid, which contains the Al(III) cation. Save this solution for the spot test for Al(III) (see following). Wash the residue with two drops of water, and remove and discard the wash liquid. The residue left in the centrifuge tube contains tin(IV) as $Sn(OH)_4$.

Spot tests for Al(III)

Aluminon test. On a spot plate, acidify (test with a litmus paper) the basic solution containing Al(III) with 6 M HCl. Add one drop of aluminon reagent followed by 6 M aqueous NH_3 until the mixture is basic to a litmus paper. Mix the solution thoroughly. Record your observations.

Thenard's blue test. Dip one-half of a piece of ashless filter paper in the Al(III) solution. Air-dry the filter paper, and then soak it again in the Al(III) solution. Add exactly two drops of 0.5 M cobalt(II) nitrate [$Co(NO_3)_2$] solution. (**NOTE: Avoid adding too much cobalt nitrate solution.**) Air-dry the filter paper, and then ignite the paper on a Bunsen burner. The blue-colored ash residue confirms the presence of aluminum. Your instructor will demonstrate how to ignite a filter paper.

Spot test for Sn(II)

Transfer the tin residue to a spot plate. Add 6 M HCl to dissolve the $Sn(OH)_4$. Stir the solution. Using 10 to 20 mg of Al powder, reduce the tin(IV) solution to tin(II). Add two drops of 0.1 M $HgCl_2$ solution. What do you observe? Enter your observations on the data sheet.

$$Sn(OH)_4(s) + 6H^+(aq) + 6Cl^-(aq) \rightarrow SnCl_6^{2-}(aq) + 4H_2O(l) + 2H^+(aq)$$

$$3SnCl_6^{2-}(aq) + 2Al^0(s) \rightarrow 3SnCl_4^{2-}(aq) + 2Al^{3+}(aq) + 6Cl^-(aq)$$

$$SnCl_4^{2-}(aq) + 2Hg^{2+}(aq) + 4Cl^-(aq) \rightarrow SnCl_6^{2-}(aq) + Hg_2Cl_2(s)$$

Borax Bead Test for Mn(II)

Evaporate 20 to 30 drops of a solution that contains Mn(II) on a watch glass over a boiling water bath inside a **HOOD**. Perform the borax bead test on the dry product by following the general procedure as described in Section 6.3c. A purple or violet bead indicates the presence of manganese.

Borax Bead Test for Fe(II) or Fe(III)

Evaporate 20 to 30 drops of a solution that contains Fe(II) or Fe(III) on a watch glass over a boiling water bath inside a **HOOD**. Perform the borax bead test on the dry product by following the general procedure as described in Section 6.3c. A light green or yellow bead indicates the presence of iron.

Section B: Development of a Qualitative Analysis Scheme for Group 2 Cations

Using the observations made during the tests on the known cations, develop a scheme for the qualitative analysis of a mixture of cations of this group. Draw a flow chart for your scheme that will enable you to separate the cations for their positive identification. Show the flow chart to your instructor for his or her approval. Using 10 to 20 drops of the unknown mixture, perform the qualitative analysis following your scheme.

NOTE: Since you are using a significant amount (10 to 20 drops) of the unknown mixture, you will have to use a larger amount of group reagents for the analysis. This fact is reflected in the procedure described below.

Guideline for the Development of a Qualitative Analysis Scheme

1. Place 10 to 20 drops of the unknown solution in a centrifuge tube. First, separate the cations of Group 1 from those of Group 2 (and other subsequent groups) by acidifying the solution with four drops of 6 M HCl. If Group 1 cations are present, they will precipitate from the solution as insoluble chlorides. The Group 2 cations are soluble in the presence of HCl. Centrifuge the solution for 45 to 60 seconds. Decant the supernatant liquid. Save the supernatant liquid (**snt #1**) for Group 2 analysis. Rinse the residue twice with four drops of deionized water, discarding the wash each time. The solid (if any) may be analyzed according to the procedure you developed for Group 1 cations in Part A.

2. Next, precipitate Group 2 cations as hydroxides by adding six drops of 6 M NH_3 (test with litmus) to the supernatant liquid from step 1 (**snt #1**). Stir the solution, then centrifuge it for two minutes. If the solution contains cations from Groups 3 or 4, save the supernatant liquid in a **labeled** test tube for future analysis (described in Parts C and D of this experiment). Otherwise, the supernatant liquid may be discarded to a proper waste container. Save the residue (**ppt #1**).

3. Wash the residue (**ppt #1**) twice with four drops of water, discarding the wash each time. Using the solubility characteristics of the Group 2 cations (**ppt #1**) in 6 M NaOH containing 3 percent H_2O_2 (use 10 drops of this solution), divide the sample into the two Subgroups, 2A and 2B (recall that the hydroxides and oxides of Subgroup 2A are insoluble in oxidizing NaOH and the hydroxides of Subgroup 2B are soluble in oxidizing NaOH).

   Centrifuge the solution, and save the supernatant liquid (**snt #2**) for Subgroup 2B analysis (Al^{3+} and Sn^{4+}/Sn^{2+}). Rinse the residue left in the centrifuge tube twice with five drops of water. Centrifuge, discard the wash liquid, and save the residue (**ppt #2**).

   Dissolve the residue (**ppt #2**) in 10 drops of 6 M HCl. Heat the mixture in a hot water bath for two minutes, stir, and centrifuge if necessary. This solution contains the cations of Subgroup 2A (ZrO^{2+}, Ti^{4+}, Fe^{3+}/Fe^{2+}, Bi^{3+}, and Mn^{2+}). Remove the supernatant liquid and divide it into *five equal parts* in five test tubes. Perform the characteristic spot tests for each of the five Group 2A cations according to your own scheme.

4. Heat the supernatant solution saved for Subgroup 2B (**snt #2**) analysis for 10 minutes in a hot water bath. Cool the mixture for two minutes. Perform the analysis for detecting the cations of Subgroup 2B by following your analysis scheme. An example appears in the accompanying flowchart.

Submit your laboratory report, which must include either of the following:

1. Prelaboratory report sheet, data sheets, prelaboratory problems, and postlaboratory problems.
2. A short written report of your experiment (see Chapter 4). Attach the prelaboratory report sheet, report sheet(s), and postlaboratory problems.

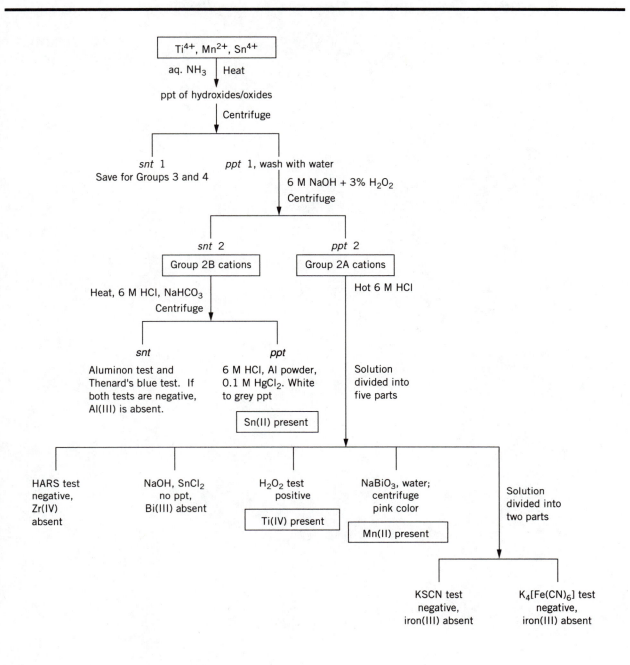

Ti^{4+}, Mn^{2+}, Sn^{4+}

aq. NH$_3$ | Heat

ppt of hydroxides/oxides

Centrifuge

snt 1
Save for Groups 3 and 4

ppt 1, wash with water

6 M NaOH + 3% H$_2$O$_2$
Centrifuge

snt 2
Group 2B cations

ppt 2
Group 2A cations

Heat, 6 M HCl, NaHCO$_3$
Centrifuge

Hot 6 M HCl

snt
Aluminon test and
Thenard's blue test. If
both tests are negative,
Al(III) is absent.

ppt
6 M HCl, Al powder,
0.1 M HgCl$_2$. White
to grey ppt

Sn(II) present

Solution
divided into
five parts

HARS test
negative,
Zr(IV)
absent

NaOH, SnCl$_2$
no ppt,
Bi(III) absent

H$_2$O$_2$ test
positive

Ti(IV) present

NaBiO$_3$, water;
centrifuge
pink color

Mn(II) present

Solution
divided into
two parts

KSCN test
negative,
iron(III) absent

K$_4$[Fe(CN)$_6$] test
negative,
iron(III) absent

Name ______________________ Section ______________

Instructor ______________________ Date ______________

PRELABORATORY REPORT—EXPERIMENT 16, PART B

Use additional sheets if necessary.

Experiment title ______________________________

Objective

Reagents, reactions/formulas to be used

Materials and equipment table

Outline of procedure

Use additional sheets if necessary.

1. All of the cations of Group 2 react with aqueous NH_3 to form the corresponding hydroxides. Write reactions for each of them.

2. Which hydroxides are soluble in oxidizing NaOH solution?

3. Mention at least one specific spot test/special test for each cation of Group 2. Write the correct procedure.

4. Suggest a scheme (flow diagram) for the separation of a mixture containing Al^{3+}, Fe^{3+}, and Sn^{2+} cations. Mention the confirmatory tests for each cation.

EXPERIMENT 16, PART B   DATA SHEET

Data Collection Table

Unknown ID number _____

1. Analysis of known individual cations with the reagents

Describe the experiments and write balanced chemical reactions for each test resulting in products different from the starting material.

Experiment performed	Observation made	Inference drawn
1.		
2.		

Use a separate sheet if necessary.

Complete the following table:

Reagents used	Fe^{3+}	Bi^{3+}	Sn^{2+}	Al^{3+}	Mn^{2+}	Zr^{4+}
Aq. NH_3						
Aq. NaOH						

Use a separate sheet if necessary.

2. Additional tests

Use a separate sheet if necessary.

3. Special Tests

Experiment performed	Observation made	Inference drawn
1. Spot tests		
2. Other tests		

Scheme for the Analysis of the Unknown Cations

Develop a scheme for analyzing the unknown solution and draw the flow diagram for the scheme. Show this to your instructor and obtain his or her approval before performing the unknown analysis.

1. What did you observe when you performed the following tests?

    Special test for Fe^{3+} _____

    Special test for Mn^{2+} _____

    Special test for Sn^{2+} _____

    Special test for Bi^{3+} _____

2. List the metal cations of this group that form insoluble hydroxides with **oxidizing** NaOH.

3. What happens when

    (*a*) ammonia is added to a mixture of $FeCl_3$ and $MnCl_2$ solutions?

    (*b*) excess NaOH solution is added to Al^{3+} solution?

    (*c*) nitric acid is added to a solution of Fe^{2+}?

Part C: Qualitative Analysis of Group 3 Cations—Investigative Approach

The Group 3 cations consist of barium, Ba^{2+}; strontium, Sr^{2+}; and calcium, Ca^{2+}.

> Complete the prelaboratory report sheet and the prelaboratory problem sheet before coming to the laboratory. For concentrations of common acids and bases, see Appendix 3.

The objectives of this experiment are (a) to perform the tests on the individual known cations and to enter your observations on the data sheet and (b) to develop a scheme for the qualitative analysis of an unknown solution.

Test Reagents

The reagent solutions are 6 M HCl, concentrated (12 M) HCl, 3 M NH_4Cl (ammonium chloride), 3 M NH_3, 6 M NH_3, solid $(NH_4)_2CO_3$ (ammonium carbonate), and 6 M CH_3COOH (HOAc, acetic acid).

Special Test Reagents

These reagents include 0.5 M $(NH_4)_2C_2O_4$ (ammonium oxalate), 1 M $(NH_4)_2SO_4$ (ammonium sulfate), $K_4Fe(CN)_6$ [potassium hexacyanoferrate(III)], and 0.1 M K_2CrO_4 (potassium chromate).

Procedure

> Microscale experiment   Estimated time to complete the experiment: 2.5 h
> Important Note: The use of a centrifuge is recommended for separating precipitates from supernatant liquids [see Chapter 6.2(g) and (h)]. After removing the supernatant liquid with a Pasteur pipet, wash the residue twice with two to four drops of water and recentrifuge the mixture. Remove the supernatant liquid and collect the residue for further testing. Always save the supernatant liquid in a marked container for further analysis unless instructed otherwise.

> Disposal procedure: Dispose of the solutions in marked containers.

Obtain six small centrifuge tubes, several test tubes ($\sim$10$\times$75 mm), litmus paper, a nichrome wire, several Pasteur pipets with a rubber bulb, a small stirring rod, and one 50 mL beaker to serve as a water reservoir for rinsing the Pasteur pipets.

Also obtain 2 to 3 mL of the unknown solution, which may contain **one, two, or three cations**. Record the ID number of the unknown solution on your record sheet. Set up a water bath in a 250 mL beaker that is three-quarters full of deionized water.

Place 0.5 to 1.0 mL of the **unknown** solution on a clean watch glass. Place the watch glass on a boiling water bath in a beaker for complete evaporation of the solution. While the evaporation is going on, proceed to the wet tests. When the solution has completely evaporated to dryness, remove the watch glass from the water bath. Collect the dry solid and perform the flame tests and other special tests as described shortly.

Section A: Analysis of Known Cations

1. Wet Tests: Separation of Cations Using Reagents

Using a Pasteur pipet, place four drops of each of the known cation solutions in three separate centrifuge tubes, labeled with the cation symbols. Add four drops of water to each of the centrifuge tubes, followed by one drop of 6 M HCl. Note whether any precipitates form. If none do, the Group 1 cations are absent. Now, add three drops of 6 M NH$_3$ to each solution. Stir to ensure that the solution is distinctly basic. Note whether any precipitate forms. If none do, the Group 2 cations are absent. Enter your observations in the data table.

Add a small amount of solid (NH$_4$)$_2$CO$_3$ (one-fourth the size of a pea) to each of the centrifuge tubes. Stir the mixture to complete the precipitation of the metal carbonates. The reaction involved is (M = Ca^{2+}, Ba^{2+}, and Sr^{2+})

$$M^{2+}(aq) + CO_3^{2-}(aq) \rightarrow MCO_3(s)$$

If precipitation has occurred, indicate it in the appropriate data box by writing *ppt*, followed by its color code (*w* for white, *y* for yellow, and so forth). Save all centrifuge tubes that contain precipitates. These precipitates will be used to perform additional tests described shortly.

2. Additional Tests

Using a Pasteur pipet, remove and discard the clear supernatant liquid from each of the centrifuge tubes, leaving behind as much residue as possible. Rinse the residue twice using six drops of deionized water, discarding the wash liquid each time. Add four drops of 6 M acetic acid (CH$_3$COOH, HOAc) to each solid and stir. Now, heat the contents of each centrifuge tube in a water bath for 1 min. Enter your observations in the data table. The observed reaction follows this pattern:

$$MCO_3(s) + 2\,HOAc \rightarrow M(OAc)_2(aq) + H_2CO_3(aq)$$

Add one drop of yellow 0.1 M K$_2$CrO$_4$ solution to each of the centrifuge tubes. Note on the data sheet which of the solutions forms a precipitate. Centrifuge and discard the supernatant liquid. Save the centrifuge tube where a precipitate has formed (call it **ct #1**) for a flame test (see next section).

$$M^{2+}(aq) + K_2CrO_4(aq) \rightarrow MCrO_4(s) + 2\,K^+(aq)$$

The contents of the remaining centrifuge tubes (where no precipitate formed—they will appear yellow because of the soluble chromate salt) should be made distinctly basic (test with red litmus paper) by adding five to six drops of 6 M NH$_3$. Now, add a few crystals of (NH$_4$)$_2$CO$_3$ to each of the solutions. Stir the solutions thoroughly and allow both tubes to stand for two minutes. Both centrifuge tubes should yield a carbonate precipitate. Centrifuge each tube, discarding the supernatant liquid. Wash the residue with a few drops of water, and discard the wash. What is the composition of each precipitate? Record your observations on the data sheet.

Dissolve each residue in four to six drops of 6 M acetic acid. Heat both centrifuge tubes in a hot water bath for 1 min to expel CO_2 gas. Add four drops of saturated $(NH_4)_2SO_4$ solution to each centrifuge tube, followed by a small crystal of sodium thiosulfate $(Na_2S_2O_3)$. Heat and stir the solutions in a water bath for two minutes, and then cool them back to room temperature. Note that one of the centrifuge tubes shows the formation of a white precipitate. Centrifuge and discard the supernatant liquid. Note on the data sheet which cation this is. Save this centrifuge tube (call it **ct #2**) for a flame test.

$$M^{2+}(aq) + (NH_4)_2SO_4(aq) \rightarrow MSO_4(s) + 2\,NH_4^+(aq)$$

To the last centrifuge tube [which did not produce a precipitate with $(NH_4)_2SO_4$], add four drops of 0.5 M $(NH_4)_2C_2O_4$ (ammonium oxalate) solution and warm the mixture in a water bath for one minute. The solution will form a precipitate. Note which cation is responsible for this precipitate, and record it on the data sheet. Centrifuge and discard the supernatant liquid. Save the residue (**ct #3**) for a flame test.

$$M^{2+}(aq) + (NH_4)_2C_2O_4(aq) \rightarrow MC_2O_4(s) + 2\,NH_4^+(aq)$$

3. Special Tests

Flame Tests

To familiarize yourself with the color of the flame characteristic of the three cations—Ba^{2+}, Sr^{2+}, and Ca^{2+}—obtain a 10 mg sample of a solid salt of each cation from your instructor. Perform the flame test as described in Chapter 6, section 3(*b*). Observe the color of the flame for each cation (see Table 6.2 to match the color of the flame with that of the cations). Record your observations on the data sheet.

Wash the residue at least twice in each of the tubes (**ct #1, ct #2,** and **ct #3**) with four drops of deionized water. Discard the wash liquids. Now, using a nichrome or platinum wire, perform the flame test as described before on each of the residues in **ct #1, ct #2,** and **ct #3**. In each case, dip the clean wire in concentrated HCl, then touch the tip of the wire into the residue inside the centrifuge tube. Heat the wire carrying a small portion of the residue in a Bunsen burner flame, noting the color of the flame. Match the color of the flame with the cation that produced it.

Use a clean wire for each residue. Dirty wire contaminated with other cation(s) may produce a different color in the flame, leading to wrong conclusions.

Section B: Development of a Qualitative Analysis Scheme for Group 3 Cations

Using the observations made during the tests on the known cations, develop a scheme for the qualitative analysis of a mixture of cations of this group. Draw a flow chart for your scheme that will enable you to separate the cations for their positive identification. Show the flow chart to your instructor for his or her approval. Using 10 to 20 drops of the unknown mixture, perform the qualitative analysis following your scheme.

Guideline for the Development of a Qualitative Analysis Scheme

1. Place 10 to 20 drops of the unknown mixture in a centrifuge tube. First, precipitate the Group 1 cations by acidifying (litmus paper) with four drops of 6 M HCl. Centrifuge the solution for 45 to 60 seconds. Decant the supernatant liquid, saving the supernatant liquid for Groups 2–4 analysis. Rinse the residue (if any) twice with deionized water, and discard the wash. The solid may be analyzed for Group 1 cations according to the procedure you developed earlier (Part A).

2. Next, precipitate any Group 2 cations that may be present by adding five to six drops of 6 M aqueous NH_3 (test with litmus) to the supernatant liquid from step 1. Stir the solution, then centrifuge it for two minutes. Decant the supernatant liquid (**snt #1**) and save it in a **labeled** test tube for the analysis of Group 3 cations (see step 3) according to the procedure already described. Wash the residue twice with five drops of water, discarding the wash. This residue may be analyzed for the Group 2 cations according to the procedure developed in Part B.

3. Finally, add 20 mg of ammonium carbonate, $(NH_4)_2CO_3$, to the supernatant liquid (**snt #1**) from step 2. Stir and heat the mixture in a boiling water bath for one minute. Decant and save the supernatant liquid for Group 4 cation analysis, as described in Part D of this experiment. The residue (**ppt #1**) is used for the analysis of Group 3 cations.

4. Wash the residue (**ppt #1**) with water. Discard the supernatant liquid. Dissolve the residue in a minimum volume of 6 M acetic acid, stir the mixture, and heat it in a water bath for two minutes. Carry out the qualitative analysis according to the scheme you developed. An example of a flow chart appears in the accompanying figure.

Flame Tests on the Dry Residue of the Unknown Mixture

Collect the residue from the watch glass (see the foregoing) and carry out the flame test for Ba^{2+}, Ca^{2+}, and Sr^{2+}. If an unknown contains a mixture of two or three of these cations, flame tests performed on this mixture cannot be used to confirm the cations. The cations must be separated before any conclusive flame test can be applied.

Submit your report in one of the following formats:

1. Prelaboratory report sheet, data sheets, prelaboratory problems, and postlaboratory problems.

2. If required by your instructor, submit a short written report of your experiment (see Chapter 4). Attach the prelaboratory report sheet and the data sheets, along with your written report including the prelaboratory and postlaboratory problems.

Example of a Flow Chart, Group 3

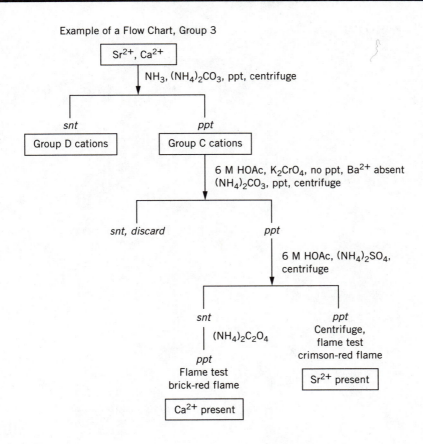

Name ______________________ Section ______________

Instructor ______________________ Date ______________

PRELABORATORY REPORT SHEET—EXPERIMENT 16, PART C

Experiment title ______________________

Objective

Reagents, reactions/formulas to be used

Materials and equipment table

Outline of procedure

1. What is the general reaction between the cations of this group and ammonium carbonate in ammonia solution?

2. Suggest a scheme (flow diagram) for the separation of a mixture containing Ba^{2+} and Sr^{2+} cations. Mention the confirmatory tests for each cation.

Name	_______________	Section	_______________
Instructor	_______________	Date	_______________

EXPERIMENT 16, PART C DATA SHEET

Data Collection Table

Unknown ID number ___________

1. Wet Analysis of Known Individual Cations

Write balanced chemical reactions for each positive test (the tests that form products different from the starting reagents).

Experiment performed	Observation made	Inference drawn
1.		
2.		

Use a separate sheet if necessary.

Summary table

Reagents used	Ba^{2+}	Sr^{2+}	Ca^{2+}
HCl			
Aq. NH_3			
NH_3 +$(NH_4)_2CO_3$			
Acetic acid			
K_2CrO_4			
$(NH_4)_2CO_3$			
$(NH_4)_2SO_4$			
$(NH_4)_2C_2O_4$			

Use a separate sheet if necessary.

2. Additional Tests

Use a separate sheet if necessary.

3. Special Tests

Experiment performed	Observation made	Inference drawn

1. Flame test

   Ba^{2+}

   Ca^{2+}

   Sr^{2+}

2. Other

DATA SHEET, EXPERIMENT 16, PART C, PAGE 3

Scheme for the Analysis of the Unknown Cations

Develop a scheme for analyzing the unknown solution, suggest the reaction sequence, and draw the flow diagram as shown in the example. Show the scheme to your instructor and obtain his or her approval before you perform the unknown analysis.

POSTLABORATORY PROBLEMS—EXPERIMENT 16, PART C

1. What did you observe when you performed the following tests?

(*a*) Flame test for Ca^{2+}

(*b*) Flame test for Sr^{2+}

2. List the metal cations of this group that form precipitates with the following ammonium salts:

(*a*) $(NH_4)_2SO_4$

(*b*) $(NH_4)_2C_2O_4$

(*c*) $(NH_4)_2CO_3$

Part D: Qualitative Analysis of Group 4 Cations— Directive Approach

> Complete the prelaboratory report sheet and the prelaboratory problem sheet before coming to the laboratory. For concentrations of common acids and bases, see Appendix 3.

Perform the qualitative analysis by following the directions given in this section. First, acquaint yourself with the scheme by performing the analysis on a mixture of known cations. Then perform the analysis of the unknown sample. The Group 4 cations consist of cadmium (Cd^{2+}), cobalt(II) (Co^{2+}), copper(II) (Cu^{2+}), magnesium (Mg^{2+}), nickel(II) (Ni^{2+}), and zinc (Zn^{2+}).

Procedure

> Microscale experiment   Estimated time to complete the experiment: 3 h
> Important Note: The use of a centrifuge is recommended for separating precipitates from supernatant liquids [see Chapter 6.2(g) and (h)]. After removing the supernatant liquid with a Pasteur pipet, wash the residue twice with two to four drops of water and recentrifuge the mixture. Remove the supernatant liquid and collect the residue for further testing. Always save the supernatant liquid in a marked container for further analysis unless instructed otherwise.

> Disposal Procedure: Dispose of all the solutions only in marked containers.

Known solutions containing the cations of this group are provided.

Test Reagents

These reagents include concentrated HNO_3, 6 M HCl, 6 M NaOH, 6 M NH_3, and solid $(NH_4)_2CO_3$. Other special test reagents used are 1 percent dimethylglyoxime (DMG), ammonium thiocyanate (NH_4SCN), amyl alcohol ($C_5H_{11}OH$), 1 M sodium hypochlorite (NaOCl), 0.5 M sodium acetate (CH_3COONa), 1 M thioacetamide (CH_3CSNH_2), 0.1 M potassium hexacyanoferrate(II) [$K_4Fe(CN)_6$], disodium hydrogen phosphate (Na_2HPO_4), potassium hexacyanocobaltate(III) [$K_3Co(CN)_6$], and sodium tetraborate decahydrate or solid borax ($Na_2B_4O_7 \cdot 10H_2O$).

Obtain seven centrifuge tubes, some test tubes ($\sim 10 \times 75$ mm), litmus paper, a nichrome or platinum wire, several Pasteur pipets with a rubber bulb, a small stirring rod, and one 50 mL beaker to serve as a water reservoir for rinsing the Pasteur pipets. Also obtain 2 to 3 mL of the unknown solution, which may contain **one, two, or three cations.** Record the ID number of the unknown on your record sheet.

Section A: Qualitative Analysis Scheme for Group 4 Cations

1. Wet Tests for Group 4 Cations

Set up a boiling water bath in a 250 mL beaker covered with an aluminum or plastic rack with holes to hold test tubes. In a clean centrifuge tube, place 1 mL of a solution that is 0.05 to 0.1 M in each of the Group 4 cations (and may contain other groups). Add five drops of 6 M HCl and stir. Then add 6 M aqueous ammonia dropwise, with stirring, until the solution tests neutral to pH paper. Add four or five additional drops of 6 M aqueous ammonia. Add a half-pea-sized portion of solid ammonium carbonate to this solution, stir it thoroughly, and heat it in a hot water bath for one minute. Centrifuge for 45 to 60 seconds. Any precipitate present will contain cations from Groups 1, 2, and 3. The supernatant liquid contains the Group 4 cations. Record all observations.

In the presence of ammonia, five of the Group 4 cations form soluble metal-ammine complexes. All the complexes have different colors, as shown below. The magnesium ion does not form a complex with ammonia, but its hydroxide is more soluble and does not precipitate.

$$Cu^{2+}(aq) + 4\,NH_3(aq) \rightarrow Cu(NH_3)_4^{2+}(aq) \qquad \text{(royal blue)}$$
$$Cd^{2+}(aq) + 4\,NH_3(aq) \rightarrow Cd(NH_3)_4^{2+}(aq) \qquad \text{(colorless)}$$
$$Ni^{2+}(aq) + 6\,NH_3(aq) \rightarrow Ni(NH_3)_6^{2+}(aq) \qquad \text{(blue)}$$
$$Zn^{2+}(aq) + 4\,NH_3(aq) \rightarrow Zn(NH_3)_4^{2+}(aq) \qquad \text{(colorless)}$$
$$Co^{2+}(aq) + 6\,NH_3(aq) \rightarrow Co(NH_3)_6^{2+}(aq) \qquad \text{(pink)}$$

The color of the supernatant liquid may provide a clue as to which cations are present.

Decant the supernatant liquid containing the Group 4 cations into a clean crucible. (**NOTE: If you are analyzing a solution as part of a continuing experiment, use the supernatant left from separating the Group 3 cations.**) Place the lid, slightly ajar, on the crucible and heat the solution to dryness over a low flame.

Cool the crucible for five minutes, and then add four to six drops of concentrated nitric acid, being sure to wash the inside of the crucible. Replace the lid and heat to dryness once again. Cool the crucible for five minutes, repeat the addition of concentrated nitric acid and heat to dryness. After cooling for five minutes, collect the solid. Divide the residue into two unequal parts in a 1:4 ratio.

2. Treatment of the Smaller Part

Flame Test for Copper
Dip a clean nichrome or platinum wire into concentrated HCl, and then touch it to the residue and perform the flame test as described in Chapter 6, section 3(*b*). A green flame shows the presence of copper cation.

Borax Bead Test for Cobalt
Using a clean nichrome or platinum wire, perform the borax bead test with the same residue by following the procedure described in Chapter 6, Section 3(*c*). A blue bead **confirms** the presence of cobalt cation.

3. Treatment of the Larger Part

Dissolve the larger part of the residue in a minimum volume of 6 M HCl (10 to 20 drops) in the crucible. Using a clean Pasteur pipet, transfer the resulting solution to a clean centrifuge tube labeled **ct #1**. Rinse the crucible with 5 drops of distilled water and add the rinse to the same centrifuge tube. Record all observations.

Spot Test for Identification of Nickel(II) Ion

Transfer one drop of the solution from **ct #1** to a spot plate. Add one drop of 1 percent dimethylglyoxime. Add 6 M NH_3 with stirring until the solution tests basic to litmus paper. A red or pink precipitate or coloration confirms the presence of nickel. Record all observations. Note that the solution must be basic for the precipitate to form.

$$Ni(H_2O)_6^{2+}(aq) + 6\,NH_3(aq) \rightarrow Ni(NH_3)_6^{2+}(aq) + 6\,H_2O(l)$$
$$Ni(NH_3)_6^{2+}(aq) + 2\,DMG^- \rightarrow Ni(DMG)_2(s) + 6\,NH_3(aq)$$

Spot Test for Identification of Copper(II) Ion

Transfer one drop of the solution from the centrifuge tube (**ct #1**) to a spot plate. Add 6 M NH_3 until the solution tests only weakly acidic. If the solution becomes basic [it will show blue color due to $Cu(NH_3)_4^{2+}$], use 6 M acetic acid to make it slightly acidic. Add two-drops of 0.1 M $K_4Fe(CN)_6$ solution. Formation of a maroon precipitate or color **confirms** the presence of copper. Record all observations.

$$2\,Cu(NH_3)_4^{2+}(aq) + 8\,H^+(aq) + Fe(CN)_6^{4-}(aq) \rightarrow Cu_2Fe(CN)_6(s) + 8\,NH_4^+(aq)$$

4. Test for Cadmium Cation and its Separation from Other Cations

Cadmium is particularly difficult to identify, as it must be isolated from virtually all other cations. Most cadmium compounds are white to pale yellow, and their presence is easily masked by the presence of even traces of copper, cobalt, and or nickel cations (which form highly colored compounds).

Cadmium and copper can be separated from cobalt, nickel, and zinc based on the solubilities of their sulfides. Cadmium sulfide and copper sulfide are less soluble under acidic conditions (low pH, 0.5 to 1.0) than are cobalt, nickel and zinc sulfides. The approximate values for the solubility product, K_{sp}, of these five sulfides are given here:

K_{sp} values

CuS	8.7×10^{-36}	CoS	5.9×10^{-21}
CdS	3.6×10^{-29}	NiS	3.0×10^{-21}
		ZnS	1.1×10^{-21}

The following equilibrium is dependent upon the hydrogen ion concentration (actually, the activity; see Experiment 28):

$$H_2S(aq) \rightleftharpoons S^{2-}(aq) + 2\,H^+(aq)$$

According to Le Chatelier's principle, at a low pH (higher $[H^+]$), the sulfide ion concentration, $[S^{2-}]$, will be low. The K_{sp} values for copper and cadmium sulfides are very low ($< 10^{-28}$, much lower than those for nickel, cobalt, and zinc sulfides).

Both these cations (~ 0.01 M in this analysis) will form insoluble sulfides even at an extremely low sulfide ion concentration ($\sim 10^{-20}$ M at a pH of 0.5). However, Co^{2+}, Zn^{2+}, and Ni^{2+} will not precipitate under these conditions; they will, however, form insoluble sulfides at a pH higher than ~ 8.

Treat the remainder of the solution in **ct #1** as follows. Adjust the pH of the solution to about 0.5 by following either of the following methods:

Separation of Copper (II) and Cadmium (II)

Method A. Add concentrated ammonia dropwise from a pipet until the solution is basic to litmus paper (red litmus turns blue). If a precipitate forms, ignore it. Now, add 6 M HCl dropwise until the mixture is just acidic (test with fresh litmus paper, blue litmus turns red). At this time, add two additional drops of 6 M HCl. If available, use pH paper to measure the pH of the solution—it should be about 0.5.

Method B. Add one drop of 1 percent methyl violet to the solution. Using a Pasteur pipet, add 6 M ammonia solution dropwise until the color of the solution changes to yellow-green. At pH 2 to 10 the color of the indicator remains violet.

CAUTION: Carry the following step out in the **HOOD**. A small amount of H_2S gas is produced in this reaction. This toxic gas has a rotten egg smell.

Add ~ 0.5 mL of freshly prepared 1 M thioacetamide solution to the solution. Suspend the centrifuge tube (**ct #1**) in a boiling water bath and heat the mixture, with stirring, for at least five minutes. These steps lead to the production of H_2S, a source of sulfide ion, in situ.

$$CH_3CSNH_2(s) + 2\,H_2O(l) + H^+(aq) \rightarrow CH_3COOH(aq) + NH_4{}^+(aq) + H_2S(aq)$$

Gradually, a precipitate will form—at first the residue is light in color, then it becomes darker, and finally it will turn black. Continue to heat the mixture for two additional minutes. Centrifuge the solution, retaining the residue (**ppt #1**). Decant the supernatant liquid into a clean centrifuge tube, labeled **ct #2**. Wash the residue in **ct #1** with deionized water and discard the wash. Save the residue for mixing with a second lot of a precipitate, obtained later.

Readjust the pH of **ct #2** to 0.5 or higher by adding 0.5 M sodium acetate solution. Add 10 drops of deionized water, followed by 6 to 10 drops of 1 M thioacetamide. Place the tube (**ct #2**) on the test tube rack and heat the solution again in the boiling water bath for five minutes. Centrifuge the solution for one minute (is it brown or yellow?). Decant the supernatant liquid into a clean centrifuge tube, labeled **ct #3**, for separating Co^{2+}, Ni^{2+}, and Zn^{2+} cations as sulfides under basic conditions (described shortly).

Wash the residue in **ct #2** with four drops of water, discarding the wash liquid. Add an additional four drops of water and, using a pipet, transfer this mixture to **ct #1**, which contains **ppt #1**. Add 10 to 20 drops of 6 M HNO_3 to **ct #1**. Heat the mixture in a hot water bath for two minutes. Almost all the residue should dissolve. Centrifuge and decant the liquid into another centrifuge tube, labeled **ct #4**. Discard the solid in **ct #1**, if any, to a waste container. Very often, light yellow sulfur separates out, which may remain in solution as a colloid. In the event of a colloidal solution, evaporate the solution almost to dryness. Using a glass rod, remove the coagulated sulfur, or centrifuge.

Add aqueous ammonia, drop by drop, to the solution in **ct #4** until the solution is basic to litmus paper. Add an additional two drops of ammonia, and stir the solution. If the solution turns blue, copper is present. (**Note: If the solution is not blue, then copper(II) cation is absent. In that case, skip to the next paragraph.**) Add 0.1 g of sodium dithionite to the blue solution. Heat the solution in a hot water bath for 1 to 2 minutes. Reduction of copper(II) to copper(I) and, finally, to dark copper metal will take place. This observation is an additional confirmation of the presence of copper. Centrifuge and collect the liquid in a clean test tube (**tt #5**). Dispose of any solid in a designated container.

Add 6 to 8 drops of 1 M thioacetamide to the supernatant liquid in **tt #5**. Heating the solution in the boiling water bath for three minutes will lead to the formation of a yellow precipitate of CdS if cadmium is present.

Separation of Nickel (II), Cobalt (II), and Zinc (II)

Centrifuge tube **ct #3** is now treated with aqueous ammonia until it is basic (red litmus turns blue, pH 9 to 10). Using a Pasteur pipet, add 15 drops of 1 M thioacetamide to the solution. Stir and heat this solution in a boiling water bath for five minutes. This process will generate the sulfides of Ni^{2+}, Co^{2+}, and Zn^{2+}.

Centrifuge the solution, decant the clear supernatant liquid into a centrifuge tube labeled **ct #6** and save it for the detection of Mg^{2+} and Zn^{2+} (see step 8, which follows). The residue **ppt #2** left in **ct #3** contains cobalt(II) sulfide (CoS, black), nickel(II) sulfide (NiS, black), and zinc sulfide (ZnS, white).

5. Separation of Cobalt, Nickel, and Zinc from Their Sulfides (ppt #2)

Stir the residue **ppt #2** with 10 drops of dilute HCl acid. Stir the mixture and allow it to stand for two minutes. Centrifuge for 45 to 60 seconds. Decant the supernatant liquid, **snt #1**, and save it for testing for the presence of zinc (see step 7). Save the solid in **ct #3** for testing for cobalt and nickel (**ppt #3**).

6. Treatment of Ppt #3 for Testing for the Presence of Cobalt and Nickel

With a tiny part of **ppt #3**, perform the borax bead test [see Chapter 6, section 3(c)]. If the bead is blue, cobalt is present. Add 10 drops of 6 M HCl and five drops of NaOCl (household bleach) to the **ppt #3** in the centrifuge tube. Heat the mixture in a hot water bath for two minutes. Centrifuge, if necessary. Collect and transfer the liquid to another test tube. Add 1 mL of water. Divide the solution into two parts.

Part 1. Add 10 drops of amyl alcohol, followed by 25 to 30 mg ammonium thiocyanate, NH_4SCN. Using a stirring rod, mix the solution thoroughly. A blue coloration of the alcohol layer confirms cobalt.

Part 2. Add a crystal or two of NH_4Cl, followed by three or four drops of 1 percent dimethylglyoxime. While stirring, add aqueous ammonia till the solution is basic (use litmus paper). A red precipitate indicates the presence of nickel.

7. Treatment of snt #1 for Testing for the Presence of Zinc

Add 6 M NaOH solution to **snt #1** until it is basic (litmus paper). Heat the mixture in a hot water bath for one minute. Render the solution acidic by adding acetic acid (litmus paper). Add three to four drops of potassium hexacyanoferrate(II) solution. A white precipitate confirms the presence of zinc.

8. Test for Magnesium and Zinc Cations

The supernatant liquid in **ct #6** contains Mg^{2+} and probably some leftover Zn^{2+} and Ni^{2+} (the presence of Ni^{2+} will give a blue color to the solution). Add 6 M ammonia until the solution is basic, and then a small portion (10 mg) of disodium hydrogen phosphate, Na_2HPO_4. A white precipitate of magnesium ammonium phosphate, $MgNH_4PO_4$, confirms the presence of Mg^{2+}.

$$Mg^{2+}(aq) + NH_4^+(aq) + HPO_4^{2-}(aq) \rightarrow MgNH_4PO_4(s)$$

Centrifuge tube **ct #6** for 45 to 60 seconds, then decant and discard the supernatant liquid. Wash the precipitate with four drops of water, centrifuge the solution, and discard the wash liquid.

Add six to eight drops of 6 M NaOH solution to the precipitate. Any zinc phosphate, $Zn_3(PO_4)_2$, that may be present will dissolve, forming colorless $Zn(OH)_4^{2-}$. Any white precipitate left is magnesium ammonium phosphate, which is insoluble in excess NaOH. Centrifuge the mixture and discard the solid, decanting the supernatant liquid into a clean test tube (**tt #7**) for zinc analysis. Add 6 M acetic acid to the supernatant solution until it is acidic (use litmus paper). Stir the solution and heat it in a water bath for one minute.

Divide this solution into two parts. One part, when treated with four drops of potassium hexacyanoferrate(II) solution, generates a white precipitate having the composition $Zn_2Fe(CN)_6$. This reaction confirms the presence of zinc cation.

9. Spot Test for Zinc Cation, Rinmann's Green Test

Concentrate the second part of the foregoing solution by heating it in a crucible until its volume is reduced by half. Neutralize the solution by adding 6 M NH_3 (litmus). Prepare a piece of filter paper for this test by dipping it in a 4 percent solution of potassium hexacyanocobaltate(III) in 1 percent aqueous potassium chlorate ($KClO_3$). Dry the filter paper before use. Add two drops of the neutralized solution to the pretreated filter paper. Dry the paper in air. On ignition in a crucible, the ash will be green if zinc is present, as a result of the formation of cobalt zincate, $CoZnO_2$, called Rinmann's green compound.

Section B: Analysis of an Unknown Containing the Group 4 Cations

Obtain 1 mL of an unknown solution containing not more than three Group 4 cations in a clean centrifuge tube. Repeat the procedures just outlined, using the unknown solution. Report which cations are present in the unknown, along with observations to verify the identification.

Name ______________________ Section ______________

Instructor ______________________ Date ______________

PRELABORATORY REPORT SHEET—EXPERIMENT 16, PART D

Experiment title ______________________________

Objective

Reagents, reactions/formulas to be used

Materials and equipment table

Outline of procedure

PRELABORATORY PROBLEMS—EXPERIMENT 16, PART D

1. What is the general reaction between the cations of this group and sodium hydroxide?

2. Suggest a scheme (flow diagram) for the separation of a mixture containing Cd^{2+}, Cu^{2+}, and Mg^{2+} cations. Mention the confirmatory tests for each cation.

3. Explain the role played by K_{sp} in the separation of CuS from NiS.

EXPERIMENT 16, PART D   DATA SHEET

Data Collection Table

Unknown ID Number _____

Develop your own table as shown in parts A, B, and C for the analysis of a mixture of known cations.

Scheme for the Analysis of the Unknown Cations

Develop a scheme for analyzing the unknown solution, suggest the reaction sequence, and draw the flow diagram.

POSTLABORATORY PROBLEMS—EXPERIMENT 16, PART D

1. What did you observe when you performed the following tests?

    (a) Add 1 M thioacetamide to an acid solution containing Cd^{2+}.

    (b) Add 1 M thioacetamide to a basic solution containing Zn^{2+}.

2. Write reactions for the following changes:

    (a) Disodium hydrogen phosphate is added to a solution of Mg^{2+} ions.

    (b) Ammonia is added to a mixture of Cu^{2+}, Ni^{2+}, and Cd^{2+} ions.

Part E: Qualitative Analysis Scheme for Anions: Directive Approach

Three sets of tests—**dry tests, wet tests**, and **confirmatory tests**—will be performed to identify a given anion. A dry test is applied to a dry sample or solid. A wet test involves the use of a solution of the sample together with a reagent that forms a precipitate or produces a visible change characteristic to the sample. In addition to these tests, the anions will also be confirmed by performing specific tests called confirmatory tests.

In this analysis, solid samples as well as solutions may be used. Thus, for dry tests, if a solution of sample has been supplied for the anion analysis, part of this solution should be evaporated to dryness so that a solid sample is obtained.

To perform wet tests, a solution of the **soluble** solid sample is prepared in water. If the sample is **insoluble** in water, an extract of the insoluble sample in sodium carbonate is prepared (see section 2b of this experiment). The analysis of anions is performed using aqueous solutions of barium nitrate $[Ba(NO_3)_2]$, silver nitrate $(AgNO_3)$, iron(III) chloride $(FeCl_3)$, iron(II) sulfate $(FeSO_4)$, and calcium chloride $(CaCl_2)$. Other specific reagents used in this section are 0.001 M potassium permanganate $(KMnO_4)$, calcium hydroxide solution, dilute hydrochloric acid, lead acetate $[Pb(CH_3COO)_2]$ solution, potassium dichromate $(K_2Cr_2O_7)$ solution, and Fe^{3+} (dissolved $FeCl_3$). The anions to be tested are the following:

Acetate	CH_3COO^-	Nitrite	NO_2^-
Borate	BO_3^{3-}	Oxalate	$C_2O_4^{2-}$
Bromide	Br^-	Phosphate	PO_4^{3-}
Carbonate	CO_3^{2-}	Sulfate	SO_4^{2-}
Chloride	Cl^-	Sulfide	S^{2-}
Fluoride	F^-	Sulfite	SO_3^{2-}
Iodide	I^-	Thiocyanate	SCN^-
Nitrate	NO_3^-		

Depending upon the tests, the anions may be classified into five groups. Note that some anions are in more than one group.

Dry Tests

Group 1 Anions: The Acid Group

Anions in Group 1, when treated with dilute or concentrated sulfuric acid, generate gases that can be identified. These anions are acetate, carbonate, fluoride, sulfide, nitrite, nitrate, sulfite, chloride, bromide, and iodide.

Wet Tests

Group 2 Anions: The Barium Nitrate Group

Basic solutions of the Group 2 anions react with an aqueous solution of barium nitrate, $Ba(NO_3)_2$, to form insoluble barium salts. The anions of this group are sulfate, phosphate, and carbonate.

Group 3 Anions: The Silver Nitrate Group

Acidic (HNO_3) solutions of the Group 3 anions form precipitates with aqueous solutions of silver nitrate, $AgNO_3$. The anions of this group are chloride, bromide, and iodide.

Group 4 Anions: The Calcium Chloride Group

When reacted with calcium chloride, $CaCl_2$, the anions of Group 4 form insoluble calcium precipitates. These anions include fluoride, oxalate, and phosphate.

Procedure: Identification of an Anion in a Sample

Obtain your unknown sample. Enter its ID number in your record sheet. Perform the tests as described below. Record all your experiments, observations, and inferences on the data sheet as shown in parts A, B, and C.

Microscale experiment	Estimated time to complete the experiment: 2.5 h

1. Dry Tests

> The following procedure is for solid samples. If you are given a solution for analysis, perform the tests using 5 to 10 drops of the solution (or obtain a residue by evaporating 0.5 mL of the solution on a watch glass over a sand bath).

Set up a boiling water bath in a 250 mL beaker. The dry tests for the anions will be performed in small test tubes (75×10 mm).

Action of Dilute Sulfuric Acid (Group 1 Anions)

Place 10 to 20 mg of the anion sample in a 10×75 mm test tube. Using a Pasteur pipet, add five to eight drops of ~6 M H_2SO_4 to the sample. Observe if any reaction takes place while the solution is at room temperature. If none occurs, heat the mixture in the boiling water bath, stirring the mixture with a glass rod. Observe any changes that take place.

Observation	Inference
A rapid effervescence of a colorless and odorless gas is observed. If a glass rod moistened with $Ca(OH)_2$ solution is held just above the surface of the solution, the drop turns milky.	Carbon dioxide gas has formed. **Carbonate (CO_3^{2-}) anion is present.**
Evolution of a colorless gas takes place. The gas turns a filter paper moistened with an acidified solution of $K_2Cr_2O_7$ green.	Sulfur dioxide gas has formed. **Sulfite anion (SO_3^{2-}) is present.**
A colorless gas with a rotten egg smell evolves. The gas turns filter paper moistened with lead acetate black.	H_2S gas has formed. **Sulfide, S^{2-}, anion is confirmed. Caution: H_2S gas is toxic.**

Observation	Inference
An odor of vinegar is detected.	Acetate, CH_3COO^-, anion is present.
A very light brown gas evolves. When placed against a white background, the brown color becomes visible.	Nitrite, NO_2^-, anion is present.

Action of Concentrated Sulfuric Acid (Group 1 Anions)

CAUTION: Concentrated sulfuric acid is highly corrosive and may cause severe burns. In the event of skin contact, wash the area with copious amounts of cold running water. Inform your instructor immediately. In the case of a spill, inform your instructor. Perform the test inside a HOOD.

Using a microspatula, place 10 to 20 mg of the anion sample in a 10×75 mm test tube. Hold the test tube with a test tube clamp. Add two to four drops of concentrated sulfuric acid to the top of the sample. Observe if any reaction has taken place at room temperature. If none has, heat the mixture with stirring in a boiling water bath.

Observation	Inference
A colorless gas with a pungent smell evolves. If a glass rod moistened with aqueous ammonia is held near the mouth of the test tube, dense white fumes are observed.	HCl gas has formed. A **chloride salt** was present.
A colorless gas evolves. It turns a drop of $Ca(OH)_2$ solution milky.	Carbon dioxide gas has formed. A **carbonate salt** was present.
Reddish brown vapors are observed. The vapor condenses on the walls of the test tube.	HBr gas (with some Br_2) has formed. A **bromide salt** was present.
Violet vapors are evolved. Confirm by adding a few specks of solid MnO_2 and heating.	HI gas (with some I_2) has formed. An **iodide salt** was present.
On slow but prolonged heating, the inside of the test tube appears "oily." A glass rod moistened with water shows milkiness due to the formation of hydrated silica.	HF gas has formed. A **fluoride salt** was present.
A light brown gas evolves. Add a small piece of copper wire. Heat. A distinctly brown gas evolves.	NO_2 gas has formed. A **nitrite** or **nitrate salt** was present.

Flame Test for Borate

Perform this test inside a well-ventilated HOOD.

Mix 10 to 20 mg of the anion with 5 mg of powdered calcium fluoride (or potassium fluoride) in a *dry* 75×10 mm test tube. Using a Pasteur pipet, add one or two drops of concentrated H_2SO_4 and stir the mixture with a glass rod. Place the end of the glass rod (carrying a small amount of the wet mixture) very close to the base of a Bunsen burner flame (avoid putting the rod inside the flame). **The green color of the flame confirms the presence of a borate salt.**

2. Wet Tests

The dry tests just described give information as to which anions **may be** present in a sample. The wet analysis of the sample must be carried out for confirmation. Wet tests are performed on **aqueous solutions** of the sample. Set up a boiling water bath in a 250 mL beaker.

If a solid sample is provided, try to dissolve 5 mg of the sample in a small amount of water. If the sample completely dissolves (the water may need to be heated), proceed with the analysis as described in **section 2a**. If the sample does not dissolve, prepare a sodium carbonate extract of the sample according to the procedure described in **section 2b**. Use this extract to perform the wet tests.

2a. Wet Tests for Soluble Anions

If the solid is soluble in water, prepare an aqueous solution of the sample by dissolving 20 to 30 mg of the solid in 3 mL of water. Using this solution, perform the following tests with barium nitrate, silver nitrate, and calcium chloride reagents.

Test with Barium Nitrate Solution (Group 2 Anions)

Place 10 drops of the anion solution in a centrifuge tube. Make the solution basic by adding 6 M ammonia (test with litmus paper). Add four or five drops of 0.1 M barium nitrate solution, and stir the solution. If a white precipitate forms, one or more of the Group 2 anions (carbonate, CO_3^{2-}; oxalate, $C_2O_4^{2-}$; sulfate, SO_4^{2-}; and phosphate, PO_4^{3-}) are present.

Centrifuge the mixture. Decant and discard the supernatant liquid. Wash the residue twice with deionized water, discarding the wash. Add five to eight drops of 6 M HNO_3 to the residue, and stir the mixture. If evolution of a colorless gas is noticed, insert a glass rod moistened with calcium hydroxide, $Ca(OH)_2$, solution inside the centrifuge tube. Do not allow the rod to touch the wall of the centrifuge tube. If the calcium hydroxide droplet at the end of the rod becomes milky, this observation **confirms the presence of carbonate, CO_3^{2-}.**

Heat the remaining solution in a hot water bath. A white precipitate that remains insoluble in 6 M HNO_3 **confirms the presence of sulfate, SO_4^{2-}** ($BaSO_4$).

Centrifuge the preceding solution. Transfer the clear supernatant liquid to another test tube, discarding the solid into a proper waste container. Add five drops of saturated ammonium molybdate, $(NH_4)_2MoO_4$, solution and stir. Heat the mixture in a hot water bath for several minutes. The slow formation of a yellow precipitate **confirms the presence of phosphate, $PO4^{3-}$.** Oxalate, $C_2O_4^{2-}$ is confirmed with $CaCl_2$ test.

Test with Silver Nitrate Solution (Group 3 Anions)

Place 10 drops of the anion solution in a centrifuge tube. Acidify the solution by adding five drops of aqueous 6 M HNO_3 followed by two or three drops of silver nitrate solution. If a precipitate forms, it is due to Group 3 anions.

A **white** precipitate of silver chloride, AgCl, shows the presence of Cl^-.

A **light yellow** precipitate of silver bromide, AgBr, shows the presence of Br^-.

A **yellow-brown** precipitate of silver iodide, AgI, indicates the presence of I^-.

Centrifuge the solution, discarding the supernatant liquid. Add 10 drops of 6 M NH_3 solution to the residue. Heat the mixture in a water bath. If most of the solid disappears, it was an AgCl residue. Acidifying the solution with HNO_3 will regenerate the *white* precipitate, confirming chloride, Cl^-. For confirming bromide or iodide anion in the solution, see the confirmatory tests in section 3.

Test with Calcium Chloride Solution (Group 4 Anions)

Place 10 drops of the test solution in a test tube. Acidify the solution with 6 M HNO_3 (test with litmus paper), then add 6 M NH_3 solution until the solution is basic. Heat the test tube in a hot water bath to expel excess NH_3. Now add five drops of 0.1 M calcium chloride solution. A white precipitate indicates the presence of one or more of the Group 4 anions: fluoride (previously confirmed by the "oily" appearance of the test tube in the dry tests), phosphate, PO_4^{3-} [previously confirmed by the $Ba(NO_3)_2$ test], and oxalate, $C_2O_4^{2-}$. Thus, only the oxalate anion needs to be confirmed.

Add 10 drops of 6 M H_2SO_4 solution to the test solution and heat it in a hot water bath. The acid will dissolve any calcium oxalate that is present. Add one drop of 0.001 M $KMnO_4$ solution, and heat the solution in a hot water bath. The pink color due to permanganate becomes colorless, due to reduction by the acidic oxalate anion. **This observation confirms the presence of oxalate.**

2b. Preparation of Sodium Carbonate Extract

If the sample is insoluble in water (see the list of insoluble salts in Chapter 6), boil a mixture of 100 mg of the solid sample, 200 mg of anhydrous, analytically pure sodium carbonate (Na_2CO_3) and 4 mL of deionized water in a 10 mL Erlenmeyer flask for five to six minutes. Cover the mouth of the flask with a 10 mL beaker. If necessary, add water to compensate for losses due to evaporation. Using a Pasteur pipet, transfer the mixture to a centrifuge tube, and centrifuge the solution. Decant the supernatant liquid to a test tube. Wash the solid left in the centrifuge tube with five to eight drops of hot water. Transfer the clear wash to the same test tube. Save this Na_2CO_3 extract for the wet tests according to the procedure described next.

Wet Tests for Anions in Sodium Carbonate Extract

Sulfate Test
Place five drops of the extract in a test tube. Acidify the solution by adding 6 M HNO_3 (test with litmus paper) and stir. Heat the solution in a hot water bath to expel carbon dioxide

gas. Add one drop of 0.1 M barium nitrate solution, and stir. **A white precipitate confirms the presence of sulfate anion.**

Halide Test
Place five drops of the extract in a test tube. Acidify the solution by adding 6 M HNO_3 (test with litmus paper) and stir. Heat the solution in a hot water bath to expel carbon dioxide gas. Add one drop of 0.1 M silver nitrate solution. A white precipitate soluble in excess aqueous NH_3 (heat if necessary) confirms the presence of Cl^-. A pale yellow precipitate shows the presence of bromide, Br^-. A yellow precipitate indicates the presence of iodide, I^-. Confirm bromide and iodide by the tests described in section 3.

Phosphate Test
Place five drops of the extract in a test tube. Acidify the solution by adding 6 M HNO_3 (test with litmus paper). Heat the solution in a hot water bath to expel carbon dioxide gas. Add five drops of saturated ammonium molybdate solution, and heat the solution in the water bath. **A yellow precipitate confirms the presence of phosphate, PO_4^{3-}.**

Oxalate Test
Place five drops of the extract in a test tube, and acidify it with 6 M sulfuric acid. Add one drop of 0.001 M $KMnO_4$ solution, stir, and heat the mixture in a water bath. If the pink color of the solution disappears, **this observation confirms the presence of the oxalate anion.**

3. Confirmatory Tests

For each of the following four tests, place five drops of anion test solution (or the sodium carbonate extract) in a 10×75 mm test tube.

Tests for Bromide and Iodide, Br^- and I^-
Add ~3 M sulfuric acid to one of the test solutions until it is acidic (test with litmus paper). Add four drops of toluene to the mixture. Toluene, being insoluble in water, forms a separate layer (upper layer). Add, **drop by drop**, 3 percent hydrogen peroxide solution, and stir the solution. Allow the solution to stand for 30 seconds. **The toluene layer becomes pink or violet, confirming the presence of iodide, I^-.** Continue the addition of 3 percent hydrogen peroxide, stirring the solution thoroughly. The toluene layer becomes brown. **This observation confirms the presence of bromide, Br^-.**

Test for Nitrate and Nitrite, NO_3^- and NO_2^-
Add three drops of freshly prepared iron(II) sulfate solution to one of the preceding solutions. Cool the test tube under running water. Hold the test tube at a 45° angle, and dropwise add concentrated sulfuric acid to this mixture. **Be careful not to shake the test tube.** Sulfuric acid, being more dense than water, will occupy the lower layer. Cool the test tube under cold running water. A brown ring at the interface of the two liquids confirms **the presence of nitrate, NO_3^-,** due to the formation of $Fe(NO)^+$. If the solution becomes brown, the color reflects the **presence of nitrite, NO_2^-.**

Test for Thiocyanate, SCN^-
Dilute one of the test solutions by adding five drops of water. Add two drops of 0.1 M iron(III) chloride solution. **If the solution turns blood-red, this observation confirms the presence of thiocyanate, SCN^-.**

Test for Acetate, CH$_3$COO$^-$

Add five drops of pure ethyl alcohol, followed by two or three drops of concentrated sulfuric acid, to one of the test solutions. Heat the mixture in a hot water bath. **A fruity smell indicates the presence of acetate, CH$_3$COO$^-$. (See Figure 1.6 for the proper way to smell a chemical.)**

Name ____________________ Section __________

Instructor ____________________ Date __________

PRELABORATORY REPORT SHEET—EXPERIMENT 16, PART E

Experiment title ____________________

Objective

Reagents and reactions to be used

Outline of procedure

PRELABORATORY PROBLEMS—EXPERIMENT 16, PART E

1. Complete and balance the following reactions. If a reaction does not occur, write "no rxn."

$NaCl(s) + \text{conc. } H_2SO_4 \rightarrow$

$NaCl(s) + \text{conc. } H_2SO_4 + MnO_2(s) \rightarrow$

$KI(aq) + AgNO_3(aq) \rightarrow$

$CaBr_2(aq) + H^+(aq) + H_2O_2(aq) \rightarrow Br_2 + Ca^{2+}(aq) + H_2O(l)$

2. Suggest a scheme for the separation of a mixture containing Br^-, BO_3^{3-}, and SO_4^{2-} anions. Mention the confirmatory tests (the tests that are conclusive for a particular anion) for each.

Name ______________________	Section ____________
Instructor ______________________	Date ____________

EXPERIMENT 16, PART E DATA SHEET

Data Collection Table

Unknown ID number ____________

Make your own data table showing dry tests, wet tests, and confirmatory tests for the anions. Indicate which tests were positive. Write reactions for all confirmatory tests.

Experiments	Observations	Inferences

Confirmatory tests

1. What observations were made for the following tests?

   Flame test for BO_3^{3-} _____

   PO_4^{3-} test with ammonium molybdate _____

   CO_3^{2-} test with $Ba(NO_3)_2$ _____

2. What color changes are observed when the following tests are performed?

   Oxalate sample solution + H_2SO_4 + $KMnO_4$ solution $\rightarrow$

   $SCN^-(aq) + Fe^{3+}(aq) \rightarrow$

   $NO_3^-(aq) + Fe^{2+}(aq) + H_2SO_4 \rightarrow$

3. List four types of observations that would lead you to believe that a chemical reaction has occurred.

Part F: Qualitative Analysis of an Unknown Solid Sample

The analysis of an unknown solid sample consisting of a cation mixture and an anion can be carried out by first dissolving the sample either in water or in 6 M hydrochloric acid (or, very rarely, in 6 M nitric acid). In either case, a general procedure for the qualitative analysis of both the cation and anion must be followed. This general procedure involves the same scheme developed for the individual cations and anions in sections A–E of this experiment.

The analysis scheme involves the following steps: (1) preparing a solution of the sample (if the sample is a solid) in a suitable solvent (water or acid), (2) performing the group analysis following the flow chart in the Figure E16.F1, (3) confirming the cation by analyzing the group precipitate (ppt), and (4) performing the anion tests for identifying the anion.

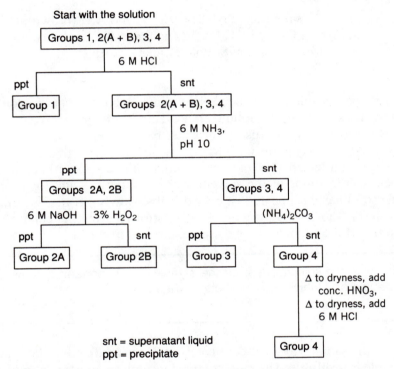

Figure E16.F1 Flow chart for the separation of cations into groups

Procedure: Analysis of an Unknown Sample

Step 1. Obtain 500 mg of an unknown sample. Record the unknown number on the data sheet. Note the physical state (crystalline or amorphous powder) and colors of the sample. The colors of some crystalline salts are as follows:

Salts	Color
Almost all Cu^{2+} salts	Blue or green
All Co^{2+} salts	Pink or purple
Many Fe^{2+} and Fe^{3+} salts	Green or yellow
Many Ni^{2+} salts	Light green
Many Mn^{2+} salts	Pink or pale pink

The colors of some other solid samples include the following:

White	All Al^{3+} salts except Al_2S_3 (yellow) and AlI_3 (brown); all Mg^{2+}, Ca^{2+}, Sr^{2+}, Ba^{2+}, Zn^{2+}, and Pb^{2+} salts except PbI_2 (yellow)
Black	CoI_2, MnO_2, CuS, NiS, PbS, FeS, and $CuBr_2$
Yellow	CdS, PbI_2, and Al_2S_3

Step 2. Prepare a solution of the sample in a suitable solvent. To do this, examine the solubility of the sample in the following solvents in the order given: (*a*) cold (room temperature) water, (*b*) hot water (use a hot water bath to heat the mixture), (*c*) cold 6 M HCl, (*d*) hot 6 M HCl, (*e*) cold or hot 6 M HNO_3. In each case, transfer 5 to 10 mg (the amount on the tip of a microspatula) of the powdered solid to a 75×10 mm test tube. Using a Pasteur pipet, add 15 drops of the solvent. Stir the mixture with a glass rod. Observe carefully if the solid has disappeared (it may dissolve slowly). If not, heat the mixture in a hot water bath for 2 to 3 minutes with constant stirring. Note the color of the solution. **Proceed to step 4.** If the solid does not dissolve, proceed to step 3.

> In connection with the solubility test for the sample, it is important to know which ionic solids are soluble and which are insoluble in water. See Experiment 15 or Chapter 6, Table 6.1.

Step 3. If the sample does not dissolve even in concentrated HCl or in concentrated HNO_3, the sample is **insoluble**. The most common insoluble compounds encountered in the laboratory are $AgCl$, $AgBr$, AgI, $SrSO_4$, $BaSO_4$, $PbSO_4$, CaF_2, Al_2O_3, Fe_2O_3, TiO_2, and WO_3. Silver salts are not given for analysis.

Take 50 mg of the insoluble sample in a 5 mL crucible. Mix the sample with 200 mg of solid Na_2CO_3, and add a flake of solid NaOH.

> CAUTION: NaOH is highly corrosive. Avoid contact with skin.

Heat the mixture strongly over a Bunsen burner flame for 3 to 5 minutes (the mixture may melt). Allow the mixture **to cool to room temperature**. Place the crucible and its contents in a 25 mL beaker. **Carefully add 5 mL of water** to the crucible. Add 6 M HCl dropwise until the solution is distinctly acidic. Transfer the mixture to a centrifuge tube, and centrifuge the mixture. Collect the supernatant liquid. **Use this supernatant liquid to perform the analysis for Groups 2 and 3 cations.** Note that if the sample contains WO_3, it will produce a yellow precipitate on being treated with an acid.

Step 4. Perform the qualitative analysis tests for identifying the cation on the sample solution. First, use the group reagents [the reagent for Group 1 cations is 6 M HCl, for Group 2 is 6 M NH_3, for Group 3 is 6 M NH_3+ solid $(NH_4)_2CO_3$, for Group 4 is thioacetamide and Na_2HPO_4] in the same order as shown in Figure E16.F1 to precipitate the cation in one of the groups. Using a centrifuge, separate the precipitate for further analysis and identify the cation. The complete procedure for the analysis of the sample was described in Parts A–D.

Step 5. Using the solid sample, perform the special tests for ammonium ion, NH_4^+, the flame test for Na^+ and K^+, and the borax bead test for Mn^{2+} and Co^{2+}.

Step 6. Perform the tests as described in Section E to identify the anion(s) present in the sample. Perform both the dry and wet tests to detect the anion.

Develop your own data sheet, and draw a flow chart similar to that in Figure E16.F1. Submit a written report of your findings according to the instructions given in Chapter 4.

Name ____________________ Section ____________

Instructor ____________________ Date ____________

PRELABORATORY REPORT SHEET—EXPERIMENT 16, PART F

Use additional sheets if necessary

Experiment title ____________________

Objective

Reagents, reactions/formulas to be used

Materials and equipment table

Outline of procedure

Use additional sheets if necessary

1. Suppose you are given an unknown solid that contains potassium aluminum sulfate, $KAl(SO_4)_2 \cdot 12H_2O$. This compound is also called alum.

    (*a*) Suggest a scheme (a chart) for qualitative analysis of this compound.

    (*b*) How would you detect the presence of water in a sample of alum?

EXPERIMENT 16, PART F   DATA SHEET

Data Collection Table

Unknown ID number _____

Analysis of an unknown solution

POSTLABORATORY PROBLEM—EXPERIMENT 16, PART F

1. After doing the whole series of qualitative analysis, how much chemistry did you learn? Write comments on your learning experience.

17 Preparation and Percent Composition of a Metal Oxide and Iodide:

Determination of the Empirical Formula of a Compound

Micro- and Macroscale Experiments

Objectives

- To determine the percent composition and empirical formula of a compound
- To use the methods of gravimetric analysis, solution reaction, and reflux

Prior Reading

- Section 3.4   Weighing
- Section 3.6   Heating methods
- Section 3.7   Stirring
- Section 3.8   Filtration and gravimetric methods

Related Experiments

- Experiments 18 (Reactions of lead and its recovery), 19 (Reactions of copper), and 20 (Oxalate complexes of iron and a photochemical reaction)

Atoms of different elements chemically combine to form different compounds following two fundamental laws. The **law of constant composition** states that a sample of a pure compound is always composed of the same elements combined in the same proportions by mass. If we analyze samples of water, we will find that the ratio of the masses of hydrogen and oxygen in each sample is the same, regardless of the source of the water. The **law of multiple proportions** states that if two elements combine to form a series of compounds, the mass ratios of the elements are small whole numbers. For example, hydrogen forms two compounds with oxygen: water (H_2O) and hydrogen peroxide (H_2O_2). In the first compound, 16.0 grams of oxygen (one mole of O) combine with 2.016 grams of hydrogen (two moles of H); and in the second compound, 32.0 grams of oxygen (two moles of O) combine with 2.016 grams of hydrogen (two moles of H). The ratio of masses of oxygen that combine with the hydrogen in these compounds is 16.0 g/32.0 g or 1:2, which supports the law of multiple proportion.

When a new compound is prepared, its formula must be determined. This task is most commonly achieved from the **percent composition** of the compound—the percent by mass of each element relative to the total mass of the compound. It is obtained by dividing the mass of each element in the compound by the molar mass of the compound, and multiplying by 100. In one mole of water (MW = 18.02 g/mol) are two moles of H (2.016 g) and one mole of O (16.00 g). Therefore, %H $= (2.016/18.016)100 = 11.19$ percent and %O $= (16.00/18.016)100 = 88.81$ percent.

The simplest whole-number ratio of the number of atoms of each element in a compound is called the **empirical formula** of the compound. The **molecular formula** shows the actual number of each type of atom present in a compound. For example, suppose that the empirical formula CH is obtained experimentally from the percent composition of a compound. This formula may represent a molecule having the molecular formula C_2H_2, C_6H_6, or any multiple of the empirical formula CH. To derive the molecular formula from the empirical formula one must know the molar mass of the compound.

The empirical formula of a compound can be determined either by chemical analysis or by synthesis. In chemical analysis, a known mass of the compound is decomposed to obtain the masses of the elements themselves or some of their known derivatives (for carbon, the derivative is CO_2 and for hydrogen it is water). From these masses, the percent composition of the compound is determined. In synthesis, known masses of the elements are allowed to form a compound and the mass of the compound is determined. From these data, it is possible to derive the empirical formula of the compound.

In Part A of this experiment, a known mass of a metal will be reacted with oxygen. The mass of the product (a metal oxide) allows the calculation of the mass of oxygen reacted with the metal. The masses of the metal and oxygen yield the percentages of the elements reacting. Finally, the empirical formula for the product is determined. The stoichiometry for the reaction is

$$x\,M(s) + \frac{y}{2}\,O_2(g) \rightarrow M_xO_y$$

Example E17.1 In an experiment, a mixture of copper and sulfur was heated to produce a sample of copper sulfide. The following data were obtained:

Weight of the empty crucible	2.077 g
Weight of the crucible + copper	2.289 g
Weight of the crucible + copper sulfide	2.396 g

Calculate the percentages of copper and sulfur in copper sulfide and derive its empirical formula.

Answer

Copper mass	$= 2.289 - 2.077 = 0.212\,g$
Copper sulfide mass	$= 2.396 - 2.077 = 0.319\,g$
Sulfur mass	$= 2.396 - 2.289 = 0.107\,g$
% copper in copper sulfide	$= (0.212/0.319)100 = 66.5\%$
% sulfur in copper sulfide	$= (0.107/0.319)100 = 33.5\%$
Moles of Cu	$= 66.5\,g(1\,mol/63.55\,g) = 1.05\,mol\,Cu$
Moles of S	$= 33.5\,g(1\,mol/32.0\,g) = 1.05\,mol\,S$
Mole ratio Cu:S	$= 1.05\,Cu{:}1.05\,S = 1{:}1$
Empirical formula $=$ CuS	

General References

1. Crossfield, A. J. *J. Chem. Educ.* **1977,** *54,* 190.
2. De Moura, J. M.; Marcello, J. A. *J. Chem. Educ.* **1987,** *58,* 565.

Experimental Section

Procedure

Part A: Formula of a Metal Oxide by Ignition (Dry Method)

Microscale method   Estimated time to complete the experiment: 2 h
Experimental Steps: The procedure involves five steps: (1) obtaining the constant mass of crucibles, (2) weighing the metal ribbons, (3) carefully igniting the metal pieces to form oxides, (4) determining the masses of metal oxides, and (5) determining the empirical formula of the metal oxide.

Obtain three porcelain microcrucibles (1 to 2 mL size) and three pieces of metal ribbon. Number the bottom of each crucible for identification. Place them on one corner of a wire gauze as shown in Figure 3.20d. The wire gauze is supported on a ring attached to a ring stand such that the tip of the nonluminous flame of a Bunsen burner just touches the base of all three crucibles. **Caution: Always handle the crucibles with tongs to prevent burns.** Moreover, fingers may leave markings on crucibles, resulting in false weighing.

Heat the crucibles until their bottoms are red-hot. Cool them to room temperature in a desiccator. Note the length of heating and cooling times. In subsequent work, use the same length of time for heating and cooling. Weigh the crucibles to ± 0.001 g. Repeat the process of heating, cooling, and weighing until you have obtained constant weights for all the crucibles. Record the data on the data sheet.

Wrap each piece of the metal ribbon around the long stem of a metal spatula, avoiding overlapping coils. Remove the coil and put it inside a crucible. Weigh each of the crucibles with the metal inside. Enter the data on the record sheet.

Place all the crucibles (with metal ribbons in them) on the wire gauze and heat them gently over a low flame. Slow heating prevents the metal from burning with a bright flame (if this happens, use a lid to cover the crucible). When all the metal has reacted with oxygen (metal turns into a white ash), heat the crucibles strongly for an additional 3 to 5 minutes. Remove the heat and allow the crucibles to cool to room temperature on the wire gauze.

When the crucibles are cool, add one or two drops of water to the metal ash (metal oxide) in each crucible to decompose any metal nitride. Water reacts with the metal nitride to form ammonia (NH_3) and the metal hydroxide. On heating, metal hydroxides lose water, generating the corresponding metal oxide. Drive off excess water and ammonia by heating the crucibles **on a very low flame**. Too strong a heating will cause the loss of the metal oxide due to spurting of the boiling water. Carefully hold a piece of moist red litmus paper over the crucibles. The litmus paper will turn blue if ammonia is produced. When the metal ash has dried completely, heat the crucibles strongly for an additional two minutes.

Finally, cool the crucibles in a desiccator for the same length of time as before. Weigh each crucible along with its contents. Repeat the process of heating, cooling, and weighing until you obtain constant weights. Record the data.

From the data collected, calculate the mass of the metal ribbon, oxygen, and the metal oxide in each crucible. From these data, calculate the percent metal and oxygen. Determine the empirical formula for the metal oxide.

Cleaning of the Crucibles[1]

Using soapy water and a test tube brush, clean the crucibles. To remove stains, put several NaOH pellets inside each crucible and heat the mixture on a flame until the mixture melts. Cool the crucible, rinse it thoroughly with copious amounts of running water, and soak it in a bath of 1 M sulfuric acid. Finally, rinse the crucibles with water, dry them, and return them to your instructor.

> CAUTION: NaOH pellets are strongly corrosive; avoid skin contact. Use a pair of tongs to handle crucibles.

Alternative Procedure

> Macroscale method   Estimated time to complete the experiment: 2.5 h

Scale up the microscale quantities by a factor of four. Use regular-sized crucibles with lids for macroscale work. Simultaneous use of three crucibles is not recommended.

Part B: Determination of the Empirical Formula of a Metal Iodide

Two experiments have been described. Either one can be performed.

Zinc Reaction (in Aqueous Medium)

> Microscale method   Estimated time to complete the experiment: 2 h
> Experimental Steps: The procedure involves three steps: (1) reacting zinc powder with a known mass of iodine crystals, (2) determining the mass of metal iodide, and (3) calculating the mass of zinc that has reacted and obtaining the empirical formula of zinc iodide.

> CAUTION: Macroscaling the following reaction is not recommended.

Obtain a 25 mL beaker, magnetic stirring hot plate, a micro stir bar, zinc powder, iodine crystals, and acetone. Clean the beaker and the stirring bar with detergent, then with deionized water, and finally rinse with acetone three times. Dry them in an oven at 110°C. Cool the beaker and the stirring bar to room temperature (preferably in a desiccator) for 10 min.

Accurately weigh the beaker containing the micro stir bar. Place 300 to 400 mg of zinc powder in the beaker, reweigh it, and record the mass. Transfer about 120 to 130 mg of I_2 crystals to the beaker and reweigh the beaker accurately. Calculate the exact amount of iodine taken. In place of a stir bar, one can use a glass rod to stir the mixture.

Add 2 to 3 mL of water to the mixture. Place the beaker on a magnetic-stirring hot plate and cover it with a watch glass. With continuous stirring, **warm** the contents of the beaker (*do not boil*). **Make sure that iodine crystals do not sublime (vaporize).**

Within a minute or two, the aqueous layer will become yellow as a result of the reaction between zinc and iodine. The reaction is exothermic and may become brisk. If that happens, remove the beaker from the heat and cool it in an ice-water bath. Continue to stir the mixture (heat if necessary) until the brown color of the aqueous layer almost disappears. Add water to make up the loss due to evaporation, washing down the sides with a few drops of water.

When the aqueous layer becomes colorless, decant the solution to a small flask (making sure that no zinc powder is transferred). Give the flask to your instructor for iodine recovery. Rinse the unreacted zinc powder five or six times with 2 mL of deionized water, decanting the aqueous layer to a waste container. **While transferring the wash liquid, you must not lose any zinc.** Finally, wash the metal with acetone four or five times, using 1 mL of acetone each time, decant the solvent, and keep all the zinc in the beaker. Collect the acetone wash in a waste container for recycling.

Place the beaker with the zinc over a *warm* sand bath. Heat *slowly* to dry the powder (to prevent spattering). When it is dry, cool the beaker in a desiccator. Weigh the beaker with the dried zinc in it. Empty the beaker by transferring the unreacted zinc into a designated container and rinse the beaker. Repeat the experiment two more times. Return the clean beaker to your instructor.

> Do not throw away the leftover zinc; it must be recovered, recycled, and reused.

Calculate the amount of zinc that has reacted with the given amount of iodine. Determine the percent zinc and percent iodine from the data and obtain the empirical formula of zinc iodide.

Antimony Reaction (Nonaqueous Solvent[2])

> Microscale method   Estimated time to complete the experiment: 2 h
> Experimental Steps: Four steps are involved: (1) weighing antimony powder, (2) taking a known mass of I_2, (3) preparing the iodide compound by heating at reflux, and (4) calculating the empirical formula of the compound

Weigh an empty clean and dry 10 mL vial or a round-bottom flask containing a micro stir bar. Add 300 mg of antimony powder and weigh the vial again. Transfer 200 ± 10 mg of I_2 to the vial or flask and reweigh it. Add 4 mL of toluene. Calculate the exact mass of iodine taken.

> CAUTION: Toluene is flammable; no open flames.

Attach the vial to a reflux condenser. Run cold water through the condenser. Heat the mixture over a sand bath to a *slow* reflux (see Figure E17.1). Continue heating until no more violet color (due to I_2) is visible. When the reaction is complete, the resulting solution becomes green-yellow. (Your instructor will demonstrate how to set up a reflux unit.)

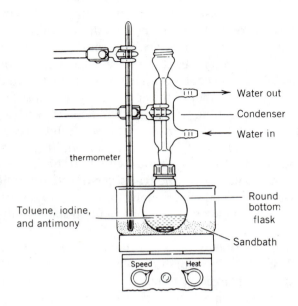

Figure E17.1   Arrangement for reflux

Filter the hot solution under suction using a Hirsch funnel (Chapter 3, section 8) or gravity filtration. Collect the filtrate in a clean and dry filter flask. Using additional hot toluene (2 to 3 mL portions), complete the transfer of all the antimony powder in the vial to the Hirsch funnel. Finally, using 1 to 2 mL acetone, rinse the leftover antimony filter cake powder several times.

Transfer the antimony powder to a 10 mL beaker, remove the filter paper, and dry the antimony over a warm sand bath. Cool to room temperature and weigh.

Calculate the mass of antimony consumed in the reaction. From this and the mass of iodine, calculate the percent antimony and percent iodine and deduce the empirical formula of antimony iodide.

CAUTION: Carry out the following step in an efficient HOOD.

Recovery and Isolation of Antimony Iodide

To recover the product of the reaction of antimony with iodine, place the filter flask containing the green-yellow solution on a warm sand bath (do not boil the solution) and reduce the volume of toluene by blowing N_2 gas over the solution. Upon cooling, red-orange antimony iodide crystals separate out from the concentrated green-yellow solution.

Collect the crystals using suction filtration (Hirsch funnel), dry them on a ceramic plate or on a filter paper, and determine the yield. Determine the melting point of the crystals and % yield of the product. Your instructor will collect the product to recover the iodine.

Additional Independent Projects

1. Simultaneous determinations of the compositions of magnesium oxide and magnesium nitride[3,4]
2. Simplest formula of copper iodide[5]

References

1. Singh, M. M.; Swallow, K. C.; Pike, R. M.; Szafran, Z. *J. Chem. Educ.* **1993,** *70,* A39.
2. Wells, N.; Boschmann, E. *J. Chem. Educ.* **1977,** *54,* 586.
3. Leary, J. J.; Gallaher, T. N. *J. Chem. Educ.* **1983,** *60,* 673.
4. Gallaher, T. N.; Moody, F. P.; Burkholder, T. R.; Leary, J. J. *J. Chem. Educ.* **1985,** *62,* 626.
5. (*a*) MacDonald, D. J. *J. Chem. Educ.* **1983,** *60,* 147.
   (*b*) Szafran, Z.; Pike, R. M.; Foster, J. C. *Microscale General Chemistry Laboratory with Selected Macroscale Experiments,* Experiment 8; Wiley; New York, 1993.

Name _____   Section _____

Instructor _____   Date _____

PRELABORATORY REPORT SHEET—EXPERIMENT 17

Experiment title _____

Objective

Reactions/formulas to be used

Chemicals and solutions—their preparation

Materials and equipment table

Outline of procedure

Waste containment and recycling procedure

PRELABORATORY PROBLEMS—EXPERIMENT 17

1. Balance the following reactions:

   ______ $Fe(s)$ + ______ $Br_2(g)$ → ______ $Fe_2Br_3(s)$

   ______ $C_6H_6(l)$ + ______ $O_2(g)$ → ______ $CO_2(g)$ + ______ $H_2O(g)$

2. A 1.275 g sample of Al was heated strongly in pure oxygen and the combustion product weighed 2.409 g. Calculate the percent Al and percent O in the product and find the empirical formula of the compound formed.

3. Cobalt forms two different chlorides. One contains 54.7 percent Cl and the other 64.4 percent Cl. Determine the empirical formulas of the two compounds.

4. What items of glassware would you need to assemble the reflux condenser for Part B (antimony + iodine reaction)? Draw a sketch of the setup.

EXPERIMENT 17   DATA SHEET

Part A: Empirical Formula of a Metal Oxide by Ignition

Data Collection Table

		Trial 1	Trial 2	Trial 3
1.	Mass of empty crucible			
	First weight, g	_____	_____	_____
	Second weight, g	_____	_____	_____
	Third weight, g	_____	_____	_____
	Average, g	_____	_____	_____
2.	Mass of crucible + metal, g	_____	_____	_____
3.	Mass of crucible + metal oxide (M_xO_y)			
	First weight, g	_____	_____	_____
	Second weight, g	_____	_____	_____
	Third weight, g	_____	_____	_____
	Average, g	_____	_____	_____

Data Manipulation Table

		Trial 1	Trial 2	Trial 3
4.	Mass of the metal, g	_____	_____	_____
5.	Mass of the oxide, g	_____	_____	_____
6.	Mass of oxygen combined with metal, g	_____	_____	_____
7.	Percent metal	_____	_____	_____
8.	Percent oxygen	_____	_____	_____
9.	Moles of metal	_____	_____	_____
10.	Moles of oxygen	_____	_____	_____
11.	Mole ratio between metal and oxygen	_____	_____	_____
12.	Formula of M_xO_y	_____	_____	_____

Show calculations:

Part B: Determination of the Empirical Formula of a Metal Iodide (Zinc or Antimony Reaction)

Data Collection Table	Trial 1	Trial 2	Trial 3
1. Mass of container, g	_____	_____	_____
2. Mass of container + metal before the reaction, g	_____	_____	_____
3. Mass of container + metal + iodine before the reaction, g	_____	_____	_____
4. Mass of container + metal after the reaction, g	_____	_____	_____

Data Manipulation Table			
5. Mass of metal taken, g	_____	_____	_____
6. Mass of leftover metal, g	_____	_____	_____
7. Mass of metal consumed, g	_____	_____	_____
8. Mass of iodine taken, g	_____	_____	_____
9. Percent metal	_____	_____	_____
10. Percent iodine	_____	_____	_____
11. Moles of metal (M)	_____	_____	_____
12. Moles of iodine (I)	_____	_____	_____
13. Mole ratio (M:I)	_____	_____	_____
14. Empirical formula of the compound	_____	_____	_____

Show calculations:

POSTLABORATORY PROBLEMS—EXPERIMENT 17

1. The iodine used in part B is weighed by difference in this experiment. Why is it not weighed directly?

2. A compound contains 90.6 percent lead and 9.40 percent oxygen. Determine the empirical formula of the compound.

3. Calculate the percent H and percent O in hydrogen peroxide (H_2O_2).

18 Preparation and Analysis of Lead(II) Iodide:

Recovery, Recycling, and Reuse of the Metal

Sequential Microscale Experiment

Objectives
- To study the chemistry of a metal
- To prepare a compound and to determine its formula by gravimetric analysis
- To recover and recycle a metal

Prior Reading
- Section 3.8   Filtration and gravimetric methods

Related Experiments
- Experiments 17 (Metal oxide and iodide), 19 (Copper compounds), 20 (Iron oxalate complexes and a photochemical reaction), and 32 (Synthesis of aspirin, iodoform, and urea)

Lead has been known since ancient times. The symbol for lead, Pb, comes from the Latin *plumbum,* and the profession that originally dealt with its applications was called plumbing. The Romans used lead to make water pipes and other plumbing fixtures. In fact, some historians believe that lead poisoning (from the lead pipes) was partially responsible for the downfall of the Roman Empire. The Hanging Gardens of Babylon, one of the Seven Wonders of the Ancient World, were floored with sheets of lead metal. Over one million tons of lead are produced every year, of which 60 percent are used for making battery plates (which are 90 percent lead).

Industrially, lead is obtained from galena (PbS) by a roasting/coking technique, which is generally applicable to the extraction of a metal from its ore:

1. *Roasting:* The ore is converted to a metal oxide by heating it in air.

$$2\,PbS(s) + 3\,O_2(g) \rightarrow 2\,PbO(s) + 2\,SO_2(g)$$

2. *Reduction:* The metal oxide is then reduced, using carbon in the form of charcoal or coke.

$$PbO(s) + C(s) \rightarrow Pb(s) + CO(g)$$

In the laboratory, this method is seldom used because of the need for extremely high temperatures. Instead, most metals are extracted from their ores by treatment with mineral acids.

In this experiment, lead(II) iodide will be prepared from galena, a lead ore (or, if galena is not available, lead carbonate). The empirical formula of the product will be determined by gravimetric analysis as $PbSO_4$. Lead will be recovered from the waste as basic lead carbonate and recycled.

General References

1.   Nechamkin, H.; Dumas, P. *J. Chem. Educ.* **1978**, *55*, 601.

Experimental Section

Procedure

> Microscale method   Estimated time to complete the experiment: 6 h
> Experimental Steps: This procedure involves the following steps: (1) preparation of lead(II) iodide and calculation of percent yield, (2) analysis of the compound by a gravimetric method, and (3) recovery, recycling, and reuse of the metal.

> Disposal Procedure: Do not dispose of any lead-containing solution in the sink. See the procedure at the end of the experimental section.

Preparation of Lead Iodide

Place approximately 100 mg of galena in a 25 mL Erlenmeyer flask, and add 2 to 3 mL of concentrated (12 M) HCl (**CAUTION:** Corrosive). Add a magnetic stir bar, and heat the mixture on a magnetic-stirring hot plate until almost all of the black material has reacted and a clear solution forms. Boil the solution gently, with stirring, for an additional five minutes. [If lead carbonate is used, add 200 mg of $PbCO_3$ and 5 mL of dilute (6 M) HCl. After the reaction is complete, add an additional 5 mL of water. Bring the mixture to a boil to dissolve all the lead chloride that forms.]

> CAUTION: The following preparation should be carried out in the HOOD. Lead is a toxic metal. Avoid contact with the skin. Do not breathe the dust. Dispose of lead-containing compounds only in a designated waste container.

Remove the flask from the hot plate, and add 10 mL of concentrated HNO_3. When effervescence ceases, heat the solution to boiling. Continue heating until the solution is colorless.

> CAUTION: Nitric acid is strongly oxidizing and corrosive. Wear gloves and avoid contact with the skin. In the event of skin contact, wash with copious amounts of cold water. Report to your instructor if you spill nitric acid.

Transfer the solution to a 50 mL Erlenmeyer flask. Rinse the original flask with a few drops of water and transfer the washings to the second flask. Allow the contents of the second flask to cool to room temperature. Using a Pasteur pipet, add concentrated NH_3 to the solution dropwise with stirring, until a voluminous white precipitate forms. Dissolve the precipitate by adding dilute nitric acid dropwise from a Pasteur pipet.

> CAUTION: Ammonia is corrosive and attacks the mucous membrane; avoid breathing the fumes.

If any solids still remain, filter the solution by gravity filtration through fluted filter paper. Collect the filtrate in a 100 mL beaker. Rinse the filter with 5 mL of hot water. Use a total of 25 mL of water. A large quantity must be used owing to the low solubility of the lead compound. Collect the wash liquid in the same beaker.

Dissolve exactly 600 mg of KI in 5 mL of deionized water, and add this solution to the solution in the beaker. A yellow precipitate of lead(II) iodide should form.

Add a magnetic stir bar to the beaker and, with stirring, heat the mixture on a magnetic-stirring hot plate until the lead(II) iodide has completely dissolved. Allow the solution to cool to room temperature, and then cool it further in an ice-water bath. This technique of dissolving a solid in a hot solvent (in which it is sparingly soluble), and then cooling the solution to effect precipitation, is known as **recrystallization.**

Collect the beautiful golden crystals of product by suction filtration using a Hirsch funnel [see Section 3.8, Figure 3.20*a*]. Wash the filter cake with two 0.5 mL portions of ice water. Allow the precipitate to air-dry on the filter. Finally, dry the crystalline product in a 110°C oven for at least 10 minutes. Weigh the precipitate.

Calculate the theoretical yield of lead iodide from the amount of galena (or lead carbonate) that you used. The following reactions apply:

$$PbS + 2\,HCl \rightarrow PbCl_2 + H_2S$$

or

$$PbCO_3 + 2\,HCl \rightarrow PbCl_2 + H_2O + CO_2$$

$$PbCl_2 + 2\,KI \rightarrow PbI_2 + 2\,KCl$$

Assume that galena is 100 percent PbS. Calculate the percent yield of the product.

Gravimetric Determination of Lead

The purity of the lead(II) iodide produced in the preceding step can be determined by comparing the percent lead obtained through gravimetric analysis with the theoretical percentage. Lead is usually characterized gravimetrically as lead sulfate, $PbSO_4$.

Accurately weigh about 200 mg of your product on a tared piece of weighing paper. Transfer the compound to a 100 mL beaker, and add a magnetic stir bar.

> CAUTION: Subsequent steps should be performed in an efficient HOOD.

Add 5 mL of water, followed by 5 mL of concentrated HNO_3.

Cover the beaker with a watch glass to prevent loss due to bubbling and bumping. Heat the mixture, with stirring, to a gentle boil in a sand bath placed on a magnetic-stirring hot plate. On heating, a brown gas (NO_2) will evolve. Iodine vapors will also be produced.

Continue to heat the beaker until the fumes become colorless. You may have to add more concentrated HNO_3 in order to complete the decomposition of the compound. At this stage, all iodide should be lost and the compound will be transformed to lead nitrate. If colorless crystals are found at the bottom of the beaker at this point, ignore them. They will dissolve when water is added in a subsequent step.

Cool the mixture to room temperature. Add 20 mL of water, and dissolve any solid left in the flask. At this point the solution should be colorless. Now, take a clean 100 mL volumetric flask, and transfer the solution from the beaker. The transfer should be as quantitative as possible. Rinse the beaker with 5 mL of water at least two times (with a total of 10mL), and transfer the washings into the same flask. Remove the stir bar.

Fill the volumetric flask to the mark with deionized water. Pipet a 25 mL aliquot of this solution into the original beaker, and add 5 mL of concentrated H_2SO_4.

Place a glass rod into the beaker to prevent bumping, and place the beaker on a sand bath. Heat the mixture to just below the boiling point of the solution. At this point, you may notice the formation of white precipitate. Continue to heat the solution until dense white fumes appear.

Carefully remove the beaker from the sand bath. Allow it to cool to room temperature, and cautiously, with stirring, add 30 mL of water. Gravity-filter the solution onto a quantitative filter paper (see Section 3.8, Figure 3.19). Use a Whatman number 42 filter paper. Transfer any white residue (lead sulfate) from the beaker quantitatively to the filter. Wash the beaker with several 2 mL portions of water containing a small amount of sulfuric acid. Transfer the washings to the filter.

While the filtration is in process, heat a crucible on a Bunsen burner to a constant weight (see Experiment 17, macroscale procedure). Do not weigh the cover along with the crucible. Place the crucible in a desiccator to cool.

When the filtration is complete, remove the wet filter paper carefully, fold it loosely and place it inside the previously weighed crucible. Place the crucible on a slant porcelain triangle (see Section 3.8, Figure 3.20c). Heat the crucible, slowly at first, to dry the filter paper. Then heat the crucible strongly in order to char the filter paper. Avoid flaming the filter paper. Immediately cover the crucible with its lid if the filter paper should catch on fire—freely burning flame will sweep away some lead sulfate, affecting your results. When all of the filter paper has been burned, heat the crucible strongly until no black residue remains, and then for an additional five minutes.

Cool the crucible to room temperature in a desiccator, and weigh it. Reheat and reweigh the crucible until you obtain a constant weight. Calculate the percentage of lead in the original sample, based upon the amount of lead sulfate produced. Rinse all your glassware and crucible. Return them to your instructor.

Recovery, Recycling and Reuse of Lead as Lead Carbonate

Do not throw any lead-containing solution or residue into the sink or the waste. Collect all the residues of lead iodide including any filtrate thereof in a beaker. Add 2 to 5 mL 6 M nitric acid. Cover the beaker with a watch glass. Heat the solution (70 to 80°C) on a sand bath (**use the HOOD**) for the oxidation of iodide to iodine, which will collect as black crystals on the surface of the solution. Add 5 mL of water and cool the solution. Filter the I_2 crystals, then wash them free of Pb^{2+} ions with water. Add solid sodium carbonate in small portions to neutralize the solution; CO_2 gas will evolve. Solid lead carbonate (basic) will precipitate. Filter the solid, wash it with water several times, and dry it in air. Return the dry solid to your instructor for recycling and reuse.

Name ______________________ Section ____________

Instructor ______________________ Date ____________

PRELABORATORY REPORT SHEET—EXPERIMENT 18

Experiment title ______________________

Objective

Reactions/formulas to be used

Materials and equipment table

Outline of procedure

Recycling procedure

PRELABORATORY PROBLEMS—EXPERIMENT 18

1. $PbCO_3$ and KI are used in this preparation. Which one is the limiting reagent (the reagent on which the yield of the product depends)?

2. Why must an excess of KI be avoided? [*Hint*: Could a complex of lead be formed?]

3. What reactions taking place in this experiment fall into the following categories?
   (*a*) Redox

   (*b*) Acid-base

   (*c*) Metathesis

   (*d*) Gas formation

EXPERIMENT 18   DATA SHEET

Data Collection and Data Manipulation

Preparation of Lead(II) Iodide

1. Weight of lead(II) iodide product _____

2. Theoretical yield of lead(II) iodide product _____

3. Percentage yield _____

Gravimetric Analysis

1. Weight of sample _____

		Trial 1	Trial 2	Trial 3
2.	Weight of empty, heated, crucible	_____	_____	_____
			Average	_____
3.	Weight of crucible plus lead sulfate	_____	_____	_____
			Average	_____
4.	Weight of lead sulfate	_____		
5.	Weight of lead in lead sulfate	_____		
6.	% lead in sample of PbI_2	_____		
7.	Theoretical % lead in PbI_2	_____		
8.	Percent error	_____		

Show calculations:

POSTLABORATORY PROBLEMS—EXPERIMENT 18

1. Although the use of lead-based paint has been banned for over 10 years, lead poisoning remains one of the greatest health hazards, especially for young children who live in tenements. Explain why.

2. Why must one cool the crucible before weighing it on a balance?

3. Write balanced reactions for the process used in the recovery and reuse of lead as lead carbonate from the laboratory waste.

Copper: Its Chemical Transformations.

Preparation of Copper(II) Compounds with Glycine and Aspirin

Sequential Microscale Experiments

Recovery and Recycling of Copper

Objectives
- To study the chemistry of copper
- To recover and recycle the metal

Prior Reading
- Section 3.6   Heating methods
- Section 3.8   Filtration

Related Experiments
- Experiments 18 (Lead Iodide), 20 (Oxalate complexes of iron and a photochemical synthesis), and 32 (Synthesis of aspirin, iodoform, and urea)
- Prepare a list of chemicals, glassware, and equipment you need for this experiment (see prelaboratory report sheet)

Activities a Week Before
- Prepare a list of chemicals, glassware, and equipment you need for this experiment (see prelaboratory report sheet)

Copper, second in commercial importance only to iron, was one of the first metals to be isolated from its ores. The process of obtaining pure copper from its ores (**metallurgy**) was known as early as 4500 B.C. It is a ductile, malleable metal, being easily pounded into various shapes for use as wire, ornaments, and implements of various types. Alloys of copper (bronze, brass) were discovered quite early in history. Pure copper is the best electrical conductor of the more abundant metals. It has good thermal conductivity and is corrosion-resistant. This experiment deals with the solution chemistry of copper, its reactions with glycine and aspirin, and the subsequent analysis of these compounds.

Glycine is a member of a biologically important group of compounds called **amino acids**. Amino acids (see Experiment 38) consist of at least one amino group ($-NH_2$) and a carboxylic acid group ($-COOH$). They are the basic structural units of proteins and therefore constitute the building blocks of living cells. Proteins are polypeptides or polyamides formed by the repeated linking of the $-NH_2$ group of one amino acid to the $-COOH$ group of another (see Figure E19.1).

Proteins play extremely diverse roles in human biochemistry. One class of proteins, the enzymes, functions as catalysts for various biological reactions. A simple protein, aspartame (NutraSweet™, L-aspartyl-L-phenylalanine methyl ester, Figure E19.2) is increasingly used as an artificial sweetener. The hydrolysis of a typical protein produces as many as 20 different amino acids. The simplest amino acid is glycine, H_2NCH_2COOH. It exists as a zwitterion (a covalent compound containing a separated positive and negative charge): $H_3N^+CH_2COO^-$.

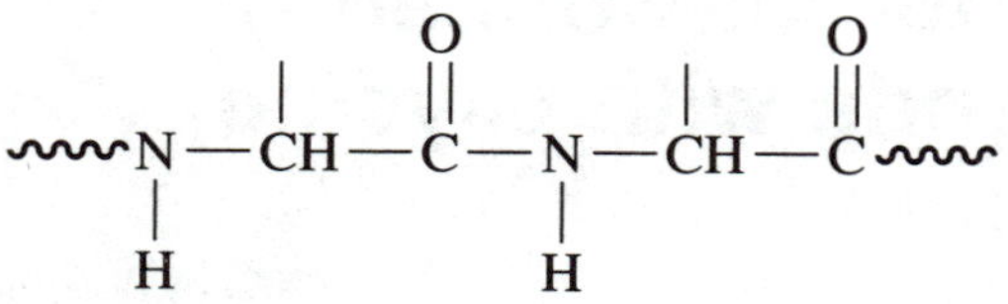

Figure E19.1   Peptide structure

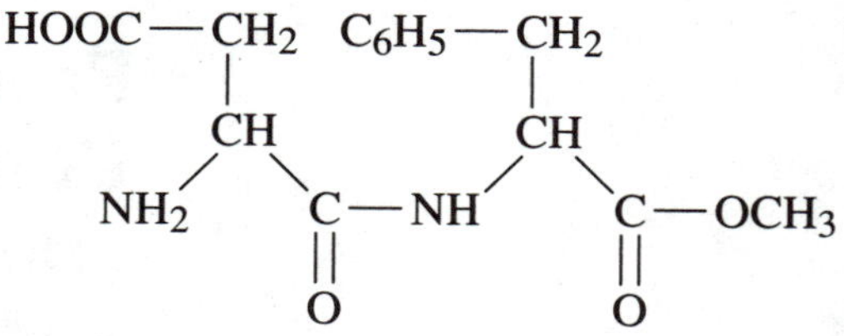

Figure E19.2   Aspartame ("NutraSweet")

Glycine contains two atoms capable of donating a pair of electrons (the N of the amine group and the O of the carboxylic acid group) and can simultaneously form two bonds with copper(II) cation. It therefore functions as a chelating ligand (from the Greek word *chele* for claw; see Experiment 20) and results in the formation of $Cu(gly)_2$. This compound exists in two isomeric forms: the *cis* isomer (like donor atoms on the same side of the copper) and the *trans* isomer (like donor atoms on opposite sides, see Figure E19.3a and b).

Aspirin (see Figure E19.4 and Experiment 32) is a well-known analgesic compound and also forms a complex, $Cu(C_9H_7O_4)_2$ where aspirin behaves as a chelating ligand.

In this experiment, a given amount of copper metal is transformed through a series of unbalanced reactions (see prelaboratory problem sheet) into two copper complexes.

$$Cu(s) + HNO_3(aq) \rightarrow Cu(NO_3)_2(aq) + NO_2(g) + H_2O(l) \tag{E19.1}$$

$$Cu(NO_3)_2(aq) + NaOH(aq) \rightarrow Cu(OH)_2(s) + NaNO_3(aq) \tag{E19.2}$$

$$Cu(OH)_2(s) \rightarrow CuO(s) + H_2O(l) \tag{E19.3}$$

$$CuO(s) + CH_3COOH(aq) \rightarrow Cu(CH_3COO)_2(aq) + H_2O(l) \tag{E19.4}$$

$$Cu(CH_3COO)_2(aq) + 2\,H_2NCH_2CO_2H(aq) \rightarrow$$
$$Cu(H_2NCH_2CO_2)_2 + 2\,CH_3COOH \tag{E19.5}$$

$$Cu(CH_3COO)_2(aq) + 2\,C_9H_8O_4 \rightarrow Cu(C_9H_7O_4)_2 + 2\,H^+ + 2\,CH_3COO^- \tag{E19.6}$$

$$Cu^{2+}(aq) + Zn(s) \rightarrow Cu(s) + Zn^{2+}(aq) \tag{E19.7}$$

The reactions involved are acid-base (with gas formation), decomposition, complex formation, and oxidation-reduction. The sequence of reactions also shows the wide variety of colors often observed for inorganic compounds. At the end of this experiment, all of the copper compounds are reduced to the metallic state for reclamation of the copper.

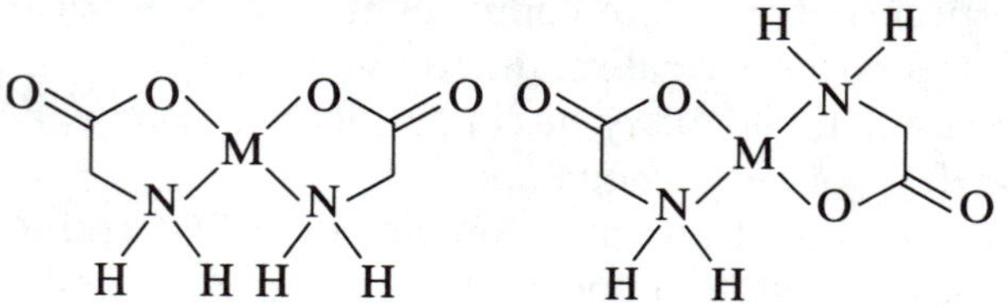

Figure E19.3   Bis(glycinato)copper(II). (*a*) *cis* structure, (*b*) *trans* structure

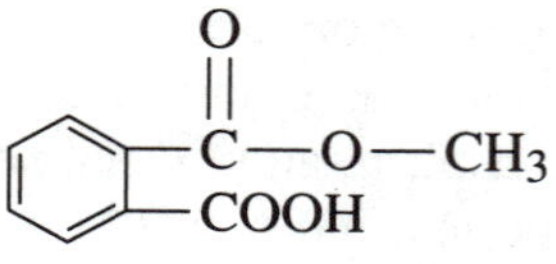

Figure E19.4   Acetyl-salicylic acid (aspirin), $C_9H_8O_4$

General References

1. Manojlovic-Muir. *Chem. Commun.* **1967**, 1057.

Experimental Section

Procedure

Part A: Conversion of Metallic Copper to Copper(II) Acetate

Microscale method   Estimated time to complete the experiment: two 2.5 h labs
Experimental Steps: Four steps are involved: (1) conversion of metallic copper to copper(II) acetate solution, (2) preparation of *cis* and *trans* glycinatocopper(II) and aspirinatocopper(II) complexes, (3) characterization of these compounds by melting point determination, and (4) regeneration of copper metal for **recycling and reuse**.

Disposal procedure: Dispose of all the solutions in marked containers. Do not throw the solids (filter papers, residual zinc or copper metal, and so forth) in the sink. Separate containers are provided for these.

Place $\sim$250 mg of copper wire (3 cm of 18 gauge wire) in a 25 mL Erlenmeyer flask. In the **HOOD**, add 4 mL of 6 M nitric acid to the flask, cover the flask with a small beaker or a watch glass, and gently warm the contents on a hot plate. Continue heating until the copper metal has completely dissolved and the evolution of brown fumes of nitrogen dioxide is no longer observed. If necessary, add an additional 1 mL of the acid. After cooling the resulting blue solution of copper(II) nitrate, add 10 mL of distilled water to the flask.

CAUTION: NO_2 is a toxic gas. Carry out the following reaction in a well-ventilated area (HOOD).

Add a magnetic stir bar (see Chapter 3, section 10 for the construction of a micro stir bar) to the Erlenmeyer flask and, with stirring, add 6 M NaOH solution dropwise from a Pasteur pipet until the solution is basic to red litmus paper. A light blue precipitate of copper(II) hydroxide forms. Add an additional 0.5 mL of 6 M NaOH. Place the flask on a sand bath heated by a magnetic-stirring hot plate. **With constant stirring**, heat the contents of the Erlenmeyer flask to boiling. During this time, the copper(II) hydroxide is transformed into black copper(II) oxide, CuO. Continue to heat the mixture for 2 to 5 minutes for proper coagulation. Allow the mixture to cool to room temperature.

NOTE: Stirring is essential to prevent bumping of the mixture and loss of CuO.

Using gravity filtration (Chapter 3, section 8), filter the black residue. Rinse the Erlenmeyer flask with 1 to 2 mL of distilled water to complete the transfer of the solid to the filter. Remove the magnetic stir bar using forceps and discard the filtrate. Wash the stir bar and the collected black residue with an additional 1 to 2 mL of distilled water.

Heat 10 mL of 6 M acetic acid to boiling in a 25 mL beaker containing a stir bar on a hot sand bath (**HOOD**). Place the Erlenmeyer flask under the stem of the funnel containing the filtered CuO.

Using a Pasteur pipet, transfer the hot acid directly onto the black solid. The copper(II) oxide will dissolve in the hot acid. Collect the filtrate in the Erlenmeyer flask. Using a pointed spatula, make a small hole through the bottom of the filter paper and recycle the hot acid filtrate through the filter paper until all the CuO has been transferred into the Erlenmeyer flask.

Finally, wash the filter paper with 2 to 3 mL of hot water to remove the last traces of the black solid. Heat the contents of the flask on a sand bath until all the solid has been dissolved, forming a clear green solution. **During heating, stir the solution constantly to avoid boiling it over.** Divide the clear solution into two equal parts in two 100 mL beakers each containing a micro stir bar (Chapter 3, section 10).

Alternate workup of CuO

After allowing the CuO mixture to cool, transfer the CuO slurry to a centrifuge tube. Centrifuge. This may have to be done in two sequential stages. After the centrifuge step, remove the supernatant liquid using a Pasteur filter pipet (PFP). Wash the CuO solid with two portions of water, centrifuging and removing the wash (PFP) after each step.

Now add 9 to 10 mL of 6 M acetic acid to the CuO and with stirring (steel spatula) warm the mixture on a sand bath until a clear green solution is obtained. Divide the clear solution into two equal parts in two 100 mL beakers, each containing a micro stir bar (Chapter 3, section 10).

Part B: Preparation of *cis*-Bis(glycinato)copper(II) Monohydrate and its *trans* Isomer

Add 1 to 2 mL of 95 percent ethanol to one portion of the foregoing solution. Heat the solution (with stirring) to a constant temperature of 70°C. Now, add ~0.350 g of glycine, and heat the solution for 5 to 6 minutes. Cool the flask to room temperature. On cooling, crystals of *cis*-bis(glycinato)copper(II) will precipitate. Complete the crystallization of the product by placing the beaker in an ice water bath for 10 minutes.

Filter the precipitate using a Hirsch funnel (see Chapter 3, section 8, Figure 3.20*a*), and dry it on a clay tile or on filter paper. When dry, determine the melting point of the *cis*-bis(glycinato)copper(II) monohydrate.

Half-fill a short melting point capillary with the product. The capillary should be sufficiently short so that any moisture expelled from the compound escapes easily to the atmosphere (cut the capillary if necessary). A longer tip of the capillary will trap the condensed moisture. Place the tube in a sand bath (a melting point apparatus or a Thiele tube may also be used), and heat to ~220°C for 5 to 10 minutes. During this time, the *cis* compound will lose its water of hydration, change color, and transform into the *trans* isomer. Remove the capillary, which now contains the *trans* compound. When dry, determine the melting point of the *trans* compound.

Part C: Preparation of Copper(II) Aspirinate

Prepare a solution of 60 mg of aspirin (acetylsalicylic acid) in 10 mL of 95 percent ethanol. Dilute the second part of the copper(II) solution from Part A by adding 50 mL of water. Add the aspirin solution to this diluted solution. With stirring, heat the mixture to 50 to 60°C on a sand bath.

Allow the alcohol to evaporate during this heating period. Remove the flask from the sand bath and set it aside to cool, preferably overnight. Dark greenish blue crystals will form. Complete the crystallization by placing the flask in an ice water bath for 20 minutes.

Suction-filter the crystals on a Hirsch funnel, and wash them with a few milliliters of 95 percent ethanol to remove the unreacted aspirin, followed by several drops of ice cold water. Allow the crystals to dry under suction, followed by further drying on a clay tile or on filter paper. Obtain the melting point of the complex.

Part D: Regeneration of Copper Metal

Collect all the compounds (and other copper waste if any) already prepared in a 30 mL beaker. Add 5 mL of water followed by several milliliters of 6 M NaOH until the solution is basic (use red litmus paper). With constant stirring, heat the solution to boiling. Continue to heat until all the copper has been transformed into a black precipitate (CuO) and has coagulated.

> CAUTION: Sodium hydroxide is corrosive. Avoid contact with skin. In the event of skin contact, wash with copious amount of cold water. Report to your instructor.

Collect the black precipitate by centrifugation. Decant off the supernatant liquid and transfer the residue into the same beaker in which the precipitation was carried out. Wash the residue with two 2 mL portions of hot water. Each time centrifuge and discard the wash liquid (use a PFP).

Add 6 mL of hot 3 M sulfuric acid to the residue. Stir the mixture with a small glass rod until the black solid has completely dissolved.

Place the beaker containing the blue copper(II) sulfate solution in the **HOOD** and add, in small portions, about 800 mg of zinc powder or dust. Vigorous evolution of hydrogen gas will occur. Stir the mixture with a glass rod, until the blue color of the original solution disappears. Heat the beaker over a sand bath to accelerate the process. A metallic precipitate of copper metal forms during this period.

> NOTE: The absence of any remaining copper(II) in the solution can be verified by adding a drop of the reaction solution to 1 mL of aqueous ammonia in a small test tube. The deep blue color of $Cu(NH_3)^{2+}$ indicates that the reaction is not yet complete. If not, it may be necessary to heat longer and add additional Zn metal.

After the reaction is complete, add 5 mL of 3 M sulfuric acid solution and stir the mixture with a glass rod. This process removes any unreacted zinc metal. The copper metal does not react under these conditions.

Allow the copper metal to settle, decant the aqueous solution, and wash the solid three times with 2 mL portions of distilled water. Decant the rinse solution between washings. Repeat this process using acetone as the wash solvent. Collect the decanted aqueous portions separately from the acetone portions. Dispose of them in separate containers. Spread the copper solid on a piece of filter paper or a clay tile, and allow it to air dry. Weigh the copper and calculate the percentage recovery.

Return all properly cleaned glassware to your instructor.

Additional Independent Projects
1. Visible spectral studies of different copper(II) compounds containing different ligands[1]
2. Determination of percent copper in different compounds[2]
3. Study of IR spectra of the complexes prepared in this experiment.

References
1. Potts, R. A. *J. Chem. Educ.* **1974**, *51*, 539.
2. Dudek, E. *J. Chem. Educ.* **1977**, *54*, 329.

Name ______________________ Section ______________

Instructor ______________________ Date ______________

PRELABORATORY REPORT SHEET—EXPERIMENT 19

Experiment title ______________________

Objective

Reactions/formulas to be used

Chemicals and solutions—their preparation

Materials and equipment table

Outline of procedure

1. Balance each of the following equations:

   (*a*) ____Cu(s)+____HNO_3(aq) → ____Cu(NO_3)$_2$(aq)+____NO_2(g)+____H_2O

   (*b*) ____Cu(NO_3)$_2$(aq) + ____NaOH(aq) → ____Cu(OH)$_2$(s) + ____$NaNO_3$(aq)

   (*c*) ____Cu(OH)$_2$(s) → ____CuO(s) + ____H_2O

   (*d*) ____CuO(s) + ____H_2SO_4(aq) → ____$CuSO_4$(aq) + ____H_2O

   (*e*) ____$CuSO_4$(aq) + Zn(s) → ____Cu(s) + ____$ZnSO_4$(aq)

2. Based on the balanced equations of Question 1, if you used 100.0 mg of copper in equation *a*, what is the exact amount of zinc required to complete the reaction in equation *e*?

EXPERIMENT 19   DATA SHEET

1.   Initial mass of copper wire   _____

2.   Mass of glycinato complex   _____

3.   Mass of aspirinato complex   _____

4.   Melting point of *cis*-glycinato complex   _____

     and *trans*-glycinato complex   _____

5.   Melting point of aspirinato complex   _____

6.   Mass of copper recovered   _____

7.   Percentage recovery (show calculation)   _____

8.   Balance the reactions given in the introduction of this experiment. Submit the report to your instructor. You may balance those reactions that are pertinent to the part of the experiment you have done.

1. In Part D of the experiment, the zinc metal is added to undergo a redox reaction with the copper(II):

$$Zn + Cu^{2+} \rightarrow Cu + Zn^{2+}$$

Hydrogen gas is also generated in this step. How is the hydrogen gas formed?

2. Why is a large excess of zinc added in the reaction of Part D of the experiment?

3. Would the percent recovery of copper metal in the experiment be high, low, or the same if insufficient NaOH was added to react with both the unreacted nitric acid and the Cu^{2+} generated in the first step of the sequence? Explain.

Preparation and Analysis of Metal-Oxalate Compounds:

A Sunlight Photolysis Experiment

Sequential Microscale Experiments— A Research Approach

Objectives

- To synthesize inorganic compounds
- To learn how to characterize the compounds using an investigative approach

Prior Reading

- Section 3.4   Weighing
- Section 3.5   Measuring liquid volumes
- Section 3.8   Filtration and gravimetric methods
- Nomenclature of complexes and an overview of the nature of bonding in these compounds

Related Experiments

- Experiments 16 (Qualitative analysis), 17 (Preparation of a metal oxide and iodide), 18 (Lead iodide: recovery and recycling), 32 (Synthesis and characterization of organic compounds)

Activities a Week Before

- Prepare a list of chemicals, equipment, and glassware needed for the experiment (see prelaboratory report)
- Prepare and standardize a 0.01 M EDTA (see Experiment 14). Dry sodium oxalate at 110° C for 2h. Store in a desiccator

The series of elements in the center portion of the periodic table are called **transition** or **d-block** elements. Transition metals have incompletely filled d subshells in their elemental forms or as ions. There are two important characteristics of these metals: (a) they exhibit variable oxidation states and (b) they are capable of forming a group of compounds called **coordination compounds** or **complexes**, most of which are colored.

A species where a metal atom or ion is surrounded by a group of neutral molecules (such as H_2O or NH_3) or ions (such as Cl^- or CN^-) is called a **complex**. Thus, $[Cu(NH_3)_4]^{2+}$, which is formed when an excess of aqueous ammonia is added to a solution of copper(II) salt, is an example of a complex cation. The neutral species $[Cu(NH_3)_4]SO_4$ is a **coordination compound**. The molecules or ions that surround the central metal are called **ligands**. Ligands usually donate one or more pairs of electrons to the central metal. The total number of bonds that a metal forms with ligands is called the **coordination number** of the metal.

Thus, in $[Cu(NH_3)_4]SO_4$ the coordination number of copper(II) is four. (The sulfate group is not a ligand; it is present to balance the charge on the complex cation.) Since ligands donate electrons to the metal, they act as **Lewis bases**. Some ligands can attach themselves to a metal by donating two or more pairs of electrons. Such ligands are called **polydentate** ligands. Complexes containing polydentate ligands are also called **chelate compounds** or **chelates** (from the Greek word *chele* for claw). For example, in $[Cu(en)_2]^{2+}$ (en = $H_2NCH_2CH_2NH_2$, ethylenediamine), en is a bidentate (two-toothed) ligand where both $-NH_2$ groups are attached to the Cu(II).

Iron, a transition metal, has an electronic configuration of $[Ar]3d^64s^2$. Iron exhibits two common oxidation states, Fe(III) (pronounced "iron three") and Fe(II) ("iron two"). Fe(II) and Fe(III) have electron configurations of $[Ar]3d^64s^04p^04d^0$ and $[Ar]3d^54s^04p^04d^0$, respectively. The Fe(III) uses its two vacant $3d$, one $4s$, and three $4p$ atomic orbitals to create a total of six vacant metal orbitals (hybridized d^2sp^3), which can receive as many as 12 electrons (6 electron pairs) from ligands.

The solution chemistry of these ions is extensive. When a solution of potassium oxalate ($K_2C_2O_4$) is added to Fe(III) in aqueous solution, a green crystalline compound forms. Here, iron is bonded to several oxalate ligands. When $CaCl_2$ is added to an aqueous solution of a simple oxalate salt such as $K_2C_2O_4$, solid calcium oxalate (CaC_2O_4) immediately precipitates. However, when $CaCl_2$ solution is added to an aqueous solution of an oxalato complex, no immediate precipitation of calcium oxalate occurs. This observation demonstrates that the oxalate ions are differently bonded to iron(III) than they are in a simple salt.

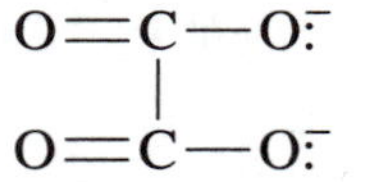

$$O{=}C{-}O^{\cdot-}$$
$$\,|$$
$$O{=}C{-}O^{\cdot-}$$

Figure E20.1
Oxalate anion

In the green oxalato compound, each oxalate anion (Figure E20.1) bonds to the iron through its two oxygen atoms, each of which donates a pair of electrons to the iron. Oxalate is therefore a bidentate ligand. Anionic ligands in coordination compounds have their endings changed to -o, so oxalate as a ligand is called **oxalato**.

In this experiment, a sample of the hydrated potassium salt (A) of "oxalato"-iron complex anion, an anhydrous oxalato derivative (B) and iron(II) oxalate dihydrate (C) will be prepared. A qualitative and quantitative analysis of the compounds will also be carried out.

General References

1. Basolo, F.; Johnson, R. C. *Coordination Chemistry*; Science Reviews; 1986.
2. Cotton, F. A.; Wilkinson, G.; Gaus, P. L. *Basic Inorganic Chemistry*, Wiley: New York, 1987.
3. McNeese, T. J.; Wierda, D. A. *J. Chem. Educ.* **1983**, *63*, 988 and its references.

Experimental Section

Procedure: Part A:

Microscale method   Estimated time to complete the experiment: three 3 h labs
Experimental Steps: This procedure involves several steps: (1) preparation of the oxalate complex, (2) standardization of $KMnO_4$ solution, (3) qualitative and quantitative analysis of the product, (4) determination of water of hydration, and (5) photolysis of the anhydrous product.

Disposal Procedure: Unless otherwise instructed, dispose of all the solutions in a marked container except those containing potassium permanganate. All wastes containing manganese must be disposed of in a separate container provided for the purpose.

Note: The later part of the experiment may be done with partners.

Standardization of 0.01 M KMnO₄ Solution

Obtain about 25 mL of 0.01 M potassium permanganate solution, two microburets, a 50 mL volumetric flask, a 10 mL graduated cylinder, and two 25 mL Erlenmeyer flasks.

Note: To conserve time, you should standardize the permanganate solution after you have started preparing the complex as described next. Use the waiting period during the synthesis of the compound for the titration.
CAUTION: Potassium permanganate is an oxidizer; avoid contact.

Prepare 50.00 mL of 0.02 M sodium oxalate solution in deionized water by dissolving 134 ± 1 mg (previously dried at 110°C for two hours) of sodium oxalate in a 50 mL volumetric flask. Calculate its molarity.

Set up two microburets, one for the $KMnO_4$ solution and the other for the standard sodium oxalate solution. Rinse and fill the permanganate buret with 0.01 M $KMnO_4$ solution. Similarly, rinse and fill the oxalate buret with the sodium oxalate solution. Record the initial volumes of solutions in both burets.

Transfer 2.00 ± 0.001 mL oxalate solution to a 25 mL Erlenmeyer flask and add 10 mL of 1 M sulfuric acid. Warm the solution to 60 to 70°C. Titrate the hot acidified oxalate solution with the permanganate solution until the first pink color appears throughout the solution and does not fade away within 30 seconds. Record the volume of $KMnO_4$ solution from the buret. Calculate the molarity of the $KMnO_4$ solution using the following reaction stoichiometry:

$$2\,MnO_4^-(aq) + 16\,H^+(aq) + 5\,C_2O_4^{2-} \rightarrow 2\,Mn^{2+}(aq) + 10\,CO_2(g) + 8\,H_2O(l) \quad (E20.1)$$

Repeat the procedure at least two more times. Find the average molarity of the $KMnO_4$ solution. Also, prepare and standardize a 0.01 M EDTA solution as described in Experiment 14.

Preparation of the Iron-Oxalate Complex (A)

Obtain all the glassware, equipment, and chemicals you need (see your prelaboratory report). Dissolve ~600 mg of potassium oxalate in 2 to 3 mL of hot deionized water in a 25 mL beaker containing a micro stir bar. While stirring, heat the solution on a sand bath placed on a magnetic-stirring hot plate. Using an Eppendorf pipet, add 200 μL of **freshly** prepared 1.5 M iron(III) chloride solution. The red-brown color of the iron(III) chloride solution should change to green as the reaction proceeds. Reduce the volume of the solution to ~1 mL by evaporation on the sand bath. Cool the solution in an ice water bath for 20 minutes. The green crystals of the complex compound should form. Adding 5 to 6 drops of absolute alcohol to the solution will accelerate the formation of crystals. Decant and discard the supernatant solution, keeping the crystals in the beaker.

For further purification, **recrystallize** the compound from hot water. Dissolve the solid in a *minimum* volume of hot water (not more than 1 to 2 mL). Add 5 to 6 drops of absolute alcohol and allow the solution to stand in an ice water bath for 10 minutes. Suction-filter the crystals on a Hirsch funnel. Collect the crystals and dry them between the folds of filter paper. Spread them on a preweighed watch glass or on a ceramic tile for further drying, *in the dark*. When the crystals are dry, weigh them along with the watch glass. Determine the yield of the compound and enter it on the report sheet.

> Note: The compound is very light-sensitive and decomposes irreversibly in light; protect the crystals from being exposed to light for a long period of time. Store them in an amber-colored vial (if you are using a colorless vial, wrap the vial with aluminum foil). On decomposition, the crystals become yellow brown due to the formation of iron(III) oxide.

For subsequent work, you may combine your product with that of your partner. Divide the product into three portions:

Portion 1 (~50 mg)	For qualitative tests of the ions
Portion 2 (~200 mg)	For oxalate and iron determination
Portion 3 (~200 mg)	For determination of the water of hydration and for photolysis

Qualitative Analysis of the Complex

To determine which ions are present in the complex, you will perform a series of chemical tests (**called qualitative analysis; see Experiment 16**) that are specific for the ions (Fe^{3+}, $C_2O_4^{2-}$, Cl^-, and K^+) used in the synthesis of the compound.

Practice Tests with Solutions Containing Known Ions

Set up a hot water bath in a 250 mL beaker. Perform the following tests (Table E20.1) with solutions of ions supplied by your instructor. Record your observations on the report sheet.

Test for potassium cation, K^+

Use any solid potassium salt (such as KCl) for this test. Insert one end of a piece of nichrome or Pt wire in a small cork. Holding only the cork, dip the free end of the wire in a drop of concentrated HCl. Heat the tip of the wire in the oxidizing flame (blue-green outer zone) of

Table E20.1.   Tests for known ions

Test	Observations	Conclusion
1. Tests for iron(III), Fe^{3+}		
(a) Take 10 drops of $FeCl_3$ solution in a test tube. Add a drop of $K_4[Fe(CN)_6]$ solution.	Dark blue color	Fe^{3+} present
(b) Add a drop of KSCN solution to another 10 drop sample of $FeCl_3$ solution in a separate test tube.	Blood red color	Fe^{3+} present
2. Tests for oxalate anion, $C_2O_4^{2-}$		
(a) Add a drop of $CaCl_2$ solution to 10 drops of potassium oxalate solution in a test tube and mix.	White precipitate	$C_2O_4^{2-}$ present
(b) Add a drop of 6 M sulfuric acid to 10 drops of the oxalate solution in a test tube, followed by 1 drop of 0.01 M $KMnO_4$. Heat the mixture on a water bath.	Pink color disappears	$C_2O_4^{2-}$ present
3. Test for chloride anion, Cl^- Place 1 drop of $FeCl_3$ solution in a test tube. Add 3 drops of 6 M HNO_3 and 1 drop of $AgNO_3$ solution. Mix.	Curdy white precipitate insoluble in HNO_3. The precipitate will appear yellow because of the yellow color of the solution.	Cl^- present

a microburner. The wire will glow but will not impart any color to the flame. Now, dip the tip of the wire in 6 M HCl, take a speck of potassium salt on the tip and place the wire in the flame. The violet color of the flame indicates the presence of K^+. If you fail to see the color, repeat the test.

Qualitative Tests for the Ions in the Complex

Prepare a sample solution by dissolving 25 to 50 mg of the complex in 4.0 mL of **deionized** water. Divide the solution into two equal portions: transfer one to a test tube for the tests in section *a* next, and the other to a centrifuge tube to be used in section *b*.

(a) Tests with the Solution of the Complex (See Table E20.1)
Perform all the tests for the ions (Fe^{3+}, $C_2O_4^{2-}$, Cl^-) as just outlined on the solution in the test tube. Tabulate your observations and draw appropriate conclusions. What ions are present in the complex? Perform the flame test for K^+ with a solid sample of the complex.

(b) Tests with the Solution of the Decomposed Complex
Add one or two drops of concentrated NH_3 (or 1 M NaOH) to the solution in the centrifuge tube. A brown precipitate will form. Centrifuge it. Using a Pasteur pipet, transfer the supernatant liquid from the centrifuged solution to a clean test tube labeled *supernatant*.

Add dilute HCl dropwise to dissolve the solid remaining in the centrifuge tube. Divide this solution into two parts in clean test tubes labeled I and II. To test tube I, add one drop of $K_4[Fe(CN)_6]$ solution. To test tube II, add a drop of KSCN solution. What conclusions can be drawn? (See Table E20.1).

Perform either of the tests described for the oxalate anion to the test tube containing the supernatant. Is oxalate ion present? (See Table E20.1).

Quantitative Analysis of the Complex

Determination of Oxalate

Accurately weigh 100 to 150 mg (± 0.001g) of the complex in a 25 mL beaker. Add 5 mL of water and 5 mL of 1 M NaOH (**CAUTION: NaOH is very corrosive. Avoid skin contact**). Heat the solution to near boiling. A gelatinous brown precipitate of iron(III) hydroxide will form.

Obtain a 50 mL volumetric flask. Set up a gravity filtration apparatus (see section 3.8). Filter the brown precipitate **quantitatively** (using Whatman 540 filter paper), collecting the filtrate directly into a 50 mL volumetric flask placed under the stem of the funnel. The filtrate contains the oxalate and potassium ions. Wash the beaker and the precipitate on the filter at least five times, using 1 mL of hot deionized water each time to remove any traces of oxalate and potassium ions. Collect all the washings in the same volumetric flask. Remove the flask from under the funnel and fill it to the mark with deionized water. Save the filter paper with the brown precipitate of iron(III) hydroxide and the beaker for iron determination.

While the filtration is in progress, set up two microburets on a buret stand. Rinse and fill one of the burets with the standardized $KMnO_4$ solution and the other with the oxalate filtrate (from the 50 mL volumetric flask). Record the initial volumes in both the burets on the data sheet.

Transfer exactly 2.0 $\pm$ 0.001 mL of the solution from the oxalate buret to a 25 mL Erlenmeyer flask. Add 10 mL of 1 M sulfuric acid and titrate it with the permanganate solution according to the procedure described before. Record the final permanganate volume from the buret.

Using the molarity of the permanganate solution, calculate the moles of $KMnO_4$ used in the titration. From this and the stoichiometry of Equation (E20.1), find the moles of oxalate present in the titrated amount of the solution taken. Convert this amount of oxalate to that present in the original 50 mL volumetric flask (by multiplying by the dilution factor). Run the titration at least two more times. Calculate the average number of moles of oxalate. The same number of moles of oxalate should be present in the amount of the complex originally taken.

Determination of Iron

Set up a 10 mL buret for 0.01 M EDTA solution. Add 1 M $NaHSO_4$ solution dropwise to the brown iron residue on the filter paper. The solution of $NaHSO_4$, being acidic, will dissolve the solid. Wash the filter paper with several 0.5 mL portions of $NaHSO_4$ until the filter paper is colorless. Collect the filtrate in the same beaker in which the precipitation was carried out. Add 5 mL of 0.01 M sulfosalicylic acid. **In a HOOD**, neutralize the acid solution by adding dilute NH_3 solution dropwise. Stop adding ammonia when the solution has just reached a red burgundy color. Titrate this solution with the standardized 0.01 M EDTA (Experiment 14) solution added from a microburet until the color of the solution changes to pure yellow. Calculate the millimoles of iron present in the sample:

$$\text{mmol Fe} = (\text{mL EDTA})(M_{\text{EDTA}})$$

Finally, determine the mol:mol ratio between iron and oxalate in the complex.

Procedure: Part B

Determination of the Waters of Hydration in the Complex (B)

Using a steel spatula, crush the dried sample of the complex to a powder on a ceramic plate. Weigh ~200 ±1 mg of the sample on a previously tared watch glass. Record the mass of the complex taken. Place the watch glass and its contents in an oven at 110 to 120°C for one hour. Cool the sample in a desiccator and weigh it again. Record the mass of the complex on the data sheet. Repeat the procedure of weighing, heating, cooling, and drying until a constant mass is obtained. **Do not discard the anhydrous product; it is used in part C.** (Note: If an oven is not available, heat the watch glass on a hot plate, of the kind used for keeping coffee hot, or on a hot sand bath at 110° C.)

Calculate the mass of the anhydrous compound and the mass of water lost. Calculate the percent water of hydration in the complex. Finally, calculate the moles of water per mole of the complex. Suggest the formula of the complex anion.

Procedure: Part C

Photolysis of the Anhydrous Product (C)

Transfer and weigh all the anhydrous complex from Part B in a tared test tube. Record the mass of the anhydrous compound on the data sheet. Dissolve the compound in 1 to 2 mL of 10 percent acetic acid solution.

Place the test tube in a rack that is exposed to direct sunlight. It can also be put under a Hg vapor lamp. This compound is photosensitive and undergoes an interesting photochemical reaction. To ensure that the reaction goes to completion, leave the sample near a window until the succeeding laboratory period.

Initially, the solution is emerald green, and some solid may remain insoluble. As the photochemical reaction proceeds, more solid goes into solution, and CO_2 gas is formed as one of the products. At the end of the reaction, a yellow precipitate of iron(II) oxalate dihydrate, $FeC_2O_4 \cdot 2H_2O$ (C), forms in quantitative yield. Filter the solution through a previously tared nail filter (see section 3.8). Wash the solid several times with 0.5 mL portions of acetone and dry under air suction. Weigh the sample. Calculate the moles of Fe and the percent Fe in the complex. Compare this result with that found before.

You have used many items of glassware in this experiment. Rinse them properly and return them to your instructor.

Extensions of the Foregoing Experiment as Independent Research Projects

1. Run the IR spectrum of the complex and of potassium oxalate in KBr and as a Nujol mull. Establish the nature of the bonding of oxalato ligands to iron(III). Detect the presence of lattice water in the spectrum of the complex.

2. If a magnetic susceptibility balance is available, determine the magnetic moment, μ_{eff}, of the complex.[1,2] Compare with that calculated for d^5 using the "spin only" formula $\mu_{eff} = g\sqrt{S(S+1)}\,BM = \sqrt{n(n+2)}\,BM$, where g is the gyromagnetic ratio (~2 for an electron), S is the total spin of the unpaired electrons (at $\frac{1}{2}$ each), and BM is the Bohr magneton (1 BM = 9.273 ergs/gauss). In the second equation, n is the number of unpaired electrons. Predict the nature of the complex: is it a high-spin or a low-spin compound?

3. Run the electronic spectrum of the compound dissolved in dilute acid media. Assign the bands observed either to d-d transitions (which are weak in the high-spin system) or to charge-transfer transitions (from oxalate to Fe^{3+}).[2]

4. Run a thermogravimetric analysis of the complex. It will show loss of three moles of water at 110°C. Further heating (260°C) results in the thermal reduction of the compound to $K_6[Fe_2(C_2O_4)_5]$. Around 400°C, this undergoes decomposition to K_2CO_3 and Fe_2O_3.[3]

Additional Independent Projects

1. Photochemical reaction between an alkane (C_7H_{16}) and a halogen (Br_2)[4]
   Brief Outline
   See Experiment 10, the additional independent project #4.

2. Synthesis and analysis of cobalt(III) complexes.[5] Develop a microscale technique for the synthesis of these compounds.

3. Preparation and analysis of acetylacetonato (2,4-pentane dionato or β-diketonato) complexes of Cr(III) and Mn(III)[1]

References

1. Szafran, Z.; Pike, R. M.; Singh, M. M. *Microscale Inorganic Chemistry: A Comprehensive Laboratory Experience*; Wiley: New York, 1991.
2. Aravamudan, G.; Gopalakrishnan, J.; Udupa, M. A. *J. Chem. Educ.* **1974**, *51*, 129, and references therein.
3. Bancroft, G. M.; Dharmawardena, K. G.; Maddock, A. G. *Inorg. Chem.* **1970**, *9*, 223.
4. Deck, E.; Deck, C. *J. Chem. Educ.* **1989**, *66*, 75.
5. (*a*) Loehlin, J. W.; Kahl, B. S.; Darlington, J. A. *J. Chem. Educ.* **1982**, *59*, 1048.
   (*b*) Wilson, L. R. *J. Chem. Educ.* **1974**, *54*, 539.

Name ______________________ Section ______________

Instructor ______________________ Date ______________

PRELABORATORY REPORT SHEET—EXPERIMENT 20
THIS PRELABORATORY REPORT MUST BE COMPLETED BEFORE DOING THIS EXPERIMENT

Experiment title ______________________

Objective

Reactions/formulas to be used

Chemicals and solutions—their preparation

Materials and equipment table

Outline of procedure

1. If 2.00 mL of a 0.0200 M solution of sodium oxalate requires 1.58 mL of $KMnO_4$ solution for the complete oxidation of the oxalate, calculate the molarity of $KMnO_4$ solution. (See equation [E20.1].)

2. Make a chart for the reactions used for the qualitative tests for Fe^{3+}, Cl^-, $C_2O_4^{2-}$, and K^+ ions.

3. A 1.201 g sample of hydrated lead acetate was heated to drive off the water of hydration. The cooled mass of the anhydrous lead acetate, $Pb(C_2H_3O_2)_2$, is 1.030 g. Calculate the value of x in the formula $Pb(C_2H_3O_2)_2 \cdot xH_2O$.

EXPERIMENT 20   DATA SHEET

Part A: Preparation and Analysis of the Complex

Concentration of KMnO₄ Solution

1. Mass of sodium oxalate taken: _____g

2. Concentration of sodium oxalate solution: _____M

		Trial 1	Trial 2	Trial 3
3.	Initial buret reading ($Na_2C_2O_4$):	_____mL	_____mL	_____mL
4.	Final buret reading ($Na_2C_2O_4$):	_____mL	_____mL	_____mL
5.	Volume of ($Na_2C_2O_4$) taken:	_____mL	_____mL	_____mL
6.	Initial buret reading ($KMnO_4$):	_____mL	_____mL	_____mL
7.	Final buret reading ($KMnO_4$):	_____mL	_____mL	_____mL
8.	Molarity of $KMnO_4$ solution:	_____mol/L	_____mol/L	_____mol/L
9.	Average molarity of $KMnO_4$:		_____mol/L	

Preparation of the Iron-Oxalate Complex (A)

Yield of the green complex   _____g

Qualitative Analysis of the Solution of the Green Complex

Test	Observations	Inference
1. Tests for iron(III), Fe^{3+}		
(*a*) Test with $K_4[Fe(CN)_6]$ solution	_____	_____
(*b*) Test with KSCN solution	_____	_____
2. Tests for oxalate anion, $C_2O_4^{2-}$		
(*a*) With $CaCl_2$ solution	_____	_____
(*b*) With 6 M H_2SO_4 + $KMnO_4$, heat	_____	_____
3. Test for chloride anion, Cl^-		
With 6 M HNO_3 + $AgNO_3$	_____	_____
4. Test for potassium cation, K^+	_____	_____

Tests with the Solution of the Decomposed Complex

Tests for Fe(III) ions
(*a*)   With $K_4[Fe(CN)_6]$ solution   _____   _____

(*b*)   With KSCN solution   _____   _____

Test for oxalate anion
(*a*)   With $CaCl_2$ solution   _____   _____

(*b*)   With 6 M H_2SO_4 + $KMnO_4$, heat   _____   _____

Quantitative Analysis of the Complex

(*a*) Determination of oxalate
1.   Mass of the green complex   _____ g

2.   Volume of the volumetric flask   _____ mL

(*b*) Titration of oxalate with standardized $KMnO_4$ solution

		Trial 1	Trial 2	Trial 3
3.	Initial oxalate buret reading, mL	_____	_____	_____
4.	Final oxalate buret reading, mL	_____	_____	_____
5.	Volume of oxalate sol'n. taken, mL	_____	_____	_____
6.	Initial $KMnO_4$ buret reading, mL	_____	_____	_____
7.	Final $KMnO_4$ buret reading, mL	_____	_____	_____
8.	Volume of $KMnO_4$ used, mL	_____	_____	_____
9.	Moles of $KMnO_4$ used	_____	_____	_____
10.	Moles of oxalate	_____	_____	_____
11.	Dilution factor	_____	_____	_____
12.	Moles of oxalate in 50 mL flask	_____	_____	_____
13.	Average moles of oxalate	_____		

(*c*) Determination of iron
14.   Molarity of EDTA solution (Experiment 14):   _____ mol/L

15.   Initial EDTA buret reading, mL   _____   _____   _____

16.   Final EDTA buret reading, mL   _____   _____   _____

17.   Volume of EDTA used, mL   _____

18.   Moles of EDTA   _____

19.   Moles of iron in the complex taken   _____

20.   Mole ratio Fe:oxalate anions   _____

21.   Percent Fe in the green complex   _____

Part B: Determination of Water of Hydration in the Complex

1. Mass of the empty watch glass ———— g

2. Mass of the watch glass + the hydrated complex ———— g

3. Mass of the hydrated green complex ———— g

4. Mass of the watch glass + anhydrous compound ———— g

5. Mass of the anhydrous compound ———— g

6. Mass of water lost ———— g

7. Percent water loss ————

8. Moles of water per mole of the complex ————

9. Formula of the complex anion _____

Part C: Photolysis of the Anhydrous Product

1. Mass of the anhydrous compound taken ———— g

2. Mass of $FeC_2O_4 \cdot 2H_2O$ after photolysis ———— g

3. Percent Fe in $FeC_2O_4 \cdot 2H_2O$ ———— %

4. Percent Fe in the original green complex ———— % (see previous page)

5. Moles of Fe present in the green complex ———— (see previous page)

6. Composition of the green complex _____

7. Name the complex _____

1. Suggest the products that may be formed when the green complex (A) is strongly heated.

2. Instead of using sunlight, can we use a light bulb for the photolysis of the compound?

3. Instead of oxalate anion, if we use ethylenediamine (en = $H_2NCH_2CH_2NH_2$) as the bidentate ligand, what would be the product? What are the donor atoms in en?

CHAPTER **8** Instrumental and Physical Methods

Introduction to Visible Spectroscopy

Introduction to Visible Spectroscopy

Many chemical compounds are colored in the solid state or in solution. In solution, the intensity of the color depends on the amount of the chemical substance present. A concentrated solution of $KMnO_4$ has a more intense purple color than a dilute solution of $KMnO_4$. This observation suggests that there is a relationship between the intensity of the color and the concentration of the chemical species present in the solution. In fact, the intensity of a color depends upon the amount of visible light absorbed by the chemical species responsible for the color of the solution. The analysis of chemical species in solution by the use of light is called **spectrophotometry**.

8.1 Interaction of Electromagnetic Radiation with Matter

Electromagnetic radiation is a form of radiant energy that is transmitted in space as a wave at the velocity of light. The wave vibrates perpendicular to the direction of propagation and produces the wave motion shown in Figure 8.1. The wave is described either in terms of its **wavelength** (λ) or its **frequency** (ν). The wavelength is the distance between crest to crest (or trough to trough). The frequency is the number of wavelengths per unit time. The inverse of the wavelength is called the **wave number** (ω). The maximum displacement of a point from the equilibrium position of the wave is called the **amplitude** (A).

The electromagnetic spectrum is subdivided into different regions depending upon the wavelength (or frequency). The various regions of the electromagnetic spectrum are shown in Figure 8.2.

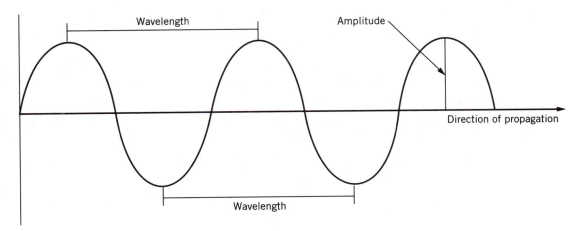

Figure 8.1 Electromagnetic radiation

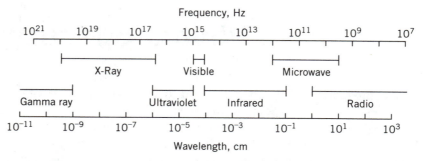

Figure 8.2   Regions of the electromagnetic spectrum

The relationships between the wavelength, frequency, and wave number are

$$\lambda = c/\nu \quad \text{and} \quad \omega = 1/\lambda = \nu/c$$

where c = the velocity of light (3.00×10^{10} cm/s); λ is in cm; ν is in s^{-1} (reciprocal seconds, also called hertz, Hz); and ω is in cm^{-1} (reciprocal centimeters). Wavelengths are commonly expressed in the following units: Å(angstrom, 10^{-10} m); nm (nanometer, 10^{-9} m) and μm (micrometer, 10^{-6} m). In the ultraviolet (UV) and visible ranges, the common unit for the wavelength is nm, whereas for the infrared range, ω in cm^{-1} is used (see Appendix 1).

Sometimes radiation is treated as a **photon** having zero mass and an energy of hν:

$$E = h\nu = h(c/\lambda) = hc\omega$$

where h is Planck's constant (6.626×10^{-34} J s or 6.63×10^{-27} erg s). From the equation, one can readily see that the greater the wavelength (or shorter the frequency), the smaller the energy.

When a solution containing analyte molecules is exposed to ordinary light, it absorbs specific wavelengths of the light, leaving the unabsorbed wavelengths to be transmitted. The transmitted wavelengths are seen as color. The observed color of the solution is the **complement** of the color absorbed by the molecules in the solution. Table 8.1 shows the complementary colors associated with those absorbed by the solution.

When molecules absorb energy, they are said to be promoted from the lowest energy level (the **ground state**) to a higher energy level (the **excited state**). This increase in energy is equal to the energy of the radiation absorbed ($h\nu$).

The energy absorbed is **quantized**—it exists at discrete levels. Molecules absorb radiation by three basic processes. Small amounts of absorbed energy are sufficient to excite **rotational transitions**. Molecules rotate around their various axes. By absorbing energy,

Table 8.1   Absorbed colors and their complements

Absorbed color	Transmitted color (complement)	Wavelength absorbed, nm
Violet	Green-yellow	380–450
Blue	Orange	450–495
Green	Violet	495–570
Yellow	Indigo (Blue)	570–590
Orange	Green-blue	590–620
Red	Blue-green	620–750

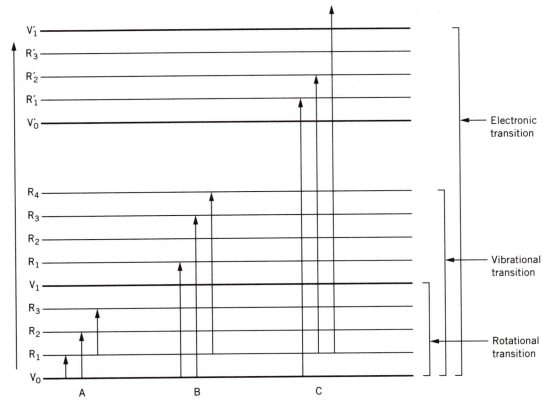

Figure 8.3 Molecular absorption transitions. (*a*) rotational; (*b*) vibrational; (*c*) electronic.

the molecules are excited to higher rotational levels. Larger amounts of absorbed energy are sufficient to excite **vibrational transitions**. The atoms in a molecule constantly vibrate. By absorbing energy, the molecules move to a higher quantized vibrational energy level. Yet higher absorptions of energy excite **electronic transitions**. Here, electrons in a lower energy orbital in a molecule are promoted to higher levels.

The visible and ultraviolet regions of the spectrum correspond to electronic transitions, whereas the infrared corresponds to vibrational transitions and the microwave region corresponds to rotational transitions. Since the energy needed to excite an electronic transition is also high enough to excite vibrational and rotational transitions, all three elements usually will be present in an electronic spectrum, as shown in Figure 8.3.

Spectroscopy is extensively used to analyze chemical samples. For a compound to be analyzed, it must absorb certain wavelengths of light. The absorption should be free from interference from absorptions due to other species present in the sample. A graph of the amount of light absorbed (*y* axis) versus wavelength (*x* axis) is called an **absorption spectrum** (Figure 8.4). Similar graphs (called **transmittance spectra**), which show the amount of light transmitted at various wavelengths, may be obtained. Transmittance and absorption spectra are obtained on instruments called **spectrophotometers**. For example, the Spectronic 20™ is a spectrophotometer that operates in the visible region of the spectrum (400 to 700 nm, the wavelengths of light that humans can see). Other types of spectrophotometers are used to investigate other regions of the electromagnetic spectrum.

An example of a visible absorption spectrum is shown in Figure 8.4. The wavelength at which the maximum absorbance occurs is referred to as λ_{max} (read as lambda max), in this case occurring at 600 nm.

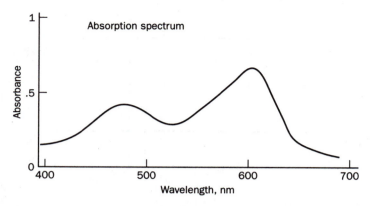

Figure 8.4 Absorption spectrum in the visible

To be detected by a visible spectrophotometer, the material being analyzed must be colored. Noncolored analytes may be detected by carrying out a reaction between them and some other material, producing a colored product. The most common visible spectrophotometer is the Spectronic 20, manufactured by Milton Roy (Figure 8.5*a*). A schematic of this unit is shown in Figure 8.5*b*.

For quantitative analysis, the Spectronic 20 is set at λ_{max}, unless an interfering species present in solution also absorbs at that wavelength. This procedure allows for maximum light absorption. It is especially useful if the absorption spectrum is reasonably flat at λ_{max}. In this case, if the wavelength drifts slightly during an analysis, the readings will be only minimally affected.

8.2 Quantitative Aspects of Spectrophotometry: Beer's Law

The relationship between the amount of radiation absorbed and the concentration of the analyte is given by **Beer's law**. When radiation having a power of P_o passes through a solution, the analyte will absorb a fraction of that light energy. As a consequence, the power of the transmitted light will be reduced from P_o to P. The **transmittance** T is defined as the ratio of the transmitted light to the incident light, or

$$T = P/P_o$$

The percent transmittance, $\%T$, is given by

$$\%T = T \times 100 = (P/P_o) \times 100$$

The absorbance A is the logarithmic inverse of the transmittance:

$$A = -\log T = -\log(P/P_o) = \log(P_o/P) \tag{8.1}$$

According to **Beer's law**, absorbance is linearly related to the concentration of the absorbing analyte and to the path length of the radiation in the solution. That is,

$$A = \epsilon bc \tag{8.2}$$

Introduction to Visible Spectroscopy

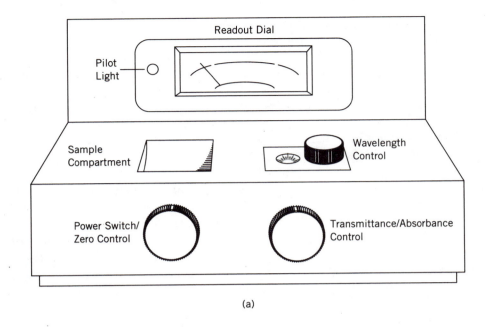

(a)

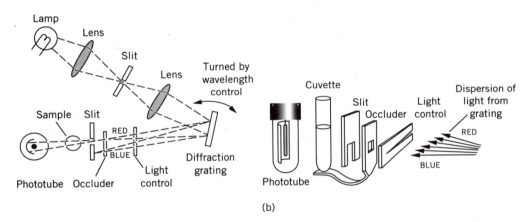

(b)

Figure 8.5 Milton Roy Spectronic 20 Spectrophotometer: (*a*) Appearance; (*b*) schematic.

where A = absorbance

ϵ = molar absorptivity (a proportionality constant, units of L cm^{-1} mol^{-1})

b = path length (cm) of light through the solution (cell width)

c = concentration of the solution in moles per liter

Since $T = \%T/100$, one can rearrange equation (8.1) and combine it with equation (8.2) as follows:

$$A = -\log(\%T/100) = \log(100/\%T) = \log 100 - \log(\%T) = 2 - \log(\%T) \quad (8.3)$$

In order to use Beer's law to determine an unknown concentration, ϵ and b must be constant. By working at a fixed wavelength, ϵ is kept constant; and if all measurements are recorded in the same cell, then b is constant. Under these conditions, the amount of light absorbed by the solution is proportional to the concentration of compound in the solution.

The first part of an experiment in quantitative analysis involves establishing the relationship between the amount of light absorbed and the concentration of the compound of interest in solution. This task is accomplished by preparing a series of solutions of known concentration, called **standards**. One of these is a **blank** solution. It should contain all the components of the standards *except* for the compound of interest.

The amount of light absorbed by each of the standard and blank solutions is read on the Spectronic 20 instrument. A Beer's law plot is prepared, that is, a graph of absorbance (y axis) versus concentration (x axis). (Graphing programs, such as Cricket Graph™, are extremely useful for this purpose.) This plot may then be used to determine the concentration of an unknown solution from the amount of light absorbed by that solution. For dilute solutions, such a calibration plot is usually linear. A typical Beer's law plot is shown in Figure 8.6.

One may readily see that the slope of the line is equal to the molar absorptivity, ϵ, in this example equal to 1.12 L/mol cm. If the absorbance of a solution of an unknown concentration is obtained, the concentration of that solution can then be read from the Beer's law plot. If the absorbance of the unknown solution was 0.08, the concentration could be read from Figure 8.6 by moving horizontally from 0.08 on the absorbance axis until the slope line is reached, and then moving vertically to the x axis, obtaining a concentration value of 0.072 mol/L. Alternatively, the absorbance value of 0.08 could be substituted for y in the slope equation

$$y = 1.12x + 0$$

Solving for x gives the concentration as 0.071 mol/L.

The Spectronic 20 instrument is capable of giving both percent transmittance and absorbance readings (some older models call the absorbance scale *optical density*). The %T scale on the instrument is linear, whereas the absorbance scale is logarithmic.

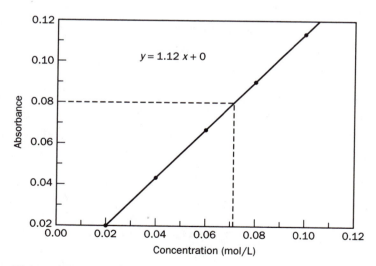

Figure 8.6 Beer's law plot

8.3 General Operating Instructions for the Spectronic 20 Spectrophotometer

The Spectronic 20 instrument is extremely easy to use and gives very accurate results. Refer to Figure 8.5a for the location of the various controls described in this section. Your instructor will provide you with more specific instructions.

1. Turn on the Spectronic 20 by turning the **power switch/zero control** knob clockwise until you feel a click. The **pilot light** will glow red at this point. Allow the unit to warm up for at least 15 minutes prior to making any measurements.
2. Set the desired wavelength (usually λ_{max}) using the **wavelength control** knob.
3. Adjust the **readout dial** to 0% T by turning the **power switch/zero control** knob. Be careful not to turn off the instrument inadvertently.
4. Rinse a matched pair of clean cells (cuvettes) with the blank solution, and wipe and blot the cells clean with a nonfibrous towel. Make sure that no fingerprints are on the cell. Fill the cells with the blank solution, and place one of them in the **sample compartment**, aligning the guide mark on the cell with the guide mark on the sample compartment. If microcuvettes are used, an insert must be placed in the sample compartment first to hold the microcells.
5. Adjust the **readout dial** to 100% T with the **transmittance/absorbance control** knob.
6. Insert the second cell of blank solution, aligning the guide marks, and take a reading. If you get something close to 100% T (zero or a small positive reading, say 0.012 on the absorbance scale), record this value as the reading for your blank solution. Record both the %T and absorbance readings. If you get a value off scale to the right (> 100%T, <0 absorbance) then you need to swap cells. Set the 100% T with this second cell. Then reinsert the first cell and you should get an on-scale reading. Record both the %T and absorbance. Note: %T values are more accurate on a nondigital instrument.
7. Retain the cell you used to set 100% T. Label it **zero cell** so you will not discard it by mistake. The zero cell will always be full of blank solution and is used only to set 100% T. The *other* cell will be referred to as the **sample cell**. The sample cell is the one for which you record the readings for all your solutions, including the blank solution. Discard the solution in the sample cell.
8. Rinse the sample cell three times with the least-concentrated standard solution. Fill the cell, wipe it dry, and insert it into the **sample holder**, once again aligning the guide marks. Record the %T and absorbance readings and the concentration of the standard.
9. Recheck the 100% T reading using the zero cell, and readjust if necessary. Repeat step 8 for the other standards, and for your unknown solution.

EXPERIMENT 21 Determination of Iron in Natural Water or in a Vitamin Tablet

Semimicro Experiment

Objectives
- To determine the concentration of a metal by visible spectroscopy

Prior Reading
- Chapter 8   Introduction to spectroscopy
- Section 2.6   Graphing of data
- Section 3.9   Solution preparation

Related Experiments
- Experiments 22 (Alcohol by spectroscopy), 23 (Manganese in steel), and 29 (Kinetics)

Apart from zinc, iron is one of the most important transition metals in biochemistry. The most familiar biochemical compound containing iron is **hemoglobin**—the oxygen carrier in red blood cells. Iron is a necessary dietary element and is found in fruits, cereals, vegetables, and some meats. In certain cases (such as for pregnant women) it may be advisable to take an iron supplement (vitamin tablet).

In this experiment, the iron in untreated natural water or in a vitamin tablet will be determined. The iron in solution must be present in the form of Fe^{2+}. To accomplish this, the iron tablet is dissolved in acid. A reducing agent (hydroquinone or hydroxylamine hydrochloride) is then added to ensure that all of the dissolved iron is in the form of Fe^{2+}. The pH of the solution is adjusted by adding a buffer (sodium citrate, pH = 3.5, sodium acetate or ammonia, pH = 6–9). Finally, an intensely colored species is formed by reacting the Fe^{2+} with 1,10-phenanthroline (also known as ferroin indicator) solution (abbreviated o-phen, see Figure E21.1). This complex is very stable and the color intensity does not change appreciably with time. Moreover, the color is stable over a wide pH range (2 to 9).

Phenanthroline has three fused rings of carbons, also containing two nitrogen atoms. Each nitrogen atom acts as a Lewis base by donating a pair of electrons to the Fe^{2+}, which acts as a Lewis acid.

General References

1. Skoog, D. A.; West, D. M.; Holler, F. J. *Analytical Chemistry*, 5th ed.; Saunders: Philadelphia, 1990.
2. Christian, C. D. *Analytical Chemistry*, 4th ed.; Wiley: New York, 1986.
3. Atkins, R. C. *J. Chem. Educ.* **1975**, *52*, 550.

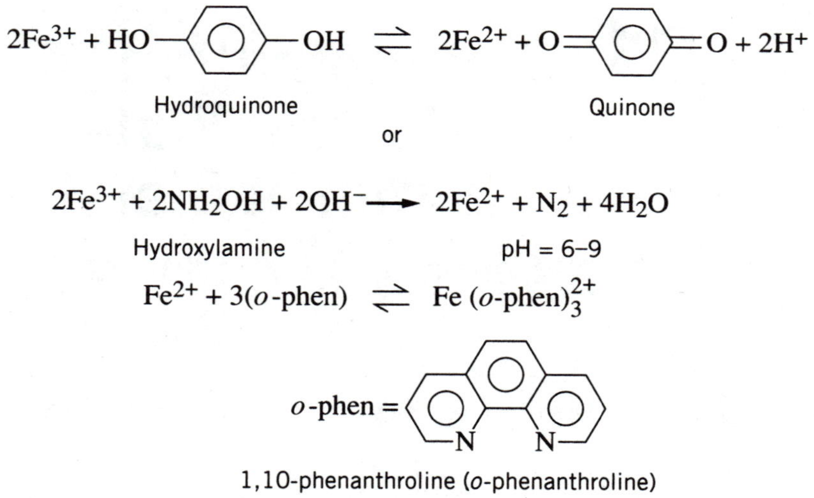

$$2Fe^{3+} + HO-\underset{\text{Hydroquinone}}{\bigcirc}-OH \rightleftharpoons 2Fe^{2+} + O=\underset{\text{Quinone}}{\bigcirc}=O + 2H^+$$

or

$$2Fe^{3+} + 2NH_2OH + 2OH^- \longrightarrow 2Fe^{2+} + N_2 + 4H_2O$$

Hydroxylamine $\qquad$ pH = 6–9

$$Fe^{2+} + 3(o\text{-phen}) \rightleftharpoons Fe(o\text{-phen})_3^{2+}$$

o-phen =

1,10-phenanthroline (*o*-phenanthroline)

Figure E21.1 Reaction scheme

Experimental Section

Procedure

Semimicroscale experiment   Estimated time to complete the experiment: 3 h
Experimental Steps: Six steps are involved in this experiment: (1) preparing a series of standard solutions of Fe^{2+}, (2) adjusting the pH of these and the unknown solution, (3) generating colored species, (4) determining a working wavelength, (5) obtaining a calibration curve, and (6) determining the concentration of Fe^{2+} in the unknown.

Disposal Procedure: Dispose of the solutions in a container or ask your instructor about alternative disposal methods. The chemicals being used in this experiment are generally not damaging to the environment.

Obtain your unknown sample of either natural water containing iron or the vitamin tablet. Procure a 125 mL Erlenmeyer flask, a funnel, a 10 mL graduated pipet, a pipet pump, a 10 mL graduated cylinder, a 25 mL beaker, a 50 mL volumetric flask, a 150 mL bottle for storing the unknown solution, six 100 mL beakers or flasks or large test tubes for storing standard solutions, cuvettes (cells), and a mortar and pestle.

The following solutions are also needed: a standard 40.00 ppm iron (Fe^{2+}) solution that should be acidic (pH = ~1.5), 0.1 M HCl, 1 percent hydroquinone in water (or 10 percent hydroxylamine hydrochloride), 2.5 percent sodium citrate solution (or 10 percent sodium acetate or ammonia solution), 0.25 percent o-phen (dissolve 0.25 g o-phen in 10 mL ethanol and dilute to 100 mL with water).

Using a 10 mL graduated pipet, transfer exactly 5.00 mL of the 40.00 ppm Fe^{2+} standard solution (supplied by your instructor) into a 25 mL beaker. Test the pH of the solution with a universal pH paper. Using a **medicine dropper**, add 2.5 percent aqueous solution of sodium citrate to the iron solution until the pH is ~3.5. **Count the drops of sodium citrate solution that are added.** Usually, 30 to 40 drops are needed if a medicine dropper is used. Very often, preliminary adjustment of pH of the solution may also be achieved by adding an aqueous solution of ammonia (or 10 percent sodium acetate).

Now, pipet 5.00 mL of the standard iron solution (40.00 ppm) into the 50 mL volumetric flask. Record the concentration of this solution on the data sheet. Using the same dropper, add the same number of drops of sodium citrate solution as before. Add 1 mL of hydroquinone followed by 1.5 mL of o-phenanthroline solution. A red color will form.

Dilute the solution to the mark with deionized water and mix well. Allow the solution to stand for at least 10 minutes to complete the reaction. Transfer this solution into a **clean and dry** 100 mL beaker. Mark this as "known 1" solution.

Use the same volumetric flask (rinse it with deionized water) in making other solutions as described next (see table). Prepare five additional standard solutions and a blank by transferring 4.0, 3.0, 2.0, 1.5, and 1.0 mL of the original Fe^{2+} standard solution (40.00 ppm) as just described.

The sodium citrate buffer (or 10 percent sodium acetate solution) and the hydroquinone solution should be added in proportion to the volume of Fe^{2+} solution taken. To each standard solution, add 1.5 mL of o-phenanthroline. In the following example, 30 drops of the citrate buffer were needed for making the "known 1" solution.

Solution	Vol. Fe^{2+}, mL	Vol. citrate, drops	Vol. hydroquinone, mL	Vol. o-Phen, mL
Known 1	5.0	30 (assumed)	1.0	1.5
Known 2	4.0	24	0.8	1.5
Known 3	3.0	18	0.6	1.5
Known 4	2.0	12	0.4	1.5
Known 5	1.5	9	0.3	1.5
Known 6	1.0	6	0.2	1.5
Blank	0.0	30	1.0	1.5

After making up each solution in turn in the 50 mL volumetric flask, dilute to the mark with deionized water, shake well, and wait 10 minutes. Transfer the solution to an appropriately labeled beaker. Calculate the concentration (in ppm) of each of the solutions. Record these concentrations on the data sheet.

Preparation of Unknown Solution from a Vitamin Tablet

NOTE: A sample of untreated water containing about 1 ppm iron may be used in place of a vitamin tablet.

Obtain an iron/vitamin tablet, and crush it to a fine powder in a mortar and pestle. Transfer (300 $\pm$ 1 mg of) the powder to a tared 125 mL Erlenmeyer flask. Add 20 mL of 0.1 M HCl to the flask. Place a small funnel (with the stem of the funnel inside the flask) on the mouth of the flask. Boil the contents of the flask gently (**HOOD**) for 15 minutes. Add additional deionized water to compensate for any loss due to evaporation.

Cool the solution. Using a quantitative filter paper (Whatman 42), gravity-filter it directly into the 50 mL volumetric flask. Rinse the Erlenmeyer flask and any residue on the filter several times with 1 mL portions of hot deionized water to complete the quantitative transfer. Now, add enough deionized water to bring the liquid level up to the mark and shake well. Label the flask "unknown solution." This solution will be sufficient for 10 students. Transfer the unknown into a clean and dry beaker or a flask.

Pipet 5.0 mL of the unknown solution (or the untreated water) to a 25 mL beaker. Using a medicine dropper, add sodium citrate solution dropwise to bring the pH of the solution to ~3.5 (use a universal pH paper). **Count the drops added.**

Using the same pipet, transfer 5 mL of the unknown solution to a 50 mL volumetric flask. Add the same number of drops of sodium citrate solution as before (medicine dropper) followed by 1.00 mL of hydroquinone solution and 1.5 mL of o-phenanthroline. Dilute to the mark with deionized water and mix thoroughly. Allow this solution to stand for 10 minutes.

Determination of the Working Wavelength at Maximum Absorbance

NOTE: Your instructor will provide you with the instructions for operating the spectrophotometer. See Chapter 8 for general instructions.

Using the blank and the "known 1" solution, determine the absorbances at various wavelengths over the interval 400 to 600 nm. For example, set the wavelength control knob at 400 nm (λ_{max}). Adjust the readout dial to 0% T by turning the power switch/zero control knob.

Next, insert the zero cuvette (cell) with the blank solution and adjust the readout dial to 100% T (or 0 absorbance). Replace the blank with the sample cuvette containing "known 1" solution and take the transmittance (%T) and the absorbance (A) readings. Enter the values in your data sheet. Now, increase the wavelength to 420 nm. Readjust the 0% T and 100% T as described before. Insert the "known 1" solution again. Read the transmittance and the absorbance values and record the values on the data table. Repeat this procedure using the "known 1" solution, taking transmittance and absorbance readings at 20 nm intervals (except in the region of maximum absorbance, where 5 nm intervals must be used). Make a plot of absorbance (y axis) versus wavelength (x axis). From the plot, determine the wavelength (called the **working wavelength**, $\lambda_{\mathbf{max}}$) at which the iron solution showed the maximum absorbance.

Construction of a Calibration Curve and Analysis of the Unknown

Set the wavelength dial to the working wavelength. Measure the absorbances for all the known solutions (1 through 6) and of the unknown at this wavelength, following the procedure just described.

Construct a calibration curve plotting the concentrations of the known iron solutions (x axis) versus the corresponding absorbances (y axis). Draw the best possible straight line through the data points. Use the absorbance of the untreated water or of the unknown solution and the calibration curve to calculate the concentration of Fe^{2+}.

Additional Independent Projects

1. Spectrophotometric investigation of the molecular complex of 2,3-dichloro-5,6-dicyanobenzoquinone with naphthalene[1]

2. Determination of phosphate in water[2,3]

   Brief Outline

   Either the ascorbic acid or tin(II) chloride method may be used.[2,3] A series of standard solutions of phosphate are prepared. The color development reagent is a mixture of ammonium molybdate solution, sulfuric acid, and either ascorbic acid or tin(II) chloride + hydrochloric acid. At first, using one of the standard solutions containing a blue phosphomolybdate complex, determine a suitable working wavelength. Next, prepare a calibration curve. Finally, obtain the concentration of the unknown phosphate solution.

   Develop the data sheets, calculate the concentration of phosphate, and write a short report according to the instructions given in Chapter 4.

References

1. Rehwaldt, R. E.; Boynton, E. *J. Chem. Educ.* **1965**, *42*, 648.
2. Mohrig, J. R. *J. Chem. Educ.* **1972**, *49*, 15.
3. Diehl-Jones, S. M. *J. Chem. Educ.* **1983**, *60*, 986.

Name ______________________ Section ____________

Instructor ______________________ Date ____________

PRELABORATORY REPORT SHEET—EXPERIMENT 21

Experiment Title: ______________________

Objective

Reactions/formulas to be used

Materials and equipment table

Outline of procedure

1. A 2.0×10^{-5} solution of $KMnO_4$ has a $\%T$ of 36.4 when measured in a 10.2 mm cell at a wavelength of 525 nm. Calculate the absorbance of the solution and the molar absorptivity of $KMnO_4$.

2. A sample consists of 10 mL of 1.00×10^{-5} M Fe^{2+}. What is the concentration of Fe^{2+} in ppm?

EXPERIMENT 21   DATA SHEET

Collection of data

1.   Mass of whole tablet _____ g

2.   Mass of the powdered tablet sample taken _____ g

Determination of working wavelength (λ_{max})

Wavelength, λ, nm	Transmittance, $\%T$	Absorbance, A instrument	Calculated absorbance, $(A = 2 - \log T)$
400	_____	_____	_____
420	_____	_____	_____
440	_____	_____	_____
460	_____	_____	_____

and so on. (Near the maximum absorbance, decrease the wavelength interval to a 5 nm increment.)

600	_____	_____	_____

Plot wavelength (x axis) versus absorbance (y axis). From the graph, determine the working wavelength, λ_{max}, at which the absorbance, A, is maximum.

3. Working wavelength   _____nm

Standard and unknown solution readings

Solution #	Conc. of Fe^{2+}, ppm	%T	Calculated Absorbance
Blank	0	_____	_____
1	_____	_____	_____
2	_____	_____	_____
3	_____	_____	_____
4	_____	_____	_____
5	_____	_____	_____
6	_____	_____	_____
Unknown	_____	_____	_____

Show calculations:

Manipulation of data

Construct a Beer's law plot as described in Chapter 8.

4. Beer's law equation (if determined)   _____

5. Unknown reading   _____

6. Concentration of unknown   _____ (use the calibration curve)

7. Concentration in original unknown solution   _____ (use the dilution factor)

8. Mass of iron in vitamin tablet   _____ (use the mass of the powdered sample taken and scale up to the mass of one whole tablet)

9. Compare content of iron in tablet to advertised value (state what brand was analyzed).

Show calculations:

POSTLABORATORY PROBLEMS—EXPERIMENT 21

1. If an Fe^{2+} solution is 40 ppm, what is its molarity?

2. What would be the effect on the calculated amount of iron in a tablet if the wash water used to rinse the beaker, filter cake, and filter flask were not added to the 10 mL volumetric flask? Explain.

3. What is the purpose of using hydroquinone? 1,10-phenanthroline? Suggest another reagent that can be used in place of hydroquinone.

22 Measurement of Alcohol Level Using Visible Spectroscopy

Microscale Experiment

Objectives

- To use visible spectroscopy and to analyze a colorless sample

Prior Reading

- Section 3.3   Handling chemicals
- Section 3.4   Weighing
- Section 3.5   Measuring liquid volumes
- Section 3.9   Solution preparation
- Chapter 8   Introduction to visible spectroscopy

Related Experiments

- Experiments 21 (Iron by visible spectroscopy), 23 (Manganese in steel), 24 (Composition of a complex), and 29 (Kinetics)

To determine whether a driver is driving under the influence of alcohol, law enforcement officers perform a Breathalyzer™ test to measure the blood alcohol content of the bloodstream. In the breath analyzer test, a breath sample (which may or may not contain alcohol) is passed through a solution containing potassium dichromate ($K_2Cr_2O_7$) in acid. Potassium dichromate, a strong oxidizing agent, is bright yellow. It oxidizes the alcohol to acetic acid (vinegar). The chromium is consequently reduced from the VI to the III oxidation state, which is not yellow. The reaction is

$$2\,Cr_2O_7^{2-} + 16\,H^+ + 3\,C_2H_5OH \rightarrow 4\,Cr^{3+} + 3\,CH_3CO_2H + 11\,H_2O$$

yellow   green

The amount of alcohol in a breath analyzer sample is therefore proportional to the amount of potassium dichromate that is used up and also therefore to the loss of yellow color (or alternatively to the amount of green color appearing). Note that alcohol is not actually detected but rather the chromium.

The Blood Alcohol Concentration (BAC) may be calculated from the equation

$$\text{BAC} = \frac{0.8\,A}{W\,R}$$

where W = body weight of the individual being tested

  A = amount of alcohol in the body (in mL)

  R = "Widmark R Factor," approximately 0.68 for men and 0.55 for women

In most states, a BAC of 0.10 percent is sufficient to be convicted for driving under the influence of alcohol; in some states the threshold BAC is even lower. The national trend is toward convictions for lower BAC values.

In this experiment, a potassium dichromate solution will be prepared and monitored using visible spectroscopy at 440 nm (a wavelength in the yellow region of the spectrum). An ethanol solution will be added (simulating a breath sample), and the amount of ethanol in the sample will be determined.[1]

General Reference

1. Timmer, W. C. *J. Chem. Educ.* **1986**, *63*, 897.

Experimental Section

Procedure

Microscale experiment Estimated time to complete the experiment: 2.5 h
Experimental Steps: Three steps are involved in this procedure: (1) preparation of a standard solution of alcohol, (2) determination of BAC in a known solution, and (3) determination of BAC in an unknown solution.

Disposal Procedure: Dispose of the solution containing chromium(VI) in a marked container.

Part A: Determination of BAC in a Known Solution

Obtain a buret, 10 mL graduated pipet, three 25 mL beakers, a 1 L volumetric flask (optional), an automatic delivery pipet (1 to 100 μL), and three cuvettes. Prepare a solution of potassium dichromate by adding 12.5 mg of $K_2Cr_2O_7$ and 12.5 mg of $AgNO_3$ to a 25 mL volumetric flask. Using a buret, add 24 mL of distilled water to the flask, and then fill to the mark using concentrated (18 M) H_2SO_4 (use a Pasteur pipet).

NOTE: Concentrated sulfuric acid is very corrosive and dehydrating. Use care when handling it. Addition of a concentrated acid to water is exothermic (the flask will get hot). Be sure to add the acid very slowly.

Prepare the known ethanol solution by adding 31 μL (3.1×10^{-2} mL) of absolute ethanol to a 1 L volumetric flask and then distilled water to the mark (your instructor may provide this solution for the whole class).

Pipet exactly 10 mL of the acidified dichromate solution into each of two 25 mL beakers. Using a graduated pipet, add 1 mL of distilled water to one of the beakers (as a blank, call this Solution A) and pipet exactly 1 mL of the ethanol solution into the other beaker (call this solution B). Determine the absorbance of the solution as described next.

Part B: Determination of BAC in an Unknown Solution

Obtain an unknown sample of ethanol solution from your laboratory instructor. Place 10 mL of the acidified dichromate solution in a 25 mL beaker. Add 1 mL of the unknown ethanol solution to the beaker (call this solution C). Determine the absorbance of the solution as described next.

Use of Spectronic 20 Visible Spectrometer

Set the Spectronic 20 (or a similar instrument) to a wavelength of 440 nm. Prepare a cuvette of a 50 percent solution of sulfuric acid (half concentrated sulfuric acid (18 M), half water), and use this to set the instrument to 100 percent transmittance. Use the blank solution (solution A) prepared in Part A to set the instrument to a full-scale reading (0 percent transmittance). Measure the absorbance of the known solution from Part A (solution B) and the unknown solution from Part B (solution C).

Calculation of Concentration

The amount of radiation absorbed by a sample from a spectrometer is described by **Beer's law**:

$$A = \epsilon \, b \, c$$

where A = amount of light absorbed (**absorbance**)

ϵ = molar absorptivity (a proportionality constant, $L \, cm^{-1} \, mol^{-1}$)

b = path length (cm) of light through the solution (cell width)

c = concentration of the solution in moles per liter

Since the constant ϵ and the path length b do not change from sample to sample, the concentration of any unknown can be easily obtained by comparing its absorbance to that of a known solution, using the relationship

$$\frac{A_1}{A_2} = \frac{\epsilon_1 b c_1}{\epsilon_2 b c_2} = \frac{c_1}{c_2}$$

In some cases, it is not the amount of radiation absorbed that is measured, but rather the fraction of radiation that is transmitted (called the transmittance, T). As noted earlier transmittance and absorbance are simply related to each other:

$$A = \log(100/\%T) = 2 - \log \%T$$

Additional Independent Project

1. Using a spectrophotometric method, determine the partition coefficient of methyl violet.[1]

   Brief Outline

   The partition coefficient of methyl violet distributed between an aqueous phase and 2-octanol phase will be determined. The concentrations of methyl violet in both phases can be determined by a spectrophotometric method. Prepare a series of standard solutions of methyl violet (1.0×10^{-5} to 1.0×10^{-6} M) in water. Using one of the standard solutions, determine the working wavelength for the absorption of methyl violet. At the working wavelength, construct a calibration curve (absorbance versus concentration). Using 2-octanol, extract methyl violet from an aqueous solution that was prepared by adding 20 mL of water to 0.5 mL stock solution of methyl violet. Determine the concentration of methyl violet in both layers and calculate the partition coefficient.

   Develop the data sheets, calculate the partition coefficient (see Experiment 8), and write a short report according to the instructions given in Chapter 4.

Reference

1. Sonnenberger, D. C.; Ferroni, E. L. *J. Chem. Educ.* **1989**, *66*, 91.

Name ______________________ Section ______________

Instructor ______________________ Date ______________

PRELABORATORY REPORT SHEET—EXPERIMENT 22

Experiment title ______________________

Objective

Reactions/formulas to be used

Materials (chemicals and solutions—their preparation) and equipment table

Outline of procedure

1. Define oxidation and reduction.

2. Balance the following redox reaction in acid solution:

$$H^+(aq) + Cr_2O_7^{2-}(aq) + Fe^{2+}(aq) \rightarrow Cr^{3+}(aq) + Fe^{3+}(aq) +$$

3. Balance the following redox reaction in base solution:

$$Cl_2(aq) + OH^-(aq) \rightarrow OCl^-(aq) + Cl^-(aq) +$$

4. A solution containing 1.00 mg of cobalt in a 100 mL volume transmits 70 percent of incident light at a particular wavelength. What fraction of the light would be transmitted for a solution containing 5.00 mg of cobalt?

EXPERIMENT 22   DATA SHEET

Collection of Data

Potassium Dichromate Solution

1. Weight of potassium dichromate, g _____

2. Moles of potassium dichromate, mol _____

3. Moles of potassium dichromate
    in 10 mL of solution, mol _____

Ethanol Solutions

4. Weight of ethanol, g ($d = 0.785$ g/mL) _____

5. Moles of ethanol, mol _____

6. Moles of ethanol in 1 mL of solution, mol _____

		Trial #1	Trial #2	Trial #3	Average %T	Average A
7.	%T of Solution A	_____	_____	_____	_____	_____
8.	%T of Solution B	_____	_____	_____	_____	_____
9.	%T of Solution C	_____	_____	_____	_____	_____

Manipulation of Data

10. Moles of potassium dichromate
     remaining in Solution B, mol _____

11. BAC of Solution B
     (use your body weight), mL/kg _____

12. Concentration of alcohol
     in Solution C, mL/L _____

13. BAC of Solution C
     (use your body weight), mL/kg _____

1. Balance the redox reaction:

$$Fe^{2+}(aq) + Cr_2O_7^{2-}(aq) + H^+(aq) \rightarrow Cr^{3+}(aq) + Fe^{3+}(aq) + H_2O(l)$$

2. The experiment described here is an example of the area of Forensic Chemistry. List three other types of chemical investigations commonly done in this area.

23 **Determination of Manganese in Steel**

Microscale Experiment

Objective
- To determine manganese in steel by spectrophotometric method (spiking method)

Prior Reading
- Section 3.5   Measuring liquid volumes
- Section 3.9   Solution preparation
- Chapter 8   Introduction to visible spectroscopy

Related Experiments
- Experiments 21 (Iron in water), 22 (Blood alcohol), and 24 (Composition of an iron complex)

In many cases, one metal will be alloyed with one or more other metals in order to change its physical properties. Iron, for example, is prone to rust in the presence of oxygen and water and to corrode in the presence of acids. If one wants a long-lasting automobile, for example, it would be desirable to modify the physical properties of iron by forming an iron alloy.

When carbon is added to iron in amounts of 1 percent or less, a strong alloy that resists rusting is obtained, called **steel**. Other materials are also frequently added to steels to render them more resistant to corrosion. The most common is chromium, which, when present from 10 to 20 percent, yields **stainless steel**. Other metals are generally present in smaller quantities, including manganese, nickel, and vanadium. **Manganese steel** (≤ 3.5 percent Mn) is used in making engine parts for automobiles and aircraft.

To analyze steel as to its various components, one must first digest the steel in an acid bath in order to form an aqueous solution. In the presence of nitric acid, the various metals will be oxidized to the metal ions Fe^{3+}, Ni^{2+}, Mn^{2+}, Cr^{3+}, and so on. Unlike the other ions, manganese(II) is only faintly colored (colorless to faint pink) and does not lend itself to analysis by absorption spectroscopy. In this experiment, periodate ion is used to oxidize Mn^{2+} to the much more intensely colored permanganate ion, MnO_4^-, for ready detection. The unbalanced reaction (see prelaboratory problem) is

$$3H_2O + 2Mn^{2+} + IO_4^- \rightarrow 2MnO_4^- + IO_3^- + H^+ \qquad \text{(E23.1)}$$

The manganese can then be detected as the dark purple permanganate ion.

Interference

One must be careful in a chemical analysis that in solving one problem (the low color intensity of Mn^{2+}) one does not create another. If one is analyzing a stainless steel, the addition

$Mn^{2+} \rightarrow MnO_4$

of periodate ion to oxidize the manganese will also oxidize any Cr^{3+} ion present to dichromate, $Cr_2O_7^{2-}$ (Cr^{6+} species). The absorbance of the dichromate ion can interfere with the manganese analysis. Thus, stainless steel cannot be analyzed for manganese in this manner.

Similarly, other materials present in solution can interfere with the measurement of the desired species. In this experiment, for example, the presence of a relatively huge quantity of the lightly colored Fe^{3+} interferes with analysis of the manganese. We can eliminate this interference by reacting the Fe^{3+} with phosphate ion, with which it forms a colorless complex.

Small amounts of nickel are also present in most steels. Ni^{2+} absorbs at the wavelength of interest for Mn^{2+}. In this case, one can compensate for the absorbance of nickel by taking the digested steel before the periodate oxidation step and using it as a blank. Since nothing happens to the Ni^{2+} during the oxidation step, the amount of Ni^{2+} present before the oxidation is the same as the amount present after the oxidation. Any increase in absorbance is therefore attributable to the Mn^{2+} being oxidized to MnO_4^-. One obtains the absorbance of the MnO_4^- ion by simply subtracting the absorbance of the blank from the absorbance of the oxidized solution:

$$\underset{\text{desired quantity}}{A_{MnO_4^-}} = \underset{\text{oxidized solution}}{A_{MnO_4^- + Ni^{2+}}} - \underset{\text{blank}}{A_{Ni^{2+}}}$$

Method of Standard Addition

In many cases, it is difficult to obtain a linear Beer's law plot. In such cases, the **method of standard addition** is often used. In this technique, two portions of the sample to be analyzed are prepared. A spike of a known amount of the material being analyzed for is added to one of the portions. The absorbances of both samples are then taken. The absorbance of the spiked sample (A_2) will be larger than that of the unspiked sample (A_1), as there is more analyte present. The difference in the absorbances is obviously due to the spike itself. Letting the absorbance of the blank solution be A_b, we may write

$$\begin{aligned}
\text{Absorbance of the } MnO_4^- \text{ in sample} &= A_1 - A_b \\
\text{Absorbance of the } MnO_4^- \text{ in sample + spike} &= A_2 - A_b
\end{aligned}$$

Since the quantity of MnO_4^- in the spike is known, the quantity of MnO_4^- can be readily calculated. A sample calculation follows.

Example E23.1   A sample of steel containing manganese is analyzed as described above. A spike consisting of 10 mL of 0.1g/mL $KMnO_4$ is added to one portion. The absorbances obtained for the blank, sample, and spiked sample are given below:

$$\begin{aligned}
\text{Absorbance of sample: } 0.24 &= A_1 \\
\text{Absorbance of sample + spike: } 0.81 &= A_2 \\
\text{Absorbance of blank: } 0.05 &= A_b
\end{aligned}$$

How much $KMnO_4$ is in the sample?

Answer

The size of the spike is as follows: 10 mL of solution at 0.1g/mL = 1 g $KMnO_4$.
Thus,

$$\text{Absorbance of } MnO_4^- \text{ in sample} = A_1 - A_b = 0.24 - 0.05 = 0.19$$
$$\text{Absorbance of } MnO_4^- \text{ in sample + spike} = A_2 - A_b = 0.81 - 0.05 = 0.76$$

Let the amount of MnO_4^- in the sample $= x$ g. Then, the amount of MnO_4^- in the sample + spike $= x + 1$ g. Set up a ratio.

$$x/0.19 = (x + 1)/0.76$$
$$0.76x = 0.19x + 0.19$$
$$0.57x = 0.19$$
$$x = 0.33 \text{ g } KMnO_4 \text{ in the sample}$$

(handwritten annotations:)
$.19 = A_{sample} = \epsilon l \cdot x$
$.76 \quad A_{sample + spike} = \epsilon l (x+1)$

General References

1. Skoog, D. A.; West, D. M.; Holler, F. J. *Fundamentals of Analytical Chemistry*, 5th ed.; Saunders College Publishing: New York, 1988.

(handwritten work:)

$A_{x+spike} = \epsilon b \cdot C_x + \epsilon b \, C_{spike} = \epsilon b (x_x + C_{spike})$

$A_x = \epsilon b C_x$

$\dfrac{A_{x+spike}}{A_x} = \dfrac{C_x + C_{spike}}{C_x} = 1 + \dfrac{C_{spike}}{C_x} = 1 + \dfrac{g_{spike}}{g_x}$

$\dfrac{\frac{n_{spike}}{L_{soln}}}{\frac{n_x}{L_{soln}}} = \dfrac{n_{spike}}{n_x} = \dfrac{g_{spike}}{g_x}$

$= \dfrac{g_{spike}}{g_x}$

$g_{spike} = C_{MnO_4} \times mL$

$\dfrac{g}{mL}$

Experimental Section

Procedure

Microscale experiment   Estimated time to complete the experiment: 2.5 h
Experimental Steps: There are three steps: (1) dissolution of steel sample, (2) preparation of sample, blank, and spiked solutions, and (3) measurements of absorbances.

Disposal procedure: Dispose of the solutions containing manganese(VII) in a marked container.

Obtain three 50 mL beakers (each containing a stir bar) a magnetic stirring hot plate, a 25 mL volumetric flask, three 25 mL beakers, a 10 mL graduated pipet, a cuvette, a sample of manganese steel powder, potassium persulfate, potassium periodate, 0.1 g/mL $KMnO_4$ solution, 4 M HNO_3, and 7 M phosphoric acid.

Place about 200 mg (± 1 mg) of a steel sample in a tared 50 mL beaker and add a magnetic stir bar. Turn on the spectrophotometer.

In a HOOD, add 10 mL of 4 M HNO_3 to the beaker. Place the beaker on a magnetic-stirring hot plate, and boil the mixture gently for 5 minutes. At this point, add 200 mg of potassium persulfate, $K_2S_2O_8$, to the digested steel, and continue to boil the solution for an additional 10 minutes. The potassium persulfate oxidizes any carbon present and should leave you with a light green solution. If any solid is present at this point, ignore it. It will not affect your analysis. Allow the solution to cool to room temperature. Transfer the solution to a 25 mL volumetric flask, and fill it to the mark with deionized water. Mix thoroughly.

Pipet a 5.0 mL aliquot of the diluted solution from the volumetric flask into each of three 25 mL beakers, labeled "sample," "blank," and "spiked." Add 2.0 mL of 7 M H_3PO_4 to each of the beakers. Add 800 mg of KIO_4 to the sample beaker. Add 800 mg of KIO_4 and 2.0 mL of 0.1g/mL $KMnO_4$ solution (freshly prepared) to the spiked beaker. Nothing should be added to the blank beaker. Place the three beakers on a magnetic stirring hot plate and add magnetic stir bars. With stirring, gently boil the solutions for 5 minutes. Allow them to cool to room temperature.

Empty and rinse the 25 mL volumetric flask from before, and transfer the solution from the sample beaker. Dilute it to the mark with distilled water, and mix thoroughly. Transfer the solution to a labeled 50 mL beaker. Repeat the procedure for the spiked and blank solutions.

Following the instructions provided by your instructor, obtain the absorbance for the three solutions at a wavelength of 525 nm. Following the method described in the example, determine the amount of $KMnO_4$ in the steel sample, and the percentage of manganese in the steel.

Additional Independent Projects

1. Phosphate determination in a detergent or in lake water

Brief Outline

Two methods—ascorbic acid or tin(II) chloride—may be used.[1,2] A series of standard solutions of phosphate are prepared. The color development reagent is a mixture of ammonium molybdate solution, sulfuric acid, and either ascorbic acid or tin(II) chloride + hydrochloric acid. At first, using one of the standard solutions containing a blue phosphomolybdate complex, determine a suitable working wavelength. Next, prepare a calibration curve. Finally, obtain the concentration of the unknown phosphate solution.

Submit your report (see Chapter 4) along with the prelaboratory report sheet, data sheets, and plots.

References

1. Diehl-Jones, S. M. *J. Chem. Educ.*, **1983**, *60*, 986.
2. Mohrig, J. R. *J. Chem. Educ.*, **1972**, *49*, 15.

Name ______________________ Section ______________

Instructor ______________________ Date ______________

PRELABORATORY REPORT SHEET—EXPERIMENT 23

Experiment title ______________________________

Objective

Reactions/formulas to be used

Chemicals and solutions—their preparation

Materials and equipment table

Outline of procedure

1. Describe the process by which iron is produced from iron ore.

2. Balance the reaction E23.1.

Name ______________________	Section ____________
Instructor ______________________	Date ____________

EXPERIMENT 23 DATA SHEET DATA

Collection Table

1. Mass of steel sample, g ____________
2. Absorbance of diluted sample ____________ (A_1')
3. Absorbance of diluted sample + spike ____________ (A_2')
4. Absorbance of blank ____________ (A_b)
5. $A_1' - A_b$ ____________ (A_1)
6. $A_2' - A_b$ ____________ (A_2)

Data Manipulation Table

Concentration of Mn standard added, mg/L ____________

Concentration of Mn in diluted steel solution, mg/L ____________

Concentration of Mn in original steel solution, mg/L ____________

Mass of Mn in the steel, g ____________

% Mn in the steel ____________

Show calculations:

1. Describe three types of interference that might occur in a visible spectroscopy experiment, and how one would overcome them.

2. Several metals other than manganese are used for producing specialty steels. Name two of these metals, and describe the desirable properties that they confer to the steel.

24 **Spectrophotometric Determination of the Composition of a Complex:**

Composition of Iron-Phenanthroline Complex by Job's Method of Continuous Variation

Semimicroscale Experiment

Objectives
- To establish the formula of a colored complex by the spectrophotometric method

Prior Reading
- Section 2.6   Graphing of data
- Section 3.9   Solution preparation: concentrations
- Chapter 8   Introduction to visible spectroscopy

Related Experiments
- Experiments 20 (Iron-oxalate complexes and photolysis), 21 (Iron determination by visible spectroscopy), 22 (BAC by visible spectroscopy), 23 (Mn in steel), and 36 (Semiconductors)

Activities a Week Before
- Prelaboratory problem # 2 must be done before performing the experiment

Transition metal ions form a large number of complex compounds. For example, the almost colorless iron(II) cation reacts with 1,10-phenanthroline (*o*-phen, colorless) to form a red complex cation:

$$x\,Fe^{2+}(aq) + y\,(o\text{-phen}) \rightarrow [Fe_x(o\text{-phen})_y]^{2+}$$

Job's method of continuous variation will be used in conjunction with visible spectroscopy to establish the formula of the complex formed by Fe(II) and 1,10-phenanthroline. Job's method gives accurate results only under these circumstances:

1. The reaction is quantitative and complete.
2. A single complex species is formed, and the species is stable under the conditions of the reaction.
3. The λ_{max} for the complex species is known and is at a different wavelength from either the ligand or the metal ion.
4. The pH and ionic strength (see Experiment 28) of the solution remain constant.

Using this method, one makes a series of absorbance measurements in which the concentrations of Fe^{2+} and o-phen are varied, while the total number of moles remains constant. This requirement is most easily achieved by preparing solutions of Fe^{2+} and o-phen at identical concentrations and then mixing them in various volume ratios, keeping the total volume constant.

For example, the mole fraction of Fe^{2+} in a series of solutions may be varied in increasing order (0, 0.10, 0.25, 0.40, 0.60, 0.75, 0.90, and 1.0) while the mole fraction of o-phen is varied in decreasing order (1.0, 0.90, 0.75, 0.60, 0.40, 0.25, 0.10, and 0). Note that the sum of the mole fractions of the metal ion and the ligand in each pair is constant.

The mole fractions of the metal and the ligand are easily calculated. By definition,

$$X_M = n_M/n_T \qquad \text{and} \qquad X_L = n_L/n_T \qquad\qquad (E24.1)$$

so that

$$X_M + X_L = 1 \qquad \text{or} \qquad X_M = 1 - X_L$$

where n_M and n_L = number of moles of the metal cation and the ligand, respectively
n_T = total number of moles
    = $n_M + n_L$
X_M and X_L = mole fractions of the metal and the ligand, respectively

Under ideal reaction conditions, the maximum amount of the complex $[Fe_x(o\text{-phen})_y]^{2+}$ will form when the mole fractions of Fe^{2+} and o-phen are in the correct stoichiometric ratio. All other combinations result in the formation of lesser amounts of the complex. Since the product complex is colored, the absorbance of the solution mixture indicates the amount of the complex that has formed. A plot of absorbance versus the mole fraction of the metal ion (X_M) and the ligand (X_L) generates a graph (Figure E24.1) whose maximum indicates the stoichiometric composition.

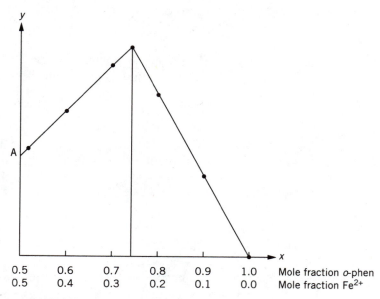

Figure E24.1   Absorption vs. mole fraction

In Figure E24.1, the absorbance maximum occurs at about $X_L = 0.75$ and $X_M = 0.25$. The ratio between them $X_L/X_M = 0.75/0.25 = 3$, indicating that the ligand-to-metal mole ratio is 3:1.

General References

1. Beck, M. T. *Chemistry of Complex Equilibria*, Van Nostrand Reinhold: New York, 1970; pp. 86–89.
2. Birk, J. P.; Ronan, T. H. *J. Chem. Educ.* **1977**, *54*, 328.
3. Christian, C. D. *Analytical Chemistry*, 4th ed.; Wiley: New York, 1986.
4. Skoog, D. A.; West, D. M.; Holler, F. J. *Analytical Chemistry*, 5th ed.; Saunders: Philadelphia, 1990.

Experimental Section

Procedure

Semimicroscale experiment   Estimated time to complete the experiment: 3 h
Experimental Steps: This procedure involves two steps: (1) preparation of solutions and (2) measurement of absorbances.

Disposal procedure: Dispose of the solutions into a suitable container.

Perform the experiment in pairs. Obtain seven clean and dry 50 mL beakers (or large test tubes) and label them "Solution #1, #2, #3, #4, #5, #6, and #7." Also procure seven watch glasses (or use waxed paper) and seven 50 mL Erlenmeyer flasks (or seven large test tubes) labeled "Blank #1, #2, #3, #4, #5, #6, and #7." Obtain one 50 mL volumetric flask, 2.5 percent sodium citrate solution (buffer), a pair of cuvettes, and four 10 mL graduated pipets. Your instructor will supply the following solutions (for each pair):

Solution A	50 mL of 1.0×10^{-3} M Fe^{2+} in 0.1 M HCl in a dry 100 mL beaker
Solution B	60 mL of 1.0×10^{-3} M o-phen in a dry 50 mL beaker
Solution C	10 mL of freshly prepared aqueous 1 percent hydroquinone solution in a dry 50 mL beaker
Solution D	10 mL of 1 M HCl solution in a dry 50 mL beaker

Preparation of Solutions

Using 10 mL graduated pipets (use a separate pipet for each solution), add the amounts of solutions A through D shown for solution #1 (in the following table) to a 50 mL volumetric flask. Add 2.5 percent sodium citrate solution to the volumetric flask to adjust the pH to ~3.5 (test this by applying a drop of solution to a piece of universal pH paper). Finally, add deionized water to the mark. Shake the solution well, and allow it to stand for 10 minutes. Transfer the solution to the 50 mL beaker labeled "Solution #1." Cover the beaker with a watch glass. Rinse the volumetric flask, and repeat the procedure for Solutions 2 through 7.

Beaker #	Solution A, mL	Solution B, mL	Solution C, mL	Solution D, mL
Solution 1	0	10	1	1
Solution 2	1	9	1	1
Solution 3	2	8	1	1
Solution 4	4	6	1	1
Solution 5	6	4	1	1
Solution 6	8	2	1	1
Solution 7	10	0	1	1

Now, prepare seven blank solutions identical to the preceding, **except for leaving out solution B**, in 50 mL volumetric flasks. Transfer the solutions to labeled Erlenmeyer flasks.

Absorbance Measurement

Follow the procedure described in Chapter 8 to obtain the spectral absorbances for each solution. Set the wavelength of the spectrophotometer to **508 nm**. Fill one with the "Blank #1" solution and the other with "Solution #1." Wipe the cuvettes clean with a Kimwipe™. Insert the "Blank #1" cell in the cavity and zero the absorbance. Remove the blank cell and insert the cell with "Solution #1." Record the transmittance, $\%T$, and the corresponding absorbance, A, on the data sheet [$A = 2 - \log(\%T)$]. Empty the cells. Rinse them twice with the second sets of solutions (Blank #2 and Solution #2), refill them, and measure the zero and the absorbance as before. Repeat the procedure with all other sets of solutions, recording the results on the data sheet.

Calculations

Determine the number of millimoles of Fe^{2+} (n_M) and o-phen (n_L) in each solution and obtain the total millimoles (n_T). Calculate the mole fractions of Fe^{2+} (X_M) and the ligand (X_L) for each solution, using equation (E24.1). Make a plot of absorbance versus the mole fractions (X_M and X_L) as shown in Figure E24.1. From the maximum of the curve, determine the mole ratio between Fe^{2+} and o-phen and establish the formula of the complex [$Fe_x(o\text{-phen})_y]^{2+}$.

Additional Independent Projects
  1. Determination of the composition of the Fe^{3+}/SCN^- system
  2. Determination of the composition of Cu^{2+}-EDTA and other complexes[1]

  Brief Outline

  Perform this project following the same general procedure as described in this Experiment. Prepare the following solutions: (a) 0.005 M EDTA (disodium salt of ethylenediaminetetraacetic acid, see Experiment 14) and (b) 0.005 M CuSO$_4$. Adjust the pH of these solutions to ~4.7 with 0.025 M HOAc/0.025 M NaOAc buffer. Using these solutions, prepare a series of 11 mixtures (10 mL) containing varying mole fractions of Cu^{2+} ($X_{Cu^{2+}} = 0.0, 0.1, 0.2, 0.3, 0.4, 0.5, 0.6, 0.7, 0.8, 0.9,$ and 1.0) and EDTA ($X_{EDTA} = 1.0, 0.9, 0.8, 0.7, 0.6, 0.5, 0.4, 0.3, 0.2, 0.1,$ and 0.0). Determine the absorbances for all the solutions at 745 nm. Plot absorbances versus $X_{Cu^{2+}}$ and X_{EDTA}. From the plot, determine the composition of the complex.

  Develop the data tables (data collection and manipulation tables) and submit a written report according to the instructions given in Chapter 4.

3. Determination of performance parameters of a spectrophotometer[2]
4. Synthesis and spectra of copper(II) complexes[3]

   Brief Outline

   This experiment is an excellent one for sequential laboratories. Synthesize a series of complexes using different ligands (EDTA, NH_3, ethylenediamine, acetylacetone, glycine, 1,10-phenanthroline, oxalic acid, and other ligands; see Experiments 19 and 20 and reference 3). Study the visible spectra of these compounds and determine the composition of the complexes.

   Develop the data tables (data collection and manipulation tables) and submit a written report according to the instructions given in Chapter 4.

References

1. Hill, Z. D.; MacCarthy, P. *J. Chem. Educ.* **1986**, *63*, 162.
2. Cope, V. W. *J. Chem Educ.* **1978**, *55*, 678.
3. Potts, R. A. *J. Chem. Educ.* **1974**, *51*, 539.

Name ______________________________ Section ______________

Instructor ______________________________ Date ______________

PRELABORATORY REPORT SHEET—EXPERIMENT 24

Experiment Title ______________________________

Objective

Reactions/formulas to be used

Materials and equipment table

Outline of procedure

PRELABORATORY PROBLEMS—EXPERIMENT 24

1. A student records a transmittance of 40 percent for a particular solution. What would the transmittance be if the solution concentration were doubled?

2. A metal cation M^{n+} reacts with a ligand L to form the complex

$$a\,M^{n+} + b\,L \rightarrow [M_aL_b]^{n+}$$

Using a 0.005 M solution of M^{n+} (*A*) and a 0.005 M solution of L (*B*), a series of solutions was prepared as shown below. Each solution was diluted to 50 mL and the absorbance was measured.

Sample #	Solution (*A*), mL	Solution (*B*), mL	Absorbance
1	0.0	10.0	0.100
2	1.0	9.0	0.238
3	3.0	7.0	0.380
4	5.0	5.0	0.510
5	7.0	3.0	0.376
6	9.0	1.0	0.240
7	10.0	0.0	0.098

(*a*) Calculate the mole fractions of the metal cation and the ligand for each solution. Show your calculations.

Solution #	$n_{M^{n+}}$	n_L	$X_{M^{n+}}$	X_L
1				
2				
3				
4				
5				
6				
7				

(*b*) Make a plot of absorbance (*y* axis) versus mole fraction of the metal cation and the mole fraction of the ligand (*x* axis). Attach the graph to this report. Determine the composition of the complex.

EXPERIMENT 24   DATA SHEET

Data Collection Table
Only solutions A and B are shown in the table. All other solution amounts remain constant.

Solution #	Solution A, mL	Solution B, mL	%T	Absorbance
1				
2				
3				
4				
5				
6				
7				

Data Manipulation Table

Solution #	mmol Fe^{2+} (n_m)	mmol o-phen (n_L)	$X_{Fe^{2+}} = \frac{n_m}{n_m+n_L}$	$X_L = \frac{n_L}{n_m+n_L}$	Absorbance
1					
2					
3					
4					
5					
6					
7					

Show calculations:

Mole fraction of Fe^{2+} ($X_{Fe^{2+}}$) at maximum absorbance _____

Mole fraction of o-phen (X_L) at maximum absorbance _____

Ratio of $X_L/X_{Fe^{2+}}$ _____

Formula of the complex _____

Show calculations:

POSTLABORATORY PROBLEMS, EXPERIMENT 24

1. What effect will a dirty cuvette have on the absorbance reading?

2. If the meter on a spectrophotometer reads 30 percent, what is the absorbance?

3. A sample of 10 mL of 1.0×10^{-3} M Cu^{2+} solution is mixed with 40 mL of 0.001 M NH_3 solution. Calculate the mole fractions of Cu^{2+} and NH_3.

25 **Chromatographic Methods:**
Part A: Column Chromatography;
Part B: Ion Exchange Chromatography;
Part C: Gas Chromatography
Microscale Experiments

Objectives
- Part A: Column chromatography: to use solvent extraction and column chromatography for the extraction, purification, and characterization of a natural product; to recover a solvent by distillation for recycling
- Part B: Ion exchange chromatography: to become familiar with the practical applications of ion exchange chromatography through the separation of a salt mixture; to do acid-base, EDTA, and pH titrations; to reclaim and recycle the resins.
- Part C: Gas chromatography: qualitative and quantitative analysis of volatile compounds

Prior Reading
- Section 3.11   Titration procedure

Related Experiments
- Experiments 1 (Determination of density); 8 (Solvent extraction), 10 to 14 (Titrations); 26, 33, and 35 (TLC).

Activities a Week Before
- Prepare labeled test tubes, Pasteur pipet column, Hickman still for distillation, and ion-exchange column. Also prepare and standardize $\sim$ 0.1 M NaOH and 0.01 EDTA solutions (see Experiment 14)
- Develop your own data tables for parts A and C

A large number of chemical compounds (drugs like taxol, quinine, morphine, cocaine, and others; perfumes; and so forth) were originally isolated from plants and are called **natural products**. One of the oldest and most widely used methods of isolating a natural product from its source is **solvent extraction**. Making a cup of tea, for example, is an extraction process where hot water is used to remove the flavoring compounds into the aqueous layer, giving a distinctive taste and aroma to the beverage. Several different types of solvent extraction procedures are known: solid-liquid, liquid-liquid, and acid-base extractions.

In **solid-liquid extraction**, a suitable solvent is added to a well-pulverized sample from which the natural product is to be isolated. If necessary, the mixture is held under reflux for a given period of time. The solvent is separated from the solid and then evaporated, leaving behind the desired chemical. In **liquid-liquid extraction** (see Experiment 8), two immiscible solvents are used, of which one carries the product and the other acts as the extracting solvent.

In **acid-base extraction**, an acidic compound reacts with a base or a basic compound reacts with an acid. In these cases an ionic salt is formed, rendering the species soluble in the aqueous phase (see Experiment 8).

Chromatographic Techniques

In 1906, the Russian scientist M. Tswett, a lecturer in botany at the University of Warsaw, reported separating the colored pigments from plants by passing an extract of leaves through a packed column. The different constituents of the extract separated out in distinct colors in the column. The technique was called **chromatography** (from the Greek words *chroma* (color) and *graphein* (writing)).

Chromatography is a general term used to cover a wide range of separation techniques. It is defined as the separation of a multicomponent mixture as it is carried through a stationary phase by the flow of a mobile phase. There are three basic types of chromatography:

1. **Solid-liquid chromatography**, which includes column, ion exchange, and thin-layer chromatography
2. **Liquid-liquid chromatography**, which includes partition chromatography
3. **Gas-liquid chromatography**, which includes gas chromatography

One of the most important features of chromatographic techniques is that they are nondestructive methods.

Part A: Column and Thin-Layer Chromatography

In these techniques, the separation of a mixture is carried out by distributing the chemical components between a stationary solid phase and a moving liquid phase. **Thin-layer chromatography** (TLC, see Experiment 26), is particularly effective in the rapid purification and identification of individual components of a mixture.

A simple **column chromatograph** consists of a narrow glass tube (such as a buret or Pasteur pipet) packed with a solid stationary phase. The solution is placed at the top of the column. The solute is immediately distributed between the liquid (mobile phase) and the solid (stationary phase). More mobile phase is added, which pushes the sample in the mobile phase farther down the column. As it reaches a new portion of the stationary phase, further partitioning occurs. This process continues throughout the entire length of the column. The rate at which a component migrates depends on the amount of time it spends in the mobile phase. Different components will spend different amounts of time in the mobile phase, thus traveling at different rates, and will therefore separate. The mobile phase is called the **eluent**, and the process of moving the sample down the column is called **elution**. The mobile phase leaving the column is called the **eluate** or effluent.

A wide variety of substances have been employed as the stationary phase in column chromatography (see Table E25.1). Many common organic solvents are used as eluents. The selection of a particular solvent or solid phase depends on the properties of the compounds being separated.

Table E25.1 Column chromatographic materials

Stationary phases	Mobile phases
Paper	Pentane
Cellulose	Cyclohexane
Magnesium sulfate	Methylene chloride
Silica gel	Diethyl ether
Alumina	Ethyl acetate
	Acetone
	Ethanol
	Water

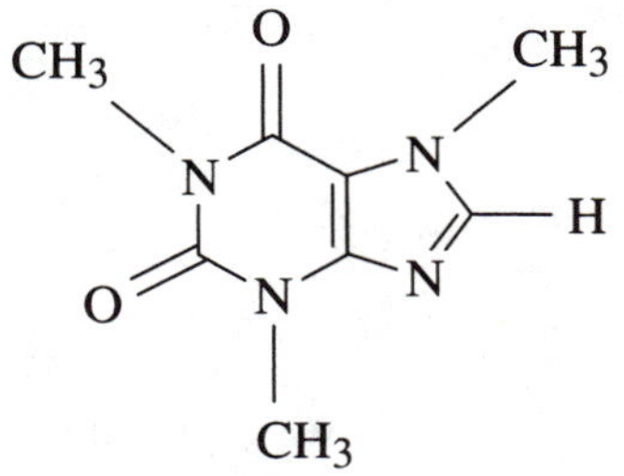

Caffeine

In Part A of this experiment, caffeine ($C_8H_{10}N_4O_2$, see the accompanying structure), will be extracted from tea. Several other chemicals (such as tannic acid) also present in tea are simultaneously extracted. To separate the tannic acids from caffeine, we add sodium carbonate to the solution, which forms water-soluble salts with the tannic acids. These salts are insoluble in methylene chloride, however, and will remain in the aqueous layer when the caffeine (which is soluble in methylene chloride) is extracted into the organic layer. Column chromatography is used to separate the caffeine from other organic impurities.

Part B: Ion Exchange Chromatography

In many practical applications one wishes to exchange one ion for another or to separate a mixture of ions. Raw water, for example, may contain ions that have undesirable properties. The presence of Mg^{2+} or Ca^{2+} ions, for example, makes water "hard" and keeps soaps from foaming in it. The presence of Fe^{2+} stains cloth that is washed in the water.

The term **ion exchange** means the exchange of ions between a solid phase (an exchange resin packed in a column) and a mobile phase (a solution or a solvent). Ion exchange chromatography is thus a special case of column chromatography. Among other things, it is used to determine the concentration of soluble salts in industrial waste, to purify water, to trap radioactive ions, and to sequence proteins.

The ion exchange resin must contain ions that can be readily and reversibly exchanged with the other ions present in a solution. Many substances, both natural (clays) and artificial, have ion exchange properties. There are two principal kinds of ion exchange resins: **cation exchange** and **anion exchange** (see Table E25.2). An example of a cation exchange resin is sulfonated polystyrene crosslinked with divinylbenzene (see Figure E25.1). In this resin, the $-SO_3^- H^+$ groups are the active parts. These are replaced in anion exchange resins by amine (or quaternary ammonium) groups. For the separation of a cation from a solution, the cation exchange resin is used in the acid (hydrogen) form, which is generated by treating the resin with aqueous HCl.

To be useful and effective, the resin must have the following properties:

1. Possess a highly crosslinked molecular structure
2. Be insoluble and chemically stable
3. Be hydrophilic, to permit free movement of ions and solvent molecules in and out of the resin structure
4. Be rich in replaceable ionic groups
5. Be able to swell when suspended in a solvent and be denser than the solvent

Table E25.2 Commercial ion exchange resin types

Type	Ion constitution	Active group	Trade name*
Cationic exchange resin (acidic)	Sulfonic acid	$-SO_3^- H^+$	Amberlite IR 120 Dowex 50W Zerolit 215 Lewatit S1020
Anionic exchange resin (basic)	Amine	$-CH_2NMe_3^+ Cl^-$	Amberlite IRA 400 Dowex 1
		Me = methyl, $-CH_3$	Zerolit FF Lewatit M5020

*Amberlite: Rohm & Haas Co., USA; Dowex: Dow Chemical Co., USA; Zerolit: Permutit Co., England; Lewatit: Merck and Bayer, Germany.

The extent of crosslinking in a resin is indicated by the number after the X in its full name. For example, Dowex 1-X4 contains 4 percent divinylbenzene, the crosslinking material. Cation exchange resins are represented by the general formula Resin–$A^- B^+$, where Resin is the basic polymer framework, A^- is the anion attached to the resin structure, and B^+ is the cation that can be replaced by another cation present in a solution. Similarly, anion exchange resins may be represented as Resin–$NMe_3^+ Cl^-$.

Suppose that an aqueous solution of NaCl is passed through a cationic exchange resin. The following quantitative exchange reaction takes place:

$$Resin–SO_3^- H^+ + NaCl(aq) \rightarrow Resin–SO_3^- Na^+ + HCl(aq)$$

The eluate now contains HCl, not NaCl. If we assume the exchange is complete, for every mole of NaCl entering the column, one mole of HCl is eluted. Similarly, if a solution of $MgCl_2$ (or any other 2+ cation) is passed through the cationic exchange column, one would get *two* equivalents of HCl in the effluent.

$$2\,Resin–SO_3^- H^+ + MgCl_2(aq) \rightarrow Mg(Resin–SO_3^-)_2 + 2\,HCl(aq)$$

The amount of acid eluting from the column can be determined by an acid-base titration.

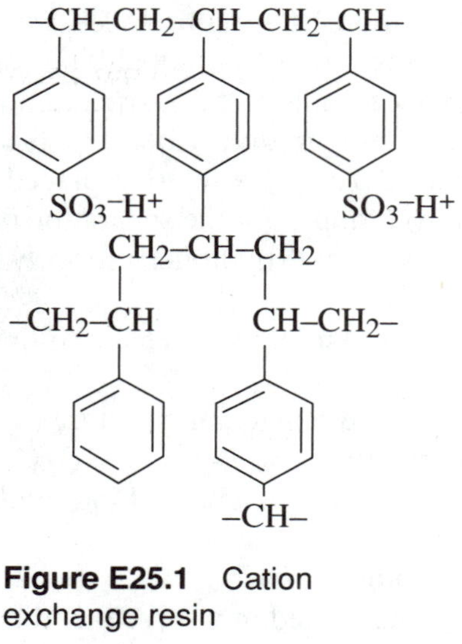

Figure E25.1 Cation exchange resin

In Part B of this experiment, a solution containing a free acid and a mixture of two salts (for example HCl, NaCl, MgCl$_2$) will be analyzed in the following way:

Step 1:   An acid-base titration of a sample of the solution determines the amount of H$^+$ present.
Step 2:   The Mg^{2+} concentration is determined by an EDTA titration.
Step 3:   A sample of the solution is passed through a cationic exchange resin.

An acid-base titration of the effluent determines the amount of H$^+$ present. Any *increase* in the amount of H$^+$ over step 1 must be due to the Na$^+$ and Mg^{2+} exchanging with the resin, producing additional H$^+$. Since the amount of Mg^{2+} is known (step 2), the amount of Na$^+$ can be calculated.

A practical application of ion exchange resins is found in softening hard water, which contains iron, calcium, and magnesium ions.

Ionic impurities in water can be removed by passing the impure water through an anion exchange resin in its OH$^-$ form and a cation exchange resin in its H$^+$ form. If a sample of laboratory water contains CaCl$_2$, it can be deionized according to the following ion exchange reactions:

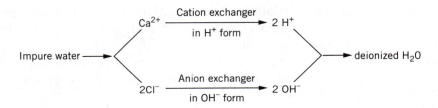

Water from which soluble ions have been removed by this treatment is called **deionized water**.

Part C: Gas Chromatography

Gas chromatography (GC) is another versatile separation technique. In gas chromatography a sample must pass through the system in the gas phase in contact with a solid or liquid stationary phase.

The basic components of a gas chromatograph are shown in Figure E25.2. The most common carrier gases are helium, argon, nitrogen, and hydrogen. The inlet pressure of the gas is generally set between 10 and 50 psi and the flow rates at 20 to 50 mL/min.

The injection port allows for the introduction of the sample onto the column. For liquid samples, the injection port is generally heated to 20 to 50°C above the boiling point of the sample to ensure that the liquid does not condense. On most GC units, a $\frac{1}{4}$ inch o.d. stainless steel column is used. These specifications normally limit sample sizes to 1 to 30 μL.

Two types of columns are commonly used in GC: capillary columns and packed columns. Capillary columns (0.3 to 0.5 mm i.d.) have their inside wall coated with a liquid stationary phase. Because of the small bore of the column, small sample sizes are necessary. Packed columns are generally 1 to 8 mm i.d. and in the range of 2 to 20 feet long. With thinner and longer columns, excellent separation efficiencies can be obtained. The stationary liquid phase must have certain properties:

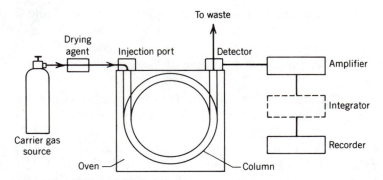

Figure E25.2 Block Diagram of a Gas Chromatograph

1. Its volatility must be low. The phase must generally have a boiling point at least 200°C above the temperature at which the column oven will operate.
2. The phase must be thermally stable over the operating temperature range.
3. It must exhibit chemical inertness with respect to samples to be analyzed.
4. As a solvent it must have good selectivity for samples that are to be separated.

A wide variety of liquids are available for use in GC. Generally, dimethylsilicone-type columns are used for nonpolar materials (separation by boiling point) whereas carbowax-treated columns are used for the polar materials.

The oven temperature is generally set close to the boiling points of the liquids being separated. With average eluent flow rates of 50 mL/min and 8-foot columns, the elution time is generally between 2 and 20 minutes.

The three most common types of detectors (which detect the gas stream) used in GC are the thermal conductivity detector (TCD), flame ionization detector (FID), and the electron capture detector (ECD). Most systems use two detectors—one sample and one reference—so that the thermal conductivity of the carrier gas (eluent) is canceled and changes in the thermal conductivity depend only on the solute material exiting from the column.

Small amounts of a sample in the carrier gas (hydrogen or helium) lead to a significant decrease in the thermal conductivity of the eluting gas. As a result the detector temperature rises in comparison to the detector temperature in the reference column. This ΔT is the recorded signal in the chromatogram. Advantages of the TCD are that it is simple and rugged in design, inexpensive, quite accurate, and nonselective; also it does not destroy the sample.

A typical chromatogram is shown in Figure E25.3. A component in the mixture can be identified by its **retention time, t_R**, which is a measure of time between injection and detection. This value is constant for a particular component as long as the conditions are fixed (flow rate of gas, temperature, column conditions). The **baseline width, W_b**, of a given peak is defined as the distance between two points where tangents to the points of inflection cross the baseline.

Column efficiency is evaluated by determining the **number of theoretical plates, N,** related to the number of equilibrium partitions established in the column. N is determined directly from the experimental chromatogram by

$$N = 16(t_R/W_b)^2$$

The units of t_R and W_b should be the same (minutes, seconds, or centimeters).

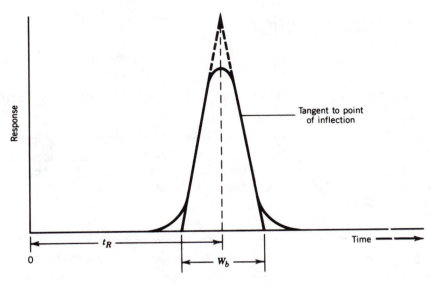

Figure E25.3   A Typical Chromatogram

The larger the number of theoretical plates, the higher the resolution of the column. Resolution of the column means the extent of separation of the peaks on the chromatogram. The **height equivalent of one theoretical plate, HETP,** is determined by

$$\text{HETP} = L/N$$

where L is the length of the column (usually in centimeters) and N is the number of theoretical plates.

General References

1. Chiong, H.-C.; Wang, P.-l. *J. Chem. Educ.* **1983,** *60,* 419.
2. Christian, G. D. *Analytical Chemistry,* 4th ed.; Wiley: New York, 1986.
3. Harris, D. C. *Quantitative Chemical Analysis,* 2nd ed.; Freeman: New York, 1987.
4. Khym, J. X. *Analytical Ion-Exchange Procedures in Chemistry and Biology*; Prentice-Hall: Englewood Cliffs, NJ, 1974.
5. Sawyer, D. T.; Heineman, W. R.; Beebe, J. M. *Chemistry Experiments for Instrumental Methods:* Wiley, New York, 1984.
6. Skoog, D. A. *Principles of Instrumental Analysis,* 3rd ed.; Saunders: New York, 1985.
7. Skoog, D. A.; West, D. M. *Analytical Chemistry: An Introduction,* 4th ed.; Saunders: New York, 1985.
8. Szafran, S,; Pike, R. M.; Singh, M. M. *Microscale Inorganic Chemistry: A Comprehensive Laboratory Experience*; Wiley: New York, 1991.
9. Taber, D. F.; Hoermer, R. S. *J. Chem. Educ.* **1991,** *68,* 73.
10. Tswett, M. *Ber. Deut. Botan. Ges.* **1906,** *24,* 235.
11. Willet, J. E. *Gas Chromatography*; Analytical Chemistry by Open Learning: London, 1987.

Experimental Section

Procedure

Part A: Solvent Extraction and Column Chromatography of Caffeine

Microscale method   Estimated time to complete the experiment: 5 h
Experimental Steps: This experiment involves two steps: (1) extraction of caffeine from tea leaves and (2) separation and characterization of caffeine by column chromatography.

Disposal procedure: Organic solvents must be disposed of in a marked container.

Obtain a tea bag, anhydrous Na_2CO_3, Na_2SO_4, a 50 mL Erlenmeyer flask, a stirring hot plate, a micro stir bar, methylene chloride, a centrifuge tube, a funnel, cotton wool or glass wool, several TLC plates, micro capillary tubes for spotting, caffeine solution, and five dry 5 mL test tubes, labeled "5%, 10%, 20%, 40%, and 100% ethyl acetate (EA)/methylene chloride (MC)."

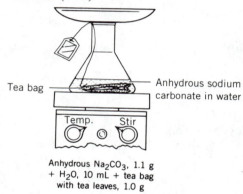

Tea bag

Anhydrous sodium carbonate in water

Temp.   Stir

Anhydrous Na_2CO_3, 1.1 g
+ H_2O, 10 mL + tea bag
with tea leaves, 1.0 g

Figure E25.4 Extraction of caffeine from tea leaves

Add 10.0 mL of deionized water followed by 1.0 g of anhydrous sodium carbonate to a 50 mL Erlenmeyer flask containing a micro stir bar (Chapter 3, section 10). Heat the flask gently with stirring to dissolve the solid. Carefully make a hole at the top of a tea bag and empty the contents into a small beaker. **Do not remove the thread from the bag.** Weigh 1.00 g of the tea leaves and place them back inside the empty bag, stapling it shut when finished. Insert the bag into the Erlenmeyer flask as shown in Figure E25.4, making sure that the bag does not rip during this procedure. Put a small watch glass on the mouth of the Erlenmeyer flask to minimize the loss of water during heating.

Boil the solution gently for 3 to 5 minutes, remove the flask and allow the solution to cool to room temperature. Remove the tea bag, squeezing it against the wall of the flask with a spatula to recover as much tea extract as possible. Discard the tea bag and its contents. Filter the deep brown solution directly into a centrifuge tube, as shown in Figure E25.5.

Extract the cooled liquid with methylene chloride according to the following procedure. Add 1.5 mL methylene chloride (**HOOD**) to the solution. **Gently** stir the mixture with a spatula.

Note: Vigorous mixing causes the formation of an emulsion, which will make separation more difficult.

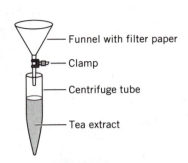

Figure E25.5 Filtration of tea extract

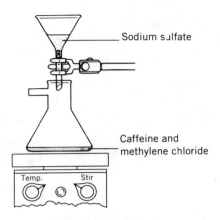

Figure E25.6 Filtration through anhydrous Na_2SO_4

The aqueous and methylene chloride layers separate slowly. Let the mixture stand for at least 15 minutes.

While waiting for the layers to separate, construct a filtration funnel by putting a small plug of cotton inside the stem of a funnel. Cover the cotton plug with a 2.0 gram portion of anhydrous sodium sulfate (see Figure E25.6). If necessary centrifuge to separate the layers. Using a 9 in. Pasteur pipet, transfer the wet lower methylene chloride layer to the filter. Transfer the dried filtrate to a 10 mL round-bottom flask. Repeat the procedure twice more, collecting the extracts in the same flask. Dispose of the remaining residue in an appropriate container.

Attach the round bottomed flask containing the CH_2Cl_2 extract to a Hickman still, fitted with an air condenser. Distill off almost all the methylene chloride solvent by **gently warming** the flask (see Figure E25.7). **Do not heat to dryness.** From time to time, remove the distillate from the collar of the Hickman still using a Pasteur pipet, and store it in a container for recycling. When about 0.3 mL of methylene chloride is left in the flask, remove the flask or the vial from the heat.

Obtain five dry 5 mL test tubes. Label the empty tubes "5%, 10%, 20%, 40%, and 100%." Transfer 1 mL each of the appropriate solutions to another set of five test tubes labeled "5%, 10%, 20%, 40%, and 100% ethyl acetate (EA)/methylene chloride (MC)." Arrange them in order on a test tube rack.

Construct a Pasteur pipet column packed with silica gel layered with a small amount of sand as described in Section 3.10. **Note: Perform all subsequent steps as quickly as possible.** Moisten the column with only a **few** drops of methylene chloride. Put the "5%" empty test tube under the tip of the column. Using a Pasteur pipet, transfer the concentrated CH_2Cl_2 residue from the round-bottom flask to the top of the column. Rinse the flask with five drops of methylene chloride and transfer the rinse to the top of the column. Allow the solution to drop to the sand layer. Using a Pasteur pipet, immediately add 500 μL of the 5 percent solvent mixture dropwise to the top of the column. (It may be necessary to let the column drain somewhat for this to fit.)

Collect the eluate in the "5%" test tube. Complete the elution by allowing the solvent to drop just to the sand layer. If the process is slow, apply gentle pressure at the mouth of

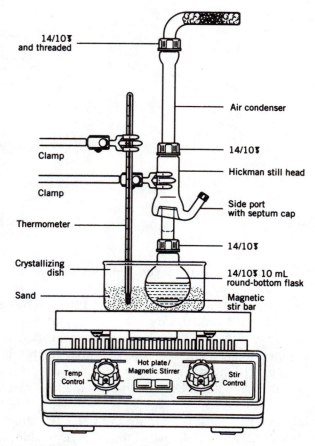

14/10T
and threaded

Air condenser

Clamp

14/10T

Hickman still head

Clamp

Side port
with septum cap

Thermometer

14/10T

Crystallizing
dish

14/10T 10 mL
round-bottom flask

Sand

Magnetic
stir bar

Hot plate/
Magnetic Stirrer

Temp
Control

Stir
Control

Figure E25.7   Distillation of methylene chloride
using a Hickman still

the column using a pipet bulb. Repeat the elution process using the other solvent mixtures (including pure ethyl acetate), collecting the eluate in the appropriate labeled test tubes.

Using silica-coated thin-layer chromatography (see Experiment 26 for procedure), check the contents of each tube for the presence of caffeine. Spot the plates with a portion of each of the eluates using micro capillary pipets. Also spot one plate with a solution of caffeine supplied by your instructor. Develop the plates in a chamber using methylene chloride as the developing solvent. If the TLC is performed correctly, the early yellow spot will be followed by a colorless spot due to caffeine. The colorless spot may be visualized by using a UV lamp or by placing the plates in a bottle containing a few crystals of iodine. Mark the spots with a pencil and calculate the R_f value for each spot. The R_f value is the ratio of the distance traveled by the solute to that traveled by the solvent (see Experiment 26 for a complete discussion of R_f values). Compare them with that for pure caffeine.

Determine which of the test tubes contain caffeine. Collect all the eluates from the test tubes that are found to contain caffeine in a *tared* round-bottom flask. Distill and collect the solvent in a Hickman still as described before. **Do not heat the flask to dryness.** Collect the solvent from the collar of the still with a Pasteur pipet and store it in the recycle container.

Remove the last traces of the solvent (250 to 500 μL) by blowing a stream of nitrogen gas gently over the solvent inside a **HOOD**. Collect the fluffy white residue and determine the yield of the caffeine. Determine the melting point (see Experiment 2) and obtain the infrared spectrum of the caffeine using a KBr pellet (your instructor will demonstrate how to prepare a sample and how to run a spectrum). Compare your results with those of an authentic sample of caffeine supplied by your instructor.

Procedure

Part B: Ion Exchange Method

Microscale method   Estimated time for completing the experiment: 4 h
Experimental Steps: There are several steps involved in the procedure: (1) preparation and standardization of NaOH and EDTA solutions, (2) construction of an ion exchange column, (3) titrations of the free acid and Mg^{2+} or Ca^{2+}, (4) separation of the ions on the column, and (5) titration of the total acid.

Perform the experiment in pairs. Obtain two Erlenmeyer flasks (25 mL and 50 mL), two 2 mL microburets (labeled *acid* and *base*), a graduated 2 mL pipet, a pipet pump, a 10 mL graduated cylinder, 0.25 percent phenolphthalein indicator, Eriochrome Black T indicator solution, blue litmus paper, an ammonia buffer (pH = 10), 10 mL of 1 M HCl solution, Bio-Rad Dowex 50W-X2, 100–200 mesh (or equivalent) ion exchange resin, 50 mL of 0.1 M NaOH solution, solid potassium hydrogen phthalate (KHP), disodium dihydrogen ethylenediaminetetraacetate (EDTA), a 10 mL and a 25 mL volumetric flask, and a section of 0.7×15 cm glass tubing or a regular buret column (15 cm, cut bottom piece) for making an ion exchange column.

Also procure an unknown solution containing ~4 mmol of H^+, ~1 mmol of Na^+ or K^+ and ~0.5 mmol of Mg^{2+} or Ca^{2+} in a 25 mL volumetric flask. Record the unknown number on the data sheet. Dilute the solution to the mark.

Construction of a Cation Exchange Chromatography Column

Using a long copper wire, place a small plug of glass wool inside the 15 cm buret column on top of the stopcock to prevent the flow of resin through the hole. Clamp the column to a ring stand as vertically as possible, close the stopcock, fill the buret with water, and test for leaks. If no leak is detected, open the stopcock and drain the water until just 1 cm of water remains in the column.

Make a slurry of ~1 g of the resin in 5 mL of water. Transfer the slurry into the column. If all the slurry cannot be poured at once, allow the resin to settle. Open the stopcock to remove excess water. Then pour in the remaining slurry. Finally, pour a 5 mm thick layer of prewashed coarse sand on top of the resin in the column.

To activate the resin, add 10 mL of 1 M HCl to the column. Open the stopcock.

CAUTION: Do not apply any pressure on the top of the column to hasten the process of HCl washing of the resin. Also, never allow the upper level of the liquid to fall below the top of the resin level.

After the meniscus has almost reached the top of the resin, add 10 mL of deionized water to the column to rinse it free from excess HCl. Wash the column two or three times more with additional 10 mL portions of water. Each time, let the first portion of water pass down almost to the top of the resin before adding the next portion. Discard the washings. Close the stopcock at the end of the process.

After the second wash, collect a few drops of the effluent from the column and test its acidity by placing a drop on a piece of blue litmus paper. If the litmus paper turns pink, continue to wash the column with deionized water until it is free from acid. Close the stopcock.

Preparation of Solutions (0.1 M NaOH and 0.01 M EDTA)

Note: If your instructor provides you with standardized NaOH and EDTA solutions, you may skip this part. Assemble two microburets for titration. Rinse and fill them with the appropriate solutions. Record the initial volume of solution in each buret.

Prepare a solution of 0.1 M KHP in a 10 mL volumetric flask and use it to standardize a 0.1 M NaOH solution according to the procedure described in Experiment 10A. Perform three titrations. Calculate the average molarity of the NaOH solution.

Prepare a standard 0.01 M EDTA solution (see Experiment 14) by adding 100 ± 1 mg of $Na_2EDTA \cdot 2H_2O$ to 15 mL of water in a 25 mL beaker equipped with a stirring bar. While stirring, heat the mixture almost to boiling to obtain a clear solution. If the solution remains turbid, add a drop or two of 0.1 M NaOH, and heat the mixture again. Transfer the clear solution to a 25 mL volumetric flask. Wash the beaker with two 2 mL portions of water and transfer the washings to the volumetric flask. Add deionized water to the mark. Calculate the molarity of the EDTA solution and enter the value on the data sheet. The EDTA solution may also be standardized according to the procedure described in Experiment 14.

Determination of H$^+$ by Acid-Base Titration

Pipet exactly 1.00 mL of the unknown solution into a 25 mL Erlenmeyer flask. Add 4 mL of deionized water followed by a **small** drop of 0.25 percent phenolphthalein indicator. Titrate the unknown solution with the standardized 0.1 M NaOH solution (Experiment 10A). The end of the titration is indicated by the appearance of a permanent pink color. Repeat this procedure at least twice more and average the results. Calculate the number of millimoles (mmol) of NaOH and, given the reaction

$$HCl(aq) + NaOH(aq) \rightarrow NaCl(aq) + H_2O(l)$$

also calculate the number of mmols of HCl in the aliquot. Finally, calculate the number of H$^+$ in the *original solution*. Calculate the amount of HCl in mg.

Determination of Mg^{2+} by EDTA Titration

Pipet exactly 1.00 mL of the unknown solution into a 25 mL Erlenmeyer flask. Add 2 mL of deionized water. Fill a 2 mL buret with 0.01 M EDTA solution. Record the initial volume of EDTA in the buret. Titrate the Mg^{2+} with the standard 0.01 M EDTA solution, following the same procedure as described in Experiment 14. You may have to refill the EDTA buret to complete the titration. Repeat the procedure at least twice more. Calculate the number of millimoles of EDTA (and hence of Mg^{2+} in the aliquot) used in the titration. Convert the millimoles of Mg^{2+} into mg. Finally, determine the mass of Mg^{2+} in the *original solution*.

Separation of the Mixture by Ion Exchange Chromatography

Pipet exactly 1.00 mL of the unknown solution into the column. Open the stopcock, and allow the solution to pass through the column followed by at least three 3 mL portions of water. Collect all the effluent in a 50 mL Erlenmeyer flask. **Note: Do not add water until the previous portion has run close to the upper level of the resin.** Continue the washings until the effluent is neutral to litmus paper. Close the stopcock.

Titration of the Effluent

Add one small drop of 0.25 percent phenolphthalein to the effluent and titrate the acid with the standardized 0.1 M NaOH. Record the volume of NaOH used. If time permits, repeat the procedure (separation of the mixture followed by titration) twice more.

Average the results. Record your data in the data table. Calculate the number of mmols of NaOH and hence of HCl. The number of mmoles of HCl corresponds to the total number of mmols of $Na^+ + Mg^{2+} + H^+$ present in the unknown mixture. Subtract the number of millimoles of H^+ originally present (determined previously) and twice the number of mmoles of Mg^{2+} present (determined before—remember, each Mg^{2+} ion exchanges with *two* H^+ ions). The remainder is the number of mmoles of Na^+ present in the original solution. Determine the concentration of Na^+ in 1 mL of the unknown, and the amount (mg) present in the *original solution*. For calculations, see the example (E25.1).

Reclamation and Recycling the Ion Exchange Column

Pass 10 mL of 1 M HCl through the column to regenerate the ion exchange resin. Wash the column with two 2 mL portions of water to remove excess acid from the column (test with blue litmus paper). Close the stopcock of the column, and add deionized water to a level at least 1 mL above the top of the resin. Seal the top of the column tightly for storage.

Procedure

Part B: Alternative Procedure Macroscale Method

Alternative macroscale method	Estimated time for completing the experiment: 6 h

Use a 50 mL buret as a column and Dowex 50-X8 cationic exchange resin (50–100 mesh). Follow the same general procedure as described for the microscale method, using a resin bed at least 30 mL in length. A 10 mL sample of the unknown should be charged onto the column. Collect the effluent in a 100 mL Erlenmeyer flask. Wash the column with two 10 mL portions of water, collecting the washings in the same flask. Continue to wash water through the column until the effluent is neutral to blue litmus paper. Perform the titration as before. To determine the free H^+ (acid-base titration) and Mg^{2+} (EDTA titration), carry out direct titrations of different aliquots of the sample using the macroscale procedures described in Experiments 10 and 14.

> **Example E25.1**   A 5.00 mL sample of an unknown solution containing HCl, NaCl, and $MgCl_2$ was diluted to 25 mL in a volumetric flask. A 1.00 mL aliquot of this solution required 0.385 mL of 0.101 M NaOH solution for complete neutralization. A second 1.00 mL aliquot of the diluted unknown needed 4.010 mL of 0.0110 M EDTA solution to completely react with the Mg^{2+}. A third 1.00 mL aliquot of diluted solution was

passed through a cationic exchange resin. The effluent and the washings collected from the column required 1.66 mL of 0.101 M NaOH for the titration. Calculate the mass in grams of each species present in the original solution.

Answer

First 1 mL aliquot

$$\text{Mol NaOH} = (0.101\,\text{mol/L})(0.385 \times 10^{-3}\,\text{L}) = 3.89 \times 10^{-5}\,\text{mol NaOH}$$
$$= 3.89 \times 10^{-5}\,\text{mol HCl/mL}$$

$$\text{Mass of HCl in 1 mL aliquot} = (3.89 \times 10^{-5}\,\text{mol HCl/mL})(36.46\,\text{g/mol})$$
$$= 1.42 \times 10^{-3}\,\text{g HCl/mL}$$

$$\text{Total HCl mass in the original solution} = (1.42 \times 10^{-3}\,\text{g HCl/mL})(25\,\text{mL})$$
$$= 0.0355\,\text{g HCl}$$

Second 1 mL aliquot

$$\text{Mol EDTA} = (0.0110\,\text{mol/L})(4.010 \times 10^{-3}\,\text{L}) = 4.41 \times 10^{-5}\,\text{mol EDTA}$$
$$= 4.41 \times 10^{-5}\,\text{mol Mg}^{2+}/\text{mL}$$
$$= (4.41 \times 10^{-5}\,\text{mol MgCl}_2/\text{mL})(2\,\text{mol HCl/mol MgCl}_2)$$
$$= 8.82 \times 10^{-5}\,\text{mol HCl}$$

$$\text{Mass of MgCl}_2/\text{mL solution} = (4.41 \times 10^{-5}\,\text{mol MgCl}_2/\text{mL})(95.22\,\text{g/mol})$$
$$= 4.20 \times 10^{-3}\,\text{g MgCl}_2/\text{mL}$$

$$\text{Total mass of MgCl}_2 \text{ in the original solution} = (4.20 \times 10^{-3}\,\text{g/mL})(25\,\text{mL})$$
$$= 0.105\,\text{g MgCl}_2$$

Third 1 mL aliquot

$$\text{Mol NaOH} = (0.101\,\text{mol/L})(1.66 \times 10^{-3}\,\text{L}) = 1.68 \times 10^{-4}\,\text{mol NaOH}$$
$$= 1.68 \times 10^{-4}\,\text{mol HCl/mL}$$

$$\text{Mol NaCl only} = 1.68 \times 10^{-4} - 3.89 \times 10^{-5} - 8.82 \times 10^{-5} = 4.09 \times 10^{-5}\,\text{mol HCl/mL}$$
$$= 4.09 \times 10^{-5}\,\text{mol NaCl/mL}$$

$$\text{Mass of NaCl in original solution} = (4.09 \times 10^{-5}\,\text{mol NaCl/mL})(58.44\,\text{g/mol})(25\,\text{mL})$$
$$= 0.0598\,\text{g NaCl}$$

Part C: Gas Chromatography

Microscale method   Estimated time to complete the experiment: 2.5 h
Experimental Steps: Two steps are involved: (1) determination of the optimum temperature of the column and (2) qualitative and quantitative analysis of a mixture.

Disposal procedure: Give the leftover organic liquids/mixtures to your instructor for proper disposal.

Obtain a clean microsyringe (25 to 50 μL size) and the unknown sample (which is a mixture of two or three liquids from the list of knowns). Known samples of organic liquids (dichloromethane, methanol, ethanol, 1-propanol, 2-propanol, ethyl acetate, and toluene) are provided. Your instructor will demonstrate how to operate the gas chromatograph and how to inject the samples into the column.

Determination of Optimum Column Temperature

> Note: In this experiment, the volume measurements must be made as accurately as possible. Handle the microsyringe very carefully; it can be damaged easily. Rinse the syringe and the graduated pipet with the same liquid each time you use them to measure the volumes of different samples.

Using a clean and rinsed graduated pipet, prepare a mixture of methanol, ethanol, and 1-propanol in a 1:2:3 volume ratio in a sealed 3 mL vial. Set the GC injector temperature at 150°C and the column temperature at 50°C. Inject 20 μL of the mixture and mark the point of injection on the recorder chart. Allow the three components to elute and obtain the chromatogram.

Bring the peak heights on scale by adjusting the attenuator. Repeat the procedure using column temperatures of about 80° and 120°C. Select the optimum temperature (good separation of peaks and peak heights) for the analysis. Determine the retention time in minutes. Similarly, determine the retention times for other known mixtures.

Density Determination

To convert the volume of a liquid to its mass, the density ρ of the liquid is determined by the micropycnometer method (see Experiment 1). The concentration of each known solution is calculated on the basis of its mass.

Qualitative Identification of the Unknown[1]

Inject 20 μL of the pure unknown and obtain a chromatogram at the optimum temperature determined as just described. By comparing this chromatogram with those obtained for the known materials, determine the identity of the components of the mixture. Then, mix pure samples (1:1 volume ratio; total volume of the mixture should not exceed 0.5 mL) of the suspected components with samples of the unknown. Inject 20 μL of the mixture and run the chromatogram again. By observing whether a new peak has developed or an old peak has grown in size, confirm the identity of each species in your unknown.

Quantitative Analysis of a Component in the Unknown

Select a component X that is well separated from the peaks of other components of the mixture. Perform a quantitative analysis of X according to the procedure described next.

From the list of the known liquids, find a liquid to use as a standard S that is not a component of the mixture of unknowns and is separated from all the peaks of the unknown. Prepare a mixture of the unknown and S (in volume ratio) such that the peak areas (or the heights, if the peaks are sharp) of X in the mixture and S remain within a factor of ~2 of each other. This ratio is accomplished by trial and error, using 20 μL injections. The volume (or mass) of S and that of the mixture must be accurately known.

Now, prepare a series of solutions by mixing accurately known volumes of pure X and pure S. Compute the ratio of the concentration of X to that of S in each solution.

Inject 20 μL of each of the mixtures into the column. Obtain the chromatogram each time. Determine the peak areas of X and of the standard S according to the method described in the next section. Calculate the ratio of the peak area of X to the peak area of S. Prepare a calibration curve by plotting the concentration ratios (x axis) versus peak ratios (y axis). The range of the plot must be such that the peak ratio measured for the unknown and for the standard is located on the curve. From your graph, calculate the concentration of X in the unknown. Using the following equation, calculate the response factor R for the species X relative to the standard S:

$$R = \left(\frac{\text{Concentration of unknown}}{\text{Concentration of standard}} \right) \times \left(\frac{\text{area of standard}}{\text{area of unknown}} \right)$$

One can calculate the concentration of any species in the unknown by using the response factor. See examples E25.2 and E25.3. An alternative method to calculate the concentration of an unknown is

$$\text{mmol of unknown sample} = \frac{(\text{peak area in mixture})(\text{volume injected})(\text{density})}{(\text{peak area of pure sample})(\text{molecular weight})}$$

$$\text{mol \% of unknown sample} = \frac{\text{mmol of sample in mixture} \times 100}{\text{total mmol in mixture}}$$

However, the accuracy of the method depends upon the exact volume measurement of the injected sample using a microsyringe.

Measurement of Peak Areas

Since the peak area is proportional to the amount of the substance present in the mixture, it is important to measure the peak areas as accurately as possible. The following methods are available to measure the peak areas:

1. If the peak is symmetrical, the product of the peak height times its width at half-height equals almost 85 percent of the total area.
2. Draw a triangle, joining two sides tangent to the inflection points on each side of the peak (see Figure E25.3). The baseline becomes the third side. Measure the area of the triangle:

$$\text{Area of the triangle} = \tfrac{1}{2}(\text{base} \times \text{height})$$

3. Make a photocopy of the chromatogram. Cut the total area along the edges of the peak. Weigh the cutout peak. This weight is proportional to the quantity of the substance present in the mixture.
4. Some GCs have a built-in capacity to integrate peaks. Many printers also have automatic integrators. Your instructor will demonstrate how to use this feature in your printer. A planimeter can also be used to measure the peak area.

Example E25.2 A mixture containing 1.20 mmol of 1-butanol and 1.00 mmol of 1-pentanol is chromatographed. The ratio of the peak areas is found to be butanol/pentanol = 1.00. Calculate the response factor, R, for butanol.

Answer

$$R = \left(\frac{\text{Concentration of unknown}}{\text{Concentration of standard}}\right) \times \left(\frac{\text{area of standard}}{\text{area of unknown}}\right)$$

$$R = (1.20 \text{mmol}/1.00 \text{mmol})(1.00/1.00) = 1.20$$

Example E25.3   In another set of experiments 0.45 mmol of pentanol is added as an internal standard to an unknown butanol solution and the mixture is chromatographed. The ratio of the peak areas for butanol/pentanol equals 0.600. What is the concentration of butanol? Use the R value obtained in example E25.1.

Answer

$$\text{mmol butanol} = 1.20 \text{ (peak area ratio)(mmol pentanol)}$$
$$= 1.20 \times 0.60 \times 0.45 = 0.32 \text{ mmol}$$

Additional Independent Projects

1. Extract and purify β-carotene from carrots[2]

2. Extract leaf pigments[3]

   Brief Outline

   Besides chlorophyll, plant leaves contain other pigments such as carotenoids, which can be seen during fall when leaves dry. Using a mortar and pestle, crush the leaves in the presence of 2 mL acetone. Using sodium bicarbonate as a solid phase suspended in petroleum ether, make a micro chromatographic column in a Pasteur pipet. Transfer the clear extract into the column. Carry out the elution process using petroleum ether, acetone, and 70/30 isopropyl alcohol/water mixture in succession. Collect each fraction separately. Explain what happens when dichloromethane is substituted for petroleum ether. Develop the data tables (data collection and manipulation tables) and submit a written report according to the instructions given in Chapter 4.

3. Determination of ion exchange capacity of an ion exchange resin[4]

   The total **ion exchange capacity** is defined as the equivalents of ion-active groups per gram of the exchanger. The ion exchange capacity of a cation exchange resin is determined by measuring the equivalents of sodium ion absorbed by 1 g of the dry resin in the hydrogen form. Load a column with ~0.5 g (W g) of the resin suspended in water. Pass 200 mL of 0.25 M sodium sulfate solution through the column. Collect the effluent and titrate the acid liberated with a prestandardized 0.1 M NaOH solution by the macroscale method. Record the volume V (milliliters). The capacity of the resin is given by the formula $0.1 \text{M} \times V(\text{mL})/W$. Submit a written report according to the instructions given in Chapter 4.

4. Separation of chloro complexes of copper(II) and iron(III) by ion exchange chromatography and spectrophotometric analysis of the elution products[5]

   This laboratory is a sequential one and can be done in two 3 hour lab periods. It uses an anion exchange column in the chloride form. In 9 M HCl solution, copper(II), Cu^{2+}, and iron(III), Fe^{3+}, exist as the following chloro complexes: $CuCl_3^-$ and $FeCl_4^-$, respectively. In the first part of the experiment, the student prepares an ion exchange column by using an anion exchange resin that has been activated by HCl. Next, the student passes a solution supplied by the instructor through the column. Extract the column first with 8 M HCl, then with 3 M HCl, and finally with deionized water. Collect all the fractions of each elution. Perform the spot tests (see Experiment 16) for copper(II) using ammonia (it will produce an intense blue color) and for iron(III) using KSCN (intense red color) solution. Collect all the fractions containing copper(II) ion and the fractions containing iron(III) ions separately. Using the procedure described in Experiment 24, determine the concentrations of both ions using a spectrophotometer (at 545 nm for copper and 508 nm for iron solutions). Generate calibration curves for known solutions of copper and iron. Develop the data tables (data collection and manipulation tables) and submit a written report according to the instructions given in Chapter 4.

5. Determination of ethanol in alcoholic beverages by GC[6]

Using a graduated pipet or automatic delivery pipet, transfer exactly 1.0 mL of ethanol into a 10 mL volumetric flask. Dilute it to the mark. Similarly, prepare other standards by diluting 2, 4, 6, and 8 mL of ethanol. Inject 1 μL of each of the standards and the unknown alcohol. Obtain the chromatograms (Gow Mac™, Porapak™ column, TCD™ only). Calculate the peak areas for both ethanol and water in each standard. For each standard, determine the ratio of the peak area of ethanol to the total peak areas of ethanol and water. Calculate the volume percent ethanol [= (volume ethanol)(100)/total volume)]. Prepare a calibration curve by plotting the ratio of the peak areas (y axis) versus volume percent ethanol (x axis). From this plot determine the concentration of the unknown. Develop the data tables (data collection and manipulation tables) and submit a written report according to the instructions given in Chapter 4.

References

1. Harris, D. C. *Quantitative Chemical Analysis,* 2nd ed.; Freeman: New York, 1987.
2. Pavia, D. L.; Lampman, G. M.; Kriz, G. S. *Introduction to Organic Laboratory Techniques,* 3rd ed.; Saunders: Philadelphia, PA, 1988.
3. Kimbrough, D. R. *J. Chem. Educ.* **1992,** *69,* 987.
4. Bassett, J.; Denney, R. C.; Jeffery, G. H.; Mendham, J. *Vogel's Textbook of Quantitative Analysis,* 4th ed.; Longman: Essex, England, 1986.
5. Foster, N.; Pestel, B.; Mease, B. *J. Chem. Educ.* **1985,** *62,* 170.
6. Leary, J. J. *J. Chem. Educ.* **1983,** *60,* 675.

Name ______________________ Section ______________

Instructor ______________________ Date ______________

PRELABORATORY REPORT SHEET—EXPERIMENT 25, PART A

Experiment title ______________________

Objective

Reactions/formulas to be used

Materials and equipment table

Outline of procedure

Develop data collection and manipulation tables for this part (see prelaboratory problem sheet, part A)

PRELABORATORY REPORT SHEET—EXPERIMENT 25, PART B

Experiment title ______________________________

and part ______________________________

Objective

Reactions/formulas to be used

Materials and equipment table

Outline of procedure

Name ______________________ Section ____________

Instructor ______________________ Date ____________

PRELABORATORY REPORT SHEET—EXPERIMENT 25, PART C

Experiment title ______________________

and part ______________________

Objective

Reactions/formulas to be used

Materials and equipment table

Outline of procedure

Develop data collection and data manipulation tables for this part (see prelaboratory problem sheet, part C)

1. Define the following:

    (*a*)   Chromatography

    (*b*)   Eluent

2. What precautions must be taken to make a column for column chromatography?

3. Write all the steps involved in the extraction and the purification of caffeine from tea leaves.

4. Using the answer from question 3, develop a Data Collection and Manipulation Table for Experiment 25, Part A. Use the blank Data Sheet provided.

PRELABORATORY PROBLEMS—EXPERIMENT 25, PART B

1. What is the purpose of adding HCl to a cationic ion exchange column prior to adding your sample?

2. A 10.0 mL sample of an unknown solution (containing H_2SO_4, NaCl, and some impurities) was diluted to 25 mL in a volumetric flask. A 2.00 mL sample of this solution was passed through a cation exchange column. The acid effluent needed 3.00 mL of 0.110 M NaOH solution for complete neutralization. In a separate direct titration of 2.00 mL of the same solution with 0.110 M NaOH solution, 2.10 mL of NaOH was consumed. How much H_2SO_4 and NaCl were present in the original sample solution?

1. Explain the following:

   (*a*) What are the differences between a thermal conductivity detector and a flame ionization detector?

   (*b*) Define retention time, R_f; theoretical plates; and HETP.

2. What precautions must one take to run a good GC chromatogram?

3. Develop your own data table.

EXPERIMENT 25   DATA SHEET

Part A: Column Chromatography

Develop your own data sheet

Part B: Ion Exchange Chromatography

Sample ID number _____

Preparation of 0.1 M Sodium Hydroxide Solution

Mass of NaOH taken or volume of NaOH _____ g (or mL)

Preparation of KHP and Standardization of Sodium Hydroxide Solution

Preparation of KHP Standard Solution (~0.1 M) in 10 mL flask

1. Initial mass of KHP, g   _____

2. Final mass of KHP, g   _____

3. Mass of KHP taken, g   _____

4. Concentration of KHP, mol/L   _____

Standardization of 0.1 M NaOH Solution using the preceding KHP standard solution

		Trial 1	Trial 2	Trial 3
1.	Initial volume of KHP, mL	_____	_____	_____
2.	Final volume of KHP, mL	_____	_____	_____
3.	Volume of KHP taken, mL	_____	_____	_____
4.	Initial volume of NaOH, mL	_____	_____	_____
5.	Final volume of NaOH, mL	_____	_____	_____
6.	Volume of NaOH used, mL	_____	_____	_____
7.	Concentration of NaOH, mol/L	_____	_____	_____

Average molarity _____ M

EDTA Solution

1. Mass of EDTA in 25 mL volumetric flask, g   _____

2. Molarity of EDTA solution, M   _____

DATA SHEET, EXPERIMENT 25, PART B, PAGE 2

Titration of the Free H$^+$ with NaOH

Molarity of NaOH _____ mol/L

		Trial 1	Trial 2	Trial 3
1.	Volume of the original solution, mL	_____	_____	_____
2.	Initial buret reading, mL (NaOH)	_____	_____	_____
3.	Final buret reading, mL (NaOH)	_____	_____	_____
4.	Volume of NaOH used, mL	_____	_____	_____
5.	mmol NaOH consumed	_____	_____	_____
6.	mmol HCl for free H$^+$	_____	_____	_____
7.	Mass of HCl, mg	_____	_____	_____
8.	Mass of HCl in original solution, mg	_____	_____	_____
9.	Average mass, mg		_____	

EDTA Titration

Molarity of EDTA _____ mol/L

		Trial 1	Trial 2	Trial 3
1.	Volume of the original solution, mL	_____	_____	_____
2.	Initial buret reading, mL (EDTA)	_____	_____	_____
3.	Final buret reading, mL (EDTA)	_____	_____	_____
4.	Volume of EDTA used, mL	_____	_____	_____
5.	mmol EDTA consumed	_____	_____	_____
6.	mmol MgCl$_2$ (Mg^{2+})	_____	_____	_____
7.	Mass of MgCl$_2$, mg	_____	_____	_____
8.	Mass MgCl$_2$ in original solution, mg	_____	_____	_____
9.	Average mass, mg		_____	

Titration of the Effluent with the Standardized 0.1 M NaOH

Molarity of NaOH _____ mol/mL

		Trial 1	Trial 2	Trial 3
1.	Volume of the original solution, mL	_____	_____	_____
2.	Initial buret reading, mL (NaOH)	_____	_____	_____
3.	Final buret reading, mL (NaOH)	_____	_____	_____
4.	Volume of NaOH used, mL	_____	_____	_____
5.	mmol of NaOH consumed	_____	_____	_____
6.	Total mmol HCl/mL (Na^+/Mg^{2+}/H^+)	_____	_____	_____
7.	mmol HCl for free H^+	_____	_____	_____
8.	mmol Mg^{2+} (from previous page)	_____	_____	_____
9.	mmol HCl equivalent to Mg^{2+}	_____	_____	_____
10.	Total mmol ($H^+ + Mg^{2+}$)	_____	_____	_____
11.	mmol NaCl equivalent to HCl (step 6 – step 10)	_____	_____	_____
12.	Mass of NaCl per mL, mg/mL	_____	_____	_____
13.	Mass of NaCl in original solution, mg	_____	_____	_____
14.	Average mass of NaCl, mg	_____	_____	_____
15.	Total mass of HCl per mL, mg/mL	_____	_____	_____
16.	Total mass of HCl in original solution, mg	_____	_____	_____
17.	Average mass of HCl, mg		_____	

EXPERIMENT 25   DATA SHEET

Part C: Gas Chromatography

Develop your own data sheet (see prelaboratory problem sheet)

1. Suggest a method to separate lycopene, a red pigment found in tomatoes, from tomato juice.

2. Why is it important to allow the level of the eluent to drop to the level of the solid phase inside the column?

POSTLABORATORY PROBLEMS—EXPERIMENT 25, PART B

1. What is deionized water? Describe the use of resins in the softening of water.

2. If a solution of Na_2HPO_4 is passed through a cation exchange column, how many moles of H^+ will be eluted from the column per mole of Na_2HPO_4?

3. Devise a procedure for determining the concentration of each species in a sample containing only NaCl and $MgCl_2$.

1. A mixture of 1.06 mmol of 1-pentanol and 0.34 mmol of 1-hexanol was chromatographed. The ratio of the peak areas was 0.33. Calculate the response factor.

2. Convert 2 μL into L.

3. What would be the order of elution of these liquids—CH_2Cl_2, methanol, and isopropanol—on a silica column?

Thin-Layer Chromatography (TLC) and Paper Chromatography

Part A: Separation of Nitrophenols on TLC Plates
Part B: Microreactions on TLC Plates
Part C: Separation of Inorganic Cations by Paper Chromatography

Microscale Experiments

Objectives
- To use thin-layer and paper chromatography to separate and identify compounds

Prior Reading
- Experiment 25

Related Experiments
- Experiments 25 (Column, ion exchange, and gas chromatography) and 33 (Amino acid mapping)

The basic theory of chromatography was presented in Experiment 25. The present experiment illustrates the utility of **thin-layer chromatography** (TLC) and **paper chromatography**. In these forms of chromatography a differential partitioning of the components in a mixture occurs by their distribution between a stationary phase (the thin layer plate or the chromatographic paper) and a mobile phase (the developing solvent).

Thin-layer Chromatography

In TLC, the stationary phase is usually a thin layer ($\sim 250\,\mu\text{m}$) of silica or alumina spread on a solid support of glass or plastic. Both types of support are now commercially available, although plates may be prepared in the laboratory if desired. It is also possible to do ion exchange TLC.

TLC is an inexpensive, fast, easy, and powerful microscale method for determining the purity of a given inorganic or organic substance. It also offers a rapid method for determining elution solvents for use in column chromatography. One potential source of trouble in inorganic analyses is that commercial silica gel usually contains some metallic impurities

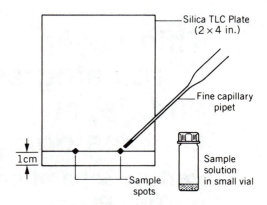

Figure E26.1 Sample application to a TLC plate

(Na^+, Mg^{2+}, Ca^{2+}, Fe^{3+}), which may interfere. Especially washed forms of silica gel are available for inorganic work.

The real advantage of TLC is that a very small amount of material is required for analysis. The disadvantage is that it is not readily amenable to large-scale separations. Another disadvantage is that the absorbent layer of silica or alumina is very thin; evaporation of the eluting solvent can occur readily.

TLC plates are purchased as plastic-backed sheets, which allows one to cut the original sheet into very economical 2×4 inch strips, or for very small plates, into 1×0.5 inch strips. Also commercially available are glass-backed plates, which can be scored with a glass cutter and carefully snapped to obtain the desired size.

One centimeter above the bottom of the thin-layer plate, a pencil line (called *sample line*) is drawn parallel to the short side of the plate. Never use a pen or ink to mark the sample line. One or two or more points, evenly spaced, are then marked on the sample line (see Figure E26.1). The sample (~ 1 mg) to be analyzed is placed in a small vial and dissolved in several drops of solvent. Next, several fine micro capillary pipets for spotting the plate are made. The micro capillary pipet is constructed by heating the center of an open-ended melting point capillary tube and pulling it quickly to form a long thin capillary. By breaking this long capillary, two fine capillary pipets may be obtained. The pipet is dipped into the sample solution, and the fine end is pressed lightly on the line (at the pencil dot) to deliver a small fraction of the solution from the pipet to the plate.

CAUTION: Since oils and moisture from the skin will interfere with the chromatographic separation, be careful not to touch the thin-layer plate with your fingers and be sure to place the plate on a clean surface when spotting the samples.

The chromatogram is developed by placing the spotted thin-layer plate (using forceps) in a developing chamber. A screw-capped widemouth jar or a beaker with aluminum foil or plastic wrap as the cover may be used for this purpose. The chamber contains a small amount of developing solvent and the inside of the chamber is lined with a piece of filter paper, which acts as a wick to saturate the atmosphere in the chamber with solvent vapor. The solvent level in the chamber must be low enough that the sample line spotted with samples will be initially positioned above the solvent level. The container should quickly be covered in order to maintain an atmosphere saturated with the developing solvent.

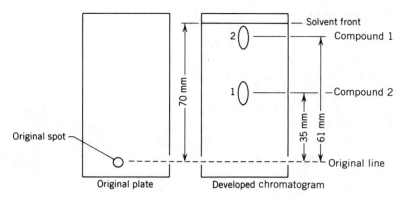

Figure E26.2 Determination of R_f values

The developing solvent will rapidly ascend the plate by capillary action, and the spotted material will elute vertically up the plate. As the solvent front nears the top of the plate, the plate should be removed from the chamber and the solvent front marked quickly with a pencil before the solvent evaporates.

If the spots themselves are colored, they can be seen immediately on the plate. They should be lightly circled with a pencil and the R_f values of the components should be determined (see following). If the compounds are colorless (and therefore do not show visually on the plate), an indirect method must be used. TLC plates containing a fluorescent indicator are now commercially available. The spots on the developed plate, when irradiated with an ultraviolet lamp, fluoresce and can thus be seen. Each spot should be outlined lightly with a pencil so that a permanent record is made of the chromatogram.

An alternative method to detect colorless substances is to place the developed plate in an iodine vapor chamber for a few seconds. Dark spots will develop in those areas containing sample material. On removal from the iodine chamber, the spots should immediately be marked with pencil. Certain colorless compounds can be detected by spraying them with special solutions containing a reagent that reacts with them to form a colored compound. This technique is very selective.

The elution characteristics of a particular species are reported as a retention factor (R_f) value. This value is a measure of the distance traveled by a substance up the plate during the development of the chromatogram, relative to the solvent front. A sample is shown in Figure E26.2. In this example, the R_f value for compound 1 would be $35/70 = 0.50$, and for compound 2 would be $61/70 = 0.87$.

In the first two parts of this experiment, TLC will be used (*a*) to analyze an unknown mixture of commercially available nitrophenols whose structures are shown here and (*b*) to study micro reactions.

2-Nitrophenol   4-Nitrophenol   2,4-Dinitrophenol

Paper Chromatography

In Part C of this experiment, paper chromatography is used to separate the cations (Zn^{2+}, Co^{2+}, Cu^{2+}, Cd^{2+}, Fe^{3+}, and Ni^{2+}). Filter paper (Whatman #1) serves as the stationary phase, and the mobile phase consists of a mixture of acetone and 6 M hydrochloric acid (90/10 by volume). Using a pencil (not a pen) draw a sample line 1 cm from the bottom of the filter paper. A single spot of the mixture to be analyzed is applied about 1 cm from the end of a strip of filter paper. A spot of solution containing each known cation that may be in the unknown mixture will also be individually placed on the sample line. The treated strip is then placed in a covered jar or beaker (which acts as a developing chamber) containing a shallow layer of the solvent mixture (see Figure E26.5). Since filter paper is very permeable to the solvent, the solvent begins to rise up the strip by capillary action. The various spots on the developed chromatogram will be highlighted by treatment with several chemical reagents to enhance their color. The reagents to be used are ammonia, 3 M HCl, ammonium sulfide, and dimethylglyoxime (DMG).

General References

1. Fono, A.; Sapse, A. M.; Ma, T. S. *Microchim. Acta* **1965**, 1098.
2. Harris, D. C. *Quantitative Chemical Analysis*, 2nd ed.; Freeman: New York, 1987.
3. Marcus, B. J.; Fono, A.; Ma, T. S. *Microchim. Acta* **1966**, 960.
4. Mayo, D. W.; Pike, R. M.; Trumper, P. K. *Microscale Organic Laboratory*, 3rd ed.; Wiley & Sons: New York, 1994.

Experimental Section

Procedure

Part A: Separation of Nitrophenols on TLC Plates

Microscale experiment   Estimated time to complete the experiment: 1 h
Experimental Steps: The procedure involves three steps: (1) making micro capillary pipets for spotting, (2) preparing TLC plates, and (3) developing the chromatogram.

Disposal procedure: Return all compounds, solutions, and solvents to your instructor for proper disposal.

The procedure is carried out using Eastman Kodak silica gel–polyethylene terephthalate plates (#13179). The plates are activated by placing them in a 100°C oven for 30 min. They are then placed in a desiccator for cooling and stored until use. Each plate should measure approximately 2.0 × 4.0 inches.

Using a pencil, draw a light line 1 cm from the bottom along the narrow edge of the plate (see Figure E26.1). Obtain four microcapillary pipets (see introduction). Use a separate pipet for each sample solution to be applied to the plate.

Select a sample of each of the four solutions (three known and one unknown) and in turn, insert a micropipet into the solution (it will fill by capillary action) and carefully apply a single small spot of the solution at the center of the pencil mark designated for the particular material (Figure E26.3). The spot should not be more than 1 mm in diameter, or else the separation will not work well.

Allow the spots to dry. Once dried, make a second application of each of the solutions directly on top of the first application. Be sure to use the same micropipet for each sample. Allow the second application to dry completely. A third application may be made if desired. **Do not overload** the spot.

Obtain a clean screw-capped jar or a beaker (aluminum foil cover) to be used as the developing chamber. Add the developing solvent (methylene chloride) to the beaker to a depth of approximately 0.5 cm. Make sure that the depth of solvent will not cover the sample spots on the TLC plate when it is immersed in the chamber.

Place the filter paper in the chamber and replace the cover. Swirl the chamber gently, and then allow it to stand for several minutes so that the air in the beaker becomes saturated with solvent vapor.

Remove the cover from the chamber and using forceps, gently lower the plate (spot end down) into the solvent. Immediately replace the cover (see Figure E26.4). Do not allow the solvent to cover the sample line.

Without disturbing the chamber, allow the system to develop until the solvent front is approximately 1.0 cm from the top of the plate. Remove the plate from the chamber, **immediately** lay the plate flat on a clean surface, and mark (pencil) the exact location of the solvent front across the width of the plate. Allow the chromatogram to dry completely (**HOOD**), using a hair dryer if it is available. Outline (pencil) the spots that are visible for the known samples as well as for your unknown mixture. Record the color of these spots.

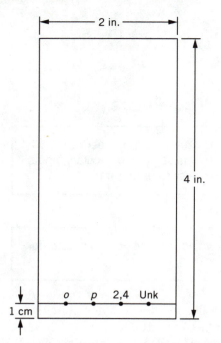

Figure E26.3 Arrangement for TLC analysis of the nitrophenols

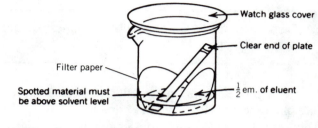

Figure E26.4 Development of a TLC plate

Analysis of the Chromatogram

If all the spots are not visible to the eye, place the plate either under a UV lamp or in a closed jar containing a few crystals of iodine.

> NOTE: Never look directly into an ultraviolet light. It can cause eye damage.

Measure the vertical distance that the approximate center of each of the spots has traveled from the original sample line, and calculate the R_f value for each known spot and for each spot in the unknown mixture. By comparing the spot colors and using the respective calculated R_f values, identify which species are in your unknown sample. Submit the chromatogram with your laboratory report.

Part B: Micro Reactions on TLC Plates

The dehydration of alcohols to yield alkenes is an important industrial reaction.

One of the common methods of carrying out this reaction is by acid catalysis. In this reaction, cyclohexanol is heated with sulfuric acid at 150°C to give cyclohexene.

Place one drop of cyclohexanol and one drop of concentrated (18 M) H_2SO_4 in a 10 × 75 mm test tube. Heat this mixture in a boiling water bath for 5 to 6 minutes. Cool the solution to room temperature, add 1 mL of diethyl ether, and loosely stopper the tube.

Spot a thin-layer plate with this mixture using a micro capillary pipet. Make a second spot with a known solution of cyclohexene (one drop) in ether (1 mL). After the ether has evaporated, develop the thin-layer plate using a solution of hexane:ethyl acetate (9:1).

Remove the developed plate, spray it in the **HOOD** with a 50 percent (9 M) H_2SO_4 solution and heat the plate for a few minutes in an oven at 150°C. Cyclohexene appears as a black spot near the solvent front. Determine the R_f value for the alkene.

Part C: Separation of Inorganic Cations by Paper Chromatography

Microscale experiment   Estimated time to complete the experiment: 1.5 h
Experimental Steps: Two steps are involved: (1) preparing and spotting the filter paper and (2) identifying the cations using specific reagents.

Disposal procedure: Return all solvents and solutions to your instructor for proper disposal.

Obtain a 10×20 cm piece of filter paper to serve as the solid phase and a 600 mL beaker to serve as a developing chamber. Using a pencil, draw a light line 1 cm from the long edge of the paper and mark the line for the Zn^{2+}, Cd^{2+}, Co^{2+}, Cu^{2+}, Fe^{3+}, and Ni^{2+} ion solutions and for the unknown as shown in Figures E26.1 and E26.5.

Obtain six micro capillary pipets (see the introduction). Using a separate micro capillary pipet for each sample solution, carefully apply a single small spot of the solution at the center of the pencil mark designated for the particular metal ion. The spot should not be more than 2 to 3 mm in diameter or else the separation will not work well.

Add the development solvent (9:1 volume ratio acetone/ 6M HCl) to the beaker to a depth of approximately $\frac{1}{4}$ inch. Make sure that the solvent level is below the sample line when the filter paper is immersed in the beaker. Cover the beaker with a watch glass (or plastic or aluminum wrap), and allow the beaker to stand for several minutes so that the air in the beaker becomes saturated with solvent vapor.

Bend the filter paper into a cylinder, without overlapping the edges. Using two flat plastic paper clips (one for the top and the other for the bottom) join the ends of the filter paper. Staple the top end as shown in Figure E26.5(b). Alternatively, fan-fold the filter paper so that each spot has its own "track" to travel up the paper. Momentarily remove the cover from the beaker and gently lower the paper (spot end down) into the solvent. Immediately replace the cover on the beaker.

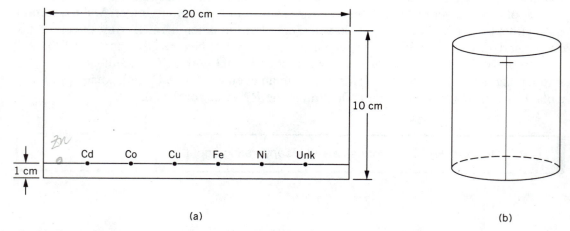

(a)   (b)

Figure E26.5   Arrangement for paper chromatographic analysis of metal ions

Without disturbing the beaker, allow the system to develop until the solvent front is approximately 1 cm from the top of the paper. By following the same procedure as described in part A, mark the solvent front with a pencil, dry the paper, and mark the various spots with a pencil. Record the color of the spots. The filter paper must be dry.

Since all the spots may not be visible at this stage, various chemical treatments will be applied to the chromatogram to make them more pronounced.

NOTE: In each of the following operations, handle the chromatogram with tongs and do each operation in the HOOD.

Hold the chromatogram over a small dish of 6 M ammonia solution for 2 to 3 minutes. The ammonia will neutralize any residual HCl remaining on the paper.

Many metal ions form stable coordination compounds with ammonia and thus new spots may appear or the original spots may change color. Mark any new spots with pencil. Record the color of all the spots, including any that change color. Now, hold the chromatogram over a petri dish of ammonium sulfide solution. As before, record any changes in the color of the spots and with pencil, outline any new spots that may appear. Sometimes, holding the paper over the NH_3 dish once again helps in detecting the spots.

Finally, hold the chromatogram so that a light spray (fine mist is best) of 1 percent dimethylglyoxime solution can be applied using an aspirator bottle. *Do not completely wet the paper*. Dimethylglyoxime forms brightly colored complexes with certain transition metal ions. If the colors do not develop, hold the paper over the NH_3 dish. As before, record any changes in color of the spots and outline any new spots that may appear after this treatment.

Dry the chromatogram completely. Measure the vertical distance that the approximate center of each of the spots has traveled from the original sample line and calculate the R_f value for each known spot and for each spot in the unknown mixture. By comparing the various colors recorded and using the respective calculated R_f values, identify which metal ions are in your unknown sample. Submit the chromatogram with your laboratory report.

Additional Independent Projects

1. Reverse phase thin-layer chromatography: Separation of plant pigments[1]
2. Distribution of salicylic acid, $C_6H_4(OH)(COOH)$ between water and 1-pentanol[2]
3. TLC of amino acids[3,4]
4. Identification of unknowns by MP and TLC[5]

   Brief Outline

   Several (two or three) organic compounds having similar melting points are grouped together. The unknown is selected from one of these groups. The lists of these compounds in groups (groups I to IV) are provided to the students. Determine the MP of the unknown given. Match the unknown with the group of compounds having similar MPs. Run the thin-layer chromatograms for the unknown and the knowns in the group. By comparing the R_f values, identify the closest match to the unknown.

   Obtain the MP of the mixture of the unknown and the known compound (see Experiment 2). If the known and unknown are the same, the mixture will have a sharp melting point. Moreover, a thin-layer chromatogram of the mixture (if the known and the unknown are the same) will produce a single spot.

   Develop your own data sheets. Write a report according to the instructions given in Chapter 4.

References

1. Cooley, J. H.; Wong, A. L. *J. Chem. Educ.* **1986**, *63*, 353.
2. Jones, M. M.; Champion, G. R. *J. Chem. Educ.* **1978**, *55*, 119.
3. Helser, T. L. *J. Chem. Educ.* **1990**, *67*, 964.
4. Hurst, M. O.; Cobb, D. K. *J. Chem. Educ.* **1990**, *67*, 978.
5. Levine, S. G. *J. Chem Educ.* **1990**, *67*, 972.

Name ______________________________ Section ______________

Instructor ______________________________ Date ______________

PRELABORATORY REPORT SHEET—EXPERIMENT 26 (FOR ALL PARTS)

Experiment title ____

Objective

Reactions/formulas to be used

Materials and equipment table

Outline of procedure

1. What would be the result of applying too much sample to the TLC plate for analysis?

2. If the solvent front moves 3.5 inches and a component in a sample being analyzed moves 1.8 inches from the baseline, what is the R_f?

3. Would you expect that changing the solvent would change the R_f value obtained for a particular unknown? Why?

4. A TLC plate is spotted with a solution containing a mixture of two components, X and Y. The chemical affinity of X for the stationary phase is less than that of Y, and the chemical affinity of X for the mobile phase is greater than that for Y. Which substance will have the larger R_f value upon analysis of the developed chromatogram? Explain your answer.

EXPERIMENT 26   DATA SHEET

Part A: Separation of Nitrophenols on TLC Plates

Sample ID number _____

Color of unknown solution _____

Average distance traveled by the solvent front _____

Species	Distance traveled by spot, cm	R_f value	Color of spot
2-Nitrophenol	_____	_____	_____
4-Nitrophenol	_____	_____	_____
2,4-Dinitrophenol	_____	_____	_____
Unknown	_____	_____	_____
	_____	_____	_____
	_____	_____	_____

Identification of the nitrophenol(s) in the unknown solution

_____

Part B: Micro Reactions on TLC Plates

Make your own data collection and data manipulation table as shown in Part A.

EXPERIMENT 26   DATA SHEET

Part C: Separation of Inorganic Cations

Unknown mixture ID number ————————————————

Average distance traveled by the solvent front ————————————————

Species	Distance traveled by spot, cm	R_f value
Zn^{2+}	————————————	————
Co^{2+}	————————————	————
Cu^{2+}	————————————	————
Fe^{3+}	————————————	————
Ni^{2+}	————————————	————
Cd^{2+}	————————————	————
Unknown	————————————	————
	————————————	————
	————————————	————
	————————————	————
	————————————	————

Effect of the various chemical reagents on the spot colors

Species	Original Color	After ammonia treatment	After sulfide treatment	After DMG treatment
Zn^{2+}	_____	_____	_____	_____
Co^{2+}	_____	_____	_____	_____
Cu^{2+}	_____	_____	_____	_____
Fe^{3+}	_____	_____	_____	_____
Ni^{2+}	_____	_____	_____	_____
Cd^{2+}	_____	_____	_____	_____
Unknown	_____	_____	_____	_____
	_____	_____	_____	_____
	_____	_____	_____	_____
	_____	_____	_____	_____

Identification of the metal ion(s) in the unknown solution

_____

POSTLABORATORY PROBLEMS—EXPERIMENT 26 (FOR ALL PARTS)

1. Why is it important to keep the spots applied to the TLC plate as small as possible?

2. Two compounds have the same R_f value (0.72) under identical conditions. Does this result show that they have identical structures? Explain your answer.

27 **Introduction to Nuclear Magnetic Resonance Spectroscopy**

Part A: Determination of Chemical Shifts and Electronegativities of Halogens in Alkyl Halides
Part B: NMR Determination of the Concentration of *n*-Butyllithium

Microscale Experiments

Objectives
- To learn the basic applications of NMR in chemistry.
- To learn to use NMR spectra to interpret properties of compounds.

Prior Reading
- Electronegativity and electronegativity scales. Read a general chemistry text.
- Section 3.5   Volume measurement using an automatic delivery pipet

Nuclear magnetic resonance (NMR) spectroscopy is a technique in which transitions of the nuclei in a sample are observed in the presence of a magnetic field. Any nucleus that does not have an even number of protons and an even number of neutrons possesses a property called **spin angular momentum** and is said to be **NMR active**. The nuclei spin leads to the generation of a magnetic field. If a sample containing these nuclei is placed in an external magnetic field (such as from a large magnet), the nuclei will interact with the external field.

In the most common example, a nucleus (^1H and ^{13}C, for example) has a spin quantum number of $\frac{1}{2}$ (called **spin**). The nuclei can either align with the external magnetic field, generating a lower energy level (called α), or align against the external magnetic field, generating a higher energy level (called β). If the nucleus absorbs energy equivalent to the difference in energy between these two levels, the nucleus will undergo a spin transition from state α to state β. This transition gives rise to the NMR signal and occurs at a specific frequency, which depends on the type of nucleus and the magnetic environment around it. Hydrogen nuclei, ^1H, for example, absorb energy at 100 MHz (in the middle of the FM radio band) in an external magnetic field strength of 23,500 gauss.

Consider an NMR spectrum of ethyl bromide (CH_3CH_2Br). A small amount of the compound is placed in an NMR tube (usually a 5 mm glass tube about the size of a pencil), a small amount of a reference compound is added [the standard reference compound for ^1H-NMR is tetramethyl-silane, $(CH_3)_4Si$, abbreviated TMS], and the tube is sealed. The tube is then placed inside an NMR spectrometer, and the spectrum shown in Figure E27.1 is obtained.

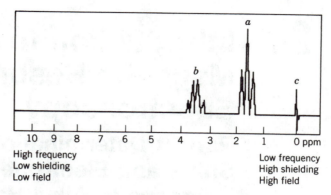

Figure E27.1 ^{1}H-NMR spectrum of CH_3CH_2Br

As one can see from the spectrum, ethyl bromide and TMS give rise to three signals, labeled a, b, and c. The signal at 0 ppm (signal at c) is the reference signal from TMS. All signals are measured relative to this one, in units of ppm, or parts per million. [Since the general frequency for ^{1}H is 100 MHz, 1 ppm = $(100 \text{ MHz})/(1 \text{ million})$ = 100 Hz.] The relative distance is called the **chemical shift** (δ). We can see that signal a is larger than signal b. Signal a is split into three lines, whereas signal b is split into four lines. Signal b is at a higher chemical shift (3.5 ppm from TMS) than is signal a (1.7 ppm). Why does a single compound, ethyl bromide, give rise to two signals, a and b? Why are the signals different, in the ways just described?

Any specific hydrogen in a compound, say the underlined H's in $CH_3\underline{CH_2}Br$, will be shifted slightly from the general frequency of 100 MHz. The reason for this is that the local magnetic field varies somewhat, depending on what other atoms are nearby. The presence of an electronegative bromine atom nearby, for example, causes a small shift to higher frequency, as the bromine withdraws electrons from the hydrogens, leaving them relatively unprotected from the magnetic field.

A weaker magnetic field (higher frequency) is therefore sufficient to excite them, and they appear at the highest chemical shift. The other hydrogens are better protected (being farther from the electronegative bromine), and appear at lower frequency.

The other differences in the NMR signals are attributable to equivalency. The three hydrogens in the $-CH_3$ group of ethyl bromide are magnetically identical—each is in the same environment, owing to free rotation about the C–C bond. As a result, each gives rise to an identical signal. Together, the signal for the three hydrogens is three times the size of the individual hydrogen signal. Signal a in the spectrum is larger than signal b, since it is caused by three hydrogens, whereas signal b is caused by only two.

The spin state of one nucleus interacts with the spin state of nuclei of other nearby atoms. Such spin-spin interaction, called **spin-spin coupling** between nuclei, is transmitted via the bonding electrons in a molecule. Spin-spin coupling gives rise to the **multiplicity, M,** of the NMR peak of a compound. The multiplicity of a signal is determined by the number of identical hydrogens, n, on the adjacent carbon, following the formula

$$M = n + 1$$

Since there are two hydrogens on the carbon adjacent to the $-CH_3$ group, the $-CH_3$ signal a will be split into $M = 2 + 1 = 3$ lines (**triplet signal**). Similarly, since there are three hydrogens on the carbon adjacent to the $-CH_2-$ group, the $-CH_2-$ signal b will be split into $M = 3 + 1 = 4$ lines (**quartet signal**).

The spacing between two adjacent peaks of the quartet or the triplet is called the **coupling constant, J**. This defines the magnitude of the coupling between the hydrogens of the $-CH_2-$ and $-CH_3-$ groups. The values of the coupling constant in a given multiplet are identical. In practice the multiplicity of a given group's signal is more complex than this, but this treatment is sufficient for simple compounds.

Part A: Determination of Chemical Shifts and Electronegativities of Halogens in Alkyl Halides

Since the different chemical shifts in different molecules in NMR spectra represent the different electron environments for the hydrogens, there is a direct relationship between the chemical shift and electronegativity. **Electronegativity** is a measure of the tendency of an atom in a molecule to attract bonding electrons to itself.

For a series of organic halides, $CH_3CH_2CH_2X$ (X = Cl, Br, I), an increase in the electronegativity of the halogens will result in a change to higher chemical shift values. In this experiment, the relation between the chemical shift and the electronegativity of halides in 1-halopropanes[1] will be studied.

It is not possible to determine absolute values for the electronegativities of atoms; however, by assigning an arbitrary value to one atom, a scale of relative electronegativity may be established. The most common scale for electronegativity is one proposed by L. Pauling. This scale is calculated on the basis of bond energies. Another method suggested by R. S. Mulliken is a direct method that defines electronegativity (EN) of an atom as the average of its ionization energy (IE) and electron affinity (EA).

$$EN = (IE + EA)/2$$

This calculation is adjusted so that fluorine, the most electronegative element, has a value of 4.0. By using NMR, it is possible to assign electronegativities to more complex groups as well. In Part A of this experiment, the electronegativities of the $-NH_2$ and $-NO_2$ groups will be determined.

Also in Part A of this experiment, NMR spectra of several organic halides (1-halopropane, C_3H_7X) will be used to determine the relative elecetronegativity of a group in an organic compound. The coupling constant J of the $-CH_2-$ multiplet (quartet) will also be determined.

Part B: NMR Determination of the Concentration of *n*-Butyllithium in a Solution

Part B of this experiment illustrates the use of NMR in quantitative determinations, specifically in determining the concentration of a commercial solution of *n*-butyllithium ($LiCH_2CH_2CH_2CH_3$, BuLi) in hexane. Solutions of BuLi are extensively used as reagents in synthetic chemistry. A solution of BuLi slowly decomposes on standing, even when it is kept in a refrigerator. It is therefore important to determine how much BuLi is actually present in solution before it is used in any reaction.

The concentration of this compound can be determined by various methods including titrimetry[2-4]. Titrimetric methods are time-consuming and need special treatment. The NMR[5,6] method is nondestructive and rapid.

The –CH$_2$– group closest to the lithium (called the terminal –CH$_2$– group) shows a characteristic chemical shift at –0.88 ppm (that is, to the right of TMS, $\delta = 0$ ppm) in hexane. A known quantity of benzene (C$_6$H$_6$) will be added to the solution to be used as an **internal standard**. An internal standard is a substance that can be used to find the ratio between a property of the standard and that of the unknown. Benzene exhibits a strong signal at around 7.40 ppm, which does not overlap with the signals shown by *n*-butyllithium. **(Note: Though benzene is toxic, its use in minute amounts in a sealed NMR tube in a well-ventilated HOOD is not considered so dangerous)**. By comparing the relative sizes (height or intensity) of the signals of benzene and of the terminal –CH$_2$– group of BuLi, the concentration of BuLi in the solution can be calculated.

Once the spectrum has been obtained, measuring the height of the signals of benzene (h_b) and the BuLi (h_l) protons allows the calculation of the concentration of the solution in the following way:

1. The height of the benzene signal (h_b) is proportional to the number of moles of hydrogen producing it. That is, the height depends on how many hydrogens there are in benzene (6 moles H), and how much benzene there is. Thus,

$$h_b \propto \left(\frac{6 \text{ moles H}}{1 \text{ mole benzene}} \right)(\# \text{ moles of benzene})$$

Since we know how much benzene was added to the sample, the number of moles of benzene is known.

2. The height of the signal due to the terminal –CH$_2$– group of BuLi ($\sim$ –0.88 ppm) is also proportional to the number of moles of hydrogen producing it. This proportionality depends on how many hydrogens there are in the signal we are looking at (2 moles of H), and how much BuLi there is. This, in turn, depends on the volume of BuLi and its concentration. That is,

$$h_l \propto \left(\frac{2 \text{ moles H}}{1 \text{ mole butyllithium}} \right)(V)(M)$$

where V is the volume of BuLi used and M is its actual molarity.

Suppose, for example, that the benzene signal was 25 mm high, and corresponded to 3.5×10^{-3} mol of hydrogen. If the BuLi signal was 5 mm high, it would correspond to $\frac{1}{5}$ that quantity of hydrogen, or 0.7×10^{-3} mol. Since the volume of the solution taken is known, one can solve for the molarity.

Example E27.1 A solution containing 1.00 mL of commercial BuLi and 25 μL of benzene (density = 0.9855 g/mL) produces an NMR spectrum with a signal height for benzene (signal at 7.4 ppm) and BuLi (signal at –0.88 ppm) equal to 2.5 cm and 4.2 cm, respectively. Calculate the concentration of BuLi.

Answer

$$\text{Benzene mass} = (25\ \mu\text{L})(1\ \text{mL}/1000\ \mu\text{L})(0.9855\ \text{g/mL}) = 0.0246\ \text{g}$$

$$\text{Moles benzene} = (0.0246\ \text{g})(1\ \text{mol}/78.1\ \text{g}) = 3.15 \times 10^{-4}\ \text{mol}$$

$$\text{Moles H in benzene signal} = (6\ \text{mol H}/\text{mol benzene})(3.15 \times 10^{-4}\ \text{mol benzene})$$
$$= 1.89 \times 10^{-3}\ \text{mol H}$$

Since the BuLi signal is larger by a factor of the ratio of heights of the signals $(4.2/2.5)$, we may write

$$\text{Moles H in BuLi signal} = 1.89 \times 10^{-3}\ \text{mol H}\ (4.2/2.5) = 3.18 \times 10^{-3}\ \text{mol H}$$
$$3.18 \times 10^{-3}\ \text{mol H} = (2\ \text{mol H}/\text{mol BuLi})(0.001\ \text{L BuLi})(\text{M of BuLi})$$
$$\text{M of BuLi} = 1.59$$

From this problem, the overall equation for the calculation of the concentration (M) of BuLi can be written as

$$\text{Molarity } (M) \text{ of BuLi} = \frac{(m_b)(H_b)(h_l)}{(MW_b)(H_l)(h_b)(V_l)}$$

where m_b = mass of benzene
H_b = # mols of H in benzene
h_l = height of the methylene ($-CH_2-$) signal in BuLi
MW_b = molar mass of benzene
H_l = # mols of H in $-CH_2-$ group
h_b = height of the signal due to benzene
V_l = volume in liters of BuLi used

This equation is known as Kasler's equation.[5]

References
1. Greever, J. C. *J. Chem. Educ.* **1978**, *55*, 538.
2. Gilman, H.; Haubein, A. H. *J. Amer. Chem. Soc.* **1944**, *66*, 1515.
3. Gilman, H.; Cartledge, F. K.; Sim, S. Y. *J. Organometal. Chem.* **1963**, *1*, 8.
4. Gilman, H.; Cartledge, F. K.; Sim, S. Y. *J. Organometal. Chem.* **1964**, *2*, 447.
5. Silveira, A.; Bretherick, H. D.; Negishi, E. *J. Chem. Educ.* **1979**, *56*, 560.
6. Irwin, J. R.; Reid, P. J. *J. Organometal. Chem.* **1968**, *15*, 1.

General References For Further Reading
1. Kasler, F. *Quantitative Analysis by NMR Spectroscopy,* Academic Press: New York, 1973.
2. Mayo, D. W.; Pike, R. M.; Trumper, P. K. *Microscale Organic Laboratory,* 3rd ed., Chapter 9; Wiley: New York, 1994.
3. Sawyer, D. T.; Heineman, W. R.; Beebe, J. M. *Chemistry Experiments for Instrumental Methods,* Wiley: New York, 1984.
4. Silverstein, R. M.; Bassler, G. C.; Morrill, T. C. *Spectrometric Identification of Organic Compounds*, 4th ed.; Wiley: New York, 1981, Chapter 4.
5. Wells, P. R. *Prog. Phys. Org. Chem.* **1968**, *6*, 111; eds. Streitwieser, A., Jr.; Taft, R. W.; Wiley: New York.

Experimental Section

Procedure

Part A: Determination of Electronegativity Using NMR[1a]

> Microscale experiment   Estimated time to complete the experiment: 1 h
> Experimental Steps: This is a one step procedure.

> Return the NMR tubes to your instructor for future use.

Obtain five previously cleaned NMR tubes and rinse them thoroughly with deionized water, followed by acetone. Dry them in an oven at 110°C for at least 30 minutes. Obtain an automatic delivery pipet (100 to 1000 μL) along with a disposable pipet tip, and transfer about 500 μL of 1-chloropropane into one of the NMR tubes. Add two drops of TMS, and replace the cap on the NMR tube. Shake the tube to mix the two, and obtain the NMR spectrum of the sample, as shown by your instructor. Repeat the process using 1-bromo-, 1-iodo-, and either 1-nitropropane or 1-aminopropane.

> CAUTION: All compounds used are flammable and irritants.

Determine the chemical shift (δ) values for the **center** of the triplet signal for a $-CH_2-$ hydrogen closest to the halogen (or $-NH_2$ or $-NO_2$ group) with respect to TMS ($\delta = 0$ ppm). This signal will be a triplet appearing farthest to the left (lowest field, $\delta \sim 3.0$ ppm) in the spectrum. Tabulate the δ values, and record the J (**coupling constant**) values for this triplet for each of the compounds. Run two trials.

Plot a graph of chemical shifts for the 1-halopropanes on the y axis versus the Pauling electronegativity (EN) of the halogen (Cl = 3.0, Br = 2.8, and I = 2.5) on the x axis. Fit the best straight line between these points. Determine the electronegativity value for the $-NH_2$ or $-NO_2$ group in 1-amino- or 1-nitropropane. The known EN values for various groups are given in references.[1a] These NMR samples may be recycled for other students. Store them in a refrigerator for future use.

Part B: Quantitative Analysis Using NMR[1b]

Microscale experiment   Estimated time to complete the experiment: 1 h
Experimental Steps: This is a two-step procedure: (1) preparing the solution of
n-butyllithium in an NMR tube and (2) obtaining the NMR spectrum of the sample.

Disposal procedure: Return your NMR tube with its contents to your instructor for reuse.

Obtain a previously cleaned NMR tube and a 10 mL Erlenmeyer flask, and rinse them thoroughly with deionized water, followed by acetone. Dry them in an oven at 110°C for at least 30 minutes. Also obtain an automatic delivery pipet (10 to 100 μL) and a disposable pipet tip. Fit the Erlenmeyer flask with a septum. Insert a syringe needle (without the syringe) through the septum. This needle will act as a vent for N_2 gas. Insert a second syringe needle connected to a Tygon tubing, that is in turn attached to a N_2 delivery system. Thoroughly flush the flask with N_2 gas for several minutes. Following the same procedure, flush the NMR tube with N_2 gas.

Momentarily remove the septum. Using an automatic delivery pipet, introduce a known volume of dry benzene (25 to 30 μL) to the Erlenmeyer flask, being sure to measure the volume as accurately as possible. Replace the septum.

CAUTION: Benzene is a carcinogen and toxic. Use gloves. Perform this part of the procedure inside a well-ventilated HOOD.

Without removing the septum from the flask, add 1.00 mL of n-butyllithium in hexane to the flask, measuring the volume of n-BuLi solution as accurately as possible. Use a separate syringe to make this transfer.

CAUTION: n-BuLi solution is highly pyrophoric (it catches fire spontaneously in air). Do not inhale the vapors. Perform this part of the work inside a well-ventilated HOOD. Use gloves. Do not remove the n-BuLi solution bottle from inside the HOOD.

Swirl the mixture inside the flask gently. Using a syringe, transfer 0.3–0.5 mL of this mixture into the NMR tube. Store the Erlenmeyer flask with the remaining solution inside the hood. Add one drop of TMS to the NMR tube, seal it, and shake the mixture well. Obtain the NMR spectrum of the solution at least two times. Identify the signals that are due to the benzene (singlet) and the terminal $-CH_2-$ group (triplet). Integrate the spectrum and calculate the height of proton signals. Calculate the molarity of the BuLi solution in the manner shown in the foregoing example. Include the NMR spectrum in your laboratory report.

Additional Independent Projects

1. Alcohol proton exchange experiment.[2]
2. Proton NMR and pK_a values for organic bases.[3]
3. Titration of an amino acid as studied by NMR.[4]

References

1a. Greever, J. C. *J. Chem. Educ.* **1978**, *55*, 538.
1b. Silveira, A.; Bretheric, H. D.; Negishi, E. *J. Chem. Educ.* **1979**, *56*, 560.
2. Pagnotta, M.; Carter, J.; Armsby, C. *J. Chem. Educ.* **1993**, *70*, 162.
3. Sawyer, D. T.; Heineman, W. R.; Beebe, J. M. *Chemistry Experiments for Instrumental Methods,* Wiley: New York, 1984.
4. Waller, F. J.; Hartman, I. S.; Kwong, S. T. *J. Chem. Educ.* **1977**, *54*, 447.

Name ______________________	Section ____________
Instructor ______________________	Date ____________

PRELABORATORY REPORT SHEET—EXPERIMENT 27 (A AND B)

Experiment title ______________________________

Objective

Reactions/formulas to be used

Chemicals and solutions—their preparation

Materials and equipment table

Outline of procedure

1. Why are internal standards used in NMR spectroscopy?

2. Why is TMS a good choice as an internal standard?

3. How many chemically different types of protons are present in an *n*-propyl halide?

Name ______________________	Section ____________
Instructor ______________________	Date ____________

EXPERIMENT 27 DATA SHEET

Part A: Determination of Electronegativity Using NMR

Sample number ______

Collection of Data	**Chemical Shift (δ)**		**Coupling constant* (J)**	
	Trial 1	**Trial 2**	**Trial 1**	**Trial 2**
1. TMS:	______	______	______	______
2. Chloro derivative	______	______	______	______
3. Bromo derivative	______	______	______	______
4. Iodo derivative	______	______	______	______
5. Nitro derivative	______	______	______	______
6. Amino derivative	______	______	______	______

* Record one value only

Manipulation of Data

1. Plot electronegativities for the halides (x axis) versus δ (y axis). Submit the graph along with your report.
2. Obtain these EN's: Amine ($-NH_2$) group ______ Nitro ($-NO_2$) group ______

Part B: Quantitative Analysis using NMR

Sample number ______
Density of benzene (C_6H_6): 0.9855 g/mL

	Trial 1	**Trial 2**	
1. Volume of benzene, mL	______	______	(convert μL to mL)
2. Mass of benzene, g	______	______	
3. Volume of *n*-BuLi, mL	______	______	
4. Molecular weight of C_6H_6, g/mol	78.1	78.1	
5. Number of protons in standard, H_b	6	6	
6. Number of protons in unknown, H_l	2	2	

DATA SHEET, EXPERIMENT 27 B, PAGE 2

7. Integral height (h_b) of NMR signal of C_6H_6, cm or mm

   (i) ______ *(ii)* ______ *(iii)* ______ *(iv)*______ *(v)* ______

   Average ______

8. Integral height (h_l) of NMR signal of $-CH_2-$ of *n*-BuLi

   (i) ______ *(ii)* ______ *(iii)* ______ *(iv)*______ *(v)* ______

   Average ______

Calculate the molarity of the unknown solution (use Kasler's equation). ____________ M

Show calculations:

POSTLABORATORY PROBLEMS—EXPERIMENT 27

1. What is electronegativity?

2. Can hexane be used as an internal standard for this experiment? Explain.

28 **Preparation of Buffers and Potentiometric Titrations:**

Determination of Ionization Constants of Acids

Micro- and Macroscale Experiments

Objectives
- To learn how to prepare buffer solutions for practical use
- To calibrate and use a pH meter
- To determine equivalent weights and ionization constants

Prior Reading
- Section 2.6   Graphing of data
- Section 3.5   Measuring liquid volumes
- Section 3.9   Solution preparation
- Section 3.10   Construction and use of a microburet
- Section 3.11   Titration procedure

Related Experiments
- Experiments 10 (Acid-base titration), 12 (Redox titrations), and 33 (Amino acid titrations)

Activities before the Experiment
- Making buffer solutions, calibrating a pH meter, and preparing solutions

One of the most important measurements in analytical chemistry is the determination of the hydrogen ion concentration, or pH, of an aqueous solution by potentiometric methods. To be able to understand the concepts on which electrochemical measurements are based, let us review some of the general concepts of acid-base equilibria in water and general principles of potentiometric methods.

Equilibria in Aqueous Solutions

Water can be considered as a species that, according to the Brønsted-Lowry concept, behaves as both an acid and a base:

$$H_2O(l) + H_2O(l) \rightleftharpoons H_3O^+(aq) + OH^-(aq)$$
$$\text{acid}_1 \qquad \text{base}_2 \qquad \text{acid}_2 \qquad \text{base}_1$$

The ion H_3O^+ is called the **hydronium ion**. The ionization constant K_w, also called the **ion product constant** or **autoprotolysis constant** of water, is given by the expression

$$K_w = [H_3O^+][OH^-] \text{ or } [H^+][OH^-] = 1.00 \times 10^{-14} \qquad (E28.1)$$

The square brackets, [ ], represent concentrations in units of molarity (mol/L) at equilibrium. When an acid is dissolved in water, it will dissociate or ionize (to some extent) to produce hydronium ions, thereby increasing the $[H_3O^+]$ in aqueous solution. Similarly, a base in water ionizes to increase the $[OH^-]$ in aqueous solution. The degree of ionization of an acid or a base into H_3O^+ and OH^-, respectively, depends upon the strength of the acid or the base. HCl, being a strong acid, dissociates completely in water,

$$HCl(aq) + H_2O(l) \rightarrow H_3O^+(aq) + Cl^-(aq)$$

For any weak acid HA or weak base B, the dissociations are written as shown below. Following the Brønsted-Lowry concept, $acid_1$ and $base_1$ or $acid_2$ and $base_2$ are **conjugate acid-base pairs**.

$$\begin{array}{cccc} HA(aq) + H_2O(l) \rightleftharpoons H_3O^+(aq) + & A^-(aq) \\ acid_1 \quad\quad base_2 \quad\quad acid_2 \quad\quad base_1 \end{array}$$

$$\begin{array}{cccc} B(aq) \;\; + H_2O(l) \rightleftharpoons BH^+(aq) \;\; + OH^-(aq) \\ base_1 \quad\quad acid_2 \quad\quad acid_1 \quad\quad base_2 \end{array}$$

The corresponding **acid and base ionization constants**, K_a and K_b, are

$$K_a = \frac{[H_3O^+][A^-]}{[HA]} \tag{E28.2}$$

$$K_b = \frac{[BH^+][OH^-]}{[B]} \tag{E28.3}$$

The concentration of water is essentially constant at approximately 55.5 moles per liter (that is, in 1.0 L of water there are 1000 g or 55.5 moles of water, assuming that the density of water is 1.0 g/mL), and it is included in the equilibrium constant K. Thus, $[H_2O]$ does not appear explicitly in the equilibrium expression. The term $[H^+]$ is used interchangeably with $[H_3O^+]$ in the discussion that follows.

The numerical values of K_a or K_b (see Appendix 4) represent the relative strengths of acids and bases. In aqueous solution, an acid whose conjugate base is weaker than water will be completely ionized. Such an acid is called a **strong acid**. For strong acids, $K_a \gg 1$. **Weak acids**, which are characterized by $K_a < 1$, are only partially ionized in aqueous solutions. A similar argument may be applied for the relative strengths of bases.

There are not many strong acids available for use in a laboratory. The commonly used strong acids are HCl, HBr, and HI (but not HF) and the oxy acids HNO_3, $HClO_4$, and H_2SO_4 (first proton ionization only).

The pH Scale

For the self-ionization of water, the ion product constant K_w is 1.0×10^{-14} at 25°C. In pure water, $[H^+] = [OH^-]$. Therefore, equation (E28.1) becomes

$$K_w = [H^+][OH^-] = [H^+]^2 = 1.0 \times 10^{-14}$$

Solving, we have

$$[H^+] = [OH^-] = 1.0 \times 10^{-7}$$

If an acid is added to pure water, the concentration of $[OH^-]$ is readily calculated if $[H^+]$ is known. The contribution to $[H^+]$ from the self-ionization of water is negligible as long as the $[H^+]$ from the acid is higher than 10^{-6} M. The concentration of H^+ or OH^- in aqueous solutions may vary over wide ranges, from greater than 1 M to less than 10^{-14} M. The common range for titration is 10^{-1} M to 10^{-13} M. It is difficult to plot a graph of $[H^+]$ against any experimental variable (such as mL of base added). This problem is avoided by expressing hydrogen ion concentration as the negative logarithm of H^+, which is called the **pH**. More precisely, the modern definition of pH is

$$pH = -\log \alpha_{H^+} = -\log[H^+]\gamma_{H^+}$$

where α_{H^+}, called the **activity** of H^+, is equal to $[H^+]\,\gamma_{H^+}$ (γ_{H^+} is known as the **activity coefficient** of H^+). For dilute solutions, γ_{H^+} is equal to 1. Thus,

$$pH = -\log[H^+]$$

Similarly,

$$pOH = -\log[OH^-]$$
$$pK_w = pH + pOH = 14$$

At 25°C, in pure water, $[H^+]$ is 10^{-7}, and the pH is 7 ($-\log 10^{-7} = 7$). Similarly, the pOH is 7. A pH of 7 represents a neutral solution. Solutions having pH < 7 are **acidic,** and those having pH > 7 are **basic**. A knowledge of pH is very important in biochemical processes. Whereas the pH of acidic gastric juice is ≈ 1.0, blood is slightly basic with a pH of about 7.4. The pH ranges from 0 to 14 for most solutions. Negative pH values are possible. They indicate $[H^+] > 1$ M. A sample of 12 M HCl has a pH of about -2.8 because the activity coefficient becomes greater than 1 at high ionic strength. A comparison of pH values of a strong acid and a weak acid having the same molarity are shown in the following example.

Example E28.1   What is the pH of a 0.025 M solution of HCl? Of acetic acid ($K_a = 1.76 \times 10^{-5}$)?

Answer
    Since HCl is a strong acid, it is completely ionized in aqueous solution. Thus, $[H^+] = 0.025$ M, and

$$pH = -\log[H^+] = -\log(0.025) = 1.60$$

Acetic acid, CH_3COOH (HOAc), is a weak acid, and therefore its ionization represents an equilibrium:

$$HOAc + H_2O \rightleftharpoons H_3O^+ + OAc^-$$

Original concentration	0.025	0	0
Change in concentration	$-x$	$+x$	$+x$
Equilibrium concentration	$0.025-x$	$+x$	$+x$

$$K_a = 1.76 \times 10^{-5} = \frac{[H^+][OAc^-]}{[HOAc]} = \frac{(x)(x)}{0.025 - x} \approx \frac{x^2}{0.025}$$

$$x = \sqrt{K_a \times 0.025}$$

Solving for x, we find

$$x = 6.63 \times 10^{-4} \text{ M} = [H^+]$$

Therefore,

$$pH = -\log(6.63 \times 10^{-4}) = 3.82$$

Thus, for the same molarity, the pH of the weak acid, HOAc, is higher than that for the strong acid, HCl, suggesting that the weak acid has a lower $[H^+]$.

Salts of Weak Acids and Bases

In aqueous solution, soluble salts of strong acids and bases (NaCl, $KClO_4$) are completely dissociated into their respective ions, which do not react with water. The pH of such solutions is ~ 7.

However, if a salt is made up of ions at least one of which is derived from a weak acid or base, then its aqueous solution is not neutral. For example, consider a solution of sodium acetate, NaOAc, in water. Na^+, being a cation of a strong base (NaOH), does not react with water. However, OAc^-, being the conjugate base of a weak acid (HOAc), establishes the following equilibrium:

$$OAc^- + H_2O \rightleftharpoons HOAc + OH^-$$

The resulting solution is basic owing to the presence of OH^- ions. Similarly, ammonium salts generate acidic solutions because of the following equilibrium:

$$NH_4^+ + H_2O \rightleftharpoons NH_3 + H_3O^+$$

Buffer Solutions

A solution that contains both substantial amounts of a weak acid (or base) and its conjugate salt will have a pH that does not change much when it is diluted or when an acid or a base is added to the solution. Such a solution is called a **buffer solution**. Solutions of acetic acid and sodium acetate or of aqueous ammonia and ammonium chloride are buffers. An acetic acid–acetate buffer contains undissociated acid, HOAc, and the acetate anion, OAc^-, from the salt. The following two equilibria are established:

$$HOAc \rightleftharpoons H^+ + OAc^-$$
$$OAc^- + H_2O \rightleftharpoons HOAc + OH^-$$

If a strong acid (such as HCl) is added to the buffer, the H^+ from the HCl will react with OAc^- (from the salt) to form more HOAc. The net effect is that a strong acid (HCl) is converted to a weak one (HOAc). This reaction minimizes changes in pH of the solution. When a strong base (such as NaOH) is added to the buffer, the OH^- from the added base reacts with HOAc to form water and OAc^-. The net effect is that a strong base (NaOH) is converted to a weak one (OAc^-). This reaction also minimizes changes in pH of the solution.

The importance of buffer solutions in all areas of chemistry and biology cannot be overemphasized. The central equation dealing with pH of solutions is the Henderson-Hasselbalch equation. For the equilibrium

$$HA + H_2O \rightleftharpoons H_3O^+ + A^-$$

we may write

$$pH = pK_a + \log \frac{[A^-]}{[HA]} \tag{E28.4}$$

Thus, if the K_a and the concentrations of HA and its conjugate base A^- are known, the pH of the buffer can be easily calculated. Similarly, for a buffer prepared from a weak base and its conjugate salt,

$$pOH = pK_b + \log \frac{[BH^+]}{[B]} \tag{E28.5}$$

The **buffer capacity b** is defined as the ability of the buffer solution to resist changes in pH. It reaches a maximum when $[A^-] = [HA]$ or, equivalently, when $pH = pK_a$. Therefore, in choosing a buffer for any experiment, one should select an acid whose pK_a is close to the desired pH. The useful range of a buffer is $pK_a \pm 1$ pH unit. Table E28.1 contains pK_a values for several useful buffers. Table E28.2 contains the pH values for buffers commonly used (NIST standards: National Institute of Standards and Technology). Note that the pH of a buffer is temperature-dependent.

General Principles of Potentiometric Methods

Potentiometry involves the measurement of a potential or voltage difference between two electrodes immersed in a solution. The electrodes and the solution constitute an electro-chemical cell. In many cases, a cell can be devised in such a way that its potential depends upon the activity of a specific ion in solution. The cell must contain an electrode whose

Table E28.1 Some common buffers and their pK_a values

Buffer name	pK_a
Phosphoric acid/potassium dihydrogen phosphate	2.15
Citric acid/sodium dihydrogen citrate	3.13
Sodium dihydrogen citrate/disodium hydrogen citrate	4.76
Acetic acid/sodium acetate	4.76
Tri(hydroxymethyl)aminomethane hydrochloride/ tri(hydroxymethyl)aminomethane	8.08
Boric acid/sodium borate	9.23
Ammonium chloride/ammonia	9.25

Table E28.2 pH of some NIST standards at different temperatures

Buffer name	pH at °C			
	20	**25**	**30**	**35**
Potassium hydrogen citrate (0.05 m)*	3.788	3.776	3.766	3.759
Potassium hydrogen phthalate (0.05 m)	4.002	4.008	4.015	4.024
Potassium dihydrogen phosphate (0.025 m) and disodium hydrogen phosphate (0.025 m)	6.881	6.865	6.853	6.844
Borax (0.01 m)	9.225	9.180	9.139	9.102

* m = molality (defined as number of moles of solute per kg of solvent)

potential varies with the activity of the ion being measured, called an **indicator electrode**. A second electrode whose potential is known and remains constant must also be present, called a **reference electrode**. In **potentiometric titrations**, an ion is titrated and the change in the potential is measured as a function of the volume of the titrant added.

The potential of an electrochemical cell is given by

$$E = E_{ind} - E_{ref} + E_{ij}$$

where E = potential of the electrochemical cell
E_{ind} = potential of the indicator electrode
E_{ref} = potential of the reference electrode
E_{ij} = liquid-junction potential

The liquid-junction potential typically develops at the interface between the reference electrode and the solution in the cell (which is negligible under the conditions of the experiment). The potential of an electrode is given by the Nernst equation. For a reversible reaction

$$A^{n+} + ne^- \rightleftharpoons A$$

the Nernst equation is

$$E = E^0 - \frac{RT}{nF} \ln \frac{[\alpha_A]}{[\alpha_{A^{n+}}]}$$

where E^0 = standard electrode potential
R = universal gas constant (8.315 VC/K mol)
n = number of electrons (eq/mol)
F = Faraday's constant (96,485 C/eq)
α_A and $\alpha_{A^{n+}}$ = activities of the species A and A^{n+}, respectively

This equation may be simplified by substituting the values for the constants R and F at T = 298 K, and by replacing ln by 2.303 log. If we assume that the activities equal the concentrations (a good assumption for dilute solutions), the Nernst equation becomes

$$E = E^0 - \frac{0.059}{n} \log \frac{[A]}{[A^{n+}]}$$

Reference and Indicator Electrodes

A reference electrode has a constant potential independent of the nature of the solution in which it is dipped. The most common reference electrodes are the saturated calomel electrode (SCE) and the silver–silver chloride electrode (SSE). The half-cell reactions and the potentials for SCE and SSE are (in saturated KCl solution)

$$SCE: \quad Hg_2Cl_2 + 2e^- \rightarrow 2Hg + 2Cl^- \quad E_{ref} = 0.241 \text{ V}$$
$$SSE: \quad AgCl + e^- \rightarrow Ag + Cl^- \quad E_{ref} = 0.197 \text{ V}$$

The indicator electrode must interact selectively with the species or ions present in the solution. Its potential should reflect the activity of the specific ion. A number of ion-selective electrodes are available. For pH measurement, the most efficient indicator electrode sensitive to H^+ activity is the glass membrane electrode, shown in Figure E28.1.

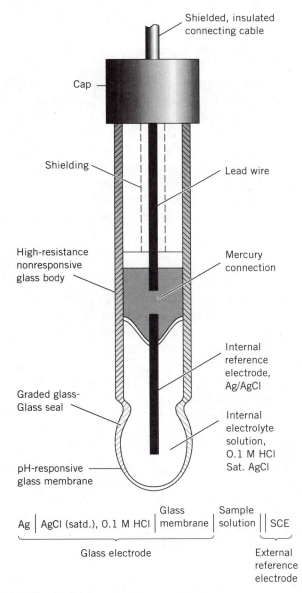

Figure E28.1 Glass membrane electrode

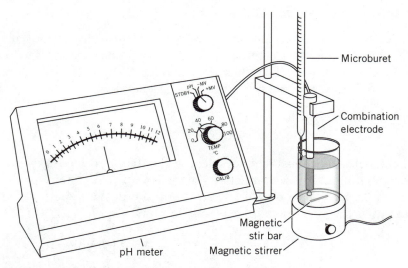

Figure E28.2   A pH meter and the setup for titration

A typical glass membrane electrode consists of a plastic or glass tube with a pH-sensitive glass membrane bulb at the end. The tube consists of an internal silver-silver chloride reference electrode, made of a silver wire coated with AgCl placed inside the bulb, which is filled with 0.1 M HCl solution saturated with AgCl. A saturated calomel electrode serves as the external electrode. The potential developed at the glass membrane is a function of the difference in the H^+ activity on either side of the membrane. This potential across the glass membrane is measured against the external electrode.

Combination pH electrodes are also available. They consist of a pH indicator electrode and a reference electrode combined in a single unit. These are designed for easier handling and smaller samples.

The pH Meter

A pH meter (Figure E28.2) is a mechanical device for measuring the potential of a glass electrode with respect to the external reference electrode. The measurement is displayed on a scale calibrated to read the pH directly. A pH meter should be standardized against two or more buffers of known pH prior to use. The first buffer of pH 7.00 is used to fix the isothermal point at 0 V. A second buffer of pH close to the solution being measured is then used to adjust the slope. Before using a pH meter, read the manufacturer's operating manual or other instructions provided by your instructor.

General References

1. Kolb, D. *J. Chem. Educ.* **1979**, *56*, 49.
2. Rossotti F. J. C.; Rossotti, H. S. *J. Chem. Educ.* **1965**, *42*, 375. This paper discusses the use of Gran's plot in pH titrations.
3. Sawyer, D. T.; Heineman, W. R.; Beebe, J. M. *Chemistry Experiments for Instrumental Methods*. Wiley & Sons: New York, 1984.
4. Starkey, R.; Norman, J.; Hintze, M. *J. Chem. Educ.* **1986**, *63*, 473.
5. Wilson, S. A.; Weber, J. H. *J. Chem. Educ.* **1977**, *54*, 513.

Experimental Section

Procedure

Part A: Preparation of Buffer Solutions

Microscale experiment   Estimated time to complete the experiment: 1 h
Experimental steps: This part involves one step: preparation of buffers.

Disposal procedure: Dispose of all the solutions in a marked container. Return the buffer to your instructor for future use.

The selection of a buffer for a given application depends on the pH range desired for the solution. Usually four primary standard solutions are used to make buffers having fixed pH values: potassium hydrogen tartrate, potassium hydrogen phthalate, potassium dihydrogen phosphate plus disodium hydrogen phosphate, and sodium tetraborate. These solutions are used to calibrate pH meters and electrodes. Standard buffer solutions are also available commercially. However, in many instances buffer solutions other than standard ones (nonstandard buffers) are prepared.

Preparation of Standard Buffer Solutions

Obtain four primary standard substances: potassium hydrogen phthalate ($KHC_8H_5O_4$—KHP—dried at 110°C for one hour); potassium dihydrogen phosphate (KH_2PO_4, dried at 120°C for two hours) and disodium hydrogen phosphate (Na_2HPO_4, dried for two hours at 120°C); and sodium tetraborate, ($Na_2B_4O_7 \cdot 10H_2O$, borax), freshly boiled and cooled deionized water (free from CO_2), three 50 mL screw-capped plastic bottles for storing buffer solutions, 50 mL pipets to transfer the deionized water, and a stirring rod.

> **pH 4 Buffer:** Dissolve 506 mg of KHP in 50.0 g of CO_2-free deionized water. This provides a 0.05 m (molal) solution of the pH 4.0 buffer at 20°C. Measure the pH of the solution with pH paper or a pH meter. Transfer the buffer to a plastic bottle.

> **pH 7 Buffer:** Dissolve 17 mg of KH_2PO_4 and 18 mg of Na_2HPO_4 in 50.0 g of CO_2-free deionized water. Measure the pH of this buffer—it should be 6.9 at 20°C. The solution is 0.025 m (molal) in both the phosphates. Transfer the buffer to a plastic container.

> **pH 9 Buffer:** Dissolve 19 mg of borax in 50.0 g of CO_2-free deionized water. Measure the pH of this buffer—it should be 9.2 at 20°C. The solution is 0.01 m (molal). Save the buffer in a plastic bottle.

Preparation of a Nonstandard Buffer Solution

The following procedure is the general method for preparing any buffer having a pH between 2 and 12. In this case, a pH 7.6 buffer of **tris** [tri(hydroxymethyl)-aminomethane hydrochloride] will be prepared.

Dissolve 800 mg of tris in about 40 mL of water in a beaker containing a stir bar set on a magnetic-stirring hot plate. Monitor the pH of the solution with a pH meter. Add 1 M NaOH (**CAUTION:** Corrosive) dropwise to the tris solution until the solution is at pH 7.6. Transfer the solution to a 50 mL volumetric flask and dilute to the mark. When done, save the buffer in a plastic bottle.

Part B: General Procedure for Potentiometric Titrations

Microscale Method

Set up a 2 mL microburet, magnetic-stirring hot plate, a tall-form 25 mL beaker containing a micro stir bar and a calibrated pH meter as shown in Figure E28.2. Calibrate the pH meter using at least two of the buffer solutions just prepared (*~7.0 pH buffer and another buffer having a pH close to that of the solution being measured*). Rinse and fill the microburet with the titrant (a solution added from the buret). Using a volumetric pipet, transfer an accurately measured volume of an analyte (or an accurately measured mass of a solid in a solution) to the beaker. Place the electrode(s) in the solution so that the solution covers the *frit* or hole at the bottom (add additional deionized water if necessary). Record the initial volume of the titrant in the data table. Stir the solution (a basic solution) and measure the pH before any titrant is added. Enter the pH value on the first line of the data table, as shown in Table E28.3.

Add one drop of indicator from a micropipet. An indicator should be chosen so as to change color over the range where a sharp rise in pH is observed. Begin the titration by adding titrant from the buret in two-drop increments, stopping for 1 to 2 minutes after each addition to allow the solution to reach equilibrium. Record the volume of titrant added and the corresponding pH in the table. Continue to titrate. Also, calculate $\Delta pH/\Delta V$ (the change in the pH value divided by the volume increment), which will be used to determine the slope of a plot of pH versus volume of titrant added. At first, the slope changes little as titrant is added. When the slope changes more rapidly, decrease the titrant addition to one drop at a time. Near the equivalence point, the pH changes rapidly. After the equivalence point is passed, resume adding titrant in two-drop increments. Continue the titration until at least an additional 0.5 mL of the titrant has been added.

Table E28.3   Representative Data Table for a Potentiometric Titration

Volume V of titrant, mL	pH	ΔpH	ΔV, mL	$\Delta pH/\Delta V$
0.00 mL	9.75	—	—	—
0.15	9.74	0.01	0.15	0.07
0.30	9.72	0.02	0.15	0.13
0.45	9.69	0.03	0.15	0.20
0.65	8.60	1.09	0.20	5.45
0.85	7.20	1.40	0.20	7.00
0.90	6.20	1.00	0.05	20.0

Macroscale Method

The procedure is the same as described for the microscale method, with the following changes. Use a 50 mL buret instead of a 2 mL microburet. Pipet 25.0 mL of the solution to be titrated into a 150 mL beaker and dilute the solution to 75 mL by adding deionized water. Add two or three drops of the indicator. Begin the titration by adding 1 mL of titrant at a time. Near the equivalence point, decrease the increment of the titrant to two or three drops at a time. After the equivalence point is passed, add 1 to 2 mL of the titrant at a time.

Determining the End Point in Potentiometric Titrations

Plot the pH (y axis) versus volume of added titrant (x axis) to generate a **titration curve** like the one shown in Figure E28.3. The equivalence point of the titration is roughly in the middle of the steep part of the titration curve. Several methods are used to determine the end point of a potentiometric titration. Some commonly used methods are described next.

Bisection Method: Figure E28.4 illustrates the bisection method for a titration curve that has a steep rise at the equivalence point. Extend the flat parts of the titration curve as shown by the dotted lines. Draw vertical lines (parallel to the pH axis) on the left and right sides of the curve. Find the point halfway up these vertical lines—this is the bisector point. Connect the bisector points. The point where the bisector line crosses the titration curve is the equivalence point.

First Derivative Method: If the titration curve does not show a steep rise at the equivalence point, or if the plot does not have straight lines before or after the end point, the first derivative method should be used. A plot of $\Delta pH / \Delta V$ (y axis) versus V (x axis) is a first derivative curve. The slope of the curve rises to a maximum at the equivalence point (Figure E28.5).

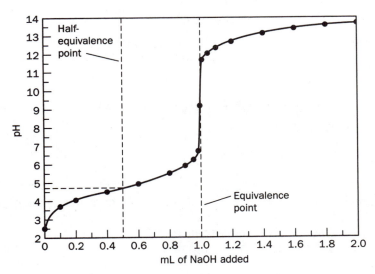

Figure E28.3 Potentiometric titration curve for acid-base titration

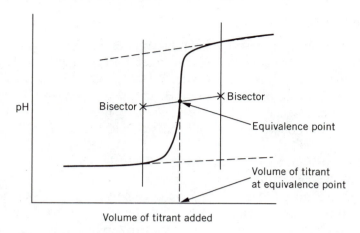

Figure E28.4 Bisection method

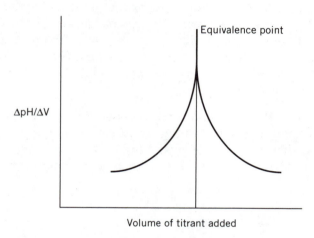

Figure E28.5 First derivative titration curve

Procedure

Part C: Determination of Equivalent Weight (EW) and K_a of a Weak Acid

Microscale Method

Microscale experiment   Estimated time to complete the experiment: 3 h
Experimental steps: There are several steps in this procedure: (1) setting up and calibration of a pH meter, (2) standardization of 0.1 M NaOH solution, (3) titration of an analyte using a pH meter, (4) plotting the data to obtain pH titration curves, and (5) calculations.

Disposal procedure: Generally, the residual solutions after titrations may be disposed of in the sink. However, ask your instructor for the proper disposal procedure.

Obtain a pH meter, a magnetic stirrer, a 25 mL tall-form beaker, a 25 mL volumetric flask, a stir bar, a volumetric pipet, two microburets (marked *acid* and *base*), a 50 mL plastic bottle for NaOH, a ruler, and a protractor.

1. Preparation of 0.1 M NaOH and KHP (Potassium Hydrogen Phthalate) Solutions

(NOTE: NaOH solutions are known to react with atmospheric carbon dioxide, forming a carbonate. It is necessary to adjust for this carbonate formation when doing a titration with a pH meter by the method shown in this section). Prepare 50 mL of ~0.1 M NaOH by diluting a supplied concentrated solution (**CAUTION: Corrosive**). Also prepare a standard 0.1 M KHP solution in a 25 mL volumetric flask according to the procedure described in Experiment 10. Calculate and record the KHP solution concentration.

2. Standardization of 0.1 M NaOH Solution by Potentiometric Titration

Rinse and fill the base buret with the 0.1 M NaOH solution. Transfer 2.00 mL of the NaOH solution into a 25 mL tall-form beaker containing a stir bar. Add 8 to 10 mL of deionized water, followed by a drop (from a fine-tip micropipet) of 0.25 percent phenolphthalein indicator. The solution will turn pink. Similarly, rinse and fill the acid buret with the KHP solution.

Set up and calibrate the pH meter (see Part B, Figure E28.2) according to the procedure given by your instructor. Wash the electrode(s) thoroughly with deionized water, and immerse them inside the beaker such that the frits in the electrodes are submerged. Titrate the NaOH solution with the KHP following the general procedure described in Part B. Collect the data in a table similar to Table E28.3. Be sure that the solution is well stirred during the titration. If time permits, repeat the procedure.

3. Potentiometric Titration of an Unknown Acid

Accurately weigh about 20 ± 1 mg of the unknown in a 25 mL tall-form beaker containing a stir bar, set atop a magnetic stirrer. Add 10 to 12 mL of deionized water and stir to dissolve the acid (the acid may not dissolve immediately; but it will subsequently react with NaOH).

Using a fine-tip Pasteur pipet, add a small drop of indicator (your instructor will tell you which indicator to use) to the solution. Fill the base buret with the standardized NaOH solution. Titrate the acid with the NaOH as before (see part B). Enter your data in the data table.

4. Manipulation of Data

Generate two titration curves: plot #1—pH (y axis) versus volume of KHP (x-axis) from the NaOH standardization and plot #2—pH (y axis) versus volume of NaOH from the unknown acid titration. **Note that the shape of plot #1 is the reverse of the shape of plot #2.** Also, construct a first derivative plot of $\Delta pH/\Delta V$ versus V of NaOH for the second titration, as shown in Figure E28.5. Determine the equivalence point either by the bisection or by the first derivative method (Figures E28.4 or E28.5) as described in Part B.

5. Calculating the Concentration of NaOH

Locate the equivalence point on curve #2 and find the pH at this point. On curve #1, find the volume of KHP that was added to the NaOH solution corresponding to this pH. Using this volume and the concentration of KHP, calculate the molarity (M) and the normality (N) of the NaOH. From the stoichiometry of the reaction

$$NaOH + KHP \rightarrow NaKP + H_2O$$

Experiment 28   Preparation of Buffers and Potentiometric Titrations   565 the molarity of the base can be calculated:

$$(M_{NaOH})(V_{NaOH}) = (M_{KHP})(V_{KHP})$$

Since NaOH has one OH^- group, its normality is equal to its molarity:

$$N_{NaOH} = M_{NaOH}$$

6. Calculation of the Equivalent Weight of the Unknown Acid

In Section 3.9, it was shown that

$$(V_{base})(N_{base}) = eq_{base}$$

From the equivalence point determined on titration curve #2, obtain the volume of NaOH needed to titrate the acid completely. Using this volume and the normality of NaOH, calculate the eq_{base}.

Since one equivalent of base always reacts with one equivalent of acid, we can calculate the number of equivalents of the unknown acid:

$$eq_{base} = eq_{acid}$$

The equivalent weight (EW) of the unknown acid is merely the mass of the unknown acid divided by the number of equivalents: $EW_{acid} = (m_{acid})/(eq_{acid})$.

7. Calculation of K_a of the Acid

At the equivalence point, the number of moles of OH^- that have been added is equal to the number of moles of H^+ in the original amount of the acid, if that acid has completely dissociated. The ionization constant of the acid, K_a, can be calculated in the following way.

The **half-equivalence point** is defined as the point where the amount of the base necessary to titrate half of the acid present has been added (see Figure E28.3). Let x be the moles of acid initially present. The amount of the acid at the half-equivalence point will be $0.5x$ moles. Since half of the acid has reacted at this point, there will be $0.5x$ moles of the conjugate base present as well. Thus, $[HA] = [A^-]$. Substituting into the acid equilibrium expression (E28.2), we may write

$$K_a = \frac{[H^+][0.5x]}{[0.5x]} = [H^+]$$

Thus, the pH at the half-equivalence point is equal to the pK_a of the acid.

Find the volume of the base needed to reach the equivalence point on curve #2. Half of this volume will be required at the half-equivalence point. Through the half-equivalence point, draw a vertical line parallel to the pH axis as shown in Figure E28.3. Draw a horizontal line to the pH axis at the point where the vertical line intersects the pH curve, as shown. The point of intersection with the pH axis is the pH value at the half-equivalence point, which is equal to the pK_a of the acid. From pK_a, find K_a, using the equation

$$K_a = 10^{-pK_a} \quad \text{or} \quad pK_a = -\log K_a$$

Example E28.2   The volume of 0.100 N NaOH needed to reach the end point in a pH titration of 21 mg of an acid in water is 1.02 mL. The pH at 0.51 mL of the base was 4.75. Calculate the equivalent weight and the K_a for the acid.

Answer

$$\text{eq}_{\text{base}} = (N_{\text{base}})(V_{\text{base}}) = (0.100 \text{ eq/L})(0.00102 \text{ L}) = 1.02 \times 10^{-4} \text{ eq}$$

$$\text{eq}_{\text{acid}} = \text{eq}_{\text{base}} = 1.02 \times 10^{-4} \text{ eq}$$

$$\text{EW}_{\text{acid}} = (m_{\text{acid}})/(\text{eq}_{\text{acid}}) = (0.021 \text{ g})/(1.02 \times 10^{-4} \text{ eq}) = 206 \text{ g/eq}$$

The volume, 0.51 mL, of the base corresponds to the half-equivalence point. At this point, $pK_a = \text{pH} = 4.75$. Therefore, $K_a = 10^{-4.75} = 1.78 \times 10^{-5}$.

Macroscale Method

Set up and standardize the pH meter as described before. Use two 50 mL burets, one labeled *acid* and the other *base*. Prepare 250 mL of ~0.1 M NaOH in a plastic bottle. Fill the acid buret with 0.100 M KHP (see Experiment 10). Pipet 25.00 mL of the NaOH solution (use a 25 mL volumetric pipet and a pipet bulb) into a 100 mL beaker containing a stir bar, and add one drop of phenolphthalein indicator. Titrate the NaOH solution with the 0.100 M solution of KHP as above. Prepare a pH titration curve for NaOH/KHP titration.

Rinse and fill the base buret with the NaOH solution. Place about 0.250 g of the unknown acid in a 250 mL beaker containing a stir bar set atop a magnetic stirrer. Add 50 mL of water to the acid, and titrate by adding NaOH as in Part B. Calculate the equivalent weight and the K_a of the acid as just shown.

Procedure

Part D: Potentiometric Titration of a Polyprotic Acid (Phosphoric Acid)

Microscale Method

Microscale experiment   Estimated time to complete the experiment: 3 to 5 h
Experimental steps: There are several steps in this procedure: (1) setup and calibration of a pH meter, (2) standardization of 0.1 M NaOH solution, (3) titration of an analyte using a pH meter, (4) plotting of the data to obtain pH titration curves, and (5) calculations.

Disposal procedure: Generally, the residual solutions after titrations may be disposed of in the sink. However, ask your instructor for the proper disposal procedure.

Note: The procedure is described for phosphoric acid, H_3PO_4. Any polyprotic acid or base can be used.

Obtain two sample solutions from your instructor. Sample #1 is 0.05 M H_3PO_4 and Sample #2 is 0.05 M H_3PO_4 + 0.05 M HCl. These samples will be provided in numbered 10 mL volumetric flasks. Dilute the samples to the mark with distilled water. Set up and calibrate the pH meter (see Part B) as described earlier.

1. Preparation of 0.1 M NaOH and KHP (Potassium Hydrogen Phthalate) Solutions

Prepare 50 mL of 0.1 M NaOH from a supplied concentrated solution. Also prepare a primary standard solution of 0.1 M KHP in a 25 mL volumetric flask according to the procedure described in Experiment 10.

2. Standardization of 0.1 M NaOH Solution by Potentiometric Titration

Rinse and fill the base buret with NaOH solution. Similarly, rinse and fill the acid buret with the standard KHP acid solution. Record the volumes from each of the burets on the record sheet. Transfer 2.00 mL of the NaOH solution from the base buret to a 25 mL tall-form beaker containing a stir bar set atop a magnetic stirrer. Add 8 to 10 mL deionized water and a drop of 0.25 percent phenolphthalein indicator (from a fine-tip micropipet). Now, following the same procedure described in Parts B and C (section 2), titrate the NaOH solution potentiometrically with the standard KHP solution added from the buret.

Collect the data in a table similar to Table E28.3. Calculate the molarity of the NaOH solution as shown below.

3. Potentiometric Titration of Samples with NaOH

Rinse and fill the acid buret with sample #1 solution. Similarly, rinse and fill the base buret with the NaOH solution. Record the initial volume of the sample #1 solution in the buret. Transfer 2.00 mL of sample #1 solution into a clean 25 mL tall-form beaker containing a stir bar and set it atop a magnetic stirrer. Add 8 to 10 mL of deionized water and immerse the thoroughly washed pH electrode(s) into the solution. Add one drop (from a fine-tip micropipet) of bromphenol blue indicator (yellow at pH 3 and blue at pH 5) to the solution. The initial pH of sample #1 should be ~2—**record this pH before adding any base**. Record the initial volume of the base.

Carry out the titration of sample #1 with the NaOH solution added from the buret, following the same procedure as described in Part B. Collect the data in a table similar to Table E28.3.

Follow exactly the same procedure with sample #2 (see Part B). In this case, note that the volume of NaOH needed to reach the first equivalence point is more than that required for the second equivalence point. Collect the data in a table similar to Table E28.3.

4. Manipulation of Data

Plot three pH titration curves (see Parts B and C):

Plot #1: pH (y axis) versus volume of KHP (x axis) for NaOH/KHP titration

Plot #2: pH (y axis) versus volume of NaOH (x axis) for NaOH/Sample #1 titration

Plot #3: pH (y axis) versus volume of NaOH (x axis) for NaOH/Sample #2 titration

Also, prepare the first derivative curves (see Figure E28.5) for samples #1 and #2.

Note: Since both samples contain a polyprotic acid, there will be two breaks in the pH titration curves resulting in the two equivalence points (two maxima in the first derivative curves). The general shapes of plots #2 and 3 are the reverse of plot #1.

Perform the necessary calculations as follows.

Sample #1 contains only H_3PO_4. In the initial stages of the titration, the addition of the base (NaOH) will not change the pH significantly. The H^+ from the acid will react with the OH^-, generating the conjugate base $H_2PO_4^-$.

$$H_3PO_4 \rightleftharpoons H^+ + H_2PO_4^- \tag{E28.6}$$

Further addition of NaOH will increase the pH of the solution sharply, resulting in the first equivalence point.

Continued addition of NaOH solution from the buret will consume the second H^+ from $H_2PO_4^-$, generating HPO_4^{2-} according to the following reaction:

$$H_2PO_4^- \rightleftharpoons H^+ + HPO_4^{2-} \tag{E28.7}$$

Since the reaction mixture contains both $H_2PO_4^-$ and HPO_4^{2-}, the solution behaves as a buffer, resisting sharp changes in pH. As the reaction proceeds, the measurement of pH will record a second rise (the second equivalence point). Further addition of OH^- will partially convert the HPO_4^{2-} to PO_4^{3-}. The pH will increase only slightly, because of the buffering action of HPO_4^{2-} and PO_4^{3-}, or

$$HPO_4^{2-} \rightleftharpoons H^+ + PO_4^{3-} \tag{E28.8}$$

Sample #2 contains both HCl and H_3PO_4. In the initial stages of the titration, the addition of OH^- will not change the pH significantly. The H^+ from HCl as well as H_3PO_4 will consume the added OH^-, generating the conjugate bases of Cl^- and $H_2PO_4^-$, or

$$HCl \rightarrow H^+ + Cl^-$$
$$H_3PO_4 \rightleftharpoons H^+ + H_2PO_4^-$$

Further addition of OH^- will increase the pH of the solution, yielding the first steep rise in the curve (the first equivalence point). Note that continued titration of the rest of the sample #2 solution with NaOH will generate a pH titration curve that is similar to that for sample #1.

5. Calculating the Concentration of NaOH Solution

Locate the first and second equivalence points on both plot #2 and plot #3 (or derivative curves), and obtain the pH values corresponding to these equivalence points. As done in Part C (see section on manipulation of data), find the volumes of KHP (from Plot #1) corresponding to these pH values.

Using the concentration of KHP, calculate the molarity for the NaOH solution for each value. A total of four molarities for NaOH will thus be determined—be sure to know which is which.

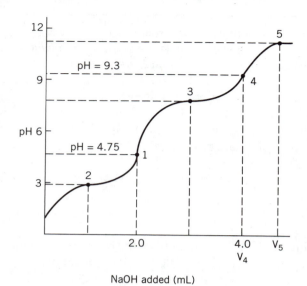

Figure E28.6   pH titration curve for sample #1 (H_3PO_4) with NaOH

6. Calculation of Molarities of the Acids

For sample #1 (H_3PO_4): Determine the location of the two equivalence points on plot #2 (see Figure E28.6). Draw a vertical line from each of the equivalence points to the x axis. The difference in volume between these two lines is the volume of NaOH that was used to titrate the **second H^+** from $H_2PO_4^-$. Note that the volume of NaOH needed to titrate the **first H^+** from H_3PO_4 is exactly the same as that required for the second H^+. Using the correct molarity of NaOH as just obtained and the volume, the molarity of H_3PO_4 in sample #1 can be easily calculated:

$$(M_{base})(V_{base}) = (M_{acid})(V_{acid})$$

 For Sample #2 (HCl + H_3PO_4): Plot #3 should resemble Figure E28.7. Determine the location of the two equivalence points. Draw a vertical line from each of the equivalence points to the x axis, as shown in the figure. The difference in volume between these two lines ($6.0 - 4.0 = 2.0$ mL in Figure E28.7) is the amount of NaOH that was used to titrate the **second H^+** from H_3PO_4. Note that the concentration of the second proton is the same as the concentration of the phosphoric acid to begin with, so the concentration of the phosphoric acid can be solved for, using

$$(M_{base})(V_{base}) = (M_{acid})(V_{acid}) \quad \text{(for phosphoric acid)}$$

The same volume of NaOH must have been used to titrate the **first H^+** from H_3PO_4 as well. The volume of NaOH needed to reach the first equivalence point (4.0 mL in Figure E28.7) was the amount needed to neutralize the **first H^+** from H_3PO_4, **plus** the amount needed to titrate the HCl. The difference between these two amounts ($4.0 - 2.0 = 2.0$ mL) is the amount of NaOH that was needed to titrate the HCl. Calculate the molarity of the HCl:

$$(M_{base})(V_{base}) = (M_{acid})(V_{acid}) \quad \text{(for hydrochloric acid)}$$

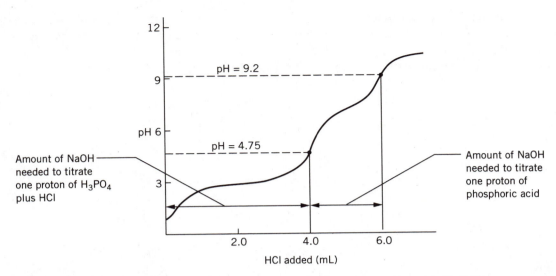

Figure E28.7 The pH titration curve for sample #2 (H_3PO_4 + HCl) with NaOH

7. Calculation of Acid Ionization Constants for H_3PO_4

This calculation is based upon the the titration data for the sample #1 (Figure E28.6).

K_1 for H_3PO_4: The first ionization constant, K_1, for H_3PO_4 [equation (E28.6)] is given by the expression

$$K_1 = \frac{[H_2PO_4^-][H^+]}{[H_3PO_4]} \qquad \text{(E28.9)}$$

From the intial pH of the solution (before adding any base), calculate $[H^+]$ for the original solution, using the equation

$$[H^+] = 10^{-pH}$$

The first ionization of H_3PO_4 forms equal moles of $H_2PO_4^-$ and H^+ (see section 7, Part C). Therefore,

$$[H_2PO_4^-] = [H^+]$$

The original concentration of H_3PO_4 was calculated in the previous section (for sample #1). The concentration of H_3PO_4 at equilibrium is the original amount, minus that which is dissociated to form H^+ (and $H_2PO_4^-$), or

$$[H_3PO_4] = [H_3PO_4]_{orig} - [H^+]$$

Since all the terms on the right side of equation (E28.9) are known, K_1 can be calculated.

K_2 for H_3PO_4: K_2 for H_3PO_4 [equation (E28.7)] can be calculated by the half-equivalence method as described in Part A. The equilibrium expression is

$$K_2 = \frac{[HPO_4^{2-}][H^+]}{[H_2PO_4^-]} \qquad \text{(E28.10)}$$

Consider Figure E28.6, which should resemble the titration curve for sample #1. Identify the equivalence points, and draw a vertical line to the x axis at these points, as shown. Point 3 in the figure represents a point halfway from the first equivalence point to the second equivalence point (halfway between 4.0 and 2.0 mL is 3.0 mL). At this half-equivalence point

$$[HPO_4^{2-}] = [H^+] = [H_2PO_4^-]$$

From (E28.10), therefore,

$$K_2 = [H^+]$$

$[H^+]$ can be calculated from the pH at point 3 (pH = 7.9 in Figure E28.6) using the equation $[H^+] = 10^{-pH}$.

K_3 for H_3PO_4: K_3 for H_3PO_4 [equation (E28.8)] is given by the equilibrium expression

$$K_3 = \frac{[PO_4^{3-}][H^+]}{[HPO_4^{2-}]} \tag{E28.11}$$

Select any point (such as point 5 in Figure E28.6) on the **flat top part** of the curve. Calculate the volume of NaOH added from the second equivalence point to reach point 5 ($V_5 - V_4$ in Figure E28.6). Obtain the number of millimoles of NaOH added by multiplying this volume by the molarity of NaOH. Finally, compute the concentration of OH^- added by dividing the millimoles of NaOH by the total volume V_t.

$$[OH^-]_{added} = \frac{(V_5 - V_4)(M_{NaOH})}{V_t}$$

The actual total concentration of $[OH^-]$ is obtained from the pH at point 5 and from recalling that pOH = 14 − pH, or

$$[OH^-]_{actual} = 10^{-pOH}$$

The difference in the concentrations ($[OH^-]_{added} - [OH^-]_{actual}$) is the measure of the $[OH^-]$ required to convert HPO_4^{2-} to PO_4^{3-}. Their concentrations may thus be determined.

Additional Independent Projects

1. Titration of a multifunctional base with an acid. Use a carbonate, soda ash, or phosphate salt. A mixture of a carbonate and a bicarbonate (see Experiment 10, Part B), or a triphosphate and diphosphate can also be used.[1]
2. Potentiometric titration of aspirin in a nonaqueous solvent.[2]
3. Determination of phosphoric acid in decarbonated cola drinks.[3]

References

1. Werner, T. C.; Werner, J. A. *J. Chem. Educ.* **1991**, *68*, 600.
2. Shen, S. Y.; Gilman, A. J. *J. Chem. Educ.* **1965**, *42*, 540.
3. Murphy, J. *J. Chem. Educ.* **1983**, *60*, 420.

Name ________________________	Section ____________
Instructor ________________________	Date ____________

PRELABORATORY REPORT SHEET—EXPERIMENT 28

Experiment title____

Objective

Reactions/formulas to be used

Chemicals and solutions—their preparation

Materials and equipment table

Outline of procedure

PRELABORATORY PROBLEMS—EXPERIMENT 28

1. Calculate the pH and pOH of a 1.0×10^{-5} M solution of HCl.

2. What is the pH of a buffer prepared by mixing 15 mL each of 0.10 M acetic acid and 0.10 M sodium acetate (K_a for acetic acid $= 1.75 \times 10^{-5}$)?

3. Calculate the pH of 0.100 M Na_3PO_4 solution ($K_3 = 7.1 \times 10^{-13}$).

4. The following data were obtained for the titration of 500 μL of 0.10 M acetic acid with 0.10 M NaOH.

Vol. NaOH, μL	0	100	200	300	400	440	480	490
pH		2.88	4.15	4.59	4.95	5.36	5.61	6.13	6.44
Vol. NaOH, μL	495	500	505	510	520	560	600	700
pH	7.05	8.70	10.32	11.05	11.30	11.80	12.00	12.02

   Plot the titration curve and first derivative curve on a sheet of graph paper. Determine the equivalence point of the titration and find the K_a of acetic acid.

EXPERIMENT 28   DATA SHEET

Part A: Preparation of Buffer Solutions

1. pH 4 Buffer

Amount of KHP taken, g   _____

Volume of water added, mL   _____

Concentration of the solution, m _____

Measured pH of the solution   _____

2. pH 7 Buffer

Amount of KH_2PO_4 taken, g   _____

Volume of water added, mL   _____

Concentration of the solution, m _____

Measured pH of the solution   _____

3. pH 9 Buffer

Amount of borax taken, g   _____

Volume of water added, mL   _____

Concentration of the solution, m _____

Measured pH of the solution   _____

4. Nonstandard Buffer

Amount of tris taken, g   _____

Volume of water added, mL   _____

Concentration of the solution, m _____

Measured pH of the solution   _____

Part C: Determination of EW and K_a of a Weak Acid

Unknown number of acid: _____

Preparation of 0.1 M NaOH and KHP solutions

1. Mass of NaOH taken or
   Volume of 50 percent NaOH taken, g or mL _____

2. Initial mass of KHP, g _____

3. Final mass of KHP, g _____

4. Mass of KHP taken, g _____

5. Concentration of KHP, M _____

Show calculation:

Standardization of NaOH by potentiometric titration

6. Volume of NaOH taken, mL _____
7. pH titration

Vol. of KHP	pH	ΔpH	ΔV	ΔpH/ΔV	Vol. of KHP	pH	ΔpH	ΔV	ΔpH/ΔV
1.					16.				
2.					17.				
3.					18.				
4.					19.				
5.					20.				
6.					21.				
7.					22.				
8.					23.				
9.					24.				
10.					25.				
11.					26.				
12.					27.				
13.					28.				
14.					29.				
15.					30.				

DATA SHEET, EXPERIMENT 28, PAGE 3

Potentiometric Titration of an Unknown Acid

1. Mass of unknown acid taken, g   _____
2. pH titration

Vol. of NaOH	pH	ΔpH	ΔV	ΔpH$/\Delta V$		Vol. of NaOH	pH	ΔpH	ΔV	ΔpH$/\Delta V$
1.						16.				
2.						17.				
3.						18.				
4.						19.				
5.						20.				
6.						21.				
7.						22.				
8.						23.				
9.						24.				
10.						25.				
11.						26.				
12.						27.				
13.						28.				
14.						29.				
15.						30.				

Show calculation:

Concentration of NaOH

1. pH at the equivalence point of plot #2 _____

2. Volume of KHP at this pH from plot #1, mL _____

3. Molarity of KHP (from page 1 of data sheet), M _____

4. Volume of NaOH (from page 2 of data sheet), mL _____

5. Molarity of NaOH, M _____

Show calculation:

Calculation of Equivalent Weight of the Acid

1. Mass of the unknown acid (from page 3 of data sheet), g _____

2. Volume of NaOH needed for titrating the acid (from plot #2 or the spike of the first derivative curve), mL _____

3. Equivalents of NaOH, eq _____

4. Equivalents of the acid, eq _____

5. Equivalent weight of the acid, g/eq _____

Calculation of K_a of the Acid

1. pH at half-equivalence point (plot #2 or derivative curve) _____

2. pK_a _____

3. K_a _____

Show calculation:

DATA SHEET, EXPERIMENT 28, PAGE 5

Part D: Potentiometric Titration of a Polyprotic Acid

Unknown number of sample #1_____

Unknown number of sample #2_____

Preparation of 0.1 M NaOH and KHP solutions

1. Mass of NaOH taken or
    Volume of 50 percent NaOH taken, g or mL   _____

2. Initial mass of KHP, g   _____

3. Final mass of KHP, g   _____

4. Mass of KHP taken, g   _____

5. Molarity of KHP, M   _____

Show calculation:

Standardization of NaOH by Potentiometric Titration

1. Volume of NaOH taken, mL   _____
2. Data Table for Potentiometric Titration of NaOH with KHP (titrant)

Vol. of KHP, mL	pH	ΔpH	ΔV, mL	ΔpH/ΔV	Vol. of KHP, mL	pH	ΔpH	ΔV, mL	ΔpH/ΔV
1.					11.				
2.					12.				
3.					13.				
4.					14.				
5.					15.				
6.					16.				
7.					17.				
8.					18.				
9.					19.				
10.					20.				

Vol. of KHP, mL	pH	ΔpH	ΔV, mL	ΔpH/ΔV		Vol. of KHP, mL	pH	ΔpH	ΔV, mL	ΔpH/ΔV
21.						26.				
22.						27.				
23.						28.				
24.						29.				
25.						30.				

Potentiometric Titration of Sample #1

1. Capacity of the volumetric flask with the unknown sample #1, mL _____

2. Data table for potentiometric titration of sample #1 with NaOH (titrant)

Volume of sample #1 taken, mL _____   Initial pH of sample #1 _____

Vol. of NaOH, mL	pH	ΔpH	ΔV, mL	ΔpH/ΔV		Vol. of NaOH, mL	pH	ΔpH	ΔV, mL	ΔpH/ΔV
1.						16.				
2.						17.				
3.						18.				
4.						19.				
5.						20.				
6.						21.				
7.						22.				
8.						23.				
9.						24.				
10.						25.				
11.						26.				
12.						27.				
13.						28.				
14.						29.				
15.						30.				

DATA SHEET, EXPERIMENT 28, PAGE 7

Potentiometric Titration of Sample #2

1. Capacity of the volumetric flask with the unknown sample #2, mL _____

2. Data table for potentiometric titration of sample #2 with NaOH (titrant)

Volume of sample #2 taken, mL _____   Initial pH of sample #2 _____

Vol. of NaOH, mL	pH	ΔpH	ΔV, mL	ΔpH/ΔV	Vol. of NaOH, mL	pH	ΔpH	ΔV, mL	ΔpH/ΔV
1.					16.				
2.					17.				
3.					18.				
4.					19.				
5.					20.				
6.					21.				
7.					22.				
8.					23.				
9.					24.				
10.					25.				
11.					26.				
12.					27.				
13.					28.				
14.					29.				
15.					30.				

Calculation of Molarities of H_3PO_4
Sample #1 (use plot #2 or its first derivative curve)

Initial pH (sample #1) _____

		At 1st equivalence pt.	At 2nd equivalence pt.
1.	pH (sample #1, plot #2, at eq. pt.)	_____	_____
2.	Volume of KHP at this pH (plot #1), mL	_____	_____
3.	Molarity of KHP, M (from previous page)	_____	_____
4.	Volume of NaOH (plot #1), mL	2.00	2.00
5.	Molarity of NaOH, M	_____	_____
6.	Volume of sample #1 taken, mL	2.00	2.00
7.	Volume of NaOH (plot #2), mL (at eq. pt.)	_____	_____
8.	Moles of NaOH, mol	_____	_____
9.	Moles of acid, mol	_____	_____
10.	Molarity of H_3PO_4 in sample #1, M	_____	_____

Calculation of Molarities
Sample #2 (use plot #3 or its first derivative curve)

Initial pH (sample #2) _____

		At 1st equivalence pt.	At 2nd equivalence pt.
1.	pH (sample #2, plot #3)	_____	_____
2.	Volume of KHP at this pH (plot #1), mL	_____	_____
3.	Molarity of KHP, M (from previous page)	_____	_____
4.	Volume of NaOH, mL	2.00	2.00
5.	Molarity of NaOH, M	_____	_____

Molarity of H_3PO_4 in Sample #2

		At 1st equivalence pt.	At 2nd equivalence pt.
1.	Volume of sample #2, mL	2.00	2.00
2.	Volume of NaOH (plot #3), (between 1st and 2nd equiv. pt.), mL	_____	_____
3.	Moles of NaOH used for H_3PO_4, mol	_____	_____
4.	Moles of acid in sample #2, mol	_____	_____
5.	Molarity of H_3PO_4 in sample #2, M	_____	_____

Show calculation:

DATA SHEET, EXPERIMENT 28, PAGE 9

Molarity of HCl in Sample #2

	At 1st equivalence pt.	At 2nd equivalence pt.
1. Total volume of NaOH (plot #3), up to 1st equivalence point, mL	_____	_____
2. Volume of NaOH used for H_3PO_4, between 1st and 2nd equiv. pt., mL	_____	_____
3. Volume of NaOH used for titrating HCl, mL	_____	_____
4. Moles of NaOH used for HCl, mol	_____	_____
5. Moles of HCl in sample #2, mol	_____	_____
6. Molarity of HCl in sample #2, M	_____	_____

Show calculation:

Calculation of K_a of H_3PO_4

1. K_1 _____

2. K_2 _____

3. K_3 _____

Show calculations:

1. If the pH of a solution is 2.03, what is the $[H^+]$?

2. The $HPO_4^{2-}/H_2PO_4^-$ couple is an effective buffer system in the blood. If the pH of blood is 7.40, what is the ratio of $[HPO_4^{2-}]/[H_2PO_4^-]$?

$$H_2PO_4^- \rightleftharpoons HPO_4^{2-} + H^+ \quad K_2 = 6.32 \times 10^{-8}$$

Chemical Kinetics

Part A: Rate of Reaction of Iodine with Acetone
Part B: Spectroscopic Study of the Reaction of Phenolphthalein with a Strong Base

Semimicroscale Experiments

Objectives
- To determine the kinetic parameters (rate, order, rate constant) of a reaction by a visual method
- To use spectroscopic methods in determining the kinetics of a reaction

Prior Reading[1]
- Chapter 2   Mathematical methods
- Section 3.5   Measuring liquid volumes
- Section 3.9   Dilutions
- Chapter 8   Introduction to spectroscopy

Related Experiments
- Experiments 1 (Density), 21 (Iron Determination), and 24 (Composition of a Complex)

Activities a Week Before
- Must complete the prelaboratory sheets for both parts and construct the solution tables

The study of kinetics involves the measurement of the rates at which chemical reactions occur and the determination of how the rates are affected by changes in the concentrations of the species, temperature, and pressure and by the presence of catalysts. Once these factors are understood, one may optimize the reaction conditions to speed up or slow down a reaction as desired. The rate at which a chemical reaction occurs in solution depends on a number of factors, the most important being the concentrations of the reactants and the temperature. For the reaction

$$x\,A + y\,B \rightarrow z\,P$$

the rate law is expressed by the following simple equation:

$$\text{rate} = k[A]^a[B]^b$$

where k is the **rate constant** (which depends only on the temperature, not on the concentrations), [A] and [B] are the **molar concentrations** of the reactants, and **a** and **b** are the **orders of the reaction** with respect to A and B.

If $a = 1$, the reaction is **first order** in terms of A; if $b = 2$, the reaction is said to be **second order** in terms of B. Note that a and b are not necessarily the reaction coefficients x and y.

There are two main difficulties in performing a kinetics experiment. The first is to find a way to simplify the complex rate law for a given reaction. The second is to devise a means by which to measure the rate of the reaction, that is, a way to signal when the reaction has progressed to a certain point.

The dependence of the rate on temperature is given by the **Arrhenius equation,**

$$k = Ae^{-E_a/RT} \tag{E29.1}$$

where A = proportionality factor (unique to each reaction)
   T = temperature (K)
   R = gas law constant (8.314 J/K mol)
   E_a = **activation energy** (J/mol) of the reaction.

The activation energy is the minimum amount of energy needed for a collision between reactants to result in the formation of product. Taking the natural logarithm and rearranging equation (E29.1), we obtain

$$\ln k = \ln A - \frac{E_a}{RT}$$
$$\ln k = -\frac{E_a}{R}\left(\frac{1}{T}\right) + \ln A$$
$$y = mx + b$$

Thus, a plot of $\ln k$ (y axis) versus $1/T$ (x axis) gives a straight line with a slope equal to $-E_a/R$ and a y intercept equal to $\ln A$.

Part A: Reaction of Iodine with Acetone. Method of Initial Rates[2]

In Part A of this experiment, the kinetics of the iodination of acetone will be studied:

$$\underset{\text{acetone}}{H_3CCOCH_3(aq)} + \underset{\text{iodine}}{I_2(aq)} \rightarrow \underset{\text{iodoacetone}}{H_3CCOCH_2I(aq)} + H^+(aq) + I^-(aq)$$

The reaction kinetics will be investigated by the **method of initial rates**, that is, by measuring the amount of time that it takes for a certain fixed amount of the reactants to react. By measuring this time interval under varying conditions of concentration and temperature, the rate law can be determined. The rate is calculated as a change in concentration of one of the reactants or products ($\Delta[I_2]$) over a time interval (Δt):

$$\text{rate} = -\Delta[I_2]/\Delta t \tag{E29.2}$$

The negative sign indicates that the $[I_2]$ is decreasing—we are measuring the rate of disappearance of iodine. The sign [ ] stands for molar concentration of a solution.

Determining the Order of Reaction and the Rate Constant k

The rate of the acetone/iodine reaction depends upon the concentrations of the reactants and on $[H^+]$. The kinetics of this reaction can therefore be represented by the general rate law,

$$\text{rate} = k[\text{acetone}]^a[H^+]^b[I_2]^c \qquad (E29.3)$$

where **a**, **b**, and **c** are the **orders of reaction** with respect to acetone, H^+, and I_2, respectively. The rate law in equation (E29.3) is fairly complex, having seven variables (the three concentrations; the orders of reaction a, b, c; and the rate constant). Knowing that the reaction does not depend upon the $[I_2]$ simplifies matters. That means the reaction is of zero order ($c = 0$, and $[I_2]^c = [I_2]^0 = 1$) in terms of the concentration of iodine.

We can use I_2 as the limiting reagent, the one that will be used up first in the reaction (as observed by the disappearance of the yellow color of I_2 as it is converted to I^-, a colorless species). We measure the time necessary for the yellow solution to become colorless. If a small concentration of I_2 and a large excess of both acetone and hydrogen ion are used, the change of concentrations of acetone and hydrogen ion will be negligible. Under these conditions, if the color of the solution has stopped changing in t seconds, when $[I_2]$ is completely used up and the reaction stops, equation (E29.2) may be written as

$$\text{rate} = \frac{-\Delta[I_2]}{\Delta t} = \text{initial}[I_2]/t \qquad (E29.4)$$

Kinetic experiments are always performed by making at least two experimental runs at constant temperature, changing one concentration (say of acetone) while keeping all other concentrations constant. Suppose in the second run, the concentration of acetone is twice that present in the first run:

For run 1: $\text{rate}_1 = k[\text{acetone}]^a[H^+]^b[I_2]^c$

For run 2: $\text{rate}_2 = k[2 \times \text{acetone}]^a[H^+]^b[I_2]^c$

Dividing the second equation by the first, we have

$$\text{rate}_2/\text{rate}_1 = (2)^a$$
$$\ln(\text{rate}_2/\text{rate}_1) = a \times \ln 2$$

Since both rate_1 and rate_2 are known, this equation can be solved for a, the order of the reaction with respect to [acetone]:

$$a = \frac{\ln(\text{rate}_2/\text{rate}_1)}{\ln 2} \qquad (E29.5)$$

The order b can be calculated in a similar manner, comparing the rates of two runs where only the $[H^+]$ changes. Once a and b are known (and recalling that $c = 0$), equation (E29.3) can be solved to determine the value of k by substituting the values for [acetone], $[H^+]$, $[I_2]$, a, b, and c.

Determination of Activation Energy, E_a

The dependency of the rate constant on the temperature may be investigated by performing similar runs (with identical concentrations of all the species involved) at several different temperatures. The Arrhenius equation is then used to determine the activation energy.

Part B: Spectroscopic Study of the Reaction of Phenolphthalein with a Strong Base

In Part A, we studied the kinetics of a reaction by a visual method. However, not many reactions can be investigated by this method. In many cases, instrumental methods are used to monitor the rates of a chemical reaction. In Part B of this experiment, a visible spectrophotometer is used to study the acid-base reactions of the indicator phenolphthalein.

Phenolphthalein (abbreviated H_2In) is extensively used as an indicator in acid-base titrations (see Experiments 10, 11, and 28). At a pH below ~ 8, it exists as the colorless acid, H_2In. In the presence of a base, at pH values from 8 to 10, it loses two protons, forming the pink ion, In^{2-}. At a pH above 10, the species In^{2-} reacts with excess base, OH^-, to form the colorless $In(OH)^{3-}$ ion. The rate law for this reaction is

$$\text{rate} = k[In^{2-}]^m[OH^-]^n \tag{E29.6}$$

In this reaction, the concentration of phenolphthalein is extremely small whereas the concentration of OH^- is very high. Only a small fraction of the OH^- reacts, so that the $[OH^-]$ remains essentially constant. The rate law, equation (E29.6), simplifies to

$$\text{rate} = k'[In^{2-}]^m$$

where k' is the **pseudo rate constant**, $k' = k[OH^-]^n$. The reaction thus *appears* to be dependent only on $[In^{2-}]$. If the reaction is first order with respect to phenolphthalein ($m = 1$), then a plot of $\log[In^{2-}]$ versus time results in a straight line with a slope equal to $-k'/2.303$. Knowing the slope, one can calculate k', the pseudo rate constant.

By measuring the reaction rate at several different values of $[OH^-]$, one can show that the reaction is first order with respect to $[OH^-]$. The n and the actual rate constant k can be evaluated in the following way. Taking the logarithm of both sides of the equation $k' = k[OH^-]^n$, one obtains

$$\log k' = n \log[OH^-] + \log k$$
$$y = mx + b$$

A plot of $\log k'$ versus $\log[OH^-]$ gives a straight line of slope of n, the order of the reaction with respect to OH^-. The intercept on the y axis will be equal to $\log k$, from which the rate constant can be easily calculated.

In Part B, the change in $[In^{2-}]$ (which is initially pink and slowly fades to a colorless species in the presence of base) is determined by measuring the absorbance of a solution with time in a spectrophotometer. The absorbance A of the solution is proportional to $[In^{2-}]$ (see Chapter 8). Thus, a plot of $\log A$ versus time will give a straight line if the reaction is first order in In^{2-}. The slope of the line will provide the rate constant, k'. *Note:* In this experiment, the total ionic concentration (the ionic strength) of the solutions is kept constant. For this purpose, a solution of NaCl is used in preparing the solutions.

References

1. Read the chapter(s) on reaction kinetics from your general chemistry textbook.
2. Slowinski, E. J.; Wolsey, W. C.; Masterton, W. L. *Chemical Principles in the Laboratory*, 5th ed.; Saunders: Philadelphia, 1989.
3. Nicholson, L. *J. Chem. Educ.* **1989**, *66*, 725.

Experimental Section

Procedure

Part A: Rate of Reaction of Iodine with Acetone

> Microscale experiment   Estimated time to complete the experiment: 2.5 h
> Experimental Steps: This experiment involves two main steps: (1) preparation of 4 M
> acetone, 1 M HCl, and 0.005 M iodine and (2) measuring the reaction time.

**Before coming to the laboratory, calculate the concentrations of the reactants, taking
into account the dilution factor. Moreover, calculate the volumes of acetone, H^+ (HCl),
and I_2 solution required to double or triple their concentrations in order to calculate
the orders of the reaction. See prelaboratory report sheet. Perform this experiment
with a partner.**

Obtain three clean test tubes (15×85 or 13×100 mm), two 10 mL graduated pipets
(one for acetone, acid, and water and the other for I_2 solution), a 25 mL and a 100 mL volumetric flask, a 250 mL beaker and a thermometer. Pure acetone, 6 M HCl, iodine crystals,
solid KI, and deionized water are provided (if the solutions are supplied by your instructor,
then collect 10 mL each of the stock solutions of 4 M acetone, 1 M HCl, and 0.005 M I_2 in
three small beakers).

Preparation of Solutions

1. *4 M acetone solution*: Using a micropycnometer (see Experiment 1) determine
    the density of 99.5 to 99.9 percent acetone (MW = 58.08 g/mol). Prepare 25 mL
    of 4 M solution by transferring the appropriate volume (using the density value
    you have determined, calculate the volume of acetone you need) of acetone to a
    25 mL volumetric flask (enter your calculation on the prelaboratory report). Add
    water to the mark, shake the flask well to mix, and transfer the solution to a labeled
    bottle.
2. *1 M HCl solution*: Dilute the proper volume of 6 M HCl in a 25 mL volumetric
    flask, mix well, and transfer the solution to a labeled bottle.
3. *0.005 M iodine solution*: Dissolve 200 to 250 mg of KI in 50 mL of deionized
    water in a 100 mL volumetric flask. Weigh 127 mg of iodine crystals, and imme-
    diately transfer them into the volumetric flask. Swirl the contents of the flask for at
    least 5 minutes to dissolve I_2 crystals. Dilute to the mark, and mix. **This amount
    makes sufficient solution for four students.**

> CAUTION: Iodine will stain the balance pan, skin, and clothing. In the event of a spill,
> rinse with a 1 percent KI solution containing 1 g of sodium thiosulfate.

Data Collection

Run #1

A room-temperature water bath should be set up in a 250 mL beaker. Pipet exactly 2 mL of 4 M acetone, 2 mL of 1 M HCl and 4 mL of deionized water into one 13 × 100 mm test tube (labeled #1). Shake the tube to mix. Using a second pipet, add exactly 2 mL of the I_2 solution to a second test tube (labeled #2). Add 10 mL of deionized water to a third test tube (labeled #3), to be used as a color reference. Allow the test tubes to stand for 3 to 4 minutes in the water bath for equilibration of temperature. Record the temperature.

Noting the time of mixing to the nearest second, pour the contents of test tube #1 into test tube #2 (total volume 10 mL). Pour back and forth between the test tubes two times to mix the solutions thoroughly. Do not spill. The mixture will appear yellow as a result of the presence of I_2. Place test tubes #2 and #3 side by side against a white background. Record the time (in seconds) that the **I_2 color disappears** (look from the top, not from the side of the test tube). Repeat the procedure, until the times for duplicate runs are within 10 percent of each other. In duplicate runs, the reacted solution in test tube #2 should be used as a reference instead of water. This is run #1. Record the time on the data sheet.

> Disposal procedure: When finished, discard the contents of test tube #2 and all reacted solutions into a proper waste container.

Runs #2 and #3: Order with Respect to Acetone

To calculate the order of the reaction with respect to acetone, perform a second run in the same manner as run #1. The concentration of acetone in run #2 should be doubled from run #1, but the concentrations of other chemicals are held constant. The total volume of the mixture (acetone + HCl + water + I_2 solution) must remain constant at 10 mL, so less water should be added to test tube #1 than before (calculate the amount). Make a duplicate run for the mixture. Record the data.

If time permits, perform run #3 using a concentration of acetone triple that of run #1 (still keeping the concentrations of other chemicals and the total volume the same). Record the time.

Runs #4 through #7: Order with Respect to H^+ and I_2

To calculate the order of the reaction with respect to H^+, perform a fourth run in the same manner as run #1. The concentration of H^+ in run #4 should be doubled from run #1, but the other concentrations are held constant. The total volume of the mixture (acetone + HCl + water + I_2 solution) must remain constant at 10 mL, so less water should be added to test tube #1 than before (calculate the amount). Transfer 2 mL of I_2 solution to test tube #2. Mix the solutions as before. If time permits, perform run #5 with triple the $[H^+]$. Similarly, perform runs #6 and #7, this time doubling and tripling $[I_2]$ in test tube #2. In each case, make duplicate trials; the time should agree within 10 percent of the previous run. Record all the data.

Runs #8 through #10: Determining the Activation Energy of the Reaction

Refer to run #1 for the amounts of reagents to be used for each temperature. Follow the same procedure as before, but keep the test tubes in a constant-temperature bath that is 10°C (use ice) below room temperature (run #8), in a constant-temperature bath

10°C above room temperature (run #9), and 20°C above room temperature (run #10). The two test tubes (#1 and #2) should sit in the water bath for at least 5 minutes prior to mixing to allow the temperatures to equilibrate. Record the data.

Data Manipulation

Before coming to the laboratory, calculate [acetone], [H$^+$], and [I$_2$] using $M_1V_1 = M_2V_2$ for all runs, where M_1 is the initial concentration of the stock solution, M_2 is the concentration to be found, V_1 is the volume of the reagent used, and V_2 is the total volume. Record these values on the data sheets.

1. Calculate *initial* [I$_2$] used in each run. *Hint*: The iodine is diluted from the initial concentration of 0.005 M to a final volume of 10 mL.
2. Calculate the rate using equation (E29.4) for each run, using the average time for the duplicate runs.
3. Calculate the order *a* (with respect to acetone) from runs #1, #2 and #3. Determine the average value. Calculate the order *b* (with respect to H$^+$) from runs #1, #4, and #5. Calculate the order *c* (with respect to I$_2$) from runs #1, #6, and #7 (use E29.5).
4. Calculate *k* (E29.3) for all runs (*k* should be approximately the same for all runs at room temperature).
5. The energy of activation, E_a, may be determined using the Arrhenius equation. From runs #1, #8, #9, and #10, plot ln *k* (*y* axis) versus $1/T$ (*x* axis). From the slope, $-E_a/R$, calculate E_a. ($R = 8.134$ J/K · mol)

Part B: Reaction of Phenolphthalein with a Strong Base

Microscale experiment   Estimated time to complete the experiment: 2.5 h
Experimental Steps: Three steps are involved: (1) setting up the instrument, (2) preparing different solutions, and (3) measuring the absorbances.

Disposal procedure: Ask your instructor if the solutions can be disposed of in the sink.

Note: Perform this experiment in pairs. Obtain two cuvettes (one for the solution and the other for the reference), a Spectronic 20 or other spectrophotometer, a Pasteur pipet, two 25 mL volumetric flasks, 1 percent phenolphthalein solution, 2 mL ethyl alcohol (absolute), solid NaOH, NaCl, a graduated 10 mL pipet, a 25 mL pipet, deionized water, four plastic bottles (100 mL), a test tube with a stopper to hold 0.25 percent phenolphthalein solution, and several 10 × 75 mm test tubes.

Preparation of Solutions

1. 0.25 percent phenolphthalein solution: Using a Pasteur pipet take five drops of 1 percent phenolphthalein in a small test tube. Add 20 drops of absolute alcohol with the same pipet. Stopper the test tube, and shake it well to mix the solution.

2. 0.30 M NaCl solution: Weigh the appropriate amount of NaCl (MW = 58.45 g/mol) to make a 0.30 M solution in a 25 mL volumetric flask.

3. 0.30 M, 0.20 M, 0.10 M, and 0.05 M NaOH solutions: Weigh enough NaOH to prepare 25 mL of 0.30 M NaOH (MW = 40.0 g/mol). Transfer the solid NaOH to a 100 mL plastic bottle labeled ("0.30M") and add 25 mL of deionized water. Shake the bottle to dissolve the NaOH. Prepare the diluted solutions in additional 100 mL plastic bottles in the following way.

To prepare the 0.20 M solution, pipet 20 mL of 0.30 M NaOH into a plastic bottle. Add 10 mL of 0.30 M NaCl (to maintain the ionic strength) solution to the bottle. Mix the solution properly and label it as 0.20 M NaOH solution. Similarly, starting with 0.20 M NaOH and 0.30 M NaCl, prepare a 20 mL sample of 0.10 M NaOH solution in the third plastic bottle. Likewise, using the 0.10 M NaOH and 0.30 M NaCl solutions, prepare a 20 mL sample of 0.050 M NaOH solution in the fourth bottle.

Spectrophotometric Method

Turn on the spectrophotometer and calibrate it according to the instructions provided in Experiment 21 (see Chapter 8). **Set the wavelength to 550 nm. Once set, do not change the setting. If a digital instrument is available, take all readings in absorbance units; otherwise all the readings must be taken on the transmittance ($\%T$) scale. Convert $\%T$ to the absorbance, A, using the equation $A = 2 - \log(\%T)$. Enter both values in the record sheet.**

Run #1
Fill the cuvette to the mark with 0.30 M NaOH solution. Add one drop of 0.25 percent phenolphthalein, invert the cuvette, and mix the solution (which will turn pink). Wipe the cuvette with a Kimwipe and place it in the sample holder. If the absorbance is above 0.60, leave the cuvette in the well until the reading drops below 0.60. If the absorbance is below 0.5, add an additional drop of phenolphthalein.

Start timing (to the second) when the absorbance is at any point between 0.6 and 0.5, and record the absorbance every 30 seconds for 3 minutes. Enter the absorbance readings and the corresponding times onto the data sheet.

Runs #2, #3, and #4
Repeat the procedure for run #1, except using 0.20 M NaOH solution. Take measurements of the absorbance and time every 1 minute for a total of 6 minutes. For run #3, perform the experiment using 0.10 M NaOH. Take absorbance readings versus time every 2 minutes for a total of 12 minutes. For run #4, perform the experiment using 0.050 M NaOH. Take absorbance readings versus time every 2 minutes for a total of 12 minutes. Enter all data on the data sheet. Finally, calculate k', the order of the reaction, and k.

Additional Independent Projects
1. Determination of the rate and the order of the iodine clock reaction, and the effect of catalysts on the reaction[1-3]
2. Determination of the rate of the reduction of potassium permanganate[4,5]
   Brief Outline
   Dilute aqueous solutions of potassium permanganate have a distinct pink color. The following reaction between permanganate and sodium oxalate in the presence of sulfuric acid takes place:

$$5\,C_2O_4{}^{2-}(aq) + 2\,MnO_4{}^-(aq) + 16\,H^+(aq) \rightarrow 10\,CO_2(g) + 2\,Mn^{2+}(aq) + 8\,H_2O(l)$$

Pink   Colorless

The rate of disappearance of the pink color depends, among other factors, on the temperature and any catalysts used. This reaction can be modified and studied in the same manner as described in Experiment 29, Part A. Develop the data sheets, calculate the orders and rate constant for the reaction, and write a short report following the directions given in Chapter 4.

3. Aquation of tris(1,10-phenanthroline)iron(II) in the presence of 2 M HCl Solution[6]

Brief Outline

In Experiments 21 and 24, the intensely colored red iron complex, $Fe(o\text{-phen})_3{}^{2+}$, is used as a probe. This complex undergoes aquation reaction in the presence of an acid, resulting in the formation of a colorless species:

$$Fe(o\text{-phen})_3{}^{2+} + 3\,H^+ + 6\,H_2O \rightarrow Fe(H_2O)_6{}^{2+} + 3\,phenH^+$$

A stock solution of $Fe(o\text{-phen})_3{}^{2+}$ is prepared by mixing an aqueous solution of iron(II) ammonium sulfate (0.066 g) in ~1.5 to 2.0 mL of water with a solution of phenanthroline. Add six drops of the red complex to ~50 mL of 2 M HCl solution in an Erlenmeyer flask. Place the flask in a water bath for the thermal equilibration. Run the visible spectra at **510 nm** by withdrawing ~3 to 4 mL portions of the solution. Start with a room-temperature (~20°C) solution and run several spectra at intervals of ~15 minutes for ~90 minutes. Following the same procedure, prepare a new sample and raise the temperature of the bath to ~30°C. Run the spectra of the new sample (~4 mL) at intervals of ~10 minutes for ~75 minutes. Repeat the procedure at ~35°C (at intervals of 10 minutes for a total of 65 to 75 minutes), ~40°C (intervals of 5 minutes for ~45 to 55 minutes), and at 45°C (intervals of 3 minutes for 30 minutes). Construct a data collection table (temperature, time, absorbance A, and log A for each series of runs). Plot 1 + log A versus time (minutes) at different temperatures. Using the plot, determine the rate constant k from the slope of the graph at the temperature concerned. If desired, the rate constants at higher temperatures can be easily obtained by following the visual half-life method (see reference 6). A plot of 3 + log k versus $1/T$ (K) will generate a straight line from which the activation energy can be calculated. Develop the data tables and submit a written report according to the instructions given in Chapter 4.

4. Second-order kinetics: dimerization of an organic compound[7]

Brief Outline

This experiment is very interesting and simple. Dissolve 30 to 50 mg of the dimer (of 2,5-dimethyl-3,4-diphenylcyclopentadienone—the formula or the structure of the compound is not important) in 10 mL of toluene. Heating the colorless solution in a steam bath (use **HOOD**) for 10 to 15 minutes will generate a colored solution of the monomer. Using an ice bath, cool the solution to room temperature. As the solution cools, the dimerization of the monomer starts. Transfer the solution to a cuvette and start measuring the absorbance at **460 nm** of the solution every 5 minutes for 30 minutes and thereafter every 10 minutes until the dimerization reaction slows down. It is known that 1.00×10^{-3} M solution of the monomer has an absorbance of 0.225. From this, calculate the concentration of the monomer in the solution corresponding to the absorbance. Plot the [monomer] versus absorbance. Draw tangents at various points on the smoothed curve and calculate the slopes. The negative slopes give the rate of disappearance of the monomer at the tangent time and concentration. Also, plot a graph of absorbance versus time in minutes. Since the absorbance is directly proportional to [monomer], the rate can be expressed as the ratio of absorbance over time in minutes. Draw three separate graphs (on the same graph paper) by plotting (A/time) versus concentration (expressed in A raised to powers 1, 2, and 3: orders of the reaction). Determine the order of the reaction from the graph that gives the best straight line. **Do not throw away the sample; it can be recycled and reused repeatedly**. Develop the data tables (data collection and data manipulation tables) and submit a written report according to the instructions given in Chapter 4.

Notes

1. Szafran, S.; Pike, R. M.; Foster, J. *Microscale General Chemistry Laboratory with Selected Macroscale Experiments*; Wiley: New York, 1993.
2. Pickering, M. *A General Chemistry Lab Manual: The Discovery Book*; Scott, Foresman/Little: Glenview, IL, 1990.
3. Slowinski, E. J.; Wolsey, W. C.; Masterton, W. L. *Chemical Principles in the Laboratory*, 5th ed.; Saunders: Philadelphia, 1989.
4. Ponraj, D. S.; Venkataraman, R.; Raghavan, P. S. *J. Chem. Educ.* **1990**, *67*, 621.
5. Datta, M.; Ameta, S. C.; Pande, P. N.; Bokadia, M. M. *J. Chem. Educ.* **1979**, *56*, 659.
6. Twigg, M. *J. Chem. Educ.* **1972**, *49*, 191.
7. Weiss, H. M.; Touchette, K. *J. Chem. Educ.* **1990**, *67*, 707.

Name ______________________	Section ____________
Instructor ______________________	Date ____________

PRELABORATORY REPORT SHEET—EXPERIMENT 29A

Experiment title ______________________________

Objective

Reactions/formulas/concentrations to be used

Solution concentration:

The total volume of each mixture is 10 mL.

	0.005 M I_2, mL	**[I_2]**	**4 M acetone, mL**	**[Acetone]**	**1 M HCl, mL**	**[H^+]**
Mixture 1	______	______	______	______	______	______
Mixture 2	______	______	______	______	______	______
Mixture 3	______	______	______	______	______	______
Mixture 4	______	______	______	______	______	______
Mixture 5	______	______	______	______	______	______
Mixture 6	______	______	______	______	______	______
Mixture 7	______	______	______	______	______	______

Materials and equipment table

Outline of procedure

PRELABORATORY REPORT SHEET—EXPERIMENT 29B

Experiment Title _____

Objective

Reactions/formulas/concentrations to be used

Materials and equipment table: prepare a table for solutions needed for part B (see part A)

Outline of procedure

PRELABORATORY PROBLEMS—EXPERIMENT 29 (FOR BOTH PARTS)

1. A solution consisting of 2 mL each of 4 M acetone and 1 M HCl is mixed with a solution containing 2 mL of 0.005 M I_2 and 4 mL of water. The color of I_2 completely disappeared after 5 minutes. Calculate the rate of the reaction. Assume that I_2 is the limiting reagent.

2. If you increase the concentration of NaOH in Part B of this experiment, how does this change affect the rate of the reaction? Why?

3. A solution containing 5.00 ppm $KMnO_4$ has a transmittance of 0.310 in a 1.00 cm cell at 520 nm. Calculate the molar absorptivity of $KMnO_4$. (See Chapter 8.)

Data Collection Table

Temperature for each run _____°C

Total volume of each mixture: $V_2 = 1.00 \times 10^{-2}$ L

Data for Order of Reaction

Run #	0.005 M I_2, mL	4 M acetone, mL	1 M HCl, mL	Water, mL	Trial 1, t, s	Trial 2, t, s	Avg. t, s
1	2	2	2	4			
2							
3							
4							
5							
6							
7							

Data for Activation Energy

Run #	Temperature, K Trial 1	Temperature, K Trial 2	Time, s Trial 1	Time, s Trial 2	Average time, s	Rate
1						
8						
9						
10						

Show calculations:

Name ______________________	Section ______________
Instructor ______________________	Date ______________

Data Manipulation Table

Calculation of Reaction Rates and Reaction Orders

Use the average time t(s) from previous page and the initial $[I_2]$ from the solution concentration table constructed in prelaboratory report sheet (use equation E29.4, rate = Initial$[I_2]/t$):

Run #1	**Run #2**	**Run #3**	**Run #4**	**Run #5**	**Run #6**	**Run #7**
$rate_1$ = ____	$rate_2$ = ____	$rate_3$ = ____	$rate_4$ = ____	$rate_5$ = ____	$rate_6$ = ____	$rate_7$ = ____

Order a with respect to acetone (use equation E29.5)

1. Using $rate_2$ and $rate_1$: _____ 2. Using $rate_3$ and $rate_1$: _____ Average a = _____

Show a sample calculation:

Order b with respect to H^+ (use equation E29.5)

1. Using $rate_4$ and $rate_1$: _____ 2. Using $rate_5$ and $rate_1$: _____ Average b = _____

Show a sample calculation:

Order c with respect to I_2 (use equation E29.5)

1. Using $rate_6$ and $rate_1$: _____ 2. Using $rate_7$ and $rate_1$: _____ Average c = _____

Show a sample calculation:

Calculation of the Rate Constant *k* for Each Run

Use equation E29.3. Substitute the values for the reaction orders calculated above, the initial concentrations of each chemical (see prelaboratory report sheet), and the corresponding observed rate for each run. Remember that the order for I_2 should be zero.

Run 1	**Run 2**	**Run 3**	**Run 4**	**Run 5**	**Run 6**	**Run 7**

Show a sample calculation:

Determination of Energy of Activation, E_a

Concentrations

$[I_2]$	[acetone]	$[H^+]$
_____	_____	_____

Run #	Temperature, K	Rate constant, k	$\ln k$	$1/T$, K^{-1}
1. At room temp.	_____	_____	_____	_____
8. 10°C below room temp.	_____	_____	_____	_____
9. 10°C above room temp.	_____	_____	_____	_____
10. 20°C above room temp.	_____	_____	_____	_____

Make a plot of $\ln k$ (y axis) vs. $1/T$ (x axis). Attach the plot to the laboratory report.

Slope $(-E_a/R)$ _____ $R = 8.31$ joules mol^{-1}K^{-1}

$E_a = $ _____ J/mol

EXPERIMENT 29B   DATA SHEET

Data Collection Table

Run #1: 0.300 M NaOH

Time t, min	Transmittance	Absorbance A	log A
_____	_____	_____	_____
_____	_____	_____	_____
_____	_____	_____	_____
_____	_____	_____	_____
_____	_____	_____	_____
_____	_____	_____	_____
_____	_____	_____	_____

Run #2: 0.200 M NaOH

Time t, min	Transmittance	Absorbance A	log A
_____	_____	_____	_____
_____	_____	_____	_____
_____	_____	_____	_____
_____	_____	_____	_____
_____	_____	_____	_____
_____	_____	_____	_____
_____	_____	_____	_____

Run #3: 0.100 M NaOH

Time t, min	Transmittance	Absorbance A	log A
_____	_____	_____	_____
_____	_____	_____	_____
_____	_____	_____	_____
_____	_____	_____	_____
_____	_____	_____	_____
_____	_____	_____	_____

Run #4: 0.050 M NaOH

Time t, min	Transmittance	Absorbance A	log A
————	————	————	————
————	————	————	————
————	————	————	————
————	————	————	————
————	————	————	————
————	————	————	————
————	————	————	————

Data Manipulation Table

Order of the Reaction with Respect to Phenolphthalein

Plot log A versus (y axis) time t (0 to 12 minutes, x axis) for each run. Determine the slope in each case and calculate k' from the slope. Are all the plots linear? Attach the graphs to the report.

Order of the Reaction in [OH⁻] where [OH⁻] = Molar Concentration of NaOH Solution

Run #	[OH⁻]	log[OH⁻]	k'	log k'
1	————	————	————	————
2	————	————	————	————
3	————	————	————	————
4	————	————	————	————

Is the reaction first order in [OH⁻]? Plot log k' (y axis) versus log [OH⁻] (x axis). Attach your plots with your report.

Calculation of the Rate Constant k

Run #	[NaOH]	k'	k	Average k
1	————	————	————	
2	————	————	————	
3	————	————	————	
4	————	————	————	————

Show calculations: use a separate sheet.

POSTLABORATORY PROBLEMS—EXPERIMENT 29A AND B

1. List five factors that affect chemical reaction rates.

2. How is the rate of a reaction affected by a temperature change?

3. **(a)** If an excess of I_2 had been used in Experiment 29A, how would it have changed the data collected? Why?

   (b) If an excess of NaOH had been used in Experiment 29B, how would it have changed the data collection? Why?

CHAPTER 9 Organic Chemistry

EXPERIMENT **30** **An Introduction to Organic Qualitative Analysis**

Objective

- To provide an introduction to organic chemistry and its major functional classes.

Prior Reading

- Inorganic qualitative analysis: Experiment 16.

Related Experiments

- Experiments 16 (Inorganic qualitative analysis) and 31 (Organic functional group analysis)

One of the exciting challenges that some chemists face regularly is the identification of organic compounds. Approximately 5 million organic compounds have been prepared or isolated to date. The identification of these materials is of utmost importance—and it appears to be an insurmountable task! The majority of these substances, however, can be grouped into a comparatively small number of classes. The chemist, working as an analyst, has at his or her disposal an enormous database of chemical and spectroscopic information that has been organized over the past years. Forensic chemistry (the identification of drugs and other evidence), environmental chemistry (the identification of, among other compounds, toxic pollutants), and industrial research and development, to name a few areas, all depend to a large extent on the ability of the chemist to isolate, purify, and identify specific organic chemicals.

The task of identification was originally based on the solubility characteristics of the compounds and on certain chemical tests that could be used to detect the presence of various functional groups. Spectroscopic techniques such as infrared, nuclear magnetic resonance, ultraviolet, visible and mass spectroscopy are now used extensively for this purpose.

In this experiment, the basic chemical and solubility tests that can be used to distinguish compounds containing the major functional groups reviewed in Experiment 31 will be explored. These groups include the alkanes, alkenes, alkyl and aryl halides, alcohols, aldehydes, ketones, carboxylic acids, and amines. In this experiment, the compounds that will be tested have previously been purified. Physical measurements are often used to determine purity. For solids, the melting point and for liquids, the boiling point (Experiment 2), refractive index (Experiment 3), and density (Experiment 1) are generally measured.

In any analysis of unknown compounds one should follow a systematic approach. One possible sequence is suggested here:

1. Perform an ignition test to determine the general nature of the compound. Does it contain double bonds? Is it aromatic or aliphatic?
2. Determine the solubility characteristics of the species. This test can often lead to valuable information related to the structural composition of an unknown organic compound.

3. Carry out chemical tests to assist in identifying elements other than carbon, hydrogen, and oxygen. Nitrogen, phosphorus, sulfur, and the halogens are frequently found in organic compounds.

4. Carry out classification tests to detect common functional groups present in the molecule. The majority of these tests may be done using a few drops of a liquid or milligrams of a solid. An added benefit, especially in relation to the chemical detection of functional groups, is that an incredible amount of chemistry can be *observed* and *learned* in performing these tests. The successful application of these tests requires that you develop the ability to think in a logical manner and that you learn to interpret the significance of each result based on your observations.

It is important to realize that *negative* findings are often as important as positive results in classifying and identifying a given compound. Cultivate the habit of following a systematic pathway or sequence so that no clue or bit of information is lost or overlooked along the way. It is also important to develop the attitude and habit of planning ahead. Outline a logical plan of attack, depending on the nature of the unknown, and follow it. As you gain more and more experience in this type of investigative endeavor, the planning stage will become easier. Record all observations and results of the tests on your data sheet. Review these data as you execute the sequential phases of your plan. This serves to keep you on the straight and narrow path to analytical success.

In this experiment, you will first be given a series of known compounds upon which to carry out the tests. In this manner you can observe and record how each species behaves under certain experimental conditions. This exercise will be followed by a series of selected unknown compounds for you to categorize into their respective classes.

Experimental Section

Procedure

Microscale experiment	Estimated time to complete the experiment: 2 h

> Disposal procedure: Dispose of all the waste in marked containers.

A. Ignition Tests

Obtain a copper wire, several 10×75 mm test tubes, soda lime, manganese(IV) oxide (MnO_2), Brilliant Yellow (or litmus) papers, test solutions and organic compounds. Calibrate a Pasteur pipet (Section 3.10). The ignition test is carried out by placing 1 to 2 mg of the sample on a spatula followed by heating with a microburner. Do not hold the sample directly in the flame—heat the spatula about 1 cm from the flat end and move the sample slowly into the flame. Important observations to be made concerning the ignition test are summarized in Table E30.1:

Table E30.1 Ignition Test Observations*

Type of Compound	Example	Observation
Aromatic and other unsaturated compounds	Toluene	Yellow, sooty flame
Lower aliphatic compounds	Hexane	Yellow, almost nonsmoky flame
Compounds with oxygen	Ethanol	Clear bluish flame
Polyhalogen compounds	Chloroform	Do not ignite until flame is applied directly to the compound
Sugars and proteins	Sucrose	Characteristic odor
Acid salts or organometallics	Sodium acetate	Residue

*Source: Cheronis, N. D.; Entrikin, J. B. *Semimicro Qualitative Analysis*; Interscience: New York, 1947; page 85.

As the heating of the sample takes place you should make the following observations:

1. Any melting or evidence of sublimation, which gives an approximate idea of the melting point by the temperature necessary to cause melting
2. The color of the flame as the substance begins to burn (see Table E30.1)
3. The nature of the combustion (flash, quiet, or an explosion). Rapid, almost instantaneous combustion indicates high hydrogen content. An explosion indicates the presence of nitrogen or N_xO_y-containing groups.
   (*a*) If a black residue remains and disappears on further heating at higher temperature, the residue is carbon.

(b) If the residue undergoes swelling during formation, the presence of a carbohydrate or similar compound is indicated.

(c) If the residue is initially black, but still remains after heating, an oxide of a metal is indicated.

(d) If the residue is white, the presence of an alkali or alkaline earth carbonate or SiO_2 from a silane or silicone is indicated.

Testing of Known Materials

1. To test the nature of hydrocarbons (aliphatic and aromatic), ignite small samples of toluene (aromatic) and hexane (aliphatic). Record your observations in the data table.

2. To test the nature of alcohols, sugars, and acid salts, ignite small samples of methanol or ethanol, glucose or sucrose, and sodium tartrate or ferrocene (iron residue) or hexamethyl disiloxane (SiO_2 residue). Record your observations in the data table.

3. The Beilstein test is used to detect the presence of halogens (Cl, Br, I). Organic compounds that contain chlorine, bromine, or iodine and hydrogen are decomposed on ignition in the presence of copper oxide to yield the corresponding hydrogen halides. These gases react to form the volatile cupric halides, which impart a green or blue-green color to a nonluminous flame. The test is very sensitive, but some nitrogenous compounds and carboxylic acids also give positive results.

   Pound the end of a copper wire to form a flat surface that can act as a spatula. Stick the other end of the wire (~4 in. long) in a cork stopper to serve as a handle. Heat the flat tip of the wire in a flame until coloration in the flame is negligible.

   Place a drop of liquid unknown or a few milligrams of solid unknown on the cooled flat surface. Gently heat the material in the flame. The carbon present in the compound will burn first (the flame will be luminous), but then the characteristic green or blue-green color will be evident. It may be fleeting, so watch carefully. Fluoride is not detected by this test since copper fluoride is not volatile.

   To observe the Beilstein test, ignite a small sample of *tert*-butyl chloride or bromobenzene as just directed. Record your observations in the data table.

B. Solubility Characteristics

Determination of the solubility characteristics of an organic compound can often give valuable information as to its structural composition. It is especially useful when correlated with spectral analysis. Several schemes have been proposed that place a substance in a definite group according to its solubility in various solvents. A simplified version covering those types of compounds is discussed shortly.

There is no sharp dividing line between soluble and insoluble, and an arbitrary ratio of solute to solvent must be selected. We suggest that a compound be classified as soluble if its solubility is greater than 15 mg/500 μL of solvent. Solubility determinations should be carried out at ambient temperature in 10×75 mm test tubes. Place 15 mg of sample in the test tube, and add a total of 0.5 mL of solvent in three separate portions from a calibrated Pasteur pipet. Between additions, stir the sample vigorously with a glass stirring rod for 1.5 to 2 minutes. If the sample is water soluble, test the solution with litmus paper to assist in classification according to the solubility scheme in Figure E30.1.

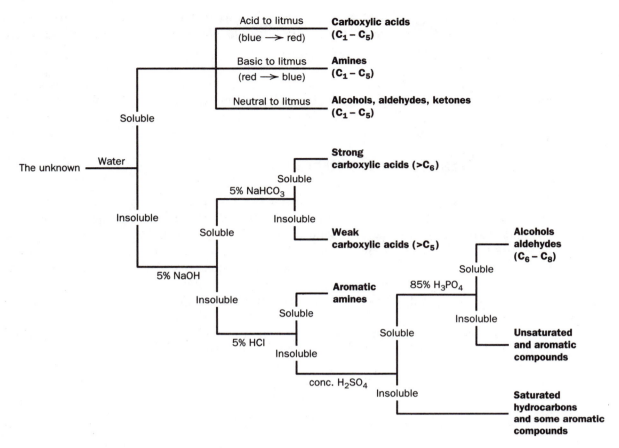

Figure E30.1 Solubility scheme for organic compounds. The subscripts under the symbol C indicate the number of carbon atoms in an organic compound.

To test the solution with litmus paper, dip the end of a small glass rod into the solution and then gently touch the litmus paper with the rod. ***Do not dip the litmus paper into the solution.***

In performing the solubility tests, follow the scheme in the order given next. ***Keep a record of all observations.***

1. Test for water solubility. If the unknown is soluble, test the acidity of the solution with litmus paper.
2. If the compound is water soluble, determine the solubility in diethyl ether. This test further classifies water-soluble materials.
3. A water-insoluble compound is now tested with 5 percent aqueous NaOH solution. If it is soluble, determine the solubility in 5 percent aqueous $NaHCO_3$. The use of the $NaHCO_3$ solution aids in distinguishing between strong (soluble) and weak (insoluble) acids.
4. Test compounds insoluble in 5 percent aqueous NaOH with 5 percent HCl.
5. Test compounds insoluble in 5 percent aqueous HCl with concentrated H_2SO_4. If they are soluble, one makes further differentiation using 85 percent H_3PO_4 as shown in the scheme.

Note that it may not be necessary to test solubility in every solvent to classify a given compound. ***Do only those tests that are required to place the compound in one of the solubility groups.*** Make all observations with care, and proceed in a logical sequence as you make the tests.

To observe solubility classifications, check the solubility of toluene, hexane, isopropyl alcohol, *tert*-butyl chloride, 2-pentene, methyl ethyl ketone, benzoic acid, and triethylamine following the scheme in Figure E30.1. Record all observations in the data table.

C. Chemical Test for Nitrogen

The soda lime test is used to detect the presence of nitrogen in an organic compound. In a 10×75 mm test tube, mix ~50 mg of soda lime and 50 mg of MnO_2. Add one drop of a liquid unknown or ~10 mg of a solid unknown. Place a moist strip of Brilliant Yellow paper (moist red litmus paper may be used if necessary) over the mouth of the tube. Using a test-tube holder, hold the tube at an angle and heat the contents gently at first and then quite strongly. Nitrogen-containing compounds will usually evolve ammonia. A positive test for nitrogen is the deep red coloration of the Brilliant Yellow paper (or blue color of the litmus paper).

To observe the soda lime test, ignite a small sample of triethylamine or N,N-dimethylaniline as just directed. Record all observations in the data table.

Name ______________________ Section ______________

Instructor ______________________ Date ______________

PRELABORATORY REPORT SHEET—EXPERIMENT 30

Experiment title ______________________

____

Objective

Reactions/formulas to be used

Materials and equipment table

Outline of procedure

1. What are aliphatic and aromatic compounds? What are alcohols, organic carboxylic acids, amines, and ketones? Give examples.

2. How can you detect the presence of nitrogen and halogen in an organic compound?

3. Describe the steps for determining the solubility of an organic compound.

Name ______________________ Section ____________

Instructor ______________________ Date ____________

EXPERIMENT 30 DATA SHEET

Observations on known and unknown compounds

A. Ignition tests:

1. Aromatic

2. Aliphatic

3. Alcohol

4. Sugar

5. Acid salt

6. Beilstein test

B. Solubility classification

Outline a solubility scheme for each of the tested compounds.

    1.   Toluene

    2.   Hexane

    3.   Isopropyl alcohol

    4.   *tert*-Butyl chloride

    5.   2-Pentene

    6.   Methyl ethyl ketone

    7.   Benzoic acid

    8.   Triethylamine

C. Chemical test for nitrogen—soda lime test

1. Triethylamine

POST LABORATORY PROBLEMS—EXPERIMENT 30

1. An unknown compound is insoluble in water, 5 percent sodium hydroxide, 5 percent hydrochloric acid, and concentrated sulfuric acid. Would you classify this material as an alcohol, ketone, saturated hydrocarbon, or an amine? Give reasons for your choice.

2. An unknown compound was observed to burn with a nonsooty yellow flame. It was insoluble in 5 percent NaOH, 5 percent HCl, and concentrated sulfuric acid solution. Indicate precisely what each of the solubility checks tells you about the unknown compound. To what class does the unknown compound belong?

EXPERIMENT **31** **Organic Functional Group Analysis**

Microscale Method

Objective
- To be able to identify several major organic functional groups

Prior Reading
- Experiment 30

Related Experiments
- Experiments 16 (Inorganic qualitative analysis) and 30 (Organic qualitative analysis)

Tests have been developed to classify organic compounds by the functional group or groups they contain. Several of these tests for the more important functional groups are outlined here and summarized in the following table.

Functional Groups	Tests
Alcohols	1. Ceric nitrate
	2. HCL/ZnCl$_2$ (Lucas)
Unsaturated hydrocarbons	1. Bromine in methylene chloride
	2. Permanganate (Baeyer)
Aldehydes and ketones	2,4-Dinitrophenylhydrazine
Carboxylic acids	1. Litmus
	2. Bicarbonate

Experimental Section

Procedure

Tests for Alcohols

Microscale experiment	Estimated time to complete the experiment: 2 h

Disposal procedure: Dispose of all the chemical compounds and solutions in marked containers.

NOTE: *For all tests given in this section, drops of reagents are measured using Pasteur pipets.*

1. Ceric Nitrate Test

In this test, primary, secondary, and tertiary alcohols having fewer than 10 carbon atoms give a positive test as indicated by a change in color from *yellow* to *red*,

$$(NH_4)_2Ce(NO_3)_6 + RCH_2OH \rightarrow [alcohol + reagent]$$
$$\text{yellow} \qquad\qquad\qquad \text{red complex}$$

Place five drops of test reagent on a white spot plate. (The reagent is prepared by dissolving 4.0 g of ceric ammonium nitrate in 10 mL of 2 M HNO_3. Warming may be necessary). Add one or two drops of the unknown sample (5 mg if a solid). Stir the mixture with a thin glass rod to mix the components, and observe any color change.

To become familiar with the ceric nitrate test, run the test on several known alcohols such as ethanol, isopropyl alcohol, and *tert*-butyl alcohol. Record all observations in the data table.

2. The HCl/ZnCl₂ Test—The Lucas Test

The Lucas test is used to distinguish between primary, secondary, and tertiary alcohols having fewer than six carbon atoms. The reaction is

$$2\,R–OH + 2\,H^+ + ZnCl_2 \rightarrow RCl\,(insol) + H_2O$$

The test requires that the alcohol initially be in solution. As the reaction proceeds, the corresponding alkyl chloride forms, which is insoluble in the reaction mixture. As a result, the solution becomes cloudy. In some cases a separate layer may be observed.

1. Tertiary alcohols react to give an immediate cloudiness to the solution. You may be able to see a separate layer of the alkyl chloride after a short time.
2. Secondary alcohols generally produce a cloudiness within 3 to 10 min. The solution may have to be heated to obtain a positive test.
3. Primary alcohols dissolve in the reagent but react very, very slowly.

Place two drops of the unknown (10 mg of solid) in a small test tube prepared by sealing a Pasteur pipet off at the shoulder (Figure E31.1), followed by 10 drops of the Lucas reagent. (The test reagent is prepared by dissolving 13.6 g of anhydrous $ZnCl_2$ in 10.5 mL of concentrated HCl, with cooling, in an ice bath.) Shake or stir the mixture with a glass rod and allow the solution to stand. Observe the results. The alcohol may be classified based on the times given above. (CAUTION: concentrated HCL is corrosive.)

To observe the Lucas test, run it on several known alcohols such as propanol, isopropyl alcohol, and *tert*-butyl alcohol. Record all observations in the data table.

Tests for Unsaturated Hydrocarbons

1. Bromine in Methylene Chloride

Unsaturated hydrocarbons (having double and triple bonds between carbon atoms) readily add bromine:

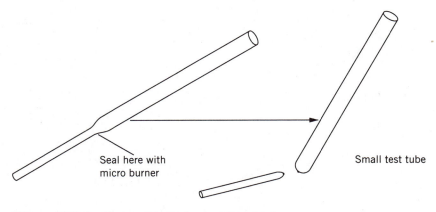

The test is based upon the decolorization of a red-brown bromine–methylene chloride solution.

> CAUTION: Perform the following tests inside a well-ventilated HOOD. Bromine is highly toxic and can cause burns. All halogenated hydrocarbons are toxic.

In a 10 x 75 mm test tube or a small tube prepared from a Pasteur pipet (see Figure E31.1), place two drops of a liquid unknown (or 15 mg if a solid) followed by 0.5 mL of methylene chloride (**HOOD**). Dropwise, with shaking, add a 2 percent solution of Br_2 in methylene chloride solvent (**HOOD**). If an unsaturated hydrocarbon is present, the solution

Seal here with micro burner

Small test tube

Figure E31.1   Preparation of a small test tube

---

will require two or three drops of the reagent before the reddish brown color of bromine persists. In the absence of an unsaturated hydrocarbon, the solution becomes reddish brown immediately.

Note that methylene chloride is used in place of the usual carbon tetrachloride since it is less toxic. Phenols, enols, amines, aldehydes, and ketones interfere with this test.

2. Permanganate Test—The Baeyer Test

Unsaturation in an organic compound can be detected by the decolorization of permanganate solution. The reaction involves the reaction of the olefin to give a 1,2-diol (glycol).

$$3 \; \text{C}=\text{C} + 2\,MnO_4^- + 4\,H_2O \longrightarrow 3 \; \underset{OH\;\;OH}{-\text{C}-\text{C}-} + 2\,MnO_2 + 2\,OH^-$$

On a white spot plate, place 0.5 mL of *alcohol-free* acetone followed by two drops of the unknown compound (or 15 mg if a solid). Dropwise, with stirring, add two or three drops of a 1 percent aqueous solution of $KMnO_4$. A positive test is the disappearance of purple permanganate color from the reagent and the precipitation of brown manganese oxides. Some functional groups that undergo oxidation with permanganate interfere with this test (phenols, aryl amines, most aldehydes, and primary and secondary alcohols).

To observe these unsaturation tests, run them on several known alkenes, such as cyclohexene or 2-pentene (mixed isomers). Record your observations in the data table.

Test for Aldehydes and Ketones

The 2,4-Dinitrophenylhydrazine Test

Aldehydes and ketones react rapidly with 2,4-dinitrophenylhydrazine to form 2,4-dinitrophenylhydrazones. These derivatives range in color from yellow to red depending on the number of double bonds in the carbonyl compound.

yellow to red precipitate

Place seven or eight drops of 2,4-dinitrophenylhydrazine reagent on a white spot plate. (Prepare the reagent by dissolving 1.0 g of 2,4-dinitrophenylhydrazine in 5.0 mL of concentrated sulfuric acid. Add this solution slowly, with stirring, to a mixture of 10 mL of water and 35 mL of 95 percent ethanol.) Add one drop of a liquid unknown. If the unknown is a solid, add one drop of a solution prepared by dissolving 10 mg of the material in 10 drops of ethanol. Stir the mixture with a glass rod. The formation of a red to yellow precipitate constitutes a positive test.

To observe the 2,4-dinitrophenylhydrazine test, run it on several known aldehydes and ketones, such as benzaldehyde and methyl ethyl ketone. Record your observations in the data table.

Tests for Carboxylic Acids

The presence of a carboxylic acid group is detected by its solubility behavior.

1. Litmus Test

An aqueous solution of the acid will be acidic to litmus paper, that is, blue litmus paper will turn red when a drop of the solution is placed on it.

2. Bicarbonate Test

A small amount of the acid added to a 10 percent sodium bicarbonate solution placed on a watch glass will evolve bubbles of carbon dioxide gas.

To observe the acid tests run them on several known acids, such as benzoic and acetic acid. Record your observations in the data table.

Reference
1. Griswold, J. R.; Rauner, R. A. *J. Chem. Educ.* **1991**, *68*, 418.
2. Mayo, D. W.; Pike, R. M.; Trumper, P. K. *Microscale Organic Laboratory,* 3rd ed., Chapter 10; Wiley: New York, 1994.

Name ______________________ Section ______________

Instructor ______________________ Date ______________

PRELABORATORY REPORT SHEET—EXPERIMENT 31

Experiment title ______________________________

Objective

Reactions/formulas to be used

Materials and equipment table

Outline of procedure

1. In each case given below, indicate the deductions that may be made as to the nature of the compound.

   (*a*) A neutral compound reacted with 2,4-dinitrophenylhydrazine reagent to form a yellow-red precipitate.

   (*b*) A compound was insoluble in water but soluble in 5 percent sodium hydroxide and 5 percent sodium bicarbonate solution. When placed in the bicarbonate solution bubbles of carbon dioxide were observed.

2. An unknown compound was observed to burn with a nonsooty yellow flame. It gave a negative ceric nitrate and Baeyer test. It did give a positive Beilstein test. Indicate precisely what each of the chemical tests tell you about the unknown compound. To what class does the unknown compound belong? (See Experiment 30)

Name ______________________ Section ______________

Instructor ______________________ Date ______________

EXPERIMENT 31 DATA SHEET

Classification tests: Known compounds

Unknown compounds

Alcohols

(*a*) Ceric nitrate test

(*b*) Lucas test

Unsaturated Hydrocarbons

(*a*) Bromine in methylene chloride

(*b*) Baeyer test

Aldehydes and Ketones

The 2,4-dinitrophenylhydrazine test

Carboxylic acids

(*a*) The litmus test

(*b*) The bicarbonate test

1. You are given five unlabeled bottles, each of which contains a colorless liquid (1-butanol, benzene, ethanoic acid, diethylamine, and 2-hexene). Explain how you would use simple chemical tests to identify the contents of each bottle.

2. How would you distinguish the compounds in each of the following sets from each other? List the reagents you would use and the expected experimental observations.

    (a) Ethanol, octanol, diethyl ketone

    (b) Cyclohexene, bromobenzene, toluene

EXPERIMENT 32 Synthesis and Analysis of Organic Compounds

Part A: Aspirin from Salicylic Acid and Acetic Anhydride

Part B: Iodoform by Electrolytic Method

Part C: Urea from Silver Cyanate and Ammonium Chloride: Recovery and Recycling of Silver

Part D: Determination of Molecular Weight of Urea by Elevation of Boiling Point; Recovery and Recycling of Ethanol and Urea

Microscale Experiments

Objectives
- To prepare organic compounds by using different methods including electrolysis
- To learn how to recrystallize and characterize organic compounds
- To determine the molecular weight of urea by the method of boiling point elevation

Prior Reading
- Sections 3.3–3.8   Handling of chemicals, weighing, volume measurement, heating, stirring, and filtration
- Experiment 2   Boiling point determination (Method 1 or 2)

Related experiments
- Experiments 9 (Electrolysis), 18 (Lead iodide), 19 (Copper), and 20 (Metal oxalates)

Activities a Week Before
- Make a chart of chemicals and glassware needed for each part

The area of chemistry that deals with compounds containing carbon and hydrogen as their primary components is known as *organic chemistry*. Carbon has a unique chemistry among the elements owing to a number of factors:

1. Carbon is a small element, capable of forming short, strong bonds, including multiple bonds.

2. Carbon has four valence orbitals (the $2s$ and the three $2p$ orbitals) and four valence electrons. It therefore is neither electron-deficient (like boron, which therefore acts as a Lewis acid) nor electron-rich (like nitrogen, which therefore acts as a Lewis base). Its compounds tend to be relatively nonreactive.

3. Carbon is capable of forming up to four strong covalent bonds *to itself* (**concatenation**). Unlike nitrogen, oxygen, and fluorine, carbon does not have lone pairs of electrons, and thus it suffers no lone pair–lone pair repulsions, which weaken single bonds. Carbon is uniquely able to form long chains, rings, and other structures of linked carbon atoms.

Organic chemistry encompasses many subdivisions, including biochemistry (the chemistry of compounds found in living systems), polymer chemistry (plastics, resins, and the like), pharmaceutical chemistry (drug synthesis), organometallic chemistry (carbon bonded to metals), and so on.

This experiment consists of three parts. In Part A, the drug aspirin is synthesized; in Part B, the compound iodoform (CHI_3) is prepared by electrolysis; and in Part C, the bio-organic compound urea (H_2NCONH_2) is synthesized from inorganic reagents. All of these compounds are organic compounds. In Part D, the molecular weight of urea is determined by elevation of the boiling point.

The **stoichiometry** (quantitative relationship between the reactants and the products) of a balanced chemical reaction is used to calculate the amount of the product formed (yield), the amount of the reactant(s) required to produce the product(s), and the amount of the reagent left (**excess reagent**) unreacted at the end of the reaction. The reactant that is completely consumed (used up) in a reaction is called a **limiting reagent**. A limiting reagent is always in short supply in a chemical reaction. The limiting reagent is the reactant that determines the yield of a particular product. **Theoretical yield** of a product is calculated on the basis of stoichiometric calculations involving the limiting reagent. The experimentally determined quantity of a product is called the **actual yield**. Knowing the theoretical yield and the experimental yield (actual yield), one can calculate the **percent yield** of a product according to the following equation:

$$\% \text{ yield} = (\text{Actual yield, g})(100)/(\text{Theoretical yield, g}) \qquad \text{(E32.1)}$$

Part A: Synthesis of Aspirin

Aspirin is the single most widely manufactured drug in the world, with more than 40 million pounds produced each year in the United States alone. Aspirin's chemical name is acetyl-salicylic acid. It is prepared by the reaction of acetic anhydride ($C_4H_6O_3$, hence, the *acetyl* root in its name) with salicylic acid.

Salicylic acid has the same analgesic and antipyretic (lowers fevers) properties as aspirin. Pure salicylic acid is more acidic than aspirin and is irritating to the stomach, mouth, and mucous membranes. It was therefore replaced by the milder acid aspirin. Some people find even aspirin to be too acidic and use various aspirin substitutes (acetaminophen, ibuprofen) instead (see Figure E32.1).

Several serious side effects are associated with the use of aspirin. High doses of aspirin have been known to damage the liver and kidneys over long periods of time. Reye's syndrome has been associated with the taking of aspirin by children with the flu or chicken pox. On the other hand, aspirin has been associated with lowering the risks of heart attacks owing to its anticoagulative properties. Obviously, care is indicated in the therapeutic use of even the most "innocuous" of drugs.[1]

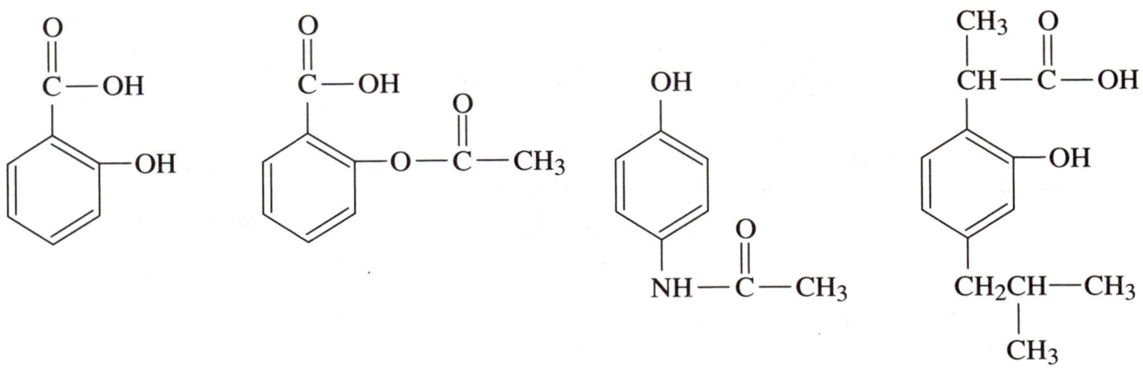

Figure E32.1   Aspirin and aspirin substitutes

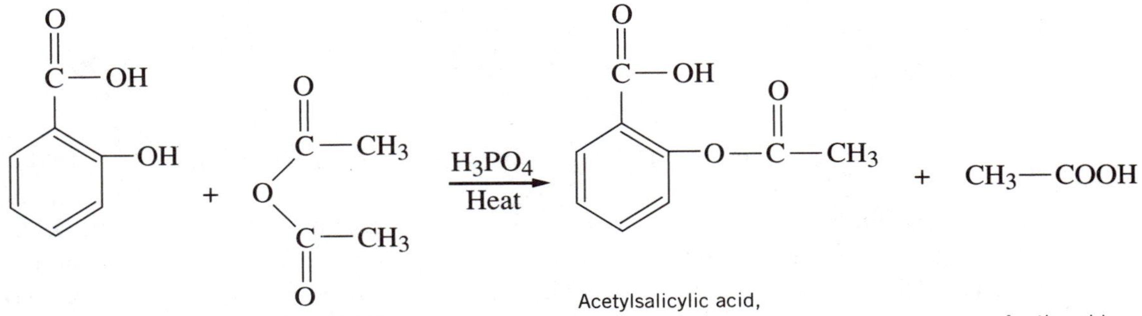

Salicylic acid

Aspirin
(Acetylsalicylic acid)

Acetaminophen
(for example, Tylenol™)

Ibuprofen
(for example, Advil™)

Figure E32.2   Synthesis of aspirin

Salicylic acid

Acetic anhydride

Acetylsalicylic acid,
or aspirin

Acetic acid

Aspirin is synthesized by reacting salicylic acid ($C_7H_6O_3$) with acetic anhydride ($C_4H_6O_3$) in the presence of phosphoric acid (which acts as a catalyst to increase the rate of reaction), as shown in Figure E32.2.[2]

Part B: Synthesis of Iodoform

Iodoform (also known as triiodomethane) is a yellow organic solid usually prepared from acetone in the presence of base as follows:[3]

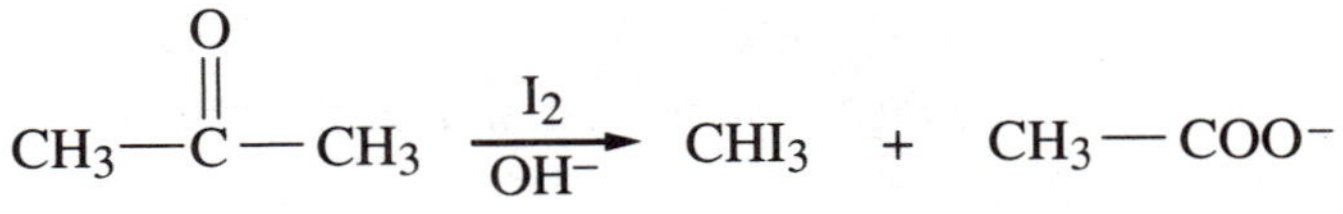

Acetone

iodoform, or
triiodomethane

Acetate anion

Figure E32.3   Iodoform reaction

In this experiment, iodoform is prepared by electrolyzing a solution of potassium iodide in acetone. When an aqueous solution of KI is electrolyzed, the following reduction half reactions are possible at the cathode:

$$K^+(aq) + e^- \rightarrow K(s) \qquad\qquad E^0 = -2.92 \text{ volt}$$
$$2\,H_2O(l) + 2\,e^- \rightarrow H_2(g) + 2\,OH^-(aq) \qquad E^0 = -0.83 \text{ volt}$$

Since the potential for H_2O reduction is far more positive, water is reduced in preference to K^+ ion.

At the anode, the following oxidation half reactions are possible:

$$2\,I^-(aq) \rightarrow I_2(s) + 2\,e^- \qquad\qquad E^0 = -0.54 \text{ volt}$$
$$2\,H_2O(l) \rightarrow O_2(g) + 4\,H^+(aq) + 4\,e^- \qquad E^0 = -1.23 \text{ volt}$$

Here, the potential for the formation of I_2 is the more positive.

The half reactions and the overall cell reaction are therefore as follows:

$$2\,H_2O(l) + 2\,e^- \rightarrow H_2(g) + 2\,OH^-(aq)$$
$$2\,I^-(aq) \rightarrow I_2(s) + 2\,e^-$$

Overall,

$$2\,I^-(aq) + 2\,H_2O(l) \rightarrow H_2(g) + I_2(s) + 2\,OH^-(aq) \qquad\qquad \text{(E32.2)}$$

Thus, I_2 and OH^- are generated **in situ** (within the reaction, instead of being added as reagents). The acetone present in the solution then undergoes the iodoform reaction in the presence of the generated I_2 and OH^-. The balanced reaction for the formation of CHI_3 is

$$H_3C–CO–CH_3 + 4\,OH^- + 3\,I_2 \rightarrow H_3C–COO^- + CHI_3 + 3\,H_2O + 3\,I^- \qquad \text{(E32.3)}$$

Using Faraday's laws, one can calculate the amount of I_2 formed during the electrolysis according to the reaction (E32.2). **For calculations, see Experiment 9**. If we know the amount of I_2, the theoretical yield of CHI_3 can be determined according to the stoichiometry of reaction (E32.3).

Part C: Synthesis of Urea

In 1828, Friedrich Wohler performed the first synthesis of an organic compound (urea, H_2NCONH_2) from purely inorganic sources [lead cyanate, $Pb(OCN)_2$, and ammonium chloride, NH_4Cl]. This synthesis was a large step in unifying organic chemistry (considered then as the chemistry of living organisms) with inorganic chemistry, the chemistry of salts and minerals. It also shattered the theory of organic **vitalism**, which held that chemists could not synthesize a molecule found in living systems.

Urea was first isolated from human urine by Roulle in 1780. It can be easily prepared by the reaction between silver cyanate, AgOCN, and ammonium chloride, NH_4Cl, according to the following reaction:[4]

$$Ag(OCN) + NH_4Cl(aq) \rightarrow AgCl(s) + H_2NCONH_2(aq) \qquad\qquad \text{(E32.4)}$$

Silver salts are both expensive and toxic to the environment. Any silver in chemical waste should be recovered and recycled, as described in the experimental section (see Part C).

General References

1. Szafran, Z.; Pike, R. M.; Foster, J. C. *Microscale General Chemistry Laboratory with Selected Macroscale Experiments;* Wiley: New York, 1993.
2. Slowinski, E. J.; Wolsey, W.; Masterton, W. L. *Chemical Principles in the Laboratory with Qualitative Analysis;* Saunders: New York, 1983.
3. Helle, K.; Rijks, J. A.; Jansenn, L. J. J.; Schuyl, J. W. *J. Chem. Educ.* **1969**, *46*, 518.
4. Tanski, S.; Petro, J.; Ball, D. W. *J. Chem. Educ.* **1992**, *69*, A128.

Experimental Section

Procedure

Part A: Preparation of Aspirin

Microscale experiment   Estimated time to complete the experiment: 2.5 h
Experimental Steps: Four steps are involved: (1) synthesis of aspirin, (2) recrystallization and characterization of the product, (3) if time permits, determination of the purity of aspirin, and (4) study of its activity as a drug.

Disposal procedure: Dispose of all the chemical wastes in designated containers.

Rinse a 10 mL Erlenmeyer flask with acetone and dry it on a sand bath or in an oven. Place a 150 mL beaker two-thirds full of water on a sand bath. Heat the water to a constant temperature of 80°C. Place about 300 mg of salicylic acid in the *dry* Erlenmeyer flask.

NOTE: If any moisture remains in the flask, the reaction will be extremely slow.

Add 700 μL of acetic anhydride to the flask, using an automatic delivery pipet. Using a medicine dropper, add two or three drops of 85 percent phosphoric acid to the mixture.

CAUTION: Both phosphoric acid and acetic anhydride are corrosive and can produce chemical burns.

Clamp the flask in the hot water bath for 10 minutes, occasionally swirling the flask to mix the contents. On heating, the solid will dissolve in acetic anhydride forming a clear solution.

Remove the flask from the water bath. While the mixture is still hot, in the **HOOD**, add 10 drops of distilled water to decompose any excess acetic anhydride to acetic acid. Warm (do not boil) the flask on a sand bath to expel the vapors of acetic acid. Add 2 to 3 mL of water all at once to the flask and allow it to cool to room temperature, plus an additional 5 to 10 minutes. During this time, a precipitate of aspirin may form. While waiting, assemble a suction filtration apparatus using a Hirsch funnel and suction flask (see Section 3.8).

To complete the crystallization of aspirin, cool the flask in an ice water bath. If the crystallization is slow, scratch the inside of the flask with a glass rod. Collect the aspirin by suction filtration, washing the product with two 1 mL portions of ice cold water. Dry the aspirin product on a clay tile or on a piece of filter paper. Weigh the aspirin, and determine the percent yield.

To recrystalize the product, transfer the aspirin crystals to a 10 mL beaker. Add 1 mL of ethyl alcohol to dissolve the crystals. Warm the solution on a sand bath and, using a Pasteur pipet, add 2 mL of water. Cover the beaker with a watch glass, and place it in an ice water bath. Filter the crystals as before and dry them under suction and then on a clay tile. Calculate the yield of the recrystallized product.

Record the melting point of the recrystallized product (see Experiment 2). Save the recrystallized product for other experiments (19, or as suggested by your instructor).

Analysis of Synthesized Aspirin

To determine the purity of the sample of aspirin you have prepared, dissolve 100 ± 1 mg of aspirin in 5 mL of 95 percent ethanol in a 50 mL volumetric flask. Add 5 mL of 0.025 M $Fe(NO_3)_3$ in 0.5 M HCl solution, and then distilled water to the mark. The concentration of aspirin can be determined following the procedure in Experiment 21, using a wavelength of 525 nm for the analysis.

Test of Aspirin Activity

Tests will now be performed to determine if aspirin becomes active in the stomach or the intestine. Remember: the salicylic acid is the active ingredient, not the aspirin itself. Salicylic acid (and all phenols) forms a violet complex with iron(III) chloride. Place an approximately 10 mg sample of salicylic acid in a small test tube, and add a few drops of water and one drop of 1 percent iron(III) chloride solution. Observe the results. Perform the same test with 10 mg of well-crushed (mortar and pestle) commercial aspirin.

Take two 10 mg portions of your aspirin product. Place one portion in a *clean* small test tube, and add five drops of 0.1 M HCl (simulating stomach acid). Add one drop of the iron(III) chloride solution. Has the aspirin become active? Place the other portion of aspirin in another small test tube. Add five drops of 0.1 M NaOH (simulating intestinal base). Add one drop of the iron(III) chloride solution. Has the aspirin become active?

Part B: Preparation of Iodoform by an Electrolytic Method

Microscale experiment   Estimated time to complete the experiment: 1.5 h
Experimental Steps: Three steps are involved: (1) setting up the electrolytic cell, (2) performing the electrolysis, and (3) collecting the product and characterizing it.

Disposal procedure: Dispose of all the chemical compounds used and the final solution in designated containers. Hand the product to your instructor.

Obtain a 30 mL wide-mouthed test tube or a 30 mL beaker containing a stir bar. Place 1.66 g of KI in the beaker (or tube), add 20 mL of water to dissolve the solid, and stir. Add 0.5 to 1.0 mL of acetone. Fit a two-hole stopper to the test tube, and insert two platinum or nichrome wires (to serve as electrodes) through the holes, making sure that they are immersed in the solution and do not touch each other (see Figure E32.4). To complete the electrolytic cell, obtain an ammeter (optional), a 6 V or 12 V battery, and two alligator clips to connect the wires from the power source to the electrodes. We recommend the use of

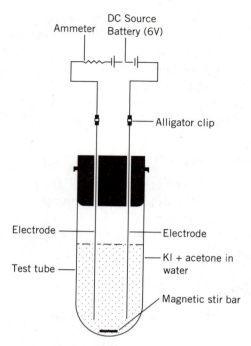

Figure E32.4 Electrolysis cell for iodoform preparation

platinum wires as the electrodes. Unlike nichrome wires, they do not corrode. These electrodes can be used over and over again.

Carry out the electrolysis for 10 to 15 minutes. During this time, observe the formation of I_2 at the anode and evolution of H_2 gas at the cathode. At the same time, the yellow iodoform compound will precipitate out of the solution. Collect the iodoform product by suction filtration using a Hirsch funnel and dry the product on a clay tile. Determine the yield and the melting point.

CAUTION: If the electrolysis is performed for a longer period, the nichrome wires are corroded by I_2, resulting in the formation of Cr^{3+} and Ni^{2+} cations. Always use freshly cut nichrome wires.

Part C: Synthesis of Urea

Microscale experiment   Estimated time to complete the experiment: 3h
Experimental Steps: Three steps are involved: (1) reacting NH_4Cl and $AgOCN$ solutions, (2) separating the product from the filtrate, and (3) recovering the silver.

Disposal procedure: Collect all the silver residue in a container for recovery and recycling.

Note: This preparation requires careful manipulation and meticulous attention to the details of the reaction conditions.

Dissolve ~16 mg of ammonium chloride, NH_4Cl, in 300 μL (automatic delivery pipet) of deionized water in a small test tube. Using a spatula, transfer ~60 mg of silver cyanate, AgOCN, to a 10 mL beaker containing a stir bar. Using the same automatic pipet as above, add 300 μL of water to the silver salt.

CAUTION: Handle all silver salts with care; they are toxic. They stain fingers and clothes.

Place the beaker containing the silver salt on a magnetic stirrer in a well-ventilated **HOOD**. Using a Pasteur pipet, add concentrated aqueous ammonia dropwise to the suspension. Continue to stir the solution. Silver cyanate should dissolve in ammonia. If the solid has not dissolved completely, add a few more drops of ammonia. **NOTE: Avoid adding too much ammonia. Excess ammonia will prevent the formation of solid silver chloride in the next step**.

Using a Pasteur pipet, transfer the ammonium chloride solution to the silver cyanate solution. A white precipitate of silver chloride should form immediately. Rinse the ammonium chloride test tube with five drops of water and add the washings to the silver cyanate solution.

Using a Hirsch funnel, filter the AgCl precipitate under suction (see Section 3.8), collecting the filtrate in a 25 mL filtration flask. **Save the silver chloride precipitate for silver recovery and recycling (see following paragraphs)**. If the filtrate is cloudy, refilter the solution. Transfer the filtrate to a 25 mL beaker. Rinse the filtration flask with five drops of water and transfer the wash solution to the beaker.

Place the beaker with the filtrate on a hot sand bath for **slow evaporation** of the ammonia and some water. The temperature of the solution should not be higher than 100°C (urea decomposes at higher temperatures). When one-fourth of the solution has been evaporated, cool the beaker in an ice bath. On cooling, the unreacted AgOCN and some solid AgCl will precipitate out. Using a Hirsch funnel, suction-filter these impurities. **Combine the residue with that of AgCl.**

Transfer the clear filtrate back to the beaker, and replace the beaker on a sand bath for further evaporation. Maintain the temperature of the solution below 100°C. When almost all of the water has been evaporated, crystals of urea will form. Cool the beaker to room temperature, decant off any water left, and collect the crystals from the bottom of the beaker. Dry the urea product on a clay tile and determine the percent yield and the melting point of the crystals.

Recovery and Recycling of Silver Chloride[1]

Collect the wet silver chloride (four to six students can collect their waste in one place) in a 250 mL beaker. Using a Pasteur pipet, add just enough concentrated ammonia to dissolve the solid. Warming the solution will enhance the rate of dissolution.

CAUTION: Ammonia is corrosive and attacks nasal mucous membranes.

With constant stirring, add an excess of 1 M ascorbic acid to reduce the silver(I) to silver metal. Silver metal will settle to the bottom of the beaker as the reduction proceeds. Decant the supernatant liquid, and then rinse the metal with water several times, decanting the liquid each time. Finally, gravity-filter the silver metal, wash it with acetone, and allow it to dry in air. The residue can be treated with nitric acid to produce technical grade $AgNO_3$ or can be melted to form silver pellets.

Additional Independent Projects

1. Potentiometric titration of aspirin in ethanol[2]
2. Determination of aspirin by ultraviolet absorption and by titrimetric methods[3]
3. Saponification of aspirin in dimethyl sulfoxide[4]
4. Use of urea as a source for ammonia in inorganic syntheses[5]

References

1. Hill, J. W.; Bellows, L. *J. Chem. Educ.* **1986,** *63,* 357.
2. Shen, S. Y.; Gilman, A. J. *J. Chem. Educ.* **1965**, *42*, 540.
3. Proctor, J. S.; Roberts, J. E. *J. Chem. Educ.* **1961**, *38*, 471.
4. Vinson, J. A.; Hocker, E. K. *J. Chem. Educ.* **1969**, *46*, 245. Also see other references in this publication for other methods.
5. Szafran, Z.; Pike, R. M.; Singh, M. M. *Microscale Inorganic Chemistry: A Comprehensive Laboratory Experience*; Wiley: New York, 1991.

Name ______________________ Section ______________

Instructor ______________________ Date ______________

PRELABORATORY REPORT SHEET—EXPERIMENT 32 (PARTS A, B, C)

Experiment title ______________________

Objective

Reactions/formulas to be used

Chemicals and solutions—their preparation

Materials and equipment table

Outline of procedure

1. (*a*) One gram of salicylic acid (138.12 g/mol) is dissolved in 460 mL of water, yielding a solution with a pH of 2.4. Determine the K_a value for salicylic acid, and compare it with that of aspirin, 3.27×10^{-4}.

   (*b*) Aspirin is prepared in 85 percent yield (on average) from salicylic acid (138.12 g/mol). How many grams of salicylic acid will be required to form 0.150 g of aspirin (180.16 g/mol)? How many milliliters of acetic anhydride (102.09 g/mol, density = 1.082 g/mL) will be required?

2. (*a*) In the synthesis of iodoform, write the balanced chemical reactions occurring at the cathode and the anode.

   (*b*) What will be the reaction in the cell if we use KBr instead of KI?

   (*c*) In the electrolytic synthesis of iodoform, 1.66 g KI and 2 mL of acetone are added to a large volume of water. Which one is the limiting reagent?

   (*d*) Make a table of chemicals and equipment you need for parts A, B, and C.

PRELABORATORY PROBLEMS—EXPERIMENT 32 (PARTS A, B, C), PAGE 2

3. (*a*) If the formula of urea is H_2NCONH_2, what is the formula of thiourea?

(*b*) Silver chloride is soluble in ammonia. What will happen if you acidify the clear solution with nitric acid?

(*c*) In the synthesis of urea, 0.219 g of silver cyanate (149.89 g/mol) is treated with 0.265 g of ammonium chloride (53.49 g/mol). What is the theoretical yield of urea (60.05 g/mol)?

Part A: Aspirin Preparation

1. Mass of salicylic acid used ———— g

2. Moles of salicylic acid used ———— mol

3. Volume of acetic anhydride used ———— mL

4. Mass of acetic anhydride used ———— g ($\rho = 1.082$ g mL^{-1})

5. Moles of acetic anhydride used ———— mol

6. Mass of aspirin product ———— g

7. Moles of aspirin product ———— mol

8. Weight percent yield ———— %

9. Melting point ———— °C

10. What was the effect of adding iron(III) chloride to the salicylic acid?

11. What was the effect of adding iron(III) chloride solution to the aspirin?

12. From your tests on the aspirin product, does the aspirin become active in the stomach or in the intestine? Explain why.

Name ______________________ Section ____________

Instructor ______________________ Date ____________

DATA SHEET EXPERIMENT 32, PAGE 2

Part B: Preparation of Iodoform

1. Mass of KI used ______ g
2. Current in amps ______ A
3. Voltage of battery ______ V
4. Electrolysis time ______ s
5. Yield of iodoform ______ g
6. % yield ______ %
7. Melting point ______ °C

Part C: Urea Synthesis

1. Mass of ammonium chloride ______ g
2. Mass of silver cyanate ______ g
3. Yield of urea ______ g
4. % yield ______ %
5. Melting point ______ °C

1. (*a*) What organic functional groups are present in salicylic acid, aspirin, acetaminophen, and ibuprofen?

(*b*) Upon standing, especially in a moist atmosphere, aspirin loses its potency and develops a distinct vinegary (acetic acid) smell. Write a balanced equation accounting for the decomposition of aspirin.

(*c*) Why is it important that the glassware used in the synthesis of aspirin be dry?

2. (*a*) Iodoform (CHI_3) forms when acetone reacts with KI in the presence of OH^-. How is the base produced in the synthesis of CHI_3 by electrolysis?

(*b*) Write the formulas of chloroform and bromoform.

(*c*) In the synthesis of iodoform, what other products form in the electrolytic cell?

POSTLABORATORY PROBLEMS—EXPERIMENT 32 (PARTS A, B, C), PAGE 2

3. (*a*) How does urea react with boiling water? Write a balanced chemical equation
for the reaction.

(*b*) What tests can be performed to detect the decomposition products formed
in the preceding reaction?

(*c*) In the synthesis of urea, addition of excess ammonia must be avoided. Why?
Explain your answer using a chemical equation.

Part D: Determination of Molecular Weight of Urea by Elevation of Boiling Point

When a pure solid is dissolved in a solvent, several physical properties of that solvent change in a way that depends only on the relative amounts of the solute and solvent present. Such properties are called **colligative properties**. There are four common colligative properties: boiling point elevation, freezing point depression, the lowering of vapor pressure and the osmotic pressure.

The normal boiling point (**bp**) of a liquid is the temperature at which its vapor pressure exactly equals atmospheric pressure. When a nonvolatile solute is added to a liquid, its boiling point increases because the addition of a solute lowers the vapor pressure of the liquid. As a result, a temperature higher than the normal boiling point is required to achieve the vapor pressure of one atmosphere.

For dilute solutions, the boiling point elevation, ΔT_b of a solution, equals $(T_b - T_b^0)°C$, where T_b is the bp of the solution and T_b^0 is the bp of the pure solvent. The boiling point elevation is proportional to the molal concentration, m, of the solution, or

$$\Delta T_b = K_b m \tag{E32.5}$$

where K_b is the **molal boiling point elevation constant** and m is the molality of the solution (m = moles of solute per kilogram of solvent). The units for K_b are °C/m. The molal boiling point elevation constants for water, ethanol, acetic acid, and cyclohexane are 0.512, 1.22, 2.93, and 2.79°C/m, respectively.

The molar mass of a solute that is a nonelectrolyte and nonvolatile, which is dissolved in a solvent, can be determined by measuring the boiling point elevation of the solution. The following example illustrates this calculation.

Example E32.1   The boiling point of a solution of 5.00 g of glucose in 95.0 g of water is 100.15°C. Calculate the molecular weight (M) of glucose ($K_b = 0.512°C/m$).

Answer

$$\Delta T_b = (100.15 - 100.00)°C = 0.15°C$$
$$m = \Delta T_b / K_b = 0.15°C / 0.512°C/m$$
$$= 0.29 \text{ m} = (0.29 \text{ mol glucose}) / (\text{kg of water})$$

$$\# \text{ mol of glucose} = [(0.29 \text{ mol glucose}) / (1000 \text{ g water})] \times (95.0 \text{ g water})$$
$$= 0.028 \text{ mol glucose}$$
$$M = 5.00 \text{ g} / 0.028 \text{ mol} = 1.8 \times 10^2 \text{ g/mol}$$

In this experiment, the molecular weight of urea as prepared in Experiment 32C will be determined. A known mass of urea will be dissolved in pure ethanol and the boiling point elevation, ΔT_b, of the ethanol solution will be determined. Using equation (E32.5), the molality m of the solution will be calculated. Since the mass of urea in the solution is known, the molecular weight can be easily determined.

General Reference

1.   See any general chemistry textbook.

Experimental Section

Procedure^[1]

Procedure[1]

> Microscale experiment   Estimated time to complete the experiment: 2.5 h
> Experimental Steps: Two steps are involved: (1) determining the bp of pure ethanol and
> (2) of the solution of urea in ethanol.

Obtain ~6 mL of ethyl alcohol, a **clean and dry** test tube (125 × 20 mm), a 250 mL beaker
for a water bath, a melting point capillary tube, a stopper fitted with an accurate thermometer
that can be read to 0.1°C, and a piece of narrow long (~20 cm) glass tubing, which acts as
an air condenser. Collect a dry sample of urea (see Experiment 32C). If you do not have
enough, your instructor will provide you with the sample.

Set up a hot water bath in the 250 mL beaker as shown in Figure E32.5. Use a sand
bath for heating water. The temperature of the bath should be around 90°C.

Weigh the empty dry test tube. Using a pipet, place 4 mL of ethanol in the test tube.
Reweigh the test tube plus ethanol. Record the weights on your record sheet. To determine
the bp of ethanol, place a 2 to 3 cm long melting point capillary tube inside the test tube,
open end down (as shown in Figure E32.5). Replace the stopper carrying the thermometer
and the air condenser. Place the test-tube assembly in the water bath so that the bulb of the
thermometer rests inside ethanol, and determine the boiling point of ethanol as described
in Experiment 2, Method 1 or 2. Record the boiling point of ethanol on your record sheet.
Avoid vigorous boiling of the alcohol. Remove the test-tube assembly, cool the system,
and repeat the procedure two more times. Determine the average boiling point.

Remove the test-tube assembly from the water bath and let it cool for 2 minutes.
Remove the stopper momentarily and add an accurately weighed sample of **dry** urea
(500 to 600 mg) to the test tube. Replace the stopper and put the test tube back in the

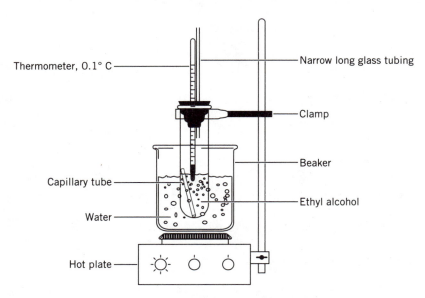

Figure E32.5   Apparatus for determining the elevation of boiling point

water bath. Let the solution boil gently for several minutes to remove any solid adhered to the side of the test tube. Determine the boiling point of the solution as before and enter the value in your data table. Using the same solution, repeat the procedure two more times. Obtain an average value for the boiling point.

Your instructor may provide you with another unknown sample. Determine the boiling point of the solution in a similar manner. Calculate ΔT_b, m, and MW of urea.

Recovery, Recycle, and Reuse of Urea and Alcohol

Collect all of the samples of urea solution in a beaker. Let the solution cool to room temperature. Urea will separate out from the solution as a solid. Cool the solution in an ice water bath. Collect the solid on a filter and dry it. Ethyl alcohol may be distilled off from the filtrate and collected for recycling. We encourage students to collect their urea samples and distill alcohol using a micro distillation column (which can be borrowed from an organic chemistry laboratory). Since urea is highly soluble in water, all the glassware can be rinsed with water.

Reference

1. Thomas, N. C.; Saisuwan, P. *J. Chem. Educ.* **1990**, *67*, 971.

Name ______________________________ Section ______________

Instructor ______________________________ Date ______________

PRELABORATORY REPORT SHEET—EXPERIMENT 32D

Experiment title ______________________________

Objective

Reactions/formulas to be used

Materials and equipment table

Outline of procedure

1. Define molarity, molality, and normality.

2. The boiling point of a solution of 2.50 g of a solid in 100 g of benzene is 80.6°C. The boiling point of pure benzene is 80.1°C. Calculate the molecular weight of the solid (K_b for benzene = 2.53°C/m).

EXPERIMENT 32D   DATA SHEET

Data Collection Table

K_b for ethyl alcohol $= 1.22°C/m$

1. Mass of empty test tube   _____ g

2. Mass of test tube + alcohol   _____ g

		Trial 1	**Trial 2**	**Trial 3**
3.	Boiling point of alcohol:	_____ °C	_____ °C	_____ °C
4.	Boiling point of the solution:	_____ °C	_____ °C	_____ °C

Data Manipulation Table

5. Average boiling point of pure alcohol   _____ °C

6. Average boiling point of the solution   _____ °C

7. Elevation of the boiling point, ΔT_b _____ °C

8. Molality m of the solution   _____ mol/kg solvent

9. Molecular weight of the solute   _____ g/mol

Show calculations:

POSTLABORATORY PROBLEMS—EXPERIMENT 32D

1. Calculate the molality of a 5 percent by mass solution of NaCl in water.

2. By mistake, you have read the boiling point of a solution of a substance in benzene as 80.6°C instead of 80.4°C. How would it affect the molecular weight calculation of the substance?

3. If we use the Kelvin scale in reporting the boiling point of a solution, will it affect our calculation? Explain your answer.

33 **Amino Acids: Qualitative and Quantitative Determination**

Part A: Amino Acid Mapping by TLC
Part B: Potentiometric Titration of Amino Acids

Micro- and Macroscale Experiments

Objectives
- To identify amino acids on TLC plates
- To determine the pK_a, molecular weight, and isoelectric point of an amino acid

Prior Reading
- Section 3.11   Titration procedure
- Experiment 28   Preparation of buffers and potentiometric titrations

Related Experiments
- Experiments   28 (pH titrations) and 26 (TLC)

Amino acids are organic acids (they contain the –COOH group) that also have a basic amino functional group (–NH$_2$). In aqueous solutions, amino acids undergo an acid-base reaction to form a dipolar ion called a **zwitterion.** This reaction is shown for the amino acid glycine:

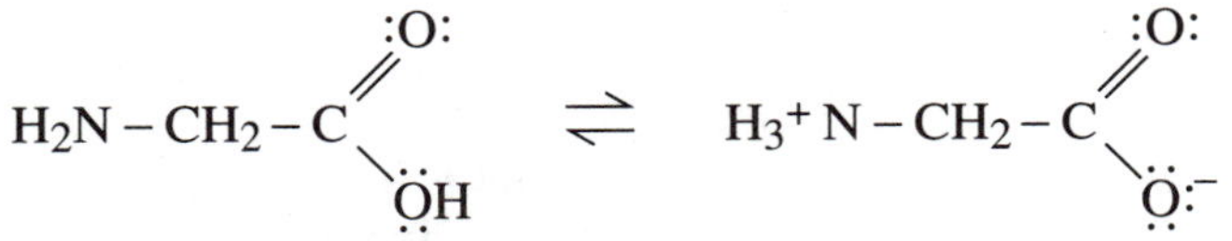

glycine   zwitterion form

The zwitterionic species predominates in neutral solutions. The forms present when the ion is treated with acid or base are shown here:

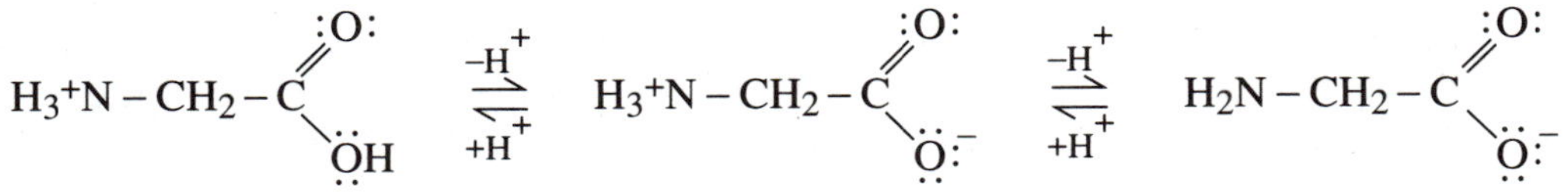

strong acid species   zwitterion form (neutral species)   strong base species

Amino acids are important because they act as biological building blocks by forming amide linkages (also called peptide linkages) that lead to the formation of proteins. An amide linkage results when the amino group of one molecule reacts with the carboxylic acid group of another molecule.

dipeptide

Polypeptides consist of many amino acid units linked in a like manner. Proteins are those polypeptides that contain more than 50 amino acids (most often 100 to 300 amino acid units). Skin, hair, muscle, and all enzymes are proteins. Twenty amino acids are commonly found in proteins.

Part A

Since amino acids exist as zwitterions, they are only slightly soluble in organic solvents but very soluble in water. Increased solubility is seen in acidic or basic solutions. **Thin-layer chromatography** is an excellent technique to use for the identification of amino acids. The solutions used in this experiment are acidic (HCl), and thus the amino acids will be in the form shown in the foregoing equations for acid solutions. Once the TLC plate is developed, the separated amino acids can be detected by generation of a blue-violet dye on treatment with **ninhydrin**, 1,3,3-triketohydrindene hydrate (except for proline and hydroxyproline, which give a yellow color). A map of R_f values will be developed for a series of known amino acids in two separate solvent systems. Using the map results, an unknown amino acid will then be identified.

Part B

At some pH, amino acids carry no net charge. This is known as the **isoelectric point**, and corresponds to the pH at which the concentration of the zwitterion is at a maximum. The pH varies between the values of 5.41 and 6.02 for those acids having no other acidic or basic functions in their structure. In more acidic solutions than the isoelectric point, the $-COO^-$ group is protonated to form the cation, $H_3N^+CH_2COOH$. In more basic solutions, the $-NH_3^+$ group loses a proton to form the anionic species, $H_2NCH_2COO^-$.

In aqueous solution, the zwitterions of glycine show the following equilibria:

$$H_3N^+CH_2COO^- + H_2O \rightarrow H_3N^+CH_2COOH + OH^- \qquad pK_a = 2.4$$
$$H_3N^+CH_2COO^- + H_2O \rightarrow H_2NCH_2COO^- + H_3O^+ \qquad pK_a = 9.6$$

or

$$H_2NCH_2COOH$$

Like all weak acids and bases, amino acids follow the Henderson–Hasselbalch equation (see Experiment 28) for example:

$$pH = pK_a + \log \frac{[A^-]}{[HA]} = pK_a + \log \frac{[H_2N^+CH_2COO^-]}{[H_2NCH_2COOH]} \qquad (E33.1)$$

From equation (E33.1), when $[A^-] = [HA]$,

$$pH = pK_a \qquad (E33.2)$$

It is under this condition that the amino acid exhibits its highest buffer capacity, b (resists changes in the pH most effectively).

The change in the pH of an amino acid solution when acid or base is added is shown in Figure E33.1. The addition of acid lowers the pH rapidly at first. When the pH reaches a value of ~2.4 (where half of the required acid has been added), exactly half of the carboxyl groups have been protonated. At this point, $[A^-] = [HA]$ (E33.1) and the solution will have formed its most effective buffer. **As shown in equation (E33.2), the pH is equal to the pK_{a1} at this point.** Further addition of acid protonates the remaining carboxyl groups, causing a smaller change in the pH. Titration of the amino acid with a base results in a similar titration curve in the *base region*. The **intersection** of the acid titration curve and the base

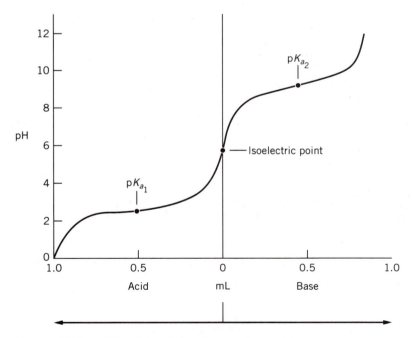

Figure E33.1   Titration curve of an amino acid

titration curve is the point where the concentrations of the cation and anion forms of the amino acid are equal—the **isoelectric point**. The isoelectric point can be determined from the following relationship:

$$\text{pH (isoelectric point)} = \tfrac{1}{2}(\text{p}K_{a1} + \text{p}K_{a2}) \qquad \text{(E33.3)}$$

In Part B of this experiment, an aqueous solution of an amino acid will be titrated with a standard acid and a standard base. The pK_a's and the isoelectric point will be determined from the pH curve.

General References

1. Buchanan, D. N.; Kleinman, R. W. *J. Chem. Educ.* **1976**, *53*, 255.
2. Helser, T. L. *J. Chem. Educ.* **1990**, *67*, 1010; and **1992**, *69*, 970.

Experimental Section

Procedure

Part A: Amino Acid Mapping by TLC Method[1]

> Microscale experiment   Estimated time to complete the experiment: 1 h
> Experimental Steps: Two steps are involved: (1) setting up the TLC plates and (2) obtaining the chromatogram and calculating the R_f values for each amino acid.

> Disposal procedure: Return the solvent mixtures and the residual amino acids to your instructor for reuse and recycle.

Note: The general procedure for paper chromatography is described in Experiment 26, Part C.

Obtain two (one for solvent A and the other for solvent B) 6 × 16 cm Whatman #1 filter papers. With a pencil (not ink), draw a very light line across the length of each of the filter papers 1 cm above the bottom edge of the papers. Put pencil dots 2 centimeters apart across the length of the line on each paper. There should be seven such dots: six for known and one for unknown amino acid samples. Number the dots 1 through 7.

Also procure a spray bottle containing 0.2 percent ninhydrin solution in ethanol or acetone. Two 250 or 400 mL beakers (marked solvent A and solvent B) with watch glass covers are used to make developing chambers. Construct or procure seven microcapillary pipets (six for known and one for unknown samples). **Do not mix up the capillary pipets after their first use.**

Using the supplied **solvents A** [an acid solvent consisting of a mixture of n-butanol (n-BuOH) + t-butanol (t-BuOH) + glacial acetic acid (HA) + water in the volume ratio of 5:5:2:4] **and B** [a basic solvent consisting of a mixture of n-butanol (n-BuOH) + t-butanol (t-BuOH) + concentrated ammonia (NH_3) + water], set up two separate developing chambers in the marked 400 mL beakers. Cover each beaker with a watch glass.

Using a separate micro capillary pipet for each sample, carefully spot a known amino acid solution (2 to 3 mg/mL of amino acid in 50 percent HCl or 1:1 concentrated HCl) on each dot on the filter paper. Thus, six spots are used for the six knowns supplied by your instructor. Spot the seventh spot with the unknown solution. Save the micro capillary pipets for later use. Keep track of which amino acid solution is spotted on each dot. Allow the spots to dry.

Construct a cylinder by folding the freshly spotted Whatman #1 paper and by stapling the slightly overlapped edges of the paper. Develop the paper chromatogram (see Experiment 26, Part C) using **solvent A. Remember to mark the solvent front when you remove the paper from the beaker.** Thoroughly dry the developed paper, and then spray it with ninhydrin solution (**HOOD**). Open the cylinder by carefully removing the staples and place the flat paper on a watch glass. Allow the paper to dry in an oven for 2 minutes at 110°C. The amino acid spot(s) should now be visible. Measure and record the R_f value for each amino acid developed in solvent A.

Using the same procedure just outlined (use the same micro capillary pipets), spot the other paper and develop it using solvent **B**. Record the R_f value for each amino acid developed in solvent B.

Enter your data (names of the known amino acids used, the distances of the solvent fronts and the distances of the amino acid spots from the baseline, the R_f values for each known acid in both the solvents) in the data table. Using the R_f data from the known amino acids, determine the identity of the amino acid(s) in the unknown solution.

Part B: Potentiometric Titration of an Amino Acid[2]

Microscale experiment   Estimated time to complete the experiment: 2.5 h
Experimental Steps: Four main steps are involved: (1) preparing and standardizing solutions of 2 M HCl and 2 M NaOH, (2) setting up and calibrating the pH meter, (3) titrating an amino acid with HCl and NaOH, and (4) plotting data and calculating MW and pK_a's of an amino acid.

Obtain and calibrate a pH meter with a combination glass electrode, standard buffer solutions (pH = 4.0, 7.0, and 10.0), a 30 mL tall-form beaker, micro stir bar (see Section 3.10), magnetic stirring motor, two 2 mL microburets, a 10 mL volumetric flask, a funnel, and 20 mL each of prestandardized ~2 M HCl and ~2 M NaOH solutions (used as titrants).

CAUTION: Both 2 M HCl and NaOH are corrosive. In case of a spill, wash with copious amounts of cold water.

Note: A better pH titration curve is obtained by using larger amounts of the amino acid, which requires the use of macroburets. In that case, use five times the quantities mentioned for the microtitration.

Using the procedure described in Experiment 28, calibrate the pH meter. Rinse and set up two microburets marked "acid" and "base." Fill the acid buret with 2 M HCl and the base buret with 2 M NaOH. Record the initial volume in each of the burets on the data sheet.

Weigh about 200 ± 1 mg of the amino acid sample into a 30 mL beaker. Add a stir bar, and set the beaker atop a magnetic stirrer. Add 20 mL of deionized water, and stir the solution until the solid has dissolved. Stir the solution and insert the combination electrode, making sure that the electrode is well immersed. Record the initial pH of the solution. Titrate the unknown sample with the 2 M HCl solution, recording the buret readings and the corresponding pH of the solution. Continue the titration until the solution shows a pH of ~1.0.

Titrate 20 mL of deionized water (as a blank) with 2 M HCl, adding the acid ***dropwise*** from the buret until the solution shows a pH of ~1.0. Record the pH and the volume of the acid after each addition.

Similarly, perform the base titration by dissolving 200 ± 1 mg of the sample in 20 mL of deionized water and titrating the solution with 2 M NaOH. Record the pH values and the buret readings on your report sheet. Continue the titration until you obtain a pH of 12.0.

Perform a blank titration of 20 mL of deionized water with the NaOH added ***dropwise*** from the buret. Enter the pH and the corresponding buret readings on the data sheet.

Table E33.1   Volume of 2 M HCl Added to the Sample and
            to Blank Water

	Volume of 2 M HCl added, mL		
pH	Sample + water	Water alone	Corrected acid volume
4.0	0.095	0.002	0.093
3.5	0.104	0.004	0.100
3.0	0.333	0.015	0.318
2.5	0.670	0.035	0.635
2.0	1.400	0.200	1.200

To determine the true titration curve (pH versus volume of titrant), one must make a volume correction for the titrant. To do this, subtract the amount of the acid (or the base) needed for the water alone to reach the designated pH from the total amount of the acid (or base) needed for the solution of the sample in water to reach the same pH. Table E33.1 illustrates the method of correcting for the dilution of the titrant (shown only for acid titration).

Draw the preliminary pH curves for both the sample and the blank (see Figure E33.2a). Construct a table similar to Table E33.1. Determine the corrected acid volume for each pH value and plot the corrected pH curve as shown in Figure 33.2b. Repeat the same procedure to prepare a graph for the base titration.

Note: If the blank microtitration does not show any significant change, the volume correctgion may be ignored.

Both plots can be drawn on the same graph paper, to obtain a graph similar to that shown in Figure E33.1. From these plots (one for the acid titration and the other for the base titration), determine the following:

Molecular Weight of the Amino Acid

The end point of the titration is indicated by the rise of the curve. Draw a vertical line through the center of the steep portion of the curve (using either the acid or the base titration curve).

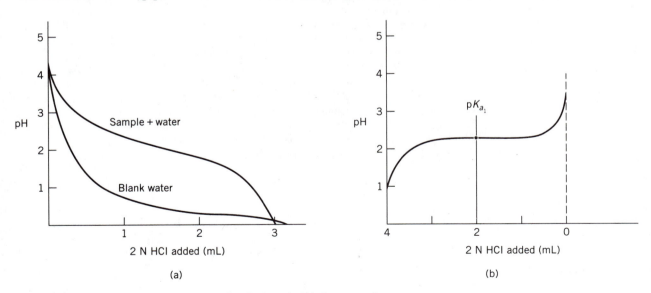

Figure E33.2   Titration curves: (*a*) Preliminary; (*b*) Corrected

The intersection of this vertical line and the x axis determines the volume of the titrant (HCl or NaOH) used to react completely with the functional group ($-COO^-$ or $-NH_3^+$).

Calculate the number of moles of titrant (which is the same as the number of moles of the amino acid). Since you know the mass of the amino acid used, determine the molecular weight of the acid.

pK_{a1} and pK_{a2} of the Amino Acid

Determine the pH value at a point corresponding to half of the total volume of titrant used for titrating the amino acid. From Equation (E33.2), pK_a = pH.

Isoelectric point from the curves

Using the pK_a's just obtained and Equation (E33.3), calculate the isoelectric point for the amino acid.

Example E33.1 A sample of 0.200 g of an amino acid is titrated with 1.33 mL of 2.00 M HCl acid. The acid contains one ionizable group ($-COO^-$). Calculate the molecular mass of the acid.

Answer

$$\text{Moles of acid} = (2.00 \text{ mol/L})(1.33 \times 10^{-3} \text{ L}) = 2.66 \times 10^{-3} \text{ mol} = \text{mol amino acid}$$

$$\text{MW} = (0.200 \text{ g})/(2.66 \times 10^{-3} \text{ mol}) = 75.2 \text{ g/mol}$$

Additional Independent Projects

1. Determination of isoelectric point of phenylglycine[3]

   Brief Outline

   The structure of phenylglycine, (C_6H_5)CH(NH$_2$)(COOH), can be derived from the structures of two individual species: benzylamine [(C_6H_5)CH$_2$(NH$_2$)] and phenylacetic acid [(C_6H_5)CH$_2$(COOH)]. Using the pH titration method (this experiment), determine the pK_1 for the protonation of benzylamine (which is titrated with a standardized 0.1 M HCl solution). Similarly, determine the pK_2 for phenylacetic acid (which is titrated with a standardized 0.1 M NaOH solution). From these pK values, calculate the isoelectric point (Iso. pt. = pK_1 + pK_2). Using a fresh sample of the amino acid given by your instructor (or synthesized by you according to the method described in reference 3), determine its isoelectric point, which can be compared with the one obtained before. The synthesis of phenylglycine from the reaction of phenylacetic acid and N-bromosuccinimide followed by the treatment of the product with an aqueous solution of ammonia takes several days.

   Develop your own data tables (data collection and data manipulation tables). Submit a written report of the experiment according to the instructions given in Chapter 4.

2. Application of the TLC technique in amino acid analysis[1,4]

References

1. Helser, T. L. *J. Chem. Educ.* **1990**, *67*, 964.
2. Clark, J. M.; Switzer, R. L. *Experimental Biochemistry*, 2nd ed.; Freeman: San Francisco, CA, 1964.
3. Barrelle, M.; Gaude, D.; Salon, M. C. *J. Chem. Educ.* **1983**, *60*, 676.
4. Hurst, M. O.; Cobb, D. K. *J. Chem. Educ.* **1990**, *67*, 978.

Name ______________________ Section ______________

Instructor ______________________ Date ______________

PRELABORATORY REPORT SHEET—EXPERIMENT 33 (PARTS A AND B)

Experiment title ______________________________

Objective

Reactions/formulas to be used

Chemicals and solutions—their preparation

Materials and equipment table

Outline of procedure

1. Draw the structures of four amino acids.

2. What steps must be taken to calibrate a pH meter?

3. If the pK_{a1} and pK_{a2} of an amino acid are 9.8 and 3.9, respectively, what is the isoelectric point of the amino acid?

4. In the development of the thin-layer plate, why must the original spots of the amino acid(s) not be immersed in the solvent when the development of the plate is initiated?

Name	____________________	Section	__________
Instructor	____________________	Date	__________

EXPERIMENT 33 DATA SHEET

Part A: Amino Acid Mapping by TLC Method

Sample ID number ________________

Complete your own data collection and data manipulation table.

Part B: Potentiometric Titration of Amino Acids

Sample ID number _____

Data Collection Table

Concentration of 2 M HCl solution   _____ mol/L

Concentration of 2 M NaOH solution   _____ mol/L

1. Mass of sample before transfer   _____ g

2. Mass of sample after transfer   _____ g

3. Mass of sample taken   _____ g

4. Initial volume in acid buret   _____ mL

5. Initial volume in base buret   _____ mL

6. pH titration

For acid titration

Sample + water

Volume of HCl, mL	pH	Volume of HCl, mL	pH
_____	_____	_____	_____
_____	_____	_____	_____
_____	_____	_____	_____
_____	_____	_____	_____
_____	_____	_____	_____
_____	_____	_____	_____
_____	_____	_____	_____

Blank water

Volume of HCl, mL	pH	Volume of HCl, mL	pH
_____	_____	_____	_____
_____	_____	_____	_____
_____	_____	_____	_____
_____	_____	_____	_____
_____	_____	_____	_____
_____	_____	_____	_____

Plot the graphs for both the sample and the water blank (pH versus volume of acid).

DATA SHEET, EXPERIMENT 33, PAGE 3

For base titration

Sample + water

Volume of NaOH, mL	pH	Volume of NaOH, mL	pH
_____	_____	_____	_____
_____	_____	_____	_____
_____	_____	_____	_____
_____	_____	_____	_____
_____	_____	_____	_____
_____	_____	_____	_____
_____	_____	_____	_____
_____	_____	_____	_____

Blank water

Volume of NaOH, mL	pH	Volume of NaOH, mL	pH
_____	_____	_____	_____
_____	_____	_____	_____
_____	_____	_____	_____
_____	_____	_____	_____
_____	_____	_____	_____
_____	_____	_____	_____

Plot the graphs for both the sample and the water blank [pH (*y* axis) versus volume of acid added (*x* axis)]. Submit your plots along with the report.

Correction from Blank Water Titration

Correction for the acid titration

	Volume of 2 M HCl added, mL		
pH	Sample + water	Water alone	Corrected acid volume

Correction for the base titration

	Volume of 2 M NaOH added, mL		
pH	Sample + water	Water alone	Corrected acid volume

Using the corrected volume of the titrant, plot the pH titration curves for both the titrations (acid and base). Submit your plots to your instructor.

Name ______________________ Section ______________

Instructor ______________________ Date ______________

DATA SHEET, EXPERIMENT 33, PAGE 5

Calculation of molecular mass of the amino acid

From acid titration ______ From base titration ______

Show calculations:

Calculation of pK_{a1} (from acid titration) and pK_{a2} (from base titration)

pK_{a1} = ______ pK_{a2} = ______

Calculation of isoelectric point

Isoelectric point = ______

Show calculations:

1. Define the following:
   (a) Buffer capacity

   (b) Zwitterion

2. In this experiment, why is it necessary to measure the pH of a blank solution?

CHAPTER **10** **Materials and Industrial Chemistry**

34 **Inorganic Polymers**

Preparation of Slime and Determination of Its Viscosity

Microscale Experiment

Objectives
- To prepare inorganic polymers and to study their properties

Prior Reading
- Section 3.3   Handling chemicals
- Section 3.6   Heating methods
- Section 3.7   Stirring

Related Experiments
- Experiments 1 (Density), 4 (Determination of viscosity), 10 (Acid-base titration), and 35 (Preparation of polystyrene)

Activities a Week Before
- Prepare a list of chemicals, equipment, and glassware you need to perform this experiment
- Develop your own data table (see Data sheet)

There are many examples of inorganic polymers. The most commonly encountered systems are **borates** ($B_xO_y^{n-}$), **silicates** ($Si_xO_y^{n-}$), and **silicones.** Organic polymers gain their strength and stability from the exceptional concatenating ability of carbon. The carbon-carbon bond is quite strong and does not undergo reaction easily. Boron-boron and silicon-silicon bonds, however, are not nearly so strong and are susceptible to oxidation or to attack by Lewis bases such as water and hydroxide. Instead of forming long chains of boron or of silicon atoms, the polymers contain alternating networks of –B–O– or –Si–O– linkages, which are much more stable. This stability comes about since boron and silicon, unlike carbon, possess empty low-energy orbitals (a p orbital on boron and d orbitals on silicon), which can effectively overlap with filled p orbitals on other elements, such as oxygen, forming extremely stable bonds.

Depending on the type of linkage and on the two- or three-dimensional nature of the bonding network, a great number of different polymers with unique properties can be prepared. The most commonly encountered polymers of borates or silicates are glasses. These materials can also be used to crosslink other polymers (usually organic), giving them greater strength through the formation of a three-dimensional network.

Experimental Section

Procedure

NOTE: Slimes of various colors can be prepared by adding colored indicator solution to the borax solution.

Microscale Experiment   Estimated time to complete the experiment: 1 h
Experimental Steps: Three steps are involved: (1) preparing 4 percent borax and 4 percent polyvinyl alcohol solutions, (2) titrating the polyvinyl alcohol solution with the borax solution, and (3) determining the viscosity of the polymer solution.

Disposal procedure: Dispose of all the solutions in marked containers.

Prepare an approximately 4 percent (by weight) solution of borax by dissolving 200 mg of borax in 5 mL of water in a 25 mL beaker. Add this solution to a micro buret. In a 100 mL beaker, prepare a 4 percent (by weight) solution of polyvinyl alcohol by slowly dissolving 1.2 g of polyvinyl alcohol in 30 mL of stirred boiling water. (If polyvinyl alcohol is added too quickly, a polymeric lump will form.) Allow the solution to cool to room temperature.

Add 0.5 mL of the borate solution to the polyvinyl alcohol solution. Stir vigorously the resulting solution with a glass rod until a consistent solution is obtained. Polymerization occurs immediately, and a slimy copolymer results from this addition. The structure of the copolymer is shown in Figure E34.1.

Determine the viscosity of this copolymer using the falling-ball method described in Experiment 4. Add an additional 0.5 mL of the borate solution, stir, and measure the viscosity. Continue this process until all the borate has been added.

Develop your own data collection and data manipulation tables. Make a plot of viscosity (*y* axis) versus milliliters of borate added (*x* axis), and comment on what the plot shows.

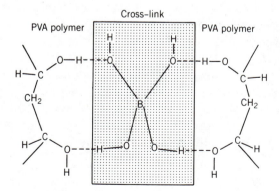

Figure E34.1 Borate crosslinking of polymers

Additional Independent Projects

1. Develop a method for the preparation of slime using polyvinyl alcohol and boric acid.

2. Determine the equilibrium volume (swelling) of a network polymer (vulcanized rubber).[1]

   ### Brief Outline

   Many crosslinked polymers swell in contact with a solvent. Swelling is the first step in the dissolution process of a polymer. Determine the density ρ_p (g/mL) of the rubber piece according to the method described in Experiment 1. Using a micropycnometer, determine the density ρ_l (g/mL) of toluene (see Experiment 1). Cut three specimens (~ 2 g each) of vulcanized rubber (or else cut pieces of a rubber stopper) and weigh each one accurately (initial mass in grams $= m_i$). Place one piece in each of three flasks containing enough toluene to cover the pieces. Stopper the flasks. The rubber specimens will absorb the solvent, causing them to swell. Periodically, remove the rubber pieces, dry them with a paper towel, and weigh them immediately. Replace them in the toluene. Repeat the procedure till the rubber pieces show constant equilibrium mass (final mass in grams $= m_f$). Using the following equation, calculate the equilibrium swelling volume V_s of the polymer:

   $$V_s = (m_i/\rho_p) + (m_f - m_i)/(\rho_l)$$

   Develop your own data collection and manipulation tables. Submit a written report according to the instructions given in Chapter 4.

References

1. Collins, E. A.; Bares, J.; Billmeyer, F. W., Jr. *Experiments in Polymer Science;* Wiley: New York, 1973. Also see Ross, J. H. *J. Chem. Educ.*, **1983,** *60,* 169.

Name ______________________ Section ______________

Instructor ______________________ Date ______________

PRELABORATORY REPORT SHEET—EXPERIMENT 34

Experiment title ______________________

Objective

Reactions/formulas to be used

Materials and equipment table

Outline of procedure

1. What function does the borate serve in the copolymers prepared in this experiment?

2. Why is the B–O bond so stable?

3. Suggest a test for detecting borate.

EXPERIMENT 34   DATA SHEET

Use this page to develop your own data table.

Data Collection Table

Data Manipulation Table

1. Define viscosity. What other method is available for measuring the viscosity of a liquid? Why can this method not be used in determining the viscosity of "slime"?

2. What is slime? Name the starting compounds used in the preparation of slime.

3. If we use boric acid instead of borax in the preparation of slime, the copolymer will not form. How can you make the polymerization reaction occur?

Objectives
- To prepare a polymer and to characterize it
- To determine molecular weight of polystyrene using reverse-phase TLC

Prior Reading
- Section 3.3   Handling chemicals
- Section 3.6   Heating methods

Related Experiments
- Experiments 25 (Column chromatography), 26 (Thin-layer chromatography), 34 (Inorganic polymers)

Polymers, both synthetic and natural, are extensively used in our daily lives. Polymers (from the Greek word *polumeres* meaning "having many parts") are extremely large molecules, termed **macromolecules**, with average molecular weights varying from 10^4 to 10^8 amu. Examples of synthetic polymers include plastics, polyethylene, PVC, Plexiglas™, textile fibers, resins, and adhesives. Natural polymers include carbohydrates, proteins (see Experiment 33), and nucleic acids.

Many small molecules can combine with each other to produce macromolecules. These smaller molecules are called **monomers**. The process of forming macromolecules from monomers is termed **polymerization**. Thus, many molecules of ethylene, $n\ H_2C=CH_2$, can combine to form polyethylene $(-CH_2-CH_2-)_n$.

Polymers are manufactured by two general methods: addition polymerization (chain-growth polymerization) and condensation polymerization (step-growth polymerization). In **addition polymerization**, monomers containing double or triple bonds join together to form macromolecules having the same elemental composition as the monomer. **Condensation polymerization** is a process where macromolecules are formed as a result of elimination of a byproduct. Frequently, catalysts are used to accelerate the rate of polymerization reactions. Straight-chain polymers are called **linear polymers**. **Cross-linked polymers** consist of infinite sheets of three-dimensional networks of monomers.

The molecular mass of a synthetic polymer depends on the mechanism of the polymerization reaction and the experimental conditions adopted for its preparation. The various macromolecules of a polymer do not have the same molecular weight, as they vary to some degree in length. The molecular weight of a polymer is the average of the molecular weights of the macromolecules (see reference 1 at the end of the experimental section). The molecular weight of a polymer determines its physical and chemical properties. For example, the strength of polyester fiber increases with molecular weight.

In this experiment, styrene, C_6H_5–CH=CH$_2$, will be polymerized in the presence of trace amounts of either 2,2'-azobisisobutyronitrile (AIBN) or benzoyl peroxide, C_6H_5–(C=O)–O–O–(O=C)–C_6H_5(BP) according to the following reaction, where Ph = C_6H_5:

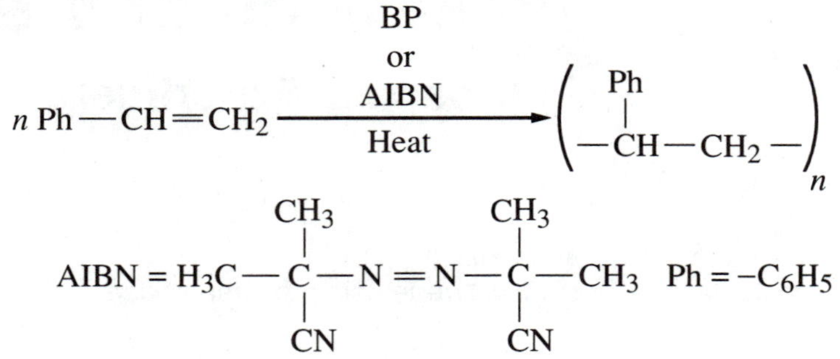

$$n\ \text{Ph}-\text{CH}=\text{CH}_2 \xrightarrow[\text{Heat}]{\substack{\text{BP} \\ \text{or} \\ \text{AIBN}}} \left(\begin{array}{c} \text{Ph} \\ | \\ -\text{CH}-\text{CH}_2- \end{array} \right)_n$$

$$\text{AIBN} = \text{H}_3\text{C}-\underset{\underset{\text{CN}}{|}}{\overset{\overset{\text{CH}_3}{|}}{\text{C}}}-\text{N}=\text{N}-\underset{\underset{\text{CN}}{|}}{\overset{\overset{\text{CH}_3}{|}}{\text{C}}}-\text{CH}_3 \quad \text{Ph}=-\text{C}_6\text{H}_5$$

In this reaction, AIBN is used as an initiator of the polymerization reaction. The value of n depends on the reaction conditions, such as length of reaction time and concentration of the initiator.

The distribution of molecular mass (which depends on the value of n) will be determined by reverse-phase thin-layer chromatography (TLC). The mobile phase is a binary mixture (71:29) of methylene chloride and methanol. The polymer is soluble in CH_2Cl_2 but insoluble in methanol. During the process of developing the TLC plate, the less polar methylene chloride is absorbed by the stationary phase, increasing the relative concentration of methanol in the remaining mobile phase. The bottom of the plate becomes richer in methylene chloride. As the chromatogram develops, the less soluble high molecular weight polymer fractions separate out first. As the solvent front moves up, the more soluble lower molecular weight polymers precipitate on the plate.

Commercially available styrene monomer contains an inhibitor (4-*tert*-butylcatechol) as a stabilizing agent. Its function is to prevent the polymerization of styrene. Prior to the polymerization reaction, the inhibitor must be removed from styrene by the micro column chromatographic method.

General References

1. Collins, E. A.; Bares, J.; Billmeyer, F. W., Jr. *Experiments in Polymer Science;* Wiley: New York, 1973.
2. Armstrong, D. W.; Marx, J. N.; Kyle, D.; Alak, A. *J. Chem. Educ.*, **1985**, *62*, 705.

Experimental Section

Procedure

> Perform this experiment inside a well-ventilated HOOD. Work in pairs.

> Microscale experiment   Estimated time to complete the experiment: 3 h
> Experimental Steps: Three steps are involved: (1) purification of styrene by column chromatography, (2) preparation of polystyrene (option 1 or 2), and (3) performance of reverse-phase thin-layer chromatography to determine the molecular weight of the polymer.

> Disposal procedure: All organic liquids must be disposed of in marked containers.

Obtain three 10×75 mm test tubes fitted with slotted corks, one 15×125 mm test tube, a micro stir bar (see Section 3.10), 2 to 4 mL of styrene, 5 mg of AIBN, and 2 to 4 g of alumina (80–200 mesh). Also procure a 5×10 cm reverse-phase thin-layer chromatographic (TLC) plate treated with an indicator (Whatman™ 4803-425 type, handle by the edge), polystyrene standards (prepared by dissolving 5 mg of each of the polystyrene standards in 1 mL of methylene chloride) with molecular mass from 2.0×10^2 to 1.0×10^5 amu, eight micro capillary pipets for spotting, a plastic ruler, and a UV lamp (254 nm).

Set up an oil bath in a 250 mL beaker containing a stir bar, atop a magnetic-stirring hot plate. Heat the oil bath ($\sim 80°$C) and maintain the bath at that temperature. If you use water to make the bath, prevent the steam from entering into the reaction vessel (test tube).

With a soft lead pencil, draw a baseline across the width of the TLC plate 1 cm from the bottom of the plate, and mark **nine spots** along the line. Also place the TLC solvent mixture consisting of methylene chloride and methanol in the **exact volume ratio of 71:29** (supplied by your instructor) in a 1000 mL beaker. The solvent must cover the bottom of the beaker, to a level of $\frac{1}{2}$ cm. Cover the beaker with a watch glass. This apparatus will be shared by a pair of students.

Purification of Styrene and Removal of Inhibitor

Prepare a microchromatographic column according to the procedure described in Experiment 25A, using alumina and a Pasteur pipet. Add 3 to 4 mL of styrene dropwise to the column and collect the eluate in a small test tube. If styrene does not elute quickly enough, use a sample from a freshly opened bottle.

Preparation of Polystyrene and Determination of its Molecular Mass

Two options are available to monitor the extent of the polymerization reaction. The reaction may be followed per unit time (at intervals of 3 minutes, 15 minutes, and 45 minutes, **option 1**) or with the change in concentration of the initiator, AIBN (use 1, 2, or 4 mg of AIBN, **option 2**). Your instructor will inform you as to which option to use.

Option 1

Obtain three 10×75 mm test tubes loosely fitted with slotted corks, labeled "3 min," "15 min," and "45 min." Add 300 μL of methylene chloride to each test tube and replace the corks. At the bottom of a fourth test tube (15×125 mm), place ~2 mL of the freshly eluted styrene and a micro stir bar. Using an automatic delivery pipet, add 2.5 mL of toluene to the styrene. Swirl the mixture, and add 30 to 40 mg of AIBN. Place a slotted cork in the test tube and clamp the test tube in the oil bath so that only the bottom fifth of the test tube is inside the oil bath.

Control the temperature of the bath to maintain a **mild reflux**. As soon as the solution begins to reflux, record the time. After 3 minutes, remove 2 or 3 drops of the solution using a Pasteur pipet, and add this to the test tube marked "3 min." Repeat this procedure at the end of 15 and 45 minutes, using the appropriate test tubes. When done, allow the reaction mixture to cool. Mix the solution in each test tube thoroughly.

While heating is going on, prepare eight microcapillary pipets for spotting solutions on the TLC plate. Using a separate capillary each time, spot five polymer standards in the first five tracks on the TLC plate. The spots should be ~2mm in diameter. Use the UV lamp to see the spot as a dark area. If the mark is not visible, spot more solution. Similarly, **spot the remaining three tracks with the unknown polymer solutions from the test tubes (3, 15, and 45 min)**.

Place the TLC plate in the developing chamber. When the solvent front has reached about 1 cm below the top of the plate, remove the plate, and mark the solvent front (see Figure E35.1). Use the UV lamp to detect and mark the locations of the bottom and the midpoint of each polymer spot.

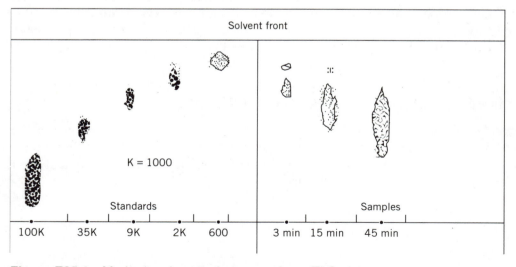

Figure E35.1 Marked and spotted reverse-phase TLC plate

Using the procedure described in Experiment 26, determine the R_f value for each standard. Obtain two R_f values for each spot: one for the bottom and the other for the midpoint of the spot. Make two calibration plots of the log of the molecular weight of the standards versus R_f values. Also, determine the bottom and the midpoint R_f values for each of the unknown polymers. Using the bottom R_f values and the calibration curve, calculate the weight average molecular weight, M_w, of the polymer formed at each time interval.

Similarly, using the midpoint R_f values, obtain the number average molecular weight, M_n (defined as the average in which the mass of a polymer sample is divided by the number of molecules present in the polymer[1]) for the sample. Calculate the polydisparity index (degree to which the molecular weight varies) for the sample by dividing M_w by M_n.

Separation of the Polymer

Pour the cooled leftover reaction mixture into a beaker containing about 5 mL of methanol. Stir the mixture vigorously. If precipitate does not form, add more methanol. Collect the precipitate by suction filtration using a Hirsch funnel (see Section 3.8). Wash the precipitate with a few drops of methanol. After it is dry, weigh the product and calculate the approximate percent yield.

Option 2

Repeat the procedure in Option 1 with the following changes. Weigh 250 mg of freshly eluted styrene in three 10×75 mm test tubes (labeled "1 mg," "2 mg," and "4 mg") and add 1, 2, and 4 mg of AIBN to the appropriate test tube. Place the test tubes in the oil bath ($\sim 80°C$), and record the time and temperature. Remove the test tubes from the oil bath after 45 minutes. Immediately pour the reaction mixture from the "1 mg" test tube into a beaker containing 20 mL of methanol. Stir the solution. Polystyrene will precipitate. Filter the residue on a Hirsch filter (see Section 3.8) and wash it several times with methanol. Dry the product at the pump, weigh it, and save the polymer. Determine the percent yield. Repeat this procedure with the contents of the other two test tubes. Plot percent yield versus milligrams of AIBN. Report the effect of the initiator.

Using the reverse-phase TLC plate, known polymer standards, and the prepared polystyrene samples, determine the weight average (M_w) and number average (M_n) molecular mass of the samples. Follow the same procedure as described in Option 1.

References

1. Pilar, F. L. *J. Chem. Educ.* **1992**, *69*, 280.

Additional Independent Projects

See Experiment 34.

Name ______________________ Section ______________

Instructor ______________________ Date ______________

PRELABORATORY REPORT SHEET—EXPERIMENT 35

Experiment title ______________________

Objective

Reactions/formulas to be used

Chemicals and solutions—their preparation

Materials and equipment table

Outline of procedure

1. Define the following terms: monomer, polymer, and copolymer.

2. Why do we use AIBN instead of benzoyl peroxide as an initiator?

EXPERIMENT 35   DATA SHEET

Complete Your Own Data Collection and Manipulation Tables

See Experiment 32 (for data involving synthesis) and Experiment 26 (for data involving TLC). Plot R_f (bottom and midpoint) versus log M_w and submit your plots. Use additional sheets if necessary.

Data Collection and Manipulation Tables

Preparation of the Polymer

Distances of the solvent front and of the spots (bottom and midpoint) from the spotting line; R_f values (bottom and midpoint) for the standards and samples

Plots and calculation of M_w and M_n of the samples

1. Define R_f.

2. Why is a mixture of methylene chloride and methanol used as the solvent in this experiment?

3. In what ways can an invisible TLC spot be rendered visible? (At least two methods, please.)

EXPERIMENT **36** Semiconductors: Preparation of Semiconducting Thin Films

Objectives
- To make semiconductor thin films and study their properties
- To use spectrophotometry in the analysis of semiconductors

Prior Reading
- Chapter 8   Introduction to visible spectroscopy

Related Experiments
- Experiments 21 (Determination of iron), 22 (Measurement of alcohol), 23 (Determination of manganese), 24 (Spectrophotometric determinations), 25 (Chromatographic techniques), 26 (Thin-layer and paper chromatography), and 37 (Analysis of a polymer by infrared spectroscopy)

Activities the Day Before the Experiment
If you are using glass slides, immerse the glass plates in a KOH/alcohol bath.

Metals are good conductors of electricity. Copper, for example, allows the flow of electrons with relatively little resistance (about 10^6 ohm cm^{-1}). It is therefore used in electrical wires. Nonmetals, like glass or diamond, generally have a large resistance to the free flow of electrons (values from 10^{10} to 10^{18} ohm cm^{-1}) and are therefore nonconductors or **insulators**. Recently, a class of chemical compounds has been developed that are called **superconductors**, because they conduct electricity with essentially no resistance. **Semiconductors** have conductivities lying between conductors and insulators (10^8 and 10^{10} ohm cm^{-1}).

The properties of conductors, semiconductors, insulators, and superconductors depend upon their electronic structures, but it is important to note that the bonding in a bulk structure (like a metal crystal) is quite different from that in a simple molecule. A solid block of a metal consists of a huge collection of atoms. A sample of 6.9 g of lithium metal (one mole of Li) will contain Avogadro's number of atoms (6.0×10^{23} atoms) and three times Avogadro's number of electrons (each Li atom contains three electrons). Each Li atom is characterized by the electronic configuration $1s^2 2s^1 2p^0$.

In the diatomic lithium molecule, Li_2, the bond between the atoms is caused by the overlap of the valence $2s$ orbitals, which forms an s (bonding) and $s*$ (antibonding) orbital set (see Figure E36.1). Each lithium has one valence electron, so the Li_2 molecule has two valence electrons, exactly filling the s bonding molecular orbital (remember—each orbital can hold two electrons) and leaving the $s*$ orbital empty.

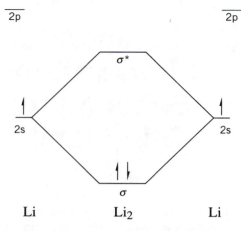

Figure E36.1 Lithium molecule and its molecular orbital diagram ($1s^2$ electrons are not shown because they do not contribute to bond formation)

Band Theory—Bonding in Crystals

We now consider the bonding found in a Li crystal. Suppose that we have n lithium atoms in the crystal and that the crystal is highly symmetric. The individual $2s$ and $2p$ orbitals on each lithium can interact with each other in many ways, forming a large number of molecular orbitals of similar energy. These individual molecular orbitals form a nearly continuous series of bands, called the **valence band** (see Figure E36.2a). For lithium, as a result of the high symmetry, the $2s$ band and $2p$ band do not exist separately; rather, they overlap and

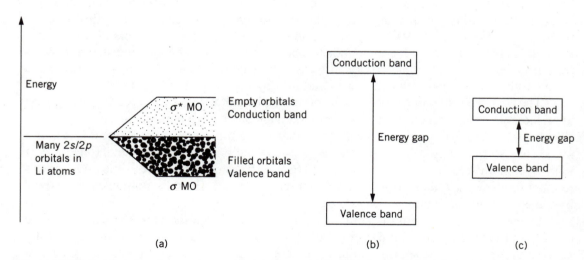

Figure E36.2 Conduction and valence bands: (a) Many atom lithium $2s/2p$ orbitals; (b) Carbon bands (energy gap 526 kJ/mol); and (c) Silicon bands (energy gap 106 kJ/mol)

merge into a $2s/2p$ band. Since there are four energy levels per atom (one $2s$ and the three $2p$ orbitals), for n atoms, there will be $4n$ orbitals, which can hold $8n$ electrons. Lithium has only one valence electron per atom, however, so there will be only n electrons in the valence band for n atoms. Thus, the valence band is only partly filled, and it is a relatively easy matter to promote an electron from a filled orbital to an empty one, as the energy needed is quite small. This ability to promote and therefore delocalize electrons is responsible for the high conductivity of metals.

In the case of nonmetals like carbon, the structures that are adopted are not so highly symmetric, and the $2s/2p$ band splits in two, as shown in Figure E36.2b. (This pattern is the equivalent of a crystal forming bonding and antibonding orbitals.) The lower energy band is still called the valence band, and the higher energy band is called the **conduction band**. The energy difference between the bands is called the **band gap energy,** E_g. Since the combined $2s/2p$ band can hold $8n$ electrons, each of the split bands can hold half of that, that is, $4n$ electrons. Since carbon has exactly four bonding electrons, a diamond crystal would have $4n$ electrons. This number exactly fills the valence band and leaves the conduction band empty. Since the band gap energy is high, free electron flow is prevented. Diamond is therefore classified as an insulator.

Semiconductors

Silicon and germanium, having the same number of electrons as carbon, might be predicted to be insulators as well. However, as the temperature is raised on these crystals, the electrons are scattered more (they are less tightly held, being farther away from the nucleus on the larger atoms), and some get promoted into the conduction band. Thus, Si and Ge conduct electricity weakly (more as the temperature rises), and they are classified as semiconductors (see Figure E36.2c).

The transition of electrons from the valence band to the conduction band can be achieved by the absorption of energy, if this energy is equal to or larger than E_g. Thus, the absorption of light occurs when

$$hc/\lambda \geq E_g$$

where the wavelength λ is expressed in nanometers, h is Planck's constant (6.63×10^{-34} J s), and c is the velocity of light (3.00×10^8 m s^{-1}). If this wavelength happens to fall in the visible region, the semiconductor is known as a photoconductor.

The Xerox™ process depends on this type of conductivity. In a Xerox machine, a positively charged plate is covered by a semiconducting film. Light is reflected by the white part of the original paper onto the film, promoting electrons from the valence band to the conduction band, and the film conducts electricity. The parts of the film that were hit by the light are no longer charged (as the promoted electrons balance the positive charge originally present), but the part of the film under the black part of the original paper is still charged. Small capsules of ink toner are then spread on the film and stick only to the charged part. A new piece of paper (the copy) then removes the toner from the film and thus acquires a copy of the black part of the original.

Example E36.1 Calculate the band gap energy of a semiconducting film that absorbs at 450 nm. What is the frequency of this light?

Answer

$$E_g = \frac{hc}{\lambda} = \frac{(6.626 \times 10^{-34}\ \text{J s})(3 \times 10^8\ \text{m s}^{-1})}{(450 \times 10^{-9}\ \text{m})} = 4.4 \times 10^{-19}\ \text{J}$$

$$\nu = \frac{c}{\lambda} = \frac{(3 \times 10^8\ \text{m s}^{-1})}{(450 \times 10^{-9}\ \text{m})} = 6.14 \times 10^{14}\ \text{s}^{-1} = 6.14 \times 10^{14}\ \text{Hz}$$

Semiconducting Compounds

Compounds that are isoelectronic (have the same number of electrons) as Si or Ge are also often semiconductors, since they have similar electronic structures. There are two common types—the III/V semiconductors (one element from Group IIIA and one from Group VA, for example, GaAs) and the II/VI semiconductors (one element from Group IIB and one from Group VIA, ZnS for example). In this experiment, we will prepare two II/VI semiconducting films.

General References

1. Wrighton, M. S. *J. Chem. Educ.* **1983**, *60, 877.*
2. Kutal, C. *J. Chem. Educ.* **1983**, *60, 882.*

Experimental Section

Procedure

Microscale experiment   Estimated time to complete the experiment: 3 h
Experimental Steps: Two main steps are involved: (1) preparing the deposits of sulfides of metals on slides and (2) obtaining the spectra of the deposits.

NOTE: All glassware including the glass slides must be meticulously cleaned. The glassware should be immersed in a KOH/alcohol bath overnight.

CAUTION: The KOH/alcohol bath is very corrosive—wear gloves.

Disposal Procedure: Dispose of all the metal salt solutions in the designated containers only.

Remove the glassware from the KOH bath using tongs. Rinse the glassware with tap water, followed by bathing them in a bath of dilute nitric acid. Rinse again with tap water and finally with deionized water (distilled water). An ultrasonic cleaning bath containing a detergent solution is an alternative to the KOH bath. Note: You may use a Kodak transparency sheet in place of a glass plate to cut out plastic slides. They are easier to handle and do not require the pretreatment with a KOH bath.

After the glass slides are clean and dry, cut them lengthwise into two halves. Your instructor may provide you with the right-sized glass slides. The slides must be cut so that they can easily fit into the spectrophotometer cuvettes.

CAUTION: Take care while handling the cut glass slides. Their edges are sharp!

Half of the slide is used for depositing the semiconducting film and the other half is used as a reference for the absorption spectrum of the film.

Preparation of a Zinc Sulfide Semiconducting Film

CAUTION: This and the following reactions generate H_2S gas, which is toxic. Carry out the deposition procedure inside a well-ventilated HOOD.

Set up a hot water bath in a 100 mL beaker on a magnetic-stirring hot plate. Place 1.0 mL of 0.1 M $ZnSO_4$ solution in a 10 mL beaker containing a stirring bar. Add 1.0 mL of 0.1 M thioacetamide (or thiourea) solution, followed by 1.5 mL of 6 M NH_3 solution. Heat the beaker in the water bath, and when the temperature has reached 85°C, begin stirring the solution. At the same time, place the glass slide halfway into the solution.

Maintain the temperature of the water bath at 85°C. A white precipitate of ZnS will form as the reaction proceeds. Remove the beaker from the water bath after 15 to 20 minutes, using tongs. Slowly and cautiously, remove the slide from the beaker using forceps. Quickly rinse the slide with deionized water. Place the slide in a desiccator for drying under an N_2 atmosphere. [It can also be dried in a side-arm test tube under vacuum or in a vacuum desiccator.] These precautions are necessary because zinc sulfide is easily oxidized to zinc sulfate in the presence of moist air.

After the slide has dried (20 to 30 min), place it in a cuvette and obtain its absorption spectrum, using the noncoated half (not coated with ZnS) as a reference. From the spectrum, determine the λ_{max} value and calculate E_g.

Preparation of a Cadmium Sulfide Semiconducting Film

> CAUTION: Cadmium compounds are toxic. Wear gloves. Perform the procedure inside a well-ventilated HOOD.

Mix 1.0 mL of 0.1 M cadmium sulfate (or cadmium acetate) solution and 1.0 mL of 0.1 M thioacetamide (or thiourea) solution in a 10 mL beaker containing a magnetic stirring bar. Add 2.0 mL of 6 M NH_3 solution. Clamp the beaker in the hot water bath. While stirring the solution, place a slide halfway into the solution. Maintain the temperature at 70 to 80°C for 10 to 15 min. During this time a yellow film will deposit on the slide.

Remove the slide, rinse it with deionized water, and allow it to air dry. Obtain the absorption spectrum of the semiconductor film as just described, using the noncoated half of the slide as the reference. Determine λ_{max} and E_g as described.

Preparation of a Bismuth Sulfide Semiconducting Film

> CAUTION: Bismuth salts and H_2S are toxic substances. Carry out the experiment in a well-ventilated HOOD.

Triturate (pulverize) a mixture of 50 mg of $Bi(NO_3)_3$, 500 μL of triethanolamine (use HOOD), and 2.5 mL of water in a mortar. Using a Hirsch funnel, filter the solution (if necessary) to obtain a clear filtrate. Using a pipet, transfer 1.0 mL of this solution to a 10 mL beaker containing a magnetic stirring bar. Add 1.5 mL of 0.1 M thioacetamide solution, followed by 1.0 mL of 6 M NH_3 solution. Insert a slide halfway into the solution, and place the beaker in a water bath maintained at 85°C. A white precipitate will soon form, and the slide will be covered with a thin film of Bi_2S_3.

After 10 minutes, remove the slide from the mixture, rinse it with deionized water, and air-dry the slide. Obtain its absorption spectrum as described earlier using the noncoated half as the reference. Determine the λ_{max} value for the film, and calculate E_g.

References

1. Gurnee, E. F. *J. Chem. Educ.* **1969**, *46*, 80.
2. Ibanez, J.; Solorza, O.; Gomez-del-Campo, E. *J. Chem. Educ.* **1991**, *68*, 872.

Name ____________________	Section __________
Instructor ____________________	Date __________

PRELABORATORY REPORT SHEET—EXPERIMENT 36

Experiment title ______________________________

Objective

Reactions/formulas to be used

Materials and equipment table

Outline of procedure

1. Explain the differences between conductors, semiconductors, insulators, and superconductors.

2. What is the band gap energy for a film that absorbs light at 510 nm? What is the
   frequency of this light?

3. What are n- and p-type conductors? What is meant by doping?
   [Hint: Consult your chemistry textbook]

EXPERIMENT 36   DATA SHEET

Semiconductor	Wavelength, nm	E_g, J
1. Zinc sulfide, ZnS	_____	_____
2. Cadmium sulfide, CdS	_____	_____
3. Bismuth sulfide, Bi_2S_3	_____	_____

Show calculations:

1. When making the Bi_2S_3 thin film, you used aqueous NH_3. Will Bi_2S_3 form if 6 M HCl solution is used instead of ammonia? (*Hint*: What is $[S^{2-}]$ at this pH?)

2. What is the purpose of using thioacetamide? Write balanced equations for the reaction that occurs when aqueous NH_3 is added to thioacetamide and the mixture is heated.

3. ZnS, CdS, and Bi_2S_3 films absorb in the range of 350, 500, and 950 nm, respectively. In which regions of the electromagnetic spectrum do these absorptions occur?

EXPERIMENT **37** Analysis of a Polymer by Infrared Spectroscopy

Microscale Experiment

Objective

- To use infrared spectroscopy to analyze a polymeric material

Prior Reading

- Chapter 8   Introduction to visible spectroscopy

Related Experiments

- Experiments   34 (Inorganic polymers), 35 (Preparation of a polymer), and 21 through 24 (Visible spectroscopy)

Spectroscopy can be used to determine the concentration of an analyte (Experiments 21 through 23) and to establish the composition of a compound using Job's method of continuous variation (Experiment 24). In this experiment, infrared (IR) spectroscopy will be used to analyze the vinyl acetate content of commercial polymer packaging films.[1] This method illustrates an application of IR in industry.

IR spectroscopy[1,2] deals with the rotational and vibrational transitions that a molecule undergoes. In a molecule, the atoms rotate and vibrate periodically about their center of mass. Each normal vibration is associated with a frequency (or wavelength) of radiation. The general requirement for a molecule to be *infrared active* is that the vibration must produce a change in the dipole moment (defined as the product of one of the charges and the charge separation). Only those vibrations that are accompanied by a change in the dipole moment are observable in the IR spectrum.

The infrared portion of the electromagnetic spectrum ranges from 4000 to 650 cm^{-1}. This range is commonly divided into two subregions: the *group frequency region* (4000 to 1300 cm^{-1}) and the *fingerprint region* (1300 to 650 cm^{-1}). Most often, the group frequency region corresponds to vibrations caused by two bonded atoms. The vibrational frequency (ν) is determined by the **reduced mass** (μ) of the atoms (m_1 and m_2 are the masses of two atoms) and the stiffness (called **force constant**, k) of the bond between them. The frequency is given by Hooke's law:

$$\nu = \left(\frac{1}{2\pi c}\right)\left(\frac{k}{\mu}\right)^{1/2}$$

where $\mu = (m_1 \times m_2)/(m_1 + m_2)$ and c is the velocity of light (3.00×10^8 m/s). From the equation, we see that as the reduced mass of the atoms in the bonds, μ, increases, the frequency decreases. That is, the lighter the atoms in a bond are, the higher its frequency is. We would therefore expect that bonds to hydrogen should appear at the highest frequencies. Similarly, as the force constant, k, increases, the frequency rises. Thus, double and triple bonds should appear at higher frequencies than single bonds.

Packaging films containing vinyl acetate groups are called ethylene vinyl acetate (EVA) **copolymers**. Copolymers are long-chain macromolecules that are produced by joining two or more different monomers (see Experiment 35). The ethylene vinyl acetate copolymer is formed according to the following reaction:

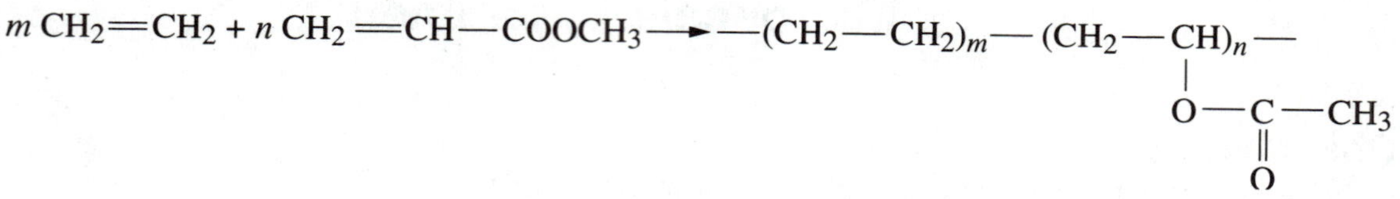

$$m\ CH_2{=}CH_2 + n\ CH_2{=}CH{-}COOCH_3 \longrightarrow {-}(CH_2{-}CH_2)_m{-}(CH_2{-}CH)_n{-}$$

(with the pendant group)

$$O{-}C{-}CH_3$$
$$\|$$
$$O$$

The percentage of vinyl acetate in the copolymer determines the grade of the polymer. Polyvinyl acetate is used as an adhesive, for textile coatings, in chewing gum, and in paints.

The IR spectrum of a sample of EVA contains bands characteristic to both the polyethylene part of the molecule and the vinyl acetate part of the molecule. The absorbances for these two bands can be measured as shown next.

In this experiment the IR absorption spectra of various samples of packaging films (with known vinyl acetate content) are obtained.[3] A correlation graph is then constructed, using the Beer-Lambert law (see Chapter 8),

$$A = \epsilon\, b\, c$$

where
A = absorbance
b = thickness of the film
c = concentration of the vinyl acetate in the film
ϵ = molar absorptivity

The method of determining A is shown in Figure E37.1.

The thickness b of the film can be measured with a micrometer. The concentration of the vinyl acetate in an unknown sample is then obtained from the calibration curve (known concentrations of vinyl acetate in films versus the absorbance/thickness).

General References

1. Szafran, Z.; Pike, R. M.; Singh, M. M. *Microscale Inorganic Chemistry: A Comprehensive Laboratory Experience;* Wiley: New York, 1991.
2. Mayo, D. W.; Pike, R. M.; Trumper, P. K. *Microscale Organic Laboratory*, 3rd ed.; Wiley: New York, 1994.
3. Allpress, K. N.; Cowell, B. J.; Herd, A. C. *J. Chem. Educ.* **1981**, *58*, 741.

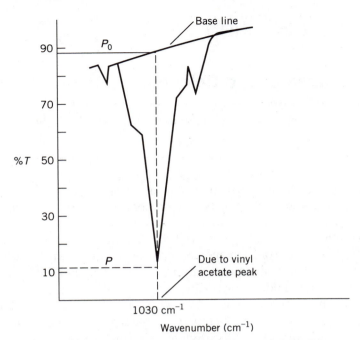

Figure E37.1 The baseline method to determine the absorbance, A [$\%T = P/P_0$ and $A = \log(1/\%T)$]

Experimental Section

Part A: Identification and Selection of IR Peaks

Your instructor will demonstrate how to operate the IR spectrophotometer.

Microscale experiment   Estimated time to complete the experiment: 2 h
Experimental Steps: Two main steps are involved: (1) obtaining the spectra of standard
vinyl acetate polymers and (2) running a spectrum of an unknown sample.

Obtain several samples of packaging film (EVA, ethylene vinyl acetate fiber) containing
known amounts of vinyl acetate and mount them on a cardboard holder (similar to the
polystyrene film standard holder used for calibration). Also obtain a strip of pure polyethylene film and mount it on a holder. Obtain an unknown EVA sample from your instructor.

   Run the IR spectrum of a known EVA sample and of the polyethylene film. Super-
impose one spectrum over the other. Determine which peak or peaks present in the EVA
sample spectrum are absent in the polyethylene spectrum. (*Hint:* There is only one such
peak present in the EVA spectrum, in the range 1100 to 600 cm^{-1}. Select this as your **target peak**.) Calculate the absorbance of the **polyethylene peak**, at 720 cm^{-1} (A_{PE}), by the
method shown in Figure E37.1.

Part B: Quantitative Determination

Obtain the spectrum of each known sample over the range 1100 to 600 cm^{-1} at a slow
scan speed. Determine the target peak (corresponding to the vinyl acetate) and measure its
absorbance A_{target}, following the baseline method shown in Figure E37.1. The absorbance
depends upon the concentration of vinyl acetate as well as the thickness of the film, b.
Measure the thickness of the film using a micrometer. Determine the thickness three times
and take the average value. The ratio of the absorbance of the target peak, A_{target}, to the
average thickness, b (μm), is proportional to the percentage of vinyl acetate present in the
polymer. Plot two calibration curves:

   Plot # 1: % vinyl acetate (x axis) versus the ratio $(A_{target})/(A_{PE})$ (y axis)

   Plot #2: % vinyl acetate (x axis) versus the ratio $(A_{target})/b$ (μm) (y axis)

Determine the percentage of vinyl acetate in the unknown sample from the calibration plots.
   Return all the polymer samples including the IR film holders to your instructor for
future use.

Additional Independent Projects:
1.   Study and interpret the IR spectra of the compounds obtained in experiments 19,
     20, and 35.
2.   Develop a qualitative IR method for identifying commercial polymer samples.[1]

Brief Outline

Use food wraps, sandwich bags made by different companies, freezer bags, and colored plastics. Run IR spectra of each sample. Compare the spectra with those of known samples (use library search process if the computer has one). Determine the characteristic group frequencies for the functional groups present in the samples. From these, determine the identity of each sample.

Develop the data collection table. Submit a written report according to the instructions given in Chapter 4. Include the spectra with your report.

3. Analyze a mixture of *ortho-*, *meta-* and *para-*xylenes by IR.[2]

References

1. Webb, J.; Rasmussen, M.; Selinger, B. *J. Chem. Educ.* **1977**, *54*, 303.
2. Veening, H. *J. Chem. Educ.* **1966**, *43*, 319.

Name ______________________________ Section ______________

Instructor ______________________________ Date ______________

PRELABORATORY REPORT SHEET—EXPERIMENT 37

Experiment title ______________________________

Objective

Reactions/formulas to be used

Chemicals and solutions—their preparation

Materials and equipment table

Outline of procedure

1. Define the following terms: monomer, polymer, and copolymer.

2. What are the IR group frequency and the fingerprint regions? What units are used to measure frequency in IR?

3. Why must the thickness of the film be used in determining the percent vinyl acetate?

EXPERIMENT 37   DATA SHEET

Part A: Identification and Selection of IR Peaks

Attach the IR spectra for the pure polyethylene sample, the knowns, and the unknown EVA copolymer sample. Mark the target peaks and the polyethylene peak at 720 cm^{-1} on each spectrum. Show the baselines and the calculation of I and I_0.

1.   Target peak in known EVA samples _____ cm^{-1}

      Polyethylene peak _720_ cm^{-1}

2.   Thickness of the polymer films, b, μm

	Trial 1	Trial 2	Trial 3	Average
Known sample 1	_____	_____	_____	_____
Known sample 2	_____	_____	_____	_____
Known sample 3	_____	_____	_____	_____
Known sample 4	_____	_____	_____	_____
Known sample 5	_____	_____	_____	_____
Unknown sample	_____	_____	_____	_____

Part B: Quantitative Determination

3.

	For target peak			
	$P_0(\%T)$	$P(\%T)$	$T = P/P_0$	$A_{target} = \log(1/T)$
Known sample 1	_____	_____	_____	_____
Known sample 2	_____	_____	_____	_____
Known sample 3	_____	_____	_____	_____
Known sample 4	_____	_____	_____	_____
Known sample 5	_____	_____	_____	_____

	Polyethylene peak			
	$I_0(\%T)$	$I(\%T)$	$T = I/I_0$	$A_{PE} = \log(1/T)$
Known sample 1	_____	_____	_____	_____
Known sample 2	_____	_____	_____	_____
Known sample 3	_____	_____	_____	_____
Known sample 4	_____	_____	_____	_____
Known sample 5	_____	_____	_____	_____
Unknown sample	_____	_____	_____	_____

4.

	Vinyl acetate $(\%)$	$(A_{target})/(A_{PE})$	$A_{target}/b\,(\mu m)$
Known sample 1	_____	_____	_____
Known sample 2	_____	_____	_____
Known sample 3	_____	_____	_____
Known sample 4	_____	_____	_____
Known sample 5	_____	_____	_____
Unknown sample	_____	_____	_____

5. Plot the calibration curves. Attach the graphs to your report.

6. Concentration of vinyl acetate in the unknown _____%

Show calculations:

POSTLABORATORY PROBLEMS—EXPERIMENT 37

1. Which part of the EVA polymer is responsible for the target peak in the IR spectrum?

2. Infrared spectroscopy is used in industry. Name some of these industries (such as the polymer industry).

CHAPTER **11** Independent Studies, Computers, and Research in Undergraduate Laboratories

Objectives

- To do independent studies through a library search
- To use computers in the laboratory
- To be able to develop and sustain undergraduate research projects

11.1 Literature Search and Independent Studies

An experiment carried out in a laboratory by a student encourages him or her to think logically and assists the individual to derive a meaningful conclusion based on the results. To a certain degree, in this process students get involved in a thinking process that enhances their ability to design independent and conceptually sound ideas. One of the important requirements for generating nascent concepts that lead to full-grown research activities is to gain knowledge about the previous contributions made by others. This process of independent inquiry is based upon a thorough library search of experimental work reported in the literature. This process constitutes what is called an **independent study**.

In chemistry (or in any other science) independent studies related to a research activity or experiment must begin with what is called a *literature search*. In chemistry, there are two basic means by which this can be efficiently done: the **Chemical Abstracts** method and the **Science Citation Index** approach.

Chemical Abstracts Method[1]

Chemical Abstracts (CA) is a journal that is published weekly by the American Chemical Society. It lists and cross-references the abstracts from all papers published in all major (and most minor) chemical journals. The abstracts are placed into one of 80 sections within CA, depending on the subject matter contained in the referenced paper. The sections cover the following material:

Sections 1–20	Biochemistry
Sections 21–34	Organic chemistry
Sections 35–46	Macromolecules
Sections 47–64	Applied chemistry and chemical engineering
Sections 65–80	Physical, inorganic, and analytical chemistry

The abstracts are cross-referenced in the index of each **weekly issue** of CA by author, patent number, and subject keyword. The weekly indices are collected into six **annual indices**, which cross-reference the abstracts in various ways:

Author index: Lists all papers in alphabetical order according to the author's name

Chemical substance index: Lists all papers according to the proper name of all chemicals used in the paper

Formula index: Lists all papers according to the formula of the chemical compounds used in the paper, in the following order: carbon first, hydrogen second, and all other elements last in alphabetical order

General subject index: Lists all papers according to subject keywords found in the paper

Index of ring systems: Useful mainly for organic chemistry, lists papers involving ring systems according to the type of ring system found

Patent index: Lists all chemical patents in numerical order, as well as subsidiary patents

Every five years, a **Collective Index** is published (in the past, it was every 10 years, and called a Decennial Index). Additionally, every 18 months an **Index Guide** is published, listing all current keywords and chemical names used in the foregoing indices.

Two commonly used methods for searching in CA are the **Formula Index** and **General Subject Index**. Your instructor will provide you with detailed guidelines as to how to proceed with the search, where the reference material is located in your institution's library, and how to find the current issue of the journal.

Suppose you are interested in knowing more about praseodymium(IV) oxide, PrO_2 (Pr is a lanthanide element). You start searching in the **Formula Index** of CA. The search will lead to a six-digit number ended with a letter (example: 110345m). The number indicates the abstract number of a paper. Besides the formula, the abstract also contains the names of authors, the journal name (abbreviated), year of publication, volume number, if any, and the page number. This information may be used for procuring the original publication for independent study.

Suppose you wish to know more about "micropycnometer." Since this is a general technique rather than a specific reaction or compound, the General Subject Index would be the logical place to start. Consulting a recent General Subject Index, we could look up "pycnometer" or "micropycnometer." If neither of these terms appeared in the General Subject Index, it might be necessary to consult the Index Guide to see the closest term to "micropycnometer" that does appear. The search, as already indicated, will lead to a six-digit number followed by a letter. This number is then used to locate the abstract and the original paper.

CAS Online[2,3]

Chemical Abstracts may also be accessed by computer database searching. The search may be conducted using various types of input, such as structures, molecular formula, CAS registry numbers, or keywords. Tutorial programs are available from STN[SM] International (2540 Olentangy River Road, P.O. Box 02228, Columbus, OH 43202).

Science Citation Index Method[4,5]

The *Science Citation Index* (SCI) is published every four months by the Institute for Scientific Information (Philadelphia, PA). Papers from all major and most minor journals are cross-referenced by use of authors, keywords, journal, location, and, most importantly, their references. The SCI is now available on compact disc, which can be accessed using a personal computer. The printed SCI consists of three major parts:

Citation index: Lists all papers as a function of the references they cite

Source index: Lists all papers alphabetically, by the first author

Permuterm subject index: Lists all papers according to subject keywords. The computer version allows searches by the following criteria:

Source author: The author of the article

Cited author: The author of the reference cited by the article

Address: The address(es) of the authors of the paper

Journal: The journal in which the paper appeared

Title word: By subject keywords appearing in the title of the paper

Assignment of an Independent Study Topic

Only one example is given here. Your instructor may assign different topics to different students.

"Carry out an independent study on superconductors and develop a report on their properties and future potential uses."

11.2 Computer Uses in General Chemistry Laboratories

The use of computers in the laboratory has been continuously increasing. In 1993, the *Journal of Chemical Education* published 43 computer-related articles. Apart from word processing, computers are being used for the following purposes:

1. For carrying out complex mathematical manipulations on data loaded into the computer. In this regard, spreadsheets (for example, MS Excel version 1.5 or Lotus 1-2-3) are extremely useful for solving simultaneous equations.[6]
2. For graphing purposes. The software program "Cricket Graph" is extremely useful in representing the data in graphical form.
3. For interfacing with other probes for the direct collection of data as the experiment is in progress. Several books have been published that describe such applications of computers in a laboratory.[7,8]
4. As learning tools.[9,10] Molecular modeling is one of the most important and powerful tools available to chemists.

Basic laboratory techniques must be learned with hands-on experience only. Computers are tools to enhance the process of doing laboratories; they cannot substitute for the benchtop experimentation and manipulations—at least not yet! The following assignments illustrate examples of the versatility of computer technology in the undergraduate laboratory. They are based upon computer programs that are readily available in the laboratory.

The description of the experimental procedure has been kept to a minimum to encourage students to do the independent library search before they begin their experiments.

Assignment 1. Use of Software Programs (Cricket Graph) in Manipulation of Data

Experiments 1 and 5: Use Cricket Graph to collect data and plot the graph for Boyle's law and Charles' law.

Experiments 28 and 33B: Use Cricket Graph to load the data and obtain the pH titration curves for each of the laboratory assignments described in these experiments.

Experiments 21 through 24: Similarly, use Cricket Graph to tabulate your data and plot the data to prepare the calibration curves wherever needed.

Assignment 2. Study of Periodic Properties of Elements Using KC? Discoverer

Using the program KC? Discoverer (available from JCE Software) or other programs, study the periodic trends among elements. Your instructor will show you how to load the database program onto the computer. To see the periodic trend, plot any selected property of elements (atomic volume, density, ionization energy, electron affinity, atomic size, and so on) versus atomic number (you may also use atomic masses) of elements. Report your results according to the instructions given in Chapter 4.

Assignment 3. Use of a Spreadsheet: Titration of Diprotic Acids and Bases

Your instructor will show you how to use the spreadsheet program (Microsoft's Excel version 1.5 or Lotus 1-2-3). This assignment is a computer-based titration procedure. Using a spreadsheet program with an iteration feature, carry out the titration of any diprotic acid or base according to the procedure described by Breneman and Parker.[6a] Generate the spreadsheet for calculating pH versus volume of base (NaOH, 0.1 M) added in titrating a diprotic acid (sulfurous acid, 10 mL of 0.1 M) solution. Submit your data and a printout of your experiment.

Assignment 4. Use of a Computer-Interfaced pH Probe. Determination of pK_a for Organic Acids by the Half-Neutralization Method

Computer interfacing with an experiment and with subsequent data collection and manipulation incorporates the constantly evolving computer technology in the chemistry laboratory. Different kinds of microprobes capable of interfacing with Apple II, IBM, and Macintosh computers are now available from commercial sources.

Set up the computer (Apple II, IBM, or Macintosh) fitted with an interface such as Universal Lab Interface or Serial Box Interface (for Macintosh), IBM game port card interface, pH amplifier and pH electrode, voltage plotter, and voltage input unit (all available from Vernier Software, Portland, OR). One can also use IBM Personal Science Laboratory (PSL) interface device.[8b] Your instructor will demonstrate how to use the computer.

Obtain 5 mL each of 0.1 M solutions of acetic acid, formic acid, and choloracetic acid. Prepare 25 mL of 0.1 M NaOH solution. Pipet 1.00 mL of the acid solution into a wide-mouthed test tube containing a micro stir bar. Add 1 mL of water followed by one drop of phenolphthalein added from a micropipet. Using a microburet, add 0.1 M NaOH dropwise to the acid solution. During the titration, stir the solution constantly. Stop adding NaOH as soon as the solution becomes pink. Using the same pipet used before, add exactly 1.00 mL of the same acid solution to the mixture. Insert the pH electrode and record the pH value on the computer. Repeat the procedure for all the acids. Under the conditions, pH = pK_a (for calculations, see Experiment 28).

Using the same equipment one can also perform the pH titration of a monoprotic or a polyprotic acid. The computer interfaced with the pH probe can be used to collect the data directly and to generate the plot of pH versus titrant volume. This plot can be used to determine the pK_a of the acid.

11.3 Independent Mini Research Projects

Research—experimental or otherwise—at the freshmen level is a rewarding experience for students. Research requires a student to think logically. This activity promotes self-learning. Further, it motivates a student to learn more than what a standard general chemistry curriculum has to offer.

Ideally, students should devise their own ideas for mini research projects. **Many ideas for independent projects are given at the end of most of the experiments described in this text.** These ideas are not the only ones; many more new concepts and topics can be chosen as research ideas. While developing a research project, take into account the following points:

1. Discuss your project with your instructor. He or she is the immediate resource person who can guide you. In fact, your instructor will schedule your research activity.
2. A research project must have a well-defined theme or problem.
3. A library search must be a component of all research projects. In fact, this search will help you to define the problem and to find how much work has been done by others on this topic. Collect as many references as you can find on the subject. Attempt to synthesize a hypothesis or a possible explanation for the problem at hand.
4. **Obtain approval of the project from your instructor or research supervisor.**
5. You and your instructor will decide what type of laboratory activities you must undertake and when to complete the project.
6. After performing the experiment(s), collect your data. Manipulate your data and explain any trends observed. If some data do not fit into a preconceived idea (hypothesis), do not get discouraged. The challenge lies in so-called negative data. In fact, negative data may provide you with new concepts and ideas that nobody thought of before.
7. Write a report according to the instructions given in Chapter 4.

The following short list of mini research projects provides illustrative examples. These ideas are open ended and can be modified as desired. For example, for GC analysis of gasoline samples, one can collect the same grade of gasoline from different gas stations or one can analyze different grades of gasoline from the same station.

1. Gas chromatographic analysis of different grades of gasoline
2. Determination of lead content in paints
3. Determination of the pH of different soil samples
4. Qualitative and quantitative analysis of common household chemicals
5. Preparation of a catalyst and its use in a synthetic reaction
6. Preparation and study of a conducting polymer
7. Preparation and study of a series of acetylacetonato complexes of different transition metals
8. Develop a method to analyze the pollution in air
9. Development of an IR technique to analyze a selected polymer material
10. Spectroscopic methods of analysis as a basis of a mini research project
11. Synthesis of a superconducting material, analysis of the composition, and evaluation of its magnetic properties
12. Many good experiments are published in *Journal of Chemical Education*. This monthly journal is published by the Division of Chemical Education of the American Chemical Society. These experiments may serve as the basis for your mini projects.

References

1. ——, *How to Search Printed CA,* American Chemical Society: Washington, DC, 1984. This 24-page pamphlet is available at no charge from the American Chemical Society.
2. Schulz, H. *From CA to CAS Online,* VCH: Weinheim, Germany, 1988.
3. Maizell, R. E. *How to Find Chemical Information,* 2nd ed.; Wiley: New York, 1987.
4. ——, *Science Citation Index*, Institute for Scientific Information: Philadelphia, PA, 1988. This 8-page pamphlet is available at no charge from ISI.
5. Garfield, E., "How to Use Science Citation Index (SCI)," *Current Contents,* **1983**, (9), 5. Reprints of this article are available at no charge from ISI.
6. For example, see (*a*) Breneman, G. L; Parker, O. J. *J. Chem. Educ.* **1992**, *69,* 46. (*b*) Parker, O. J.; Breneman, G. L. *J. Chem. Educ.* **1990**, *67*, A5. (*c*) Simpson, J. M. *J. Chem. Educ.* **1994**, *71*, A88 and references therein. (*d*) Bushey, M. M. *J. Chem. Educ.* **1994**, *71*, A90 and references therein.
7. Holmquist, D. D; Volz, D. L. *Chemistry with Computers,* Vernier Software: Portland, OR, 1993.
8. (*a*) Krause, D. C. *The Computer Based Laboratory,* JCE:Software, Vol. IA, Number 2, **1988**, *65,* 875.
    (*b*) Moore, J. W.; Hunsberger, L. R.; Gammon, S. D. *J.Chem. Educ.* **1994**, *71*, 403.
9. Chem 1 Ware Ltd. *Instructional Software for General Chemistry,* Burnaby, Canada, 1994.
10. Borman, S. "Electronic Laboratory Note Books May Revolutionize Research Record Keeping," *Chem. Eng. News* **1994**, *72*, No. 21, 10.

APPENDIX 1 Units, Definitions, and Conversion Factors

SI Units

In 1960, the General Conference on Weights and Measures, the international authority on units, introduced a revised and more modernized version of the metric system, called the **International System of Units** or **SI units** (for *Système International*). There are seven SI base units for seven physical quantities (see Table A1.1).

For many practical purposes, SI base units are modified by using decimal factors and prefixes. Most commonly used prefixes (symbols) with SI base units are these: tera (T) — 10^{12}, giga (G) — 10^9, mega (M) — 10^6, kilo (k) — 10^3, deci (d) — 10^{-1}, centi (c) — 10^{-2}, milli (m) — 10^{-3}, micro (μ) — 10^{-6}, nano (n) — 10^{-9}, and pico (p) — 10^{-12}. See any general chemistry textbook for details. For example, a kilometer (km) = 10^3 meters (m), a millimeter = 10^{-3} meter (m), and a picometer (pm) = 10^{-12} m.

For many years, the centimeter, gram, and second (cgs) system has been used to define derived units. In many modern publications these units are still in use. However, we emphasize the use of SI units as far as practicable. In Table A1.2 a complete list of common units and corresponding conversion factors is given. It may be noted that, for the sake of practicality, other common units are frequently used in chemistry laboratories; fortunately, there are not many of them.

Table A1.1 The Seven Base Units

Physical quantity	Name	Symbol
Base units		
Mass	kilogram	kg
Length	meter*	m
Time	second	s
Temperature	kelvin	K
Amount of substance	mole	mol
Luminous intensity	candela	cd
Electric current	ampere	A
Derived units		
Volume	cubic meter	m^3
Energy	joule	J

*In the United States of America, this unit is spelled as shown; in all other English-speaking countries the unit is spelled as *metre*.

Table A1.2 Units and Conversion Factors

Physical Quantity	Common units	Conversion to SI units
Length	1 inch = 2.540 centimeters (cm)	1 inch = 0.02540 meter (m)
	12 inches = 1 foot (ft)	1 ft = 0.3048 m
	3 feet = 1 yard (yd)	1 yd = 0.9144 m
	1760 yds = 1 mile	1 mile = 1609.3 m = 1.6093 km
	1 Ångstrom (Å) = 10^{-8} cm	1 Ångstrom (Å) = 10^{-10} m
		= 10^{-1} nm
Mass	1 pound (lb) = 453.6 grams (g)	1 lb = 0.4536 kilogram (kg)
	1 lb = 16 ounces (oz)	1 oz = 0.02835 kg
	1 ton = 2000 lb	1 metric ton = 10^3 kg = 10^6 g
	1 Dalton (D) = 1.66×10^{-24} g	1 avoirdupois oz = 28.3495 g
	1 g = 10^{-3} kg	1 troy oz = 31.1035 g
Volume	1 US quart = 946.33 milliliters (mL) = 0.94633 liter (L)	
	1 US gallon = 4 quarts	1 US gallon = 3.78532 L
	1 mL = 1 cm^3 = 10^{-3} L	1 μL = 10^{-3} mL = 10^{-6} L
	10^3 L = 10^3 dm^3 = 1 m^3	
Pressure	1 atmosphere (atm) = 760 torr	1 atm = 1.01325×10^5 pascal (Pa)
	= 760 mm Hg	= 101.325 kilopascal (kPa)
	= 14.70 pounds/inch^2 (psi)	1 Newton (N)/m^2 = 1 Pa
Force	1 dyne = 1 gm · cm/s^2	10^5 dynes = 1 N = 1 kg · m/s^2
Energy	1 erg = 1 dyne · cm = 1 g cm^2/s^2 =	1 kg · m^2/s^2 = 1 N · m = 1 J
		= 10^7 ergs
	1 calorie (cal) = 4.184×10^7 ergs	1 cal = 4.184 joule (J)
	1 watt · s = 1 volt (V) coulomb (C)	1 V C = 1 J = 1×10^7 erg
	1 electron volt (eV) = 23.06 kcal/mol	1 eV = 96.485 kJ/mol
	1 kcal/mol = 350 cm^{-1}	1 kJ/mol = 1.46×10^5 m^{-1}
Viscosity	1 poise = 1 dyne · s/cm^2	0.1 N · s/m^2

The following conversion factors can be used in any unit system:

Charge	1 coulomb (C) = 3.00×10^9 esu (electrostatic units)
Current	1 ampere (A) = 1 C/s
Resistance	1 ohm = 1 V/A
Power	1 watt = 1 A V
Potential	1 volt (V)
Frequency	1 hertz (Hz) = 1/s
Radioactivity	1 curie (Ci) = 3.7×10^{10} disintegrations/s

Name	Symbol	Value
Speed of light	c	2.998×10^8 m · s^{-1}
Planck's constant	h	6.626×10^{-34} J · s
Avogadro's number	N	6.023×10^{23} mole^{-1}
Faraday's constant	F	9.65×10^4 coulombs mole^{-1}
Charge on an electron	e	1.602×10^{-19} coulomb
Atomic mass unit or Dalton	amu, D	1.660×10^{-24} g
Gas constant	R	0.08206 L atm mol^{-1}K^{-1}
		8.315 J K^{-1} mol^{-1}
Acceleration due to gravity	g	9.806 m · s^{-2}
Bohr magneton	BM	9.274×10^{-24} J/T (tesla)

3 **Common Acids and Bases**

Acetic acid (concentrated, "glacial," CAS 64-19-7), commercially available in up to 99.99 percent purity. FW = 60.05 g/mol, bp = 116–118°C, density = 1.05 g/mL. Available from lab supply houses in concentration of 17.6 M. It is a corrosive acid. It has a strong odor. Vinegar is ~5 percent acetic solution; it has a pickle-like odor. **Do not inhale the vapors.**

Ammonium hydroxide (CAS 1336-21-6), commercially available as a 30 percent solution, FW = 35.05 g/mol, density = 0.900 g/mL. Concentration is ~15 M. It is a corrosive and toxic base. It has a strong odor. **Do not inhale the vapors.**

Hydrochloric acid (CAS 7647-01-0), commercially available as a 37 percent solution, density = 1.20 g/mL. FW = 36.46 g/mol. Concentration is ~12 M. The acid is highly toxic and corrosive. **Do not inhale the vapors.**

Nitric acid (CAS 7697-37-2), commercially available as 70 percent solution, density = 1.40 g/mL. FW = 63.01 g/mol. Concentration is 15.6 M. It is a highly toxic acid and a strong oxidizer. It is not compatible with any reducing agent.

Phosphoric acid (orthophosphoric acid, CAS 7664-38-2), commercially available as 85 weight percent solution, density = 1.68 g/mL. FW = 98.00 g/mol. Concentration is 14.6 M. The acid is corrosive and hygroscopic (water-absorbent).

Potassium hydrogen phthalate (CAS 877-24-7), commercially available as a 100 percent pure primary standard substance. FW = 204.2 g/mol. It is used as an acid to standardize NaOH solution.

Potassium hydroxide pellets (CAS 1310-58-3), commercially available as a ~100 percent pure substance. FW = 56.11 g/mol. It very corrosive and toxic.

Sodium carbonate (CAS 497-19-8), commercially available as 99 percent powder, density = 2.532 g/mL. FW = 105.99 g/mol. It is an irritant and a hygroscopic base.

Sodium hydroxide pellets (CAS 1310-73-2), commercially available with purity up to 97–99 percent. FW = 40.00 g/mol. The base is highly corrosive and toxic.

Sulfuric acid (CAS 7664-93-9), commercially available as a 98 percent oily substance, density = 1.84 g/mL. FW = 98.08 g/mol. Concentration is 18 M. It is a corrosive, strongly dehydrating, and oxidizing agent.

CAS # = Chemical Abstract Service Registry Number

Other Chemicals

Hydrogen peroxide (CAS 7722-84-1), commercial product is 30 percent, density = 1.11 g/mL. FW = 34.02 g/mol. It is a strong oxidizer and corrosive.

APPENDIX 4 Acid (K_a, pK_a) and Base (K_b, pK_b) Ionization Constants at 25°C

For K_a and pK_a, the general reaction is

$$\underset{\text{acid}}{\text{HA}} + H_2O \rightleftharpoons H_3O^+ + A^-$$

For K_b and pK_b, the general reaction is

$$\underset{\text{base}}{B^-} + H_2O \rightleftharpoons HB + OH^-$$

Acids	Formula	K_a	pK_a	Comments
Acetic acid	CH_3COOH	1.75×10^{-5}	4.76	
Ammonium ion	NH_4^+	5.60×10^{-10}	9.26	
Benzoic acid	$C_6H_5CO_2H$	6.30×10^{-5}	4.20	
Boric acid	H_3BO_3	5.81×10^{-10}	9.24	K_{a1}
		1.82×10^{-13}	12.7	K_{a2}
		1.58×10^{-14}	13.8	K_{a3}
Carbonic acid	H_2CO_3	4.45×10^{-7}	6.35	K_{a1}
		4.69×10^{-11}	10.3	K_{a2}
Formic acid	$HCOOH$	1.80×10^{-4}	3.75	
Glycine	NH_2CH_2COOH	4.47×10^{-3}	2.35	for $-CO_2H$
		1.67×10^{-10}	9.78	for NH_3
Hydrogen sulfide	H_2S	9.50×10^{-8}	7.02	K_{a1}
		1.30×10^{-14}	13.9	K_{a2}
Oxalic acid	$H_2C_2O_4, 2\,H_2O$	5.90×10^{-2}	1.23	K_{a1}
		6.40×10^{-5}	4.19	K_{a2}
Phosphoric acid	H_3PO_4	7.11×10^{-3}	2.15	K_{a1}
		6.32×10^{-8}	7.20	K_{a2}
		7.10×10^{-13}	12.2	K_{a3}
Sulfuric acid	H_2SO_4	1.02×10^{-2}	1.99	K_{a2}, K_{a1} is large

Bases	Formula	K_b	pK_b	
Acetate	$C_2H_3O_2^-$	5.60×10^{-10}	9.26	
Ammonia	NH_3	1.80×10^{-5}	4.74	
Bicarbonate	HCO_3^-	2.20×10^{-8}	7.66	
Carbonate	CO_3^{2-}	2.30×10^{-4}	3.64	

APPENDIX 5 Solubility Product Constants at 25°C

Compounds	K_{sp}	Compounds	K_{sp}
Aluminum hydroxide, $Al(OH)_3$	1.90×10^{-33}	Aluminum phosphate, $AlPO_4$	9.83×10^{-20}
Barium carbonate, $BaCO_3$	2.58×10^{-9}	Barium fluoride, BaF_2	1.84×10^{-7}
Barium chromate, $BaCrO_4$	1.17×10^{-10}	Barium phosphate, $Ba_3(PO_4)_2$	1.30×10^{-29}
Barium sulfate, $BaSO_4$	1.07×10^{-10}	Barium iodate, $Ba(IO_4)_2$	4.01×10^{-9}
Bismuth sulfide, Bi_2S_3	1.82×10^{-99}	Cadmium sulfide, CdS	3.60×10^{-29}
Calcium carbonate, $CaCO_3$	6.18×10^{-9}	Calcium iodate, $Ca(IO_4)_2$	6.47×10^{-8}
Calcium iodate, $Ca(IO_4)_2 \cdot 6\,H_2O$	7.54×10^{-7}	Calcium fluoride, CaF_2	1.46×10^{-10}
Calcium oxalate, $CaC_2O_4 \cdot H_2O$	2.34×10^{-9}	Calcium phosphate, $Ca_3(PO_4)_2$	2.07×10^{-33}
Cobalt(II) carbonate, $CoCO_3$	8.00×10^{-13}	Cobalt(II) sulfide, CoS	5.90×10^{-21}
Copper(II) hydroxide, $Cu(OH)_2$	2.20×10^{-20}	Copper(I) iodide, CuI	1.27×10^{-12}
Copper(II) sulfide, CuS	8.70×10^{-36}	Copper(I) thiocyanate, $CuSCN$	1.77×10^{-13}
Gold(I) chloride, $AuCl$	2.00×10^{-13}	Gold(III) iodide, AuI_3	1.00×10^{-46}
Iron(III) hydroxide, $Fe(OH)_3$	2.64×10^{-39}	Iron(II) sulfide, FeS	1.59×10^{-19}
Lead(II) bromide, $PbBr_2$	6.60×10^{-6}	Lead(II) carbonate, $PbCO_3$	1.46×10^{-13}
Lead(II) chloride, $PbCl_2$	1.17×10^{-5}	Lead(II) hydroxide, $Pb(OH)_2$	1.42×10^{-20}
Lead(II) iodide, PbI_2	8.49×10^{-9}	Lead(II) oxalate, PbC_2O_4	8.51×10^{-10}
Lead(II) sulfate, $PbSO_4$	1.82×10^{-8}	Lead(II) sulfide, PbS	9.04×10^{-29}
Lithium carbonate, Li_2CO_3	8.15×10^{-4}	Magnesium carbonate, $MgCO_3$	6.82×10^{-6}
Magnesium hydroxide, $Mg(OH)_2$	5.61×10^{-12}	Magnesium oxalate, $MgC_2O_4 \cdot 2\,H_2O$	4.83×10^{-6}
Magnesium phosphate, $Mg_3(PO_4)_2$	9.86×10^{-25}	Manganese(II) carbonate, $MnCO_3$	2.24×10^{-11}
Manganese(II) sulfide, MnS	4.65×10^{-14}	Manganese(II) hydroxide, $Mn(OH)_2$	2.06×10^{-13}
Mercury(I) chloride, Hg_2Cl_2	1.45×10^{-18}	Mercury(I) iodide, Hg_2I_2	5.33×10^{-29}
Mercury(II) sulfide, HgS	6.44×10^{-53}	Mercury(II) hydroxide, $Hg(OH)_2$	3.13×10^{-26}
Nickel(II) carbonate, $NiCO_3$	1.42×10^{-7}	Nickel(II) sulfide, NiS	3.00×10^{-21}
Silver bromide, $AgBr$	5.35×10^{-13}	Silver carbonate, Ag_2CO_3	8.45×10^{-12}
Silver chloride, $AgCl$	1.77×10^{-10}	Silver chromate, Ag_2CrO_4	1.12×10^{-12}
Silver iodide, AgI	8.51×10^{-17}	Silver cyanide, $AgCN$	5.97×10^{-17}
Silver sulfate, Ag_2SO_4	1.20×10^{-5}	Silver sulfide, Ag_2S	6.69×10^{-50}
Strontium carbonate, $SrCO_3$	5.60×10^{-10}	Strontium sulfate, $SrSO_4$	3.44×10^{-7}
Tin(II) hydroxide, $Sn(OH)_2$	5.45×10^{-27}	Tin(II) sulfide, SnS	3.25×10^{-28}
Zinc carbonate, $ZnCO_3$	1.19×10^{-10}	Zinc hydroxide, $Zn(OH)_2$	6.86×10^{-17}
Zinc sulfide, ZnS	1.10×10^{-21}		

Source: Most of the values are from Lide, D. R., ed. *Handbook of Chemistry and Physics,* 73rd ed.; CRC Press: Boca Raton, FL, 1992–93.

APPENDIX 6 Standard Reduction Potentials, E^o in Volts, V

Reaction	E^o	Reaction	E^o
(aq) + e⁻ ⇌ Li(s)	−3.04	$AgBr + e^- \rightleftharpoons Ag + Br^-$	0.071
(aq) + e⁻ ⇌ K(s)	−2.93	$[Co(NH_3)_6]^{3+} + e^- \rightleftharpoons [Co(NH_3)_6]^{2+}$	0.108
⁺(aq) + 2 e⁻ ⇌ Ba(s)	−2.91	$Sn^{4+} + 2 e^- \rightleftharpoons Sn^{2+}$	0.151
⁺(aq) + 2 e⁻ ⇌ Sr(s)	−2.89	$Cu^{2+} + e^- \rightleftharpoons Cu^+$	0.153
⁺(aq) + 2 e⁻ ⇌ Ca(s)	−2.87	$AgCl + e^- \rightleftharpoons Ag + Cl^-$	0.197
(aq) + e⁻ ⇌ Na(s)	−2.71	Calomel electrode (SCE)	0.236
²⁺(aq) + 2 e⁻ ⇌ Mg(s)	−2.37	$IO_3^- + 3 H_2O + 6 e^- \rightleftharpoons I^- + 6 OH^-$	0.260
⁺(aq) + 3 e⁻ ⇌ Ce(s)	−2.34	$Hg_2Cl_2 + 2 e^- \rightleftharpoons 2 Hg + 2 Cl^-$	0.241
⁺(aq) + 3 e⁻ ⇌ Al(s)	−1.66	$Bi^{3+} + 3 e^- \rightleftharpoons Bi$	0.308
⁺(aq) + 2 e⁻ ⇌ Ti(s)	−1.63	Calomel electrode, 0.1 M KCl	0.334
⁺(aq) + 4 e⁻ ⇌ Zr(s)	−1.45	$Cu^{2+} + 2 e^- \rightleftharpoons Cu$	0.342
²⁺(aq) + 2 e⁻ ⇌ Mn(s)	−1.19	$O_2 + 2 H_2O + 4 e^- \rightleftharpoons 4 OH^-$	0.401
(aq) + 2 e⁻ ⇌ V(s)	−1.18	$I_2 + 2 e^- \rightleftharpoons 2 I^-$	0.536
O_3^{2-}(aq) + 2 H₂O + 2 e⁻ ⇌ $S_2O_4^{2-}$ + 2 OH⁻	−1.12	$I_3^- + 3 e^- \rightleftharpoons 3 I^-$	0.536
₂O(aq) + 2 e⁻ ⇌ H_2(g) + 2 OH⁻	−0.828	$MnO_4^- + 2 H_2O + 2 e^- \rightleftharpoons MnO_2 + 4 OH^-$	0.600
²⁺(aq) + 2 e⁻ ⇌ Zn(s)	−0.762	$BrO_3^- + 3 H_2O + 6 e^- \rightleftharpoons Br^- + 6 OH^-$	0.610
⁺(aq) + 3 e⁻ ⇌ Cr(s)	−0.744	$ClO_3^- + 3 H_2O + 6 e^- \rightleftharpoons Cl^- + 6 OH^-$	0.620
O_4^{3-}(aq) + 2 H₂O + 2 e⁻ ⇌ AsO_2^- + 4 OH⁻	−0.710	$O_2 + 2 H^+ + 2 e^- \rightleftharpoons H_2O_2$	0.695
O_2^-(aq) + 2 H₂O + 3 e⁻ ⇌ As + 4 OH⁻	−0.680	$Fe^{3+} + e^- \rightleftharpoons Fe^{2+}$	0.771
2 e⁻ ⇌ S^{2-}	−0.476	$Hg_2^{2+} + 2 e^- \rightleftharpoons 2 Hg$	0.797
₂⁻(aq) + H₂O + e⁻ ⇌ NO + 2 OH⁻	−0.460	$Ag^+ + e^- \rightleftharpoons Ag$	0.800
⁺(aq) + 2 e⁻ ⇌ Fe(s)	−0.447	$2 NO_3^- + 4 H^+ + 2 e^- \rightleftharpoons N_2O_4 + 2 H_2O$	0.803
²⁺(aq) + 2 e⁻ ⇌ Cd(s)	−0.403	$Hg^{2+} + 2 e^- \rightleftharpoons Hg$	0.851
SO_4(s) + 2 e⁻ ⇌ Pb(s) + SO_4^{2-}	−0.359	$2 Hg^{2+} + 2 e^- \rightleftharpoons Hg_2^{2+}$	0.920
⁺(aq) + 2 e⁻ ⇌ Ni(s)	−0.257	$NO_3^- + 4 H^+ + 3 e^- \rightleftharpoons NO + 2 H_2O$	0.957
⁺(aq) + e⁻ ⇌ V^{2+}(aq)	−0.255	$Br_2 + 2 e^- \rightleftharpoons 2 Br^-$	1.066
+ 2 H₂O + 2 e⁻ ⇌ H_2O_2 + 2 OH⁻	−0.146	$2 IO_3^- + 12 H^+ + 10 e^- \rightleftharpoons I_2 + 6 H_2O$	1.195
²⁺(aq) + 2 e⁻ ⇌ Sn(s)	−0.138	$O_2 + 4 H^+ + 4 e^- \rightleftharpoons 2 H_2O$	1.229
²⁺ + 2 e⁻ ⇌ Pb(s)	−0.126	$Cr_2O_7^{2-} + 14 H^+ + 6 e^- \rightleftharpoons 2 Cr^{3+} + 7 H_2O$	1.232
³⁺(aq) + 3 e⁻ ⇌ Fe(s)	−0.037	$Cl_2 + 2 e^- \rightleftharpoons 2 Cl^-$	1.358
H⁺(aq) + 2 e⁻ ⇌ H₂(g)	**0.0000**	$MnO_4^- + 8 H^+ + 5 e^- \rightleftharpoons Mn^{2+} + 4 H_2O$	1.507
O_3^-(aq) + H₂O + 2 e⁻ ⇌ NO_2^- + 2 OH⁻	0.010	$F_2 + 2 e^- \rightleftharpoons 2 F^-$	2.866

urce: Most of the values are from Lide, D. R., ed. *Handbook of Chemistry and Physics*, 73rd ed.; CRC Press: Boca Raton, FL, ⎻2–93.

Safety Data for Common Solvents

Acetone: Acetone is an extremely flammable liquid. It is not normally considered dangerous, but normal precautions should be employed. ORL-RAT LD50: 5800 mg/kg.

Chloroform: Chloroform is a potent narcotic agent. It may be fatal if inhaled, swallowed, or absorbed through the skin. It is classified as a carcinogen. ORL-RAT LD50: 908 mg/kg.

Cyclohexane: Cyclohexane is harmful if inhaled or swallowed. It is extremely flammable. ORL-RAT LD50: 1215 mg/kg.

Diethyl ether: Ether is an extremely flammable solvent. Exposure to moisture tends to form peroxides, which may be explosive. The solvent is a potent narcotic. ORL-RAT LD50: 1215 mg/kg.

Dimethylsulfoxide: Dimethylsulfoxide (DMSO) is harmful if swallowed, inhaled, or absorbed through the skin. Overexposure has been found to have effects on fertility. ORL-RAT LD50: 14,500 mg/kg.

Ethyl alcohol: Ethyl alcohol (ethanol) may be fatal if inhaled, swallowed, or absorbed through the skin in large amounts. It has been shown to have effects on fertility and on embryo development. The vapor may travel considerable distances to the source of ignition and flash back. ORL-RAT LD50: 7060 mg/kg.

Hexane: Hexane is harmful if inhaled, swallowed, or absorbed through the skin. It is a flammable liquid. ORL-RAT LD50: 28,710 mg/kg.

Isopropyl alcohol: Isopropyl alcohol (rubbing alcohol) is not normally considered dangerous, but the usual precautions should be followed. ORL-RAT LD50: 5045 mg/kg.

Methyl alcohol: Methyl alcohol (methanol) may be fatal if swallowed. It is harmful if inhaled or absorbed through the skin. It is a flammable liquid. ORL-RAT LD50: 5628 mg/kg.

Methylene chloride: The compound is harmful if swallowed, inhaled, or absorbed through the skin. ORL-RAT LD50: 1600 mg/kg. It is a possible carcinogen.

Toluene: Toluene is a flammable liquid. ORL-RAT LD50: 5000 mg/kg.

All safety data in this appendix are derived from the Sigma-Aldrich Material Safety Data Sheets on CD-ROM, Aldrich Chemical Co., Inc., Milwaukee, WI.